es in Solid-State Sciences features fundamental scientific books
ng and up-and-coming researchers in the field. They strive to
systematic and comprehensive way, the basic principles as well
nts in theoretical and experimental solid-state science.
submissions for monographs or edited volumes from scholars
lomain. Topics of current interest include, but are not limited to:

s and superconductors
omena

sulators

d nanophotonics
tron systems and strongly correlated materials
l electronic properties of solids
nd magnetic resonance

Springer Series in Soli

Volume 209

The Springer
prepared by
communicate
as new develo
We welco
across this bro

- Semicondu
- Quantum p
- Spin physi
- Topologica
- Multiferroi
- Nano-optic
- Correlated
- Vibrational
- Spectrosco

Jean Bornarel

Ferroic Crystal Domains, Walls, and Phase Fronts

Springer

Jean Bornarel
Department of Physics
Grenoble Alpes University
Venon, France

ISSN 0171-1873 ISSN 2197-4179 (electronic)
Springer Series in Solid-State Sciences
ISBN 978-3-031-98098-5 ISBN 978-3-031-98099-2 (eBook)
https://doi.org/10.1007/978-3-031-98099-2

The original submitted manuscript has been translated into English. The translation was done using artificial intelligence. A subsequent revision was performed by the author(s) to further refine the work and to ensure that the translation is appropriate concerning content and scientific correctness. It may, however, read stylistically different from a conventional translation.

This Springer imprint is published by the registered company Springer Nature Switzerland AG
The registered company address is: Gewerbestrasse 11, 6330 Cham, Switzerland

Acknowledgements

First and foremost, I would like to thank the students, doctoral candidates, and colleagues who have pushed me to seek as simple explanations as possible for the observed phenomena. Thank you to those who convinced me to undertake the creation of this book rather than the multiplication of publications, which was obstructed by various activities, and thus to take some distance from the thousands of accumulated results.

On this theme of interfaces, the bibliography continually provides evidence of their impacts. The richness of theoretical models and hypotheses has often motivated the desire to see what it is all about, and we have rarely been disappointed. Thank you to all the researchers who contribute to this richness. The most free and exhaustive interaction remains the indispensable condition of relevant work and more generally of any progress.

Special mention goes to those who have contributed to the experiments, explanations, and models, whose names are cited, notably in the figures, but insufficiently. Thanks also to the technicians and engineers with whom we have designed and improved such a cryostat, a setup that did not exist before the experiment and seemed so useful afterwards.

I would also like to thank the photographers, companies, and organizations that allowed the dissemination and valorization of scientific work and established links between phenomena, models, and lines of explanation. Thanks to Alice Giraud, graphic designer, without whom the visuals would not have existed with this beauty, and to the team of Édition Diffusion Presse Sciences led by Agnès Henri, who invested in this project for the publication of the French version of the book, and to that of Springer Nature for the English version.

My thanks go particularly to the members of the reading committee, Pierre Bastie, Richard Cach, Olivier Isnard, Zdenek Kvitek, and Jean François Legrand, whose reactions and suggestions have greatly contributed to improving the strategy and content of the book.

Finally, thanks to my loved ones for their constant support in all situations and especially to my wife, Nicole Sauval, without whom I would not have reached the end of the project.

Jean Bornarel

Competing Interests The author has no competing interests to declare that are relevant to the content of this manuscript.

Introduction

The Objectives

Ferroic crystals are characterized by hysteresis cycles linking magnetization to the magnetic field, polarization to the electric field, and deformation to the mechanical stress. These cycles are mainly traced at the macroscopic level, and changes at the nanoscopic level are actively studied. The study of microscopic phenomena, particularly with regard to domains and their boundaries, key elements of the properties, and performances of these materials during "reversals" ("switchings"), remains modest if not neglected. The interest in applications at the thin-film level and the increasing ease of performing calculations and simulations with large databases could explain this trend in research for over a decade. Nevertheless, a better understanding of microscopic phenomena remains a necessity and is one of the motivations of this book.

Moreover, unlike ferromagnetism, ferroelasticity and ferroelectricity both necessarily involve objects at the level of cells, atoms, and ions. Linking the two phenomena, the specificities of the energies involved (mechanical and electrostatic) and their competition are another motivation, particularly in situations such as phase transitions.

Finally, the ambition to address not only researchers to propose some lines of thought but also science students with simple presentations is constant. The opportunity to produce visual demonstrations, with photographs of reality, was an opportunity to seize.

To go as far as possible in understanding the phenomena, the choice of crystals was restricted to simple symmetry changes. Thus, the reader easily appropriates the deformations (shear, compression, expansion), with the simplest possible deformation–polarization coupling. It will be easy to transpose the results obtained, situations, phenomena, and reasoning to other contexts. With the concern to provide experimental evidence for the proposed models, particular attention is paid to the conditions of relevance of the experiments, the main causes of errors, and the precautions to be taken to avoid or, at least, limit them.

Reading Guide

The chosen presentation allows various readings and consultations:

- The text alone should suffice for the reader who does not wish to delve into the detailed study of the experimental evidence often provided by the figures.
- Two hundred figures propose either results of experiments or simulations for the researcher or diagrams accessible to any student. Their captions are often detailed, so that those who choose to explore through visuals and figures can follow their path with good autonomy.
- Memos at the end of each chapter summarize what should be remembered and list outstanding questions. Some will receive answers in the following chapters or are also fields of research to explore.
- Annexes provide additional information adapted to the diversity of the potential audience.
- The bibliography remains limited compared to the thousands of publications whose results have been integrated. The choice of references values the most significant experiences and the decisive contributions to the subject. The author apologizes in advance to the non-cited colleagues who would have been in a comprehensive monograph, which is not the objective of this book.

Contents

The first chapter presents the necessary foundations for reading the book for a student and the characteristics of the chosen crystals such as the proportionality between polarization and the main deformation, here chosen as the order parameter. For the classical thermodynamic models presented, the often implicit assumptions for physicists are highlighted: crystal of infinite dimensions, homogeneous temperature, crystals assumed to be continuous without interface, etc. The terms "para" and "ferro-ferro" used recall for the second limiting context chosen among ferroic crystals.

The chosen problem is based on a prior presentation of the phenomena and so-called natural properties, which depend on the specificities of the sample (quality of the crystal, shape, preparation) and the laws in temperature (history, spatial distribution). Thus, Chaps. 2 for domains and walls and 5 for phase fronts demonstrate the role of microscopic interfaces in relation to the macroscopic properties of the single-domain crystal, a role that is sometimes subtle and sometimes major.

Then comes the application of fields conjugated to the order parameter (electric field or mechanical stress) in two contexts: either with sufficient values to change the overall situation of the crystal, for example the density of domain textures, or with the smallest possible value in a susceptibility measurement, to better understand a situation without (theoretically) modifying it. Chapters 3 and 6 involve fields sufficient to trace a macroscopic hysteresis cycle and study what happens at the microscopic level at any temperature of the ferro phase (Chap. 3) and during phase coexistence for a first-order transition (Chap. 6). In Chap. 3, two reversal modes are studied in detail, that of lateral displacements of plane domain walls and that of "needle" domains which change longitudinally. This allows to understand, by combining these two modes, the reversals with other forms of domains. In Chap. 6, the interplay between electric field and thermal context (history, gradient, etc.) allows to explore forms of phase fronts and their correlation with domains. The interpretation of many macroscopic results at phase transitions is thus facilitated.

Throughout the book, macroscopic experimental results of dielectric constant and loss constant are correlated with the microscopic phenomena of interfaces. A complete chapter (Chap. 4) dedicated to susceptibility highlights the phenomena that can exist depending on the temperature in the ferro-ferro crystals studied, in other ferroelectric and/or ferroelastic crystals, or in solid physics. The phenomena that concern the entire sample, those related to sets of domains, others related to domain walls with or without interaction, and finally those internal to the walls are highlighted.

The annexes are either elements known by the researcher specializing in the theme and summarized for those who are not or syntheses with a small bibliography on a subject that may interest many people.

About This Book

Hysteresis cycles symbolize nonlinear dynamics between stable situations. In solid-state physics, the phenomenon is usually detected at the macroscopic level and explained at the nanoscopic ones. In the example of crystals, simultaneously ferroelectric and ferroelastic, with very simple symmetry variation, the importance of the microscopic level, walls, and domains is demonstrated. The role of interfaces (phase fronts and domains) is also important during the phase coexistence in first-order transitions. In this book, the experimental results are firstly described, which allow realistic models. The phenomena due to the temperature variations, called "natural," are explained with regard to those induced by applications of conjugate fields (electric fields, mechanical stresses). Some conclusions could be transposed in solid-state physics and, more generally, in other scientific fields.

This book is intended to scientists concerned by interfaces in materials science, particularly in ferroic crystals, but also to high-level students. Different working processes, with 250 figures and photographic sequences, memos at the end of chapters, and appendices for students, allow its use as a reference book or for consulting.

Contents

About the Author

Jean Bornarel Professor of the Joseph Fourier University (Honorary Professor of the Grenoble Alpes University), Jean Bornarel has been among the first to understand the ferroelasticity existence at a moment where only ferroelectricity was known. Activities in different fields persuaded him to propose bridges, bonds between on the one hand ferroelectrics and ferroelastics and on the other hand at nano, micro, and macro levels.

Chapter 1
Ferroelectricity and Ferroelasticity

This chapter contains the necessary elements for an easy reading of the work. Other insights and supplements can be consulted in the appendices. Thus, a list of symbols, constants, and definitions of quantities can be found in Annex A1.

1.1 Ferroic Crystals and Hysteresis Loops

The hysteresis loop is the symbol of the ferroic properties of a material. It is the graph of magnetization as a function of the applied magnetic field for a ferromagnetic material, of polarization as a function of the applied electric field for a ferroelectric, and of deformation as a function of the applied external mechanical stress for a ferroelastic. Its existence proves that one can, by applying an external excitation, "reverse" the magnetization, polarization, or deformation and make the crystal transition from one stable state to another stable state of opposite (or different) magnetization, polarization, or deformation to that of the first state. These two stable states correspond to the horizontal parts of the hysteresis loop: the susceptibility, the ratio between the variation of the characteristic quantity (magnetization, polarization, or deformation) and that of the excitation (magnetic field, electric field, or stress) is almost constant as a function of the excitation. A stable state corresponds, for example, for a ferroelectric, to a crystal fully polarized in a single direction. The other stable state will correspond in this example to a crystal also fully polarized but with polarization opposite to the previous one.

Figure 1.1 presents the classic experiment of plotting a hysteresis loop for a ferroelectric using the Sawyer and Tower setup (1930). The characteristic parameters are defined: coercive field E_c value of the electric field when the polarization is zero, saturation polarization P_S obtained at the maximum values of the applied periodic electric field, and remanent polarization P_r obtained when the electric field passes through the zero value. Figure 1.1a presents a loop with horizontal tangents for the

J. Bornarel, *Ferroic Crystal Domains, Walls, and Phase Fronts*, Springer Series in Solid-State Sciences 209, https://doi.org/10.1007/978-3-031-98099-2_1

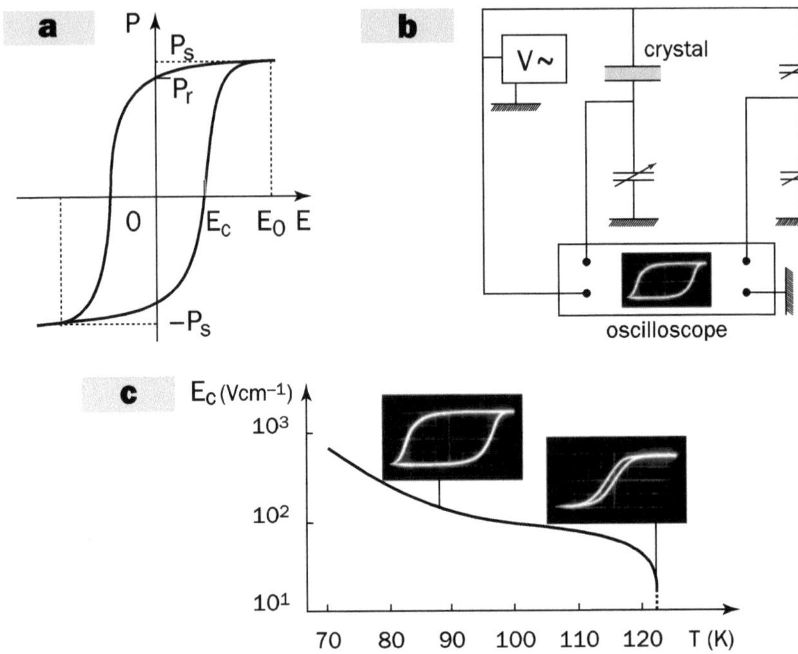

Fig. 1.1 Hysteresis cycle $P(E)$ plotted with the Sawyer and Tower (1930) setup (**a**) A ferroelectric crystal is cut into a platelet with the large faces perpendicular to the possible polarization axis are metallized; the capacity thus created is placed in the electrical circuit represented in (**b**), in series with a linear capacitor of very large capacity. This results in a proportional voltage to the electric field applied to the crystal on the horizontal sweep of the oscilloscope, and a proportional voltage to the induced charge on the electrodes, proportional to the polarization, on the vertical sweep. If the crystal was a classic insulator, the observed graph would be a straight line. If it is ferroelectric, the hysteresis cycle of (**a**) is observed with its characteristic parameters, coercive field E_c, saturation polarization P_s, and remanent polarization P_r. The variation of E_c as a function of temperature for a KDP crystal is presented in (**c**) (external alternating electric field of 50 Hz frequency and maximum amplitude of 2 kV/cm (Guyon and Lajzerowicz, 1966)). The branches of the cycle corresponding to saturation appear horizontal on the photographs of the cycles thanks to unrepresented electronic assemblies that compensate for the conduction phenomena of the saturated crystal (Diamant et al., 1951; Schubling et al., 1964)

maximum values E_0 of the applied electric field, hence a constant polarization for the single-domain sample. The experimenter imposes the shape of the applied periodic field (sinusoidal, triangular, etc.), its frequency, and the maximum value of the amplitude. The result obtained in this forced system obviously depends on these parameters.

In a ferroelastic material, one can also plot the hysteresis loop of deformation as a function of the applied stress, modified periodically (Salje and Hoppman, 1976) as illustrated in Fig. 1.2. Thus, the values of coercive stress, saturation deformation, and remanent deformation are obtained. Mechanical measurements are sometimes delicate to perform with precision given the tensorial nature of the quantities

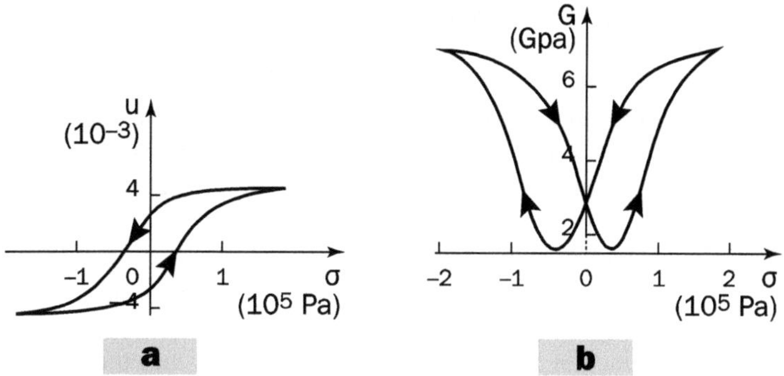

Fig. 1.2 Hysteresis cycle of a ferroelastic crystal (**a**) Deformation(u)-stress(σ) cycle of a $KH_3(SeO_3)_2$ crystal at 201 K obtained by the torsion pendulum method (sample in the form of an elongated rod along c and frequency of 10^{-1} Hz); (**b**) variation during this cycle of the shear modulus (measured by successive steps with stabilization of one minute at each σ value (Gridnev et al., 1982; Shuvalov et al.,1984))

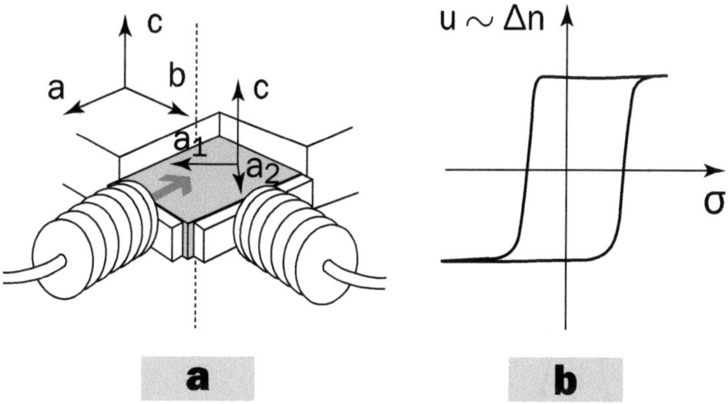

Fig. 1.3 (**a**) A parallelepiped sample of a ferroelastic crystal of orthorhombic symmetry can be subjected to shear stresses using judiciously applied normal stresses: in the case of KDP, at 45 degrees of arc from the tetragonal axes a_1 and a_2. Bellows are used for this purpose, in which the pressure of a helium gas is varied; (**b**) the birefringence of the crystal for light propagating along the c axis is proportional in the chosen crystals to the shear deformation of interest u_{xy}. Thus, the graph is comparable to a deformation-stress hysteresis cycle

(deformation and stress). Various strategies are used to achieve the result: direct measurements of stress and deformation (Pakulski et al., 1987) or using a process, for example, optical, to achieve deformation (Salje, 1990). The setup in Fig. 1.3 provides an example of obtaining results used in the work. The measurement of deformation is deduced from a measurement of birefringence Δn_c (birefringence for light passing through the crystal propagating along the ferroelectric axis c) performed using a Berek compensator or by any other method: the birefringence, in this crystal (again KDP), is proportional to the deformation u_{xy}. Measuring one quantity to deduce another can present great advantages but also be a source of additional uncertainties that must be considered.

A ferroic material whose response to an external excitation is a hysteresis loop within a certain temperature range can exhibit linear responses in other temperature ranges. Figure 1.4 shows that at room temperature, the application of an electric field in the direction of the c-axis on a KDP crystal induces a deformation and polarization along the *c*-axis that vary linearly with the amplitude of the field: the crystal is paraelastic and paraelectric (Sedlakova et al., 1975).

Figure 1.5 complements, again for KDP, the results of the disappearance of the coercive field from Fig. 1.1, above 120 K: if the temperature is above 123 K, the response of P as a function of the electric field E_z is linear, thus the dielectric constant is constant as a function of the field: the crystal is paraelectric. Conversely, if the temperature approaches 121 K, the response becomes non-linear, and one transitions into the ferroelectric phase (Baumgartner, 1950).

If one calculates the energies necessary for the transition from a ferroelectric state to a paraelectric state, a few millielectronvolts (meV) are obtained. Therefore, an increase in temperature of a few tens of degrees Kelvin may suffice to transition a crystal from a "ferro" phase to a "para" phase; this is the case in most studied crystals, except obviously if they reach their melting point before the transition.

We will now study what symmetry relations exist, for a given crystal, between its "ferro" phase and its "para" phase, which we will henceforth refer to by these abbreviations without using quotation marks.

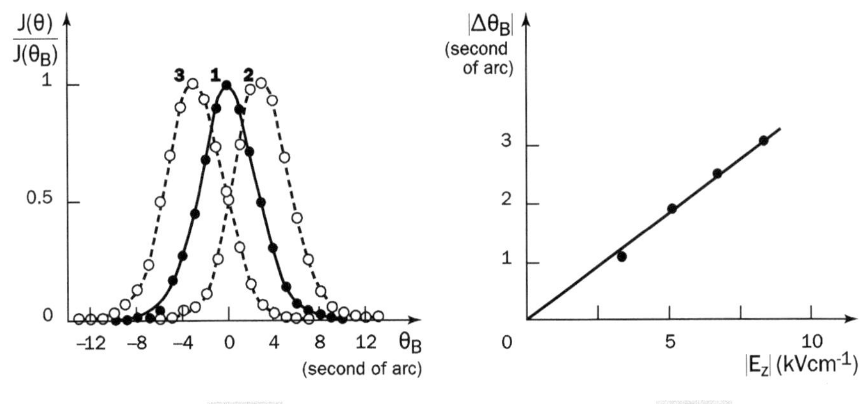

Fig. 1.4 "Para" Crystal: The neutron diffraction allows very precise measurements of lattice deformation as a function of an applied external electric field (or applied stress). For a KDP crystal in the para phase, at room temperature, the diffraction pattern moves linearly with the applied electric field E_z (the distance between the crystallographic planes varies as ($d_{36}\,E_z/2$), d_{36} being the piezoelectric module): (**a**) diffraction pattern with $E_z = 0$ (**1**), $E_z = +E_0$ (**2**) = 6.25 kV cm^{-1}, and $E_z = -E_0$ (**3**); (**b**) displacement of the diffraction pattern as a function of the magnitude of the applied electric field (Sedlakova et al., 1975)

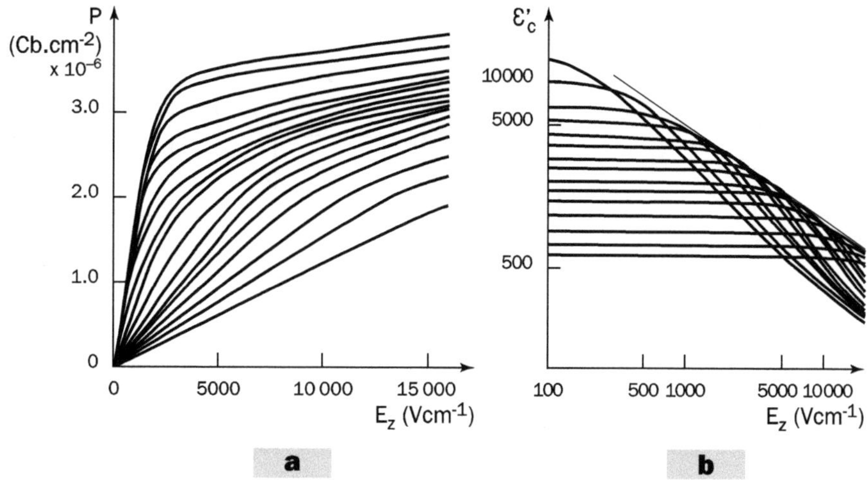

Fig. 1.5 Dielectric measurements carried out at temperatures close to the transition (Baumgartner, 1950). The author defined a temperature θ assimilated to the Curie temperature, and each curve corresponds to a value of $T - \theta$. The results of (**a**) come from hysteresis cycles (60 Hz, E_z maximum on the abscissa) for $P(E_z)$ with $(T - \theta)$ between 4 K (lower linear curve) and 0.1 K (upper curve); (**b**) the results of dielectric constant ε_c' are obtained using a measurement bridge (1 kHz, 40 V cm^{-1}) with $(T - \theta)$ between 5.32 K (high curve) and 0.14 K (low curve)

1.2 Transitions and Symmetry

1.2.1 Unit Cells and Symmetry

The ferro state and the para state of a crystal correspond to different unit cells (primitive cells or multiple cells) and often to different symmetries. To delve deeper into this question, one can consult the numerous works dedicated to symmetry changes during transitions (Tomashewski, 1992a, b; Tonkov, 1992; Tagantsev et al., 2010). We will limit our discussion to a few illustrative examples.

Four comparisons of unit cells between para and ferro phases of the same crystal are described in Figs. 1.6, 1.7, 1.8, and 1.9. A first example illustrates a ferroelastic transition (Fig. 1.6). The example in Fig. 1.7 corresponds to the symmetry change considered in this work: materials that are both ferroelectric and ferroelastic with fully coupled properties (dielectric and elastic) to possess the best means of investigation and thus understanding of phenomena (domains, walls). This seemingly simple situation is not general, as illustrated by the example in Fig. 1.8 of a crystal that is both ferroelectric and ferroelastic, in which these properties are only partially coupled. Finally, comparisons in Fig. 1.9 of tetragonal para and orthorhombic ferro unit cells in the crystal $Ba_2NaNb_5O_{15}$ remind us that comparing unit cells without considering transitions is not sufficient.

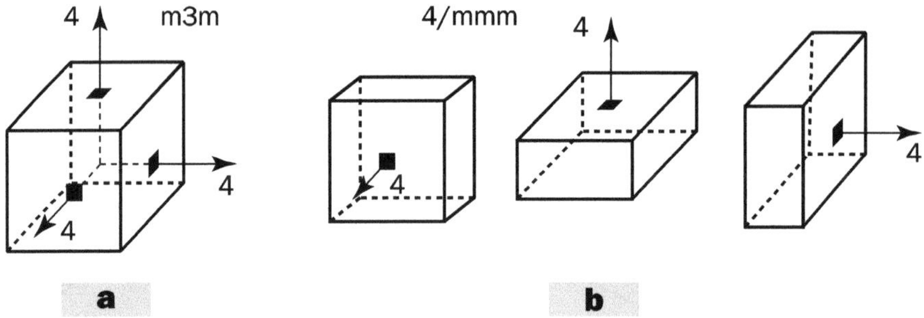

Fig. 1.6 Unit cells in a ferroelastic transition: (**a**) The symmetry of strontium titanate $SrTiO_3$ is cubic m3m in the paraelastic phase at room temperature: we observe the three fourfold axes. (**b**) The symmetry becomes tetragonal 4/mmm in the ferroelastic phase with only one fourfold axis and therefore three possibilities of orientation for the lattice and the unit cells. We can switch from one orientation to another by applying an appropriate stress (Barsh and Krumhansl, 1984)

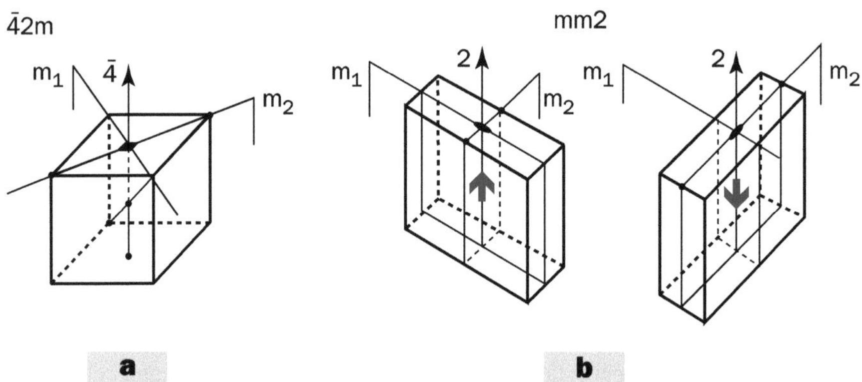

Fig. 1.7 Unit cells in a ferroelastic and ferroelectric transition: The situation widely used in the rest of this book is that of crystals from the KDP family (de Quervain, 1944), that of the gadolinium molybdate family $Gd_2(MoO_4)_3$ often referred to as GMO (Cross et al., 1968) as well as that of other compounds such as Tanane. (**a**) In the "para" phase, these crystals are of tetragonal symmetry $\bar{4}$ 2 m with one inversion centre, one fourfold axis, and two perpendicular mirror planes. (**b**) In the "ferro" phase of orthorhombic symmetry mm^2, the fourfold axis is replaced by one twofold axis along which an electric polarization appears. Two elements of symmetry have been lost at the transition: the 90° rotation, around the vertical fourfold axis, and the symmetry with respect to the plane perpendicular to it (loss of inversion); using these two lost symmetry elements, we can deduce the unit cells of the mm^2 symmetry phase from each other. We are dealing with a crystal that is both ferroelastic as shown by the two primitive cells forms and ferroelectric with the two opposite orientations of the polarization. Polarization and deformation are, in the case of these crystals, linearly coupled. We can switch from one "ferro" lattice to the other by applying either an electric field E_3 (or E_z) along the twofold axis or a shear stress σ_{12} (or σ_{xy}) in the plane perpendicular to this axis

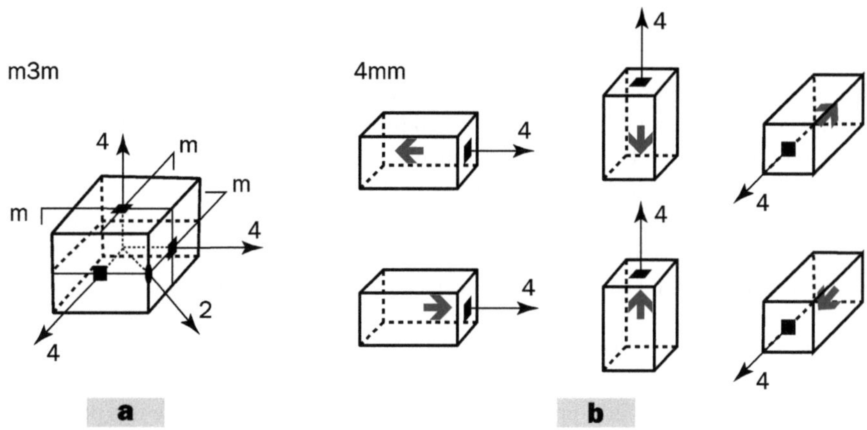

Fig. 1.8 Partial coupling between ferroelectricity and ferroelasticity. In this transition of barium titanate $BaTiO_3$ with a change of symmetry from m3m to 4 mm (Megaw, 1946; von Hippel et al., 1946), ferroelectricity and ferroelasticity are only partially coupled: three deformations for the unit cells and six polarizations are possible in the "ferro" phase; a simple inversion of polarization does not change the deformation but a change in deformation results in a 90° rotation for the direction of the polarization

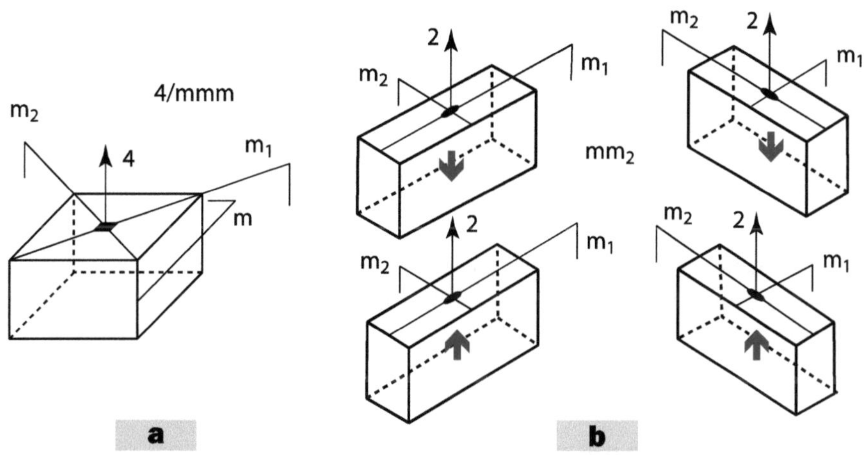

Fig. 1.9 Different phases tetragonal and orthorhombic: The barium and sodium niobate $Ba_2NaNb_5O_{15}$ has a phase of tetragonal symmetry 4/mmm (**a**) and another phase of orthorhombic symmetry mm^2 (**b**) (Singh et al., 1970). We note that the symmetry of the latter is the same as that of the ferro phase of KDP, but the number of possible unit cells is different: four stable states instead of two. Indeed, the symmetry elements lost during the transitions are not the same. Here, we obtain this result by going through two transitions: the symmetry first goes from 4/mmm to 4 mm during a first transition at 843 K not presented here, then from 4 mm to mm^2 during a transition at 533 K

1.2.2 Number of Possible Domains

The examples concern situations in which relations exist between the symmetry groups of the para and ferro phases. We will remain in this context (there are transitions without relation between the symmetry groups of the two phases). The comparisons of unit cells in Figs. 1.7 and 1.9 show that knowing the symmetry of a ferro phase is not sufficient to determine the possible unit cells, and thus the domains. On the other hand, the symmetry elements lost during the transition between the para phase and the ferro phase seem to be decisive. Is this a general result?

The answer to the question concerning the symmetry elements lost during the transition is given by Curie's principle stated more than a century ago (Curie, 1894): "when certain causes produce certain effects, the symmetry elements of the causes must be found in the effects produced". The above examples illustrate this principle. More generally, consider a crystal whose symmetry in the para phase conforms to the point symmetry group G_P. If at the phase transition, a physical quantity appears whose symmetry is described by the group K, the phase transition will be accompanied by a reduction in symmetry, with the point symmetry group of the ferro phase G_F being a subgroup of the group G_P. G_P also contains the group K. By considering the symmetry elements lost during the transition, one thus obtains the number of possible unit cells n in the ferro phase. It is equal to the ratio between the order g_P of the symmetry group G_P of the para phase and the order g_F of the group G_F of the ferro phase, a ratio often called the index of F in G:

$$n = \text{order of} G_P / \text{order of} G_F = g_P / g_F \tag{1.1}$$

n is necessarily equal to or >2. For example, in the previously proposed examples, the value of n is found to be 3 in the case of Fig. 1.6 (48/16), 2 in that of Fig. 1.7 (8/4), 6 in that of Fig. 1.8 (48/8), and 4 in that of Fig. 1.9 (16/4) (referring to the complete table on the order of groups, subgroups, and supergroups of the 32-point symmetry groups (Tagantsev et al., 2010)). Knowing n, one can predict how many pairs of domains N can exist:

$$N = n(n-1)/2 \tag{1.2}$$

The two domains of a pair will be related by symmetry operations of the group G_P. For example, in the case of Fig. 1.8 (barium titanate), one can predict $N = 15$. It should be noted that one can predict the possible numbers of domains and the domains for a determined symmetry change (Tagantsev et al., 2010).

1.2.3 Order Parameter

The transition from the para phase to the ferro phase is characterized by the appearance of a polarization for a ferroelectric and a deformation for a ferroelastic in a crystal subjected only to a temperature variation, without an applied electric field (or applied mechanical stress). But, how can these so-called spontaneous polarizations and deformations be rigorously defined (Tagantsev, 2008)?

If the paraelectric phase is non-polar, the polarization that appears at the transition is the spontaneous polarization, a vector (rank one tensor). A more general definition is necessary in the case of a polar para phase to differentiate the quantity that appears due to the transition from the variations that would naturally occur as a function of temperature if the crystal were still in the para phase. In Toledano's formalism (1974), the elements P_{Si} of the spontaneous polarization of a symmetric tensor (Nye, 1960, 1961) at a temperature T in the ferro phase are defined as follows:

$$P_{\mathrm{Si}}\left(T\right)=r_{\mathrm{F}}P_{\mathrm{i}}\left(T\right)-r_{\mathrm{P}}P_{\mathrm{i}}\left(T_{\mathrm{P}}\right) \tag{1.3}$$

T_{P} is a temperature above the transition, and r_{F} and r_{P} are the projectors associated with the symmetric irreducible representations of the point groups G_{F} and G_{P} (Hammermesh, 1964). The projector r_{X} is the sum of all the symmetry elements belonging to the group X divided by the order g_{X} of this group. The second term of the relation (1.3) is obviously zero if the para phase is non-polar.

One can also use the pyroelectric coefficients * $p = \delta P/\delta T$ (Tagantsev et al., 2010) to define the spontaneous polarization:

$$P_{\mathrm{S}}\left(T\right)=\int_{T_P}^{T}\left(\frac{\delta P}{\delta T}\right)dT \tag{1.4}$$

These definitions are in agreement with the values of P_{S} used to present experimental results or thermodynamic and phenomenological models.

In the same way, one can define a spontaneous deformation. It is more delicate, since deformation is a rank two tensor. As with the spontaneous polarization of a polar ferroelectric in the para phase, the authors propose definitions of spontaneous deformation relative to what would exist if the crystal were still in the para phase. Thus, Toledano (1974) proposes a formulation for the elements of the deformation tensor that dissociates these two elements:

$$e_{Sij}=r_{\mathrm{F}}e_{ij}-r_{\mathrm{P}}e_{ij} \tag{1.5}$$

$r_{\mathrm{F}}e_{ij}$ is the most general deformation tensor of the ferro phase, and $r_{\mathrm{P}}e_{ij}$ represents the thermal expansion of the prototype para phase extrapolated to the temperature under consideration. A comparable formulation is found in Tagantsev et al. (2010):

$$e_{Sij}\left(T\right)=\int_{T_{\mathrm{P}}}^{T}\left(\alpha_{ij}\left(T\right)-\alpha_{ij}\left(T_{\mathrm{P}}\right)\right)\mathrm{d}T \tag{1.6}$$

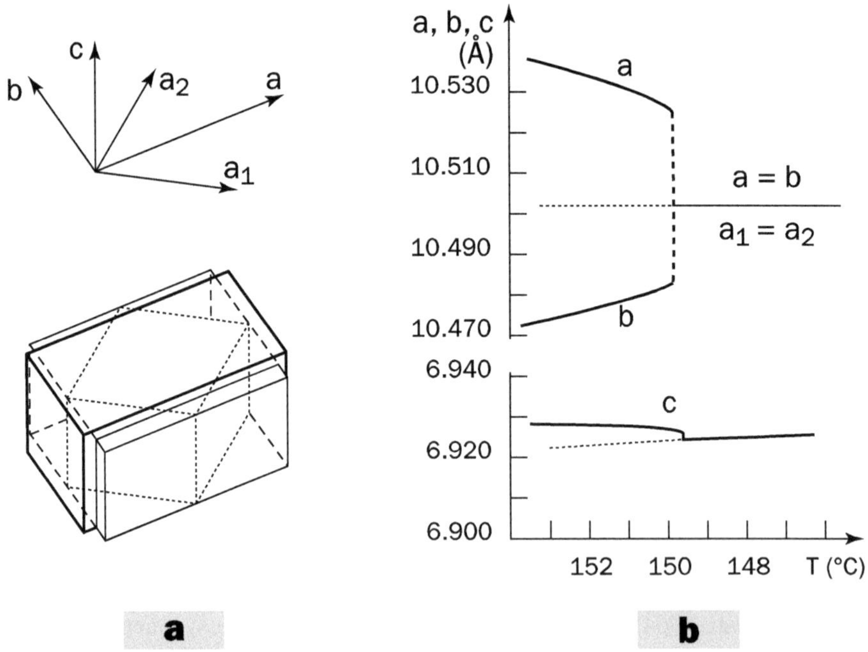

Fig. 1.10 Variation of lattice parameters during the KDP transition: (**a**) the tetragonal primitive cell is represented in dotted lines (a_1, a_2, c), the multiple tetragonal cell (with multiplicity 2) in thin lines (a, b, c), and the modification for the latter in the orthorhombic phase in bold lines. The deformations are considerably exaggerated for a better understanding; (**b**) variations of the lattice parameters for the double tetragonal cell with at the transition an increase of a and c and a decrease of b; the volume of this double primitive cell increases when passing from the para phase to the ferro phase (Kobayashi et al., 1970; Cook, 1967)

T is the temperature of interest in the ferroelectric phase, and T_P is a reference temperature in the paraelectric phase; $\alpha_{ij}(T)$ is the element of the crystal's thermal expansion tensor at temperature T (see Annex A1). The second term of the second member is obtained by extrapolating the temperature variations observed in the paraelectric phase. Figure 1.10b illustrates these temperature variations of the lattice parameters for KDP, along with a schematic representation of these in the para and ferroelectric phases Fig. 1.10a and the extrapolation of the lattice parameters from the tetragonal phase into the orthorhombic phase. This method can obviously pose some precision issues if one strays too far from the transition.

The choice of the reference system for writing the e_{Sij} elements of the strain tensor in both phases is very important: we want to identify the spontaneous parameters, such as spontaneous polarization, spontaneous strain, and other properties represented by non-zero values in the ferroelectric phase and zero values in the paraelectric phase. To do this, the reference system of the paraelectric phase is usually retained. Figure 1.11 presents the tensors of the properties of most interest concerning the type of transition considered here: elastic, dielectric, piezoelectric,

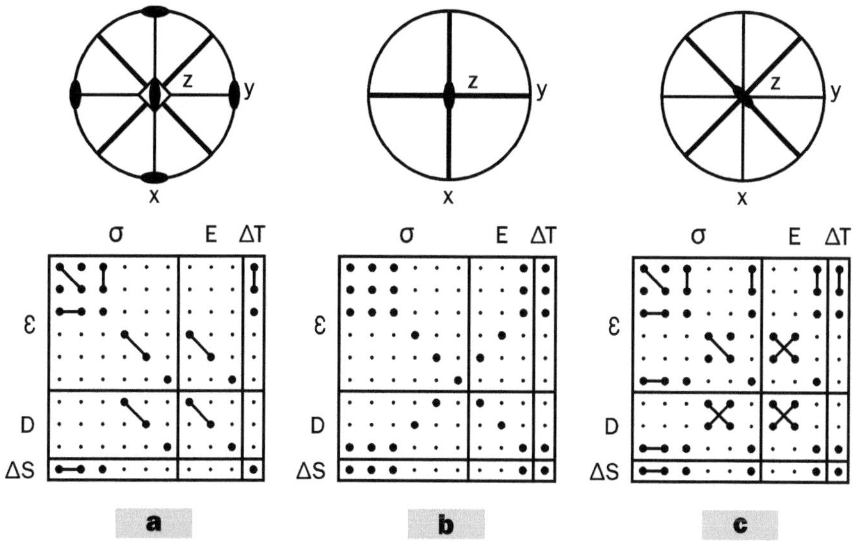

Fig. 1.11 Elements of symmetry (at the top) and matrices of the tensorial physical properties for the corresponding point symmetry group (at the bottom) (Nye, 1992): (**a**) group $\bar{4}2$ m; (**b**) mm2 group in the usual reference system of this group (axes **a** and **b** at 45° from the tetragonal axes $\mathbf{a}_1$ and $\mathbf{a}_2$); (**c**) mm2 group considered as a subgroup of the group $\bar{4}2$ m

pyroelectric, and calorific properties. As illustrated in Fig. 1.11c, all properties characterized by a symmetric second-rank tensor have a spontaneous quantity at the transition. This is the case for birefringence Δn_c for light propagating along the ferroelectric *c*-axis. Figures 1.10 and 1.11 illustrate the choice of definitions retained in this work: extrapolation of the values observed in the paraelectric phase to define the quantities that will exhibit spontaneous variations at the transition and writing the tensorial physical properties in the reference system of the paraelectric phase (see Annexes A1.1 and A1.2). Tagantsev et al. (2010) highlight the differences between this commonly accepted definition, Aizu's (1969, 1970) definition, Wadhawans (1982) definition, and the ambiguities that should be avoided. In the same work (Tagantsev et al., 2010), one can find summaries concerning changes in the symmetries of ferroelectric crystals that integrate previous works (Aizu, 1969; Toledano and Toledano, 1980, 1988), etc. For example, if one considers a tensorial property that appeared in the ferroelectric phase with symmetry described by group K, one can calculate, using the order g_K of its group, the quantity

$$n_K = g_P / g_K \tag{1.7}$$

which gives the number of lattices and domains involved; n_K obviously varies with property *K*. It is also possible to define a degeneracy factor using relation (1.1),

$$d_K = n / n_K \tag{1.8}$$

If, for example, the property defined by group K is polarization, n_K will give the number of ferroelectric domains; if the property is strain, n_K will give the number of ferroelastic domains. It is thus possible in this context of transitions with symmetry reduction to introduce a parameter describing the difference between the low-symmetry phase, which is usually more ordered at low temperatures, and the high-symmetry phase, which is more disordered. This "order parameter" is an extensive quantity that allows for a thermodynamic analysis of the transition (Annex A2).

In summary, the physical quantity chosen as the order parameter must be zero in the reference phase at any temperature above the transition and non-zero below the transition temperature in the ferro phase. It is determined by the symmetry characteristics and the change in physical properties observed at the transition.

1.3 Thermodynamic and Phenomenological Approach to the Transition

1.3.1 The Different Orders of Structural Phase Transitions

Transitions between the para state and the ferro state can be described within the general framework of structural transitions, as illustrated by an extensive bibliography. We differentiate between transitions for which the two phases can coexist, called first-order transitions since Ehrenfest (1933), and those for which the material continuously transitions from one phase to another without the two phases being in equilibrium in each other's presence. These transitions are called second-order transitions. In this section, we limit ourselves to a general and phenomenological presentation without delving into the significance of theoretical models with free energy developments, based on a chosen parameter.

Initially, the "Landau development" addressed second-order transitions, excluded fluctuations, and corresponded to an asymptotic situation at absolute zero well illustrated by static masses and springs (Landau, 1965; Landau and Lifshitz, 1967). The works of Ginzburg, Mueller and Cady, Devonshire (1954) allowed progress, which would explain the success of these models, by estimating that the transition temperatures are small compared to "atomic temperatures," thus relatively close to absolute zero (Vaks, 1968). A publication summarizes the situation (Levanyuk et al., 2020). For our part, we will follow the widespread approach of using a model that works phenomenologically while avoiding overly bold exploitations.

As shown previously, if the transition results in a reduction of symmetry (from the para phase to the ferro phase), the difference between the high symmetry phase and the low symmetry phase accompanies the introduction of an order parameter (a term that effectively means "symmetry-breaking parameter"). It seems quite logical to choose as the order parameter for a ferroelectric transition the polarization that appears, and the deformation for a ferroelastic transition. This is the case for so-called "proper" ferroelectric crystals (as "proper ferroelastics" for ferroelastics).

However, transitions can also occur by minimizing energy through the appearance of another physical quantity that, by coupling, induces the variation of polarization and/or deformation. We are dealing with so-called improper ferroelectrics and/or improper ferroelastics. Attention must be paid to the number of different unit cells in the ferro phase and the number of possible domains, using relations (1.1), (1.7) and (1.8): for example, if K corresponds to the polarization in a "proper" ferroelectric crystal, we will find d_K equal to 1, whereas it will be <1 for an "improper" ferroelectric (and similarly for the deformation of a ferroelastic crystal).

To illustrate, typical experimental results with a simple model are proposed to account for them. In the case of reference crystals, spontaneous polarization and spontaneous deformation are fully coupled, along with birefringence for light propagating along the c-axis (quantities represented by a rank 2 tensor as illustrated in Fig. 1.7). Figure 1.12 experimentally confirms the proportionality between polarization, deformation, and birefringence for different crystals of these symmetries. The ease offered to the experimenter is understood since it suffices to measure the variation of one of these quantities to know the variation of another. This choice of crystal families clearly highlights the competition between mechanical energy and electrostatic energy of crystals that are simultaneously ferroelectric and ferroelastic.

But how do these spontaneous quantities vary with temperature? Let us choose the one that expresses the ferroelastic aspect, the deformation u_{xy} (or e_6) for crystals of the KDP family. A small-angle gamma-ray diffraction experiment (description of

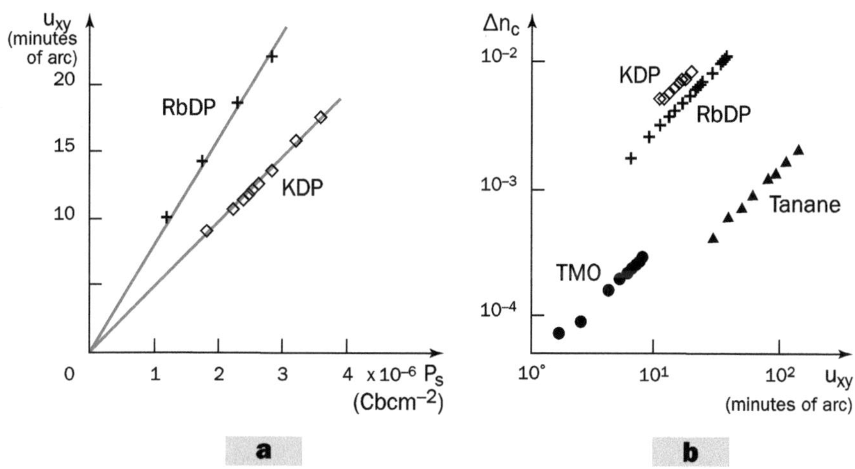

Fig. 1.12 Linear relationships are observed in crystals with the transition $\bar{4}2$ m to mm2 between spontaneous deformation, spontaneous polarization, and spontaneous birefringence in ferro phase: (**a**) relationship between deformation and polarization in KDP (Zwicker and Scherrer, 1944; Benepe and Reese, 1971; Bastie et al., 1978) and RbDP (Strukov et al., 1973; Bastie et al., 1978), (**b**) relationship between birefringence Δn_c and deformation u_{xy} in KDP (Bastie et al., 1975), RbDP (Troussaut and Vallade, 1980), TMO (Bastie and Bornarel, 1982), and Tanane (Bastie et al., 1976a, b)

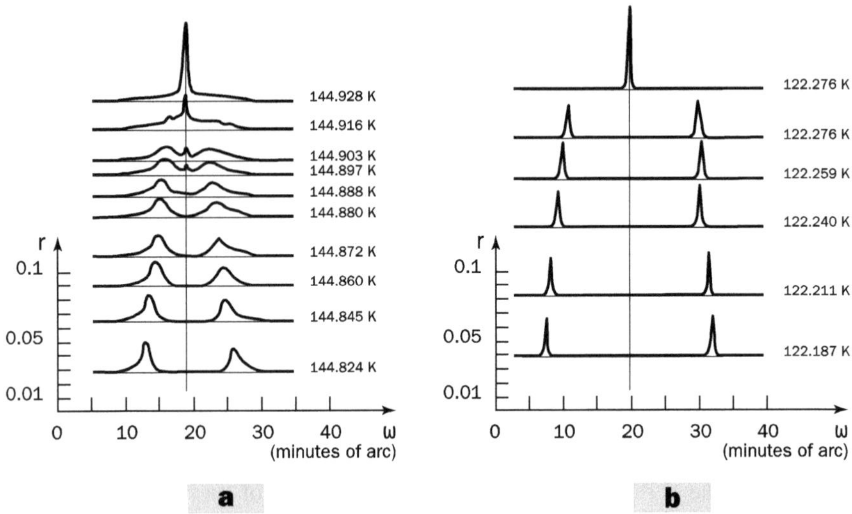

Fig. 1.13 Spontaneous deformation in the KDP family: Gamma diffraction experiments as a function of temperature allow very precise measurements of shear deformation u_{xy} during transitions (see Annex A3); thus, we obtain illustrations of the second-order transition of RbDP (**a**) and the first-order transition of KDP (**b**), even if these two transitions are close to tri-critical situations (Bastie et al., 1978)

the experiment in Annex A3), very sensitive to the rotations of the reticular crystallographic planes, illustrates (Fig. 1.13) the difference in results between crystals of the same family: the deformation appears regularly from a zero value for the second-order transition of RbDP (Bastie et al., 1978) and instantaneously with a non-zero value for the KDP transition (Bastie and Becker, 1984).

This difference between types of transitions is also illustrated in Fig. 1.14 by the variation of u_{xy} with temperature for crystals concerned by the same symmetry change. During the para-ferro transition, a regular increase from a zero value is observed for RbDP, a significant jump at the transition for DKDP, and a less marked jump for KDP. The result concerning the Tanane crystal is also presented, illustrating a second-order transition with very significant shear deformation values (2 arc degrees at 0 °C!).

This experimental illustration continues in Fig. 1.15a with a characteristic of first-order transitions: the measurements taken at increasing temperature and those at decreasing temperature do not systematically overlap: there may be thermal hysteresis. This type of transition will be studied in Chaps. 5 and 6 of this book with the coexistence of the two phases within a certain temperature range for the DKDP crystal. Figure 1.15b shows that this crystal, placed in an enclosure where the temperature evolves regularly and uniformly, maintains a constant temperature throughout the coexistence between the para and ferro phases and the exchange of latent heat. Each of us uses this phenomenon with the water-ice transition by putting a few pieces of ice in our glass of water to keep a drink cool.

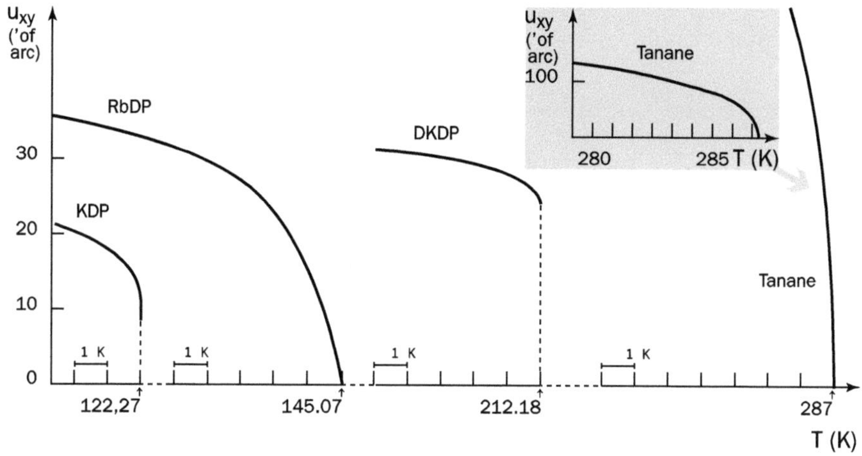

Fig. 1.14 Spontaneous shear deformation as a function of the temperature of ferro-ferro crystals with tetragonal-orthorhombic transition ($\bar{4}2$ m to mm2). The DKDP crystal is a good illustration of a first-order transition (Reese and May, 1968; Zeyen and Meister, 1976), RbDP of a second-order transition (Bastie et al., 1976a, b), and KDP of a first-order transition close to the tri-critical situation (Bastie et al., 1975; Zeyen and Meister, 1976). The Tanane transition is evidently of the second order; it should be noted that the large value of deformation obtained, of 2 degrees of arc at 0 °C (Bordeaux et al., 1973)

1.3.2 Development of Landau and Second-Order Transition

To describe such macroscopic phenomena, one generally uses an expression of free energy or another appropriate thermodynamic function. Annex A2 summarizes the most common expressions of thermodynamic functions, recalling that they concern stable states: parameter modifications should not deviate the system too far from these states for differentials such as dU for energy, dW for work done, etc., to be relevant, which highlights the importance of quasi-static and reversible measurements. Some relations will remain valid only if the function is expressed with an appropriate choice of intensive parameters and some of the extensive parameters that are conjugated to them. By respecting these conditions, very different situations of continuous transitions and even those of discontinuous transitions, without too significant discontinuities, have been described in a generally satisfactory manner using the Landau formalism (1967) taken up for ferroelectricity by Devonshire (1954).

Suppose that the transition of interest is continuous, correctly described by a very simple scalar order parameter called q, and that the energy is not altered if the sign of this scalar is changed. The value of q is assumed to be homogeneous throughout the crystal in a determined experimental situation. Furthermore, since stable equilibria are sought, the curves representing the free energy as a function of q must have a positive concavity at equilibrium points, including that at $q = 0$ in the para phase. The free energy can then be described at a temperature close to that of the transition using an expansion in even powers q^{2n} of the order parameter with the first

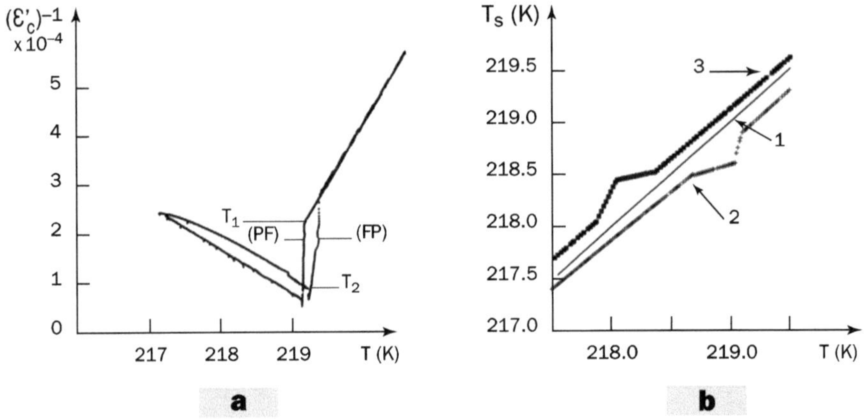

Fig. 1.15 First-order transition of the DKDP crystal: (**a**) the variation of the inverse dielectric constant along the *c* axis shows the existence of a thermal hysteresis during this transition; (**b**) the temperatures indicated correspond for the continuous line 1 to the temperature *T* measured using a platinum resistance in the gas exchange cell (helium), very close to the crystal; *T* evolves regularly over time. The curves 2 and 3 correspond to the measurement of sample temperature T_S by a thermocouple on the upper face of the sample during, respectively, a para-ferro (PF) transition and a ferro-para one (Bornarel and Cach, 1999). The temperature of the crystal remains constant during the phase coexistence, with exchange of latent heat and volume change between the two phases. Even more precise demonstrations have been experimentally produced by high-precision calorimetric measurements at about 10^{-4} K (Reese, 1969) and by neutron diffraction at about 10^{-3} K (Zeyen et al., 1976)

term, in q^2, necessarily positive in the para phase. Using the Gibbs thermodynamic function (see Annex A2), one can write (Landau, 1967), for a second-order transition and for a hydrostatic pressure considered constant:

$$\left\{G(T,q)-G_0=(a/2)q^2+(b/4)q^4+(c/6)q^6+\ \ldots\right\} \tag{1.9}$$

*G*0 is the part of *G* independent of *q*. The intensive parameter *X* conjugated to *q* (such as *E* and *σ*) is assumed to be zero, so the temperature *T* is the only variable parameter. For the coefficient a, the simplest form that vanishes at the transition temperature *T*0 is chosen, i.e., $a = A(T - T0)$ with *a*, *b*, and *c* assumed independent of temperature and hydrostatic pressure. If the transition is continuous, *q* remains small when *T* is close to *T*0, and the term in q^6 can be neglected. The coefficient *b* must then be positive for the energy to increase as the order parameter increases. A stable state *G* minimum as a function of *q* is defined by:

$\delta G/\delta q = 0 = aq + bq^3 + cq^5$, not much different from $aq + bq^3 = 0$,

and $\delta^2 G/\delta q^2 = a + 3bq^2 > 0$

The stable solutions are evident on the graph of Fig. 1.16a:

$$\text{For}\, T > T_0, q = 0\, \text{and for}\, T < T_0, q = \pm(-a/b)^{1/2}$$

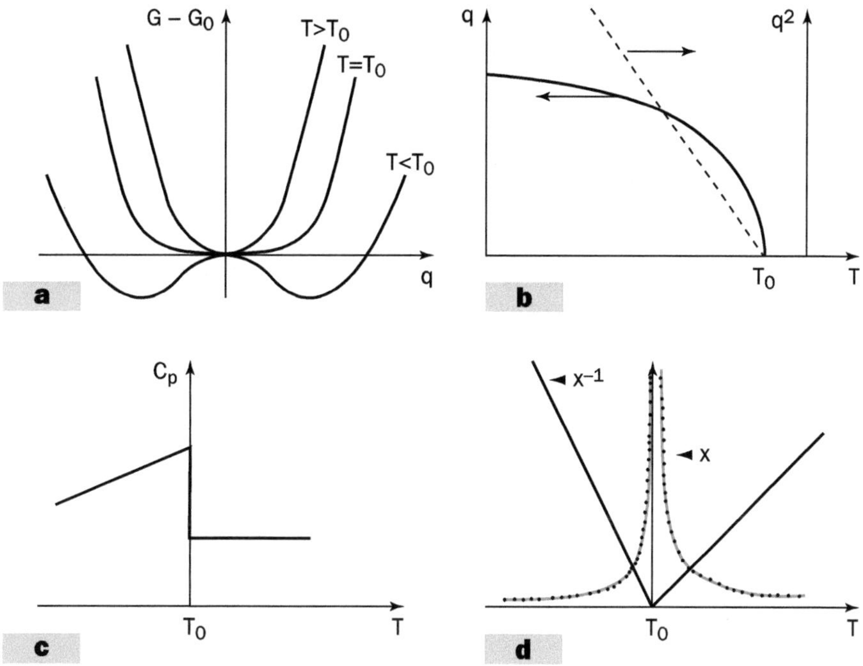

Fig. 1.16 Second-order transition within the framework of the phenomenological model of Landau with positive b constant: (**a**) Variation of the Gibbs free energy as a function of the order parameter q (and as a function of temperature T). (**b**), (**c**), and (**d**): Variations as a function of T of the order parameter q and of q^2 (b), of the specific heat at constant pressure $C_P(c)$, of the susceptibility $\chi = \delta q/\delta X$ and its inverse (**d**)

In summary, in this simple phenomenological model of a second-order transition, in the ferro phase, two stable states are characterized by two opposite values of the order parameter q, whose square will vary linearly with the temperature as illustrated in Fig. 1.16b:

$$q^2 = (A/b)(T_0 - T) \tag{1.10}$$

we will also have:

$$G_0 = G_0 - A^2 (T_0 - T)^2 / 4b \tag{1.11}$$

The entropy in the vicinity of this transition without latent heat can be written by neglecting the higher powers of q

$$S = -\delta G/\delta T = S_0 - (\delta a/\delta T) q^2 = S_0 - (A^2/b)(T_0 - T) \tag{1.12}$$

in which $S_0 = -\delta G(T,0)/\delta T$. With these simple assumptions, in the vicinity of the transition (T close to T_0), the specific heat at constant pressure is written as:

$$C_P = T(\delta S/\delta T)_P = C_{P0} + (A^2\ T)/b \tag{1.13}$$

Since in the para phase, S is equal to S_0 and C_P to C_{P0}, a discontinuity of the specific heat at the transition appears in this model as represented in Fig. 1.16c. In the same way, one can predict discontinuities of the specific heat at constant volume C_v, the thermal expansion coefficient, the compressibility, etc. and express the values of these discontinuities in terms of each other.

The quantity q, an extensive quantity, like the polarization **P** for a dielectric, can be modified by the application of a conjugate intensive quantity X (the electric field **E** in the cited case) which adds $-Xq$ in the second term of relation (1.9), i.e., $-EP$ for the electrostatic aspect. The adapted thermodynamic function in the electrostatic case (Annex A2) is the function G_2:

$$G_2 = G - EP \tag{1.14}$$

To remain within the conditions recalled above, one must not stray too far from the equilibrium states and one will write $dG_2/dq = 0$

$$aq + bq^3 - X = 0, \text{i.e. } aP + bP^3 - E = 0 \tag{1.15}$$

hence for the inverse susceptibility:

$$\delta X/\delta q = a + 3bq^2, \text{ i.e. } \delta E/\delta P = a + 3bP^2 \tag{1.16}$$

If T is greater than T_0, the inverse susceptibility is equal to $A\ (T - T_0)$, which corresponds to the well-known Curie-Weiss law, and, if T is less than T_0, it is equal to $2A\ (T_0 - T)$. This concerns the inverse dielectric permittivity ($\delta E/\delta P$) for a (proper) ferroelectric and the elastic stiffness constant ($\delta\sigma/\delta u$) for a (proper) ferroelastic as illustrated in Fig. 1.16d. In this second-order transition model, the susceptibility should take an infinite value at the transition temperature T_0.

To account for the situation when the crystal is subjected to a significant conjugate field X (E in electrostatics), one seeks stable solutions by continuity with the previous case ($X = E = 0$) using linear approximations such as $P_+ = P_S + \chi E$ (and $P_- = -P_S + \chi E$). Thus, for a ferroelectric,

$$G_2 = G_0 + (1/2)A(T - T_0)P^2 + (1/4)b\,P^4 + (1/6)cP^6 - EP \tag{1.17}$$

a function whose minima are the solutions of

$$A(T - T_0)P + bP^3 + cP^5 - E = 0 \tag{1.18}$$

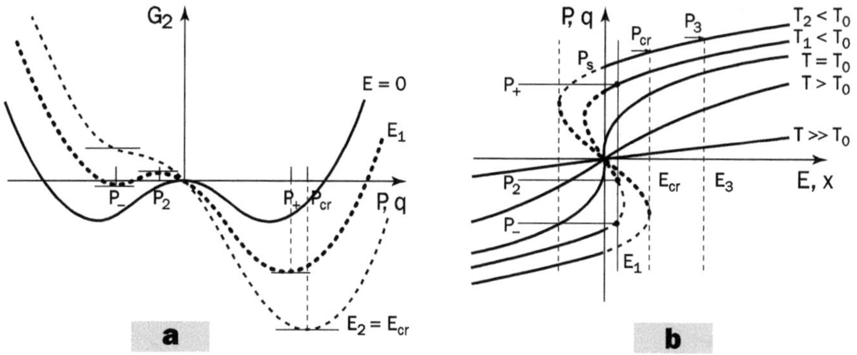

Fig. 1.17 Second-order transition of a crystal subjected to the field conjugated to the order parameter: (**a**) Representation of the function $G_2 = G - EP$ (or $-qX$) (relation 1.14) as a function of P(or q) for various values of the parameter E (or X): curve for $E = 0$ in solid line; curve for $E = E_1$ whose minima occur in P_- and P_+ (in dotted line); we have a maximum in P_2 situation of instability; curve $E = E_{cr}$ critical field for which only one stable P_{cr} state is possible at a temperature lower than T_0 (in dashed lines); the values of E_{cr} and P_{cr} are given by the relation (1.19) of the text; (b) variation of P as a function of E with the temperature T as a parameter, according to the relation (1.18). The parts of curves in continuous lines correspond to stable situations, the dashed lines to metastable situations, and the dotted lines to non-physical situations from a thermodynamic point of view. We find the values P_-, P_+ from (**a**) for $E = E_1$. For the critical field E_{cr}, the metastable situation cannot persist. When E is greater than E_{cr} as for E_3, only one value of P can exist: the crystal is mono-domain

Figure 1.17 illustrates the effect of T and E with the continuous transition between the para phase and a mono-domain ferro state. One obtains situations with stable, metastable, unstable states and the infinite tangent at $E = 0$ and $P = 0$ (infinite susceptibility already mentioned) for the curve at $T = T_0$ in Fig. 1.17b. Similarly, the critical situation (E_{cr}, P_{cr}) in Fig. 1.17a (thin dashed lines) corresponds to the existence of only one possible polarization. The "critical field" E_{cr} corresponds to the relations

$$\delta G_2 / \delta P = \delta^2 G_2 / \delta P^2 = \delta^3 G_2 / \delta P^3 = 0 \text{ and } \delta^4 G_2 / \delta P^4 \text{ positive.}$$

If we still neglect in G_2 the term in P^6, we find

$$E_{cr} = 2A^{3/2} \left(T_0 - T\right)^{3/2} / \left(3^{3/2} b^{1/2}\right) \text{ and, } P_{cr} = A^{1/2} \left(T_0 - T\right)^{1/2} / \left(3^{1/2} b^{1/2}\right) \quad (1.19)$$

A remark is necessary: these reasonings use a thermodynamic, macroscopic approach and conclude the existence of only one polarization if the applied conjugate field is greater than the critical field. This assertion poses no problem. But what about the simultaneous existence of stable and unstable states? The $P(E)$ representation in Fig. 1.18a, a simple transcription of the results from Fig. 1.17b, reveals the limits of the exercise: the branches AH and DE correspond to simple mono-domain situations. But what happens in the branches HC and DG? The assumptions made do not suppose the existence of domains, and their limits are illustrated by this theoretical "hysteresis cycle" of a mono-domain crystal which should be drawn with an

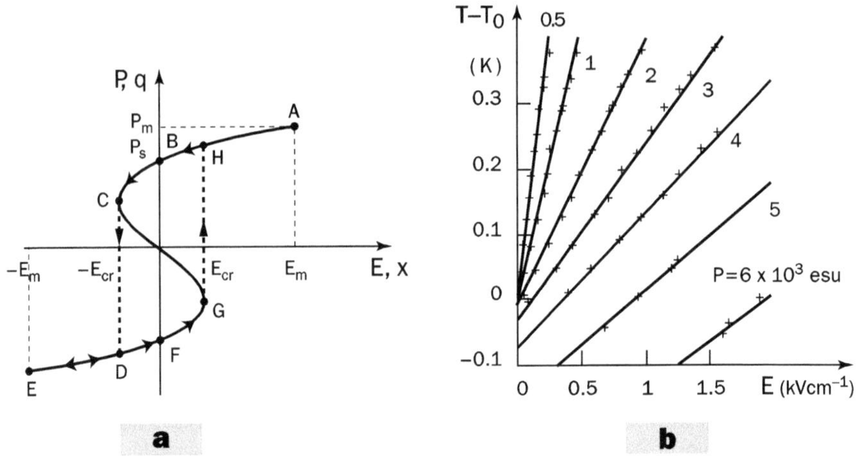

Fig. 1.18 Second-order transition of a crystal subjected to the field conjugated to the order parameter: (**a**) The hysteresis cycle illustrates a quasi-static switching: starting from situation A under an applied field E_m with a polarization P_{m+}, the field E is decreased very slowly to remain in quasi-stable conditions; in B, with E equal zero, the polarization is, by definition, the spontaneous polarization P_S. If E becomes negative, the polarization P_+ corresponds to a metastable state. In C under an applied electric field $-E_{cr}$, only the polarization P_- can exist, and the positively polarized system becomes unstable and flips discontinuously as expressed by the dashed lines on the representation. If the modulus of the field grows, we will arrive at E. The reverse process to return to a positive mono-domain state will occur following EDGHA; (**b**) the isopols defined by the relation (1.20): $T - T_0 = (AP)^{-1} E - (bP^2 + cP^4) A^{-1}$ are straight lines. As P increases, their slope and their y-intercept (for $E = 0$) decrease. These variations constitute a mark of second-order transition. The results presented were obtained with KDP under an electrostatic pressure of 3 kbar (Western et al., 1978). The polarization is expressed in units esu ×10 (reminder: 1 $Cm^{-2} = 3.10^5$ ues)

infinite period. This theoretical approach provides no information on the phenomena that allow a mono-domain crystal of a certain polarity to transition to the mono-domain situation of opposite polarity. Moreover, the critical field values obtained by relation (1.19) are very large compared to the values of coercive fields measured experimentally. E_{cr} is the maximum value of the field E that would be necessary to apply in the thermodynamic model for the entire crystal to have the same polarization with a macroscopic polarization reversal occurring as a whole. As often in solid-state physics, microscopic processes (at the domain and wall level) and nanoscopic ones (at the unit cells level) will offer possibilities of reversal with applied fields much weaker than this critical field as we will see in Chaps. 3 and 6. Ordinarily, the experimenter performs measurements at constant temperature (isotherms) or at constant applied electric field (iso-fields), but the Landau formalism also allows presentations of results at constant polarization useful for quickly detecting the order of transitions. Such "isopols" are plotted for one of the typical crystals in the context of a second-order transition in Fig. 1.18b (Western et al., 1978). From relation (1.18) written in the form

$$T - T_0 = (AP)^{-1} E - (bP^2 + cP^4) A^{-1} \tag{1.20}$$

one obtains the representation of lines $T - T_0 = f(E)$ from which the values of A are deduced by their slope, and of the values of b and c by the ordinate value for $E = 0$ which moves downwards as the parameter P of the curves increases. This mode of representation has been effectively used to clarify the order of transitions including that of KDP in experiments under hydrostatic pressure (Schmidt et al., 1976).

1.3.3 From the Phenomenological Model to a Microscopic Model

The success of Landau's development (1935) and his phenomenological approach, using macroscopic quantities, has motivated the search for descriptions at the nanoscopic or microscopic level (Salje et al., 1991; Salje, 1992). The starting point is to write a Hamiltonian with three terms in a particle scheme illustrated by Fig. 1.19:

- A contribution of kinetic energy from particles of mass m and momentum p
- A potential energy illustrated by a field in which the particle is located, with potential wells of a depth U_0
- Interaction energies of the particle with other particles of type $v_{ij}(q_i - q_j)^2$, q_i and q_j linked to the order parameter

In summary, if the sum of the interaction energies is much greater than U_0, the system is close to a "displacive" transition, whereas in the opposite situation, it is close to the order-disorder limit. The thermodynamic limit of the system is found in the quasi-harmonic approximation of Einstein with a characteristic frequency Ω. Within the framework of the mean field model, one can relate the Hamiltonian to the Gibbs free energy and thus propose a formulation of the excess energy G in the form

$$G = (A\theta_S / 2)(\coth(\theta_S / T) - \coth(\theta_S / T_C))q^2 + (B/4)q^4 + \ldots \tag{1.21}$$

with θ_S proportional to Ω, and T_C the Curie temperature in this second-order transition. This relation works well for fairly low temperatures, and when T becomes

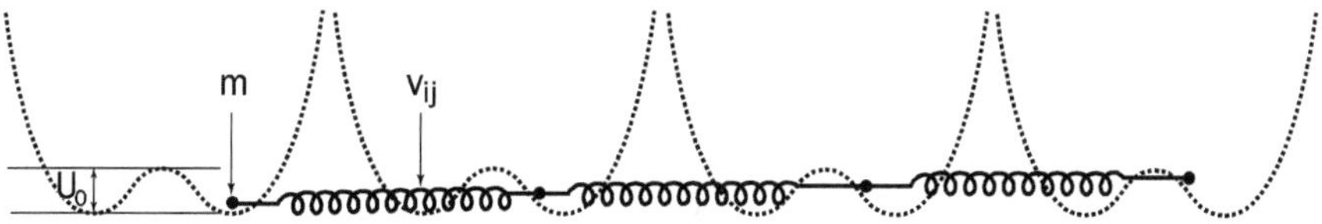

Fig. 1.19 Mechanical analogy of the Hamiltonian to describe the phenomenon of phase transition using particles of mass m in a potential with two successive wells separated by an energy barrier U_0; each particle interacts with others located in nearby potentials with the parameter v_{ij} as an indicator of stiffness. The thermodynamic limit of the model allows to find the Gibbs free energy in its usual form as a function of the order parameter, here related to the position of the particle

much greater than θ_S, the well-known development is recovered, including over hundreds of degrees for crystals whose phase transitions occur at high temperatures (Salje et al., 1991):

$$G = (A/2)(T - T_C)q^2 + (B/4)q^4 + \dots$$

1.3.4 Landau Development and First-Order Transition

The previous phenomenological description can also be used regarding first-order transitions even if the order parameter is no longer a continuous function. It is necessary, strictly speaking, to limit oneself to small discontinuities and to temperatures close to the transition; one also works as close as possible to the equilibrium states that appear in the ferro phase. The coefficient b of relation (1.9) without applied conjugate field is chosen to be negative and c to be positive; the curves plotted in Fig. 1.20a account for the results obtained by this phenomenological

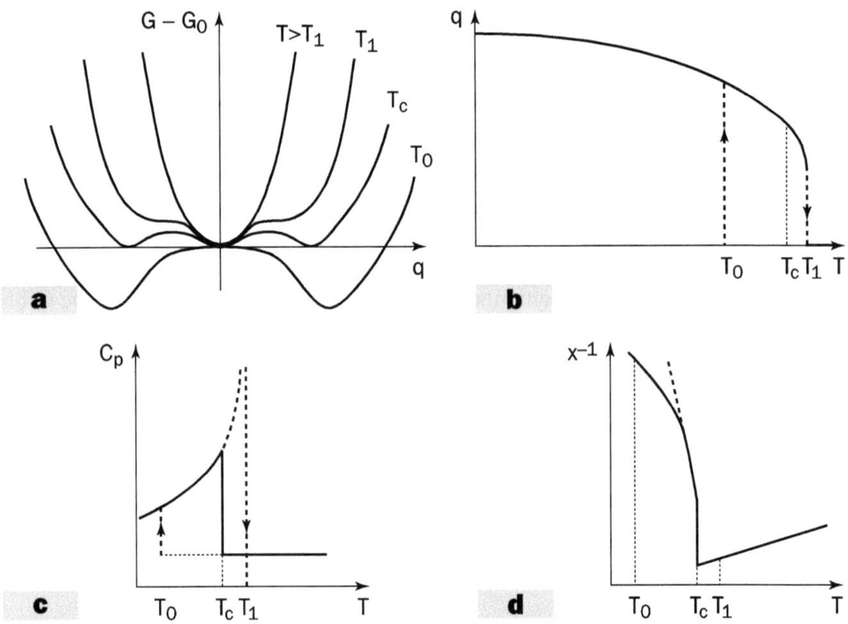

Fig. 1.20 First-order transition in the Landau model: (**a**) Variation of the Gibbs free energy as a function of the order parameter q; (**b**) variation of the order parameter q with the temperature T: phase coexistence is possible between the temperatures T_0 and T_1, the Curie temperature T_C corresponds to energy minima at identical G values; (**c**) variation with T of the specific heat at constant pressure C_P: the dotted paths illustrate the thermal hysteresis that can exist between a para-ferro temperature cycle and a ferro-para cycle. (**d**) Inverse susceptibility $\chi^{-1} = (\delta q/\delta X)^{-1}$ as a function of T: the graph corresponds to the theoretical passage from one phase to another at the Curie temperature T_C

description. It is also noted that for a high temperature in the para phase, there exists a single stable state of G at $q = 0$. But, unlike the second-order transition, when the temperature decreases, from a temperature T_1 higher than T_0:

$$T_1 = T_0 + b^2 / 4Ac \tag{1.22}$$

Two metastable states appear for

$$q_{\mathrm{MS}} = \pm(-b/2c)^{1/2} \text{ and } G - G_0 = -b^3 / 48c^2 \tag{1.23}$$

(b is negative).

The most stable state remains at $q = 0$ with $G - G_0 = 0$ for T greater than T_C. At temperature T_C, the three possible states ($-q_S$, 0, $+q_S$) are at the same level concerning the value of the Gibbs free energy:

$$T_C = T_0 + 3b^2 / 16Ac \tag{1.24}$$

$$q_S = \pm\ (-3b/4c)^{1/2} \tag{1.25}$$

with barriers between wells of height ($-b^3/96c^2$).

For a temperature between T_C and T_0, the stable states correspond to the two values of q different from zero, and the situation at $q = 0$ is metastable. Finally, at the temperature T_0, the state at $q = 0$ becomes unstable and only the two stable ferro states remain:

$$q_S = \pm(-b/c)^{1/2} \text{ with } G - G_0 = b^3 / 12c^2 \tag{1.26}$$

For temperatures below the value T_0, the stable states of the ferro phase are defined for the values $\pm\ q_S$ with $q_S\ (T)$ represented in Fig. 1.20b:

$$q_S = \pm\left(\left(b^2 - 4cA\left(T_0 - T\right)\right)^{1/2} - b\right)^{1/2} / (2c)^{1/2} \tag{1.27}$$

The entropy and specific heat are calculated as for the second-order transition:

$$S - S_0 = (A/4c)\left(b - \left(b^2 - 4Ac\left(T - T_0\right)\right)^{1/2}\right) \tag{1.28}$$

and

$$C_P - C_{P0} = \left(A^2\ T/2\right)/\left(b^2 - 4Ac\left(T - T_0\right)\right)^{1/2} = A^{3/2}\, T / 4c^{1/2}\ \left(T_1 - T\right) \tag{1.29}$$

Thus, a divergence of the specific heat C_p is observed at T_1 as illustrated in Fig. 1.20c, whereas it only included a discontinuity for a second-order transition.

The susceptibility can be calculated as previously; the result concerning the inverse susceptibility represented in Fig. 1.20d as a function of T differs from that of the second-order transition: indeed, the Curie law in $A(T - T_0)$ is found again for temperatures above T_C, but at the temperature T_C, a discontinuity appears between the value $3b^2/16c$ in the para phase and $3b^2/4c$ in the ferro phase. For temperatures below T_C, the inverse susceptibility is written (see Fig. 1.20d):

$$1/\chi = 4A\left(T_1 - T\right) + 2bA^{1/2}c^{-1/2}\left(T - T_1\right)^{1/2} \tag{1.30}$$

with a quasi-linear temperature variation near the transition but more complicated when moving away from it, with a negative concavity. This characteristic of concavity has indeed been used by some researchers as an argument to conclude the existence of a first-order transition based on experimental results of dielectric constant (Tsunekawa et al., 1969). The greater caution is recommended in the use of this argument since the contribution of the domains to the dielectric permittivity can have the same consequence on the shape of the curves. As with the second-order transition, one can study the effect of a conjugate field, namely E in the ferroelectric case: Fig. 1.21a presents the functions $G_2 = G - EP$ as a function of P for some values of E, and Fig. 1.21b provides the representation $P(E)$ for a fixed temperature higher than T_0; only the solid line parts of the curves have physical significance. The possible existence of double hysteresis loops, specific to the first-order transition, is understood. As previously, one can calculate the value of the critical applied field beyond which there exists throughout the crystal a polarization of the same orientation as that of E. The calculation follows the same approach as before with a function G_2 of the form (1.14) with zero hydrostatic pressure. By neglecting the terms in powers of P higher than 6, the following critical values are obtained:

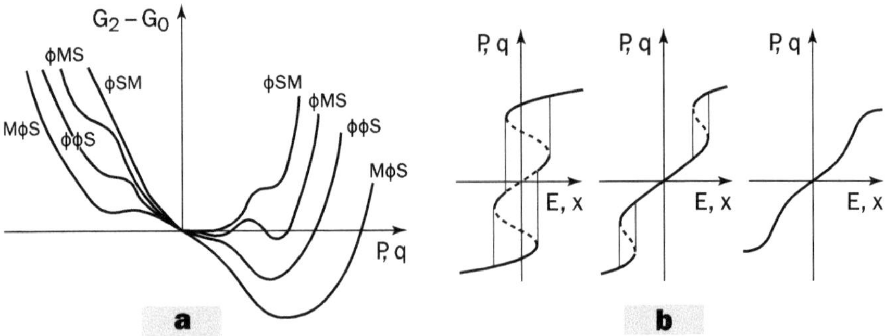

Fig. 1.21 First-order transition with applied conjugate field: During a first-order transition, there are stable (S), metastable (M), and unstable (ϕ) situations: (**a**) Representation of the $G_2(P)$ function (relation 1.14) at various temperatures and for a determined applied E field. The proximity to $G_2(P)$ curves from Fig. 1.20a is recognized. The different situations are indicated as ϕMS (unstable, metastable, stable). (**b**) The P (or q) variations as a function of E (or X) are comparable to those of a second-order transition (Figs. 1.17a, b) when $T >> T_1$ and $T << T_0$. The coexistence of phases is illustrated by double cycles for $T > T_1$ (right curve), $T_1 > T > T_C$ (middle), and $T_C > T > T_0$. Stable situations are indicated by solid lines

$$P_{cr} = \left(-3b/10c\right)^{1/2},$$
$$E_{cr} = \left(8c/3\right)P_{cr}^{\;5} = 2^{1/2}\left(-b/5\right)^{5/2}\left(3/c\right)^{3/2}, \quad (1.31)$$
$$T_{cr} = T_0 + \left(9b^2/20Ac\right)$$

As with a second-order transition, the value obtained for E_{cr} is usually much higher than those of the fields measured experimentally: these calculations yield maximum values derived from a homogeneous macroscopic model without any assumption about what may occur at the microscopic and nanoscopic levels, for example at the interfaces between two regions of opposite value of the order parameter.

The representations of *P*(*E*) at constant temperature, like the curves of *P*(*T*) at constant applied electric field, are frequently produced in the literature, but the isopols introduced previously are also of great interest; some are plotted in Fig. 1.22a for KDP at zero pressure (Western et al., 1978). Unlike the second-order transition,

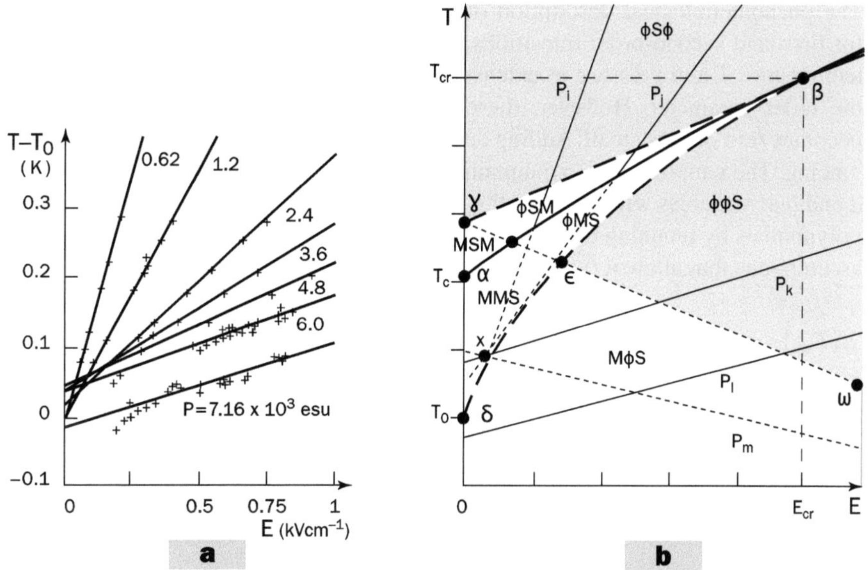

Fig. 1.22 Isopols and first-order transition: (**a**) Representation of isopols in accordance with relation (1.18), in a *T*(*E*) plane, for KDP under a pressure of 0 kbar (Western et al., 1978). The polarization is indicated in esu ×10^3 units. The slopes of the isopol lines decrease with increasing *P*; on the other hand, the *y*-intercept increases then decreases with increasing *P* which results in the crossing of lines, characteristic of a first-order transition; (**b**) phase diagram in the *T*(*E*) plane with isopols (relation (1.20)) ($P_i < P_j < P_k < P_l$) in continuous lines (stable situations) and discontinuous (metastable situations). The first-order transition line joins point α ($E = 0$, $T = T_C$) to the critical point β (E_{cr}, T_{cr}). One can continuously pass from a stable para phase ($\phi S \phi$) to a stable ferro phase ($\phi\phi S$) by bypassing point β. Between the caustics of the isopols ($\gamma\beta$ and $\delta\varepsilon\beta$) various situations are encountered as illustrated by Fig. 1.21a: MSM, ϕSM, ϕMS, MMS. At a point such as χ, the isopols P_j, P_k, and P_m (extension of an isopol with $P < 0$) intersect, which illustrates the discontinuity of *P* in this *P* and *E* situation

crossings of lines are observed; at a crossing point, two values of polarization are possible, hence the discontinuities observed in the variations $P(T)$. All the crossing points lie in the (E, T) plane within a region defined by caustics whose most extreme point in T and E corresponds to the values T_{cr} and E_{cr} given previously. Figure 1.22a published by Western et al. (1978) thus indisputably demonstrated that at zero hydrostatic pressure, the KDP transition is of the first order. Figure 1.22b qualitatively summarizes the situations that can be encountered during the coexistence of a first-order transition between the para phase and the two types of polarization (up and down for an applied up field), situations that can be stable, metastable, or unstable. This working diagram will be used in the rest of the book, notably in Chap. 6 to understand the first-order transition of DKDP under an applied electric field.

1.3.5 The Tri-critical Transition

The phenomenological description of Landau's development is quite satisfactory for first- and second-order transitions, with parameters a, b, and c, independent of temperature T and pressure p, extensive parameters external and not conjugate to the order parameter. However, these assumptions are no longer sufficient if b becomes zero or very small. Adding terms to the development, in P^8, P^{10} ... is unconvincing. The simplest way to maintain the same type of development is to consider a and b as variables with T and p; this variation can still be simply expressed using polynomials by retaining only the first terms of a development without omitting the assumptions that allow it (proximity of stable states); then, a and b can be written:

$$a(T,p) = a_T\left(T - T^*\right) + a_p\left(p - p^*\right) + \text{higher order terms in} \left(T - T^*\right)^n \text{and} \left(p - p^*\right)^n$$
$$b(T,p) = b_T\left(T - T^*\right) + b_p\left(p - p^*\right) + \text{higher order terms}$$
$$\text{with}\, a_T, a_p, b_T, b_p \text{ constants in } T \text{ and in } p \tag{1.32}$$

Under the assumption of linear approximation and by writing $a(T, p) = 0$ and $b(T, p) = 0$, one obtains the temperature T_0 by extrapolating the variation of inverse susceptibility with temperature, and the temperature T_b at which the coefficient b vanishes:

$$T_0\left(p\right) = T^* - \left(a_p / a_T\right)\left(p - p^*\right), \text{ and } T_b\left(p\right) = T^* - \left(b_p / b_T\right)\left(p - p^*\right) \tag{1.33}$$

Thus, the Gibbs potential is written (without application of conjugate fields) with T_0 and T_b function of p:

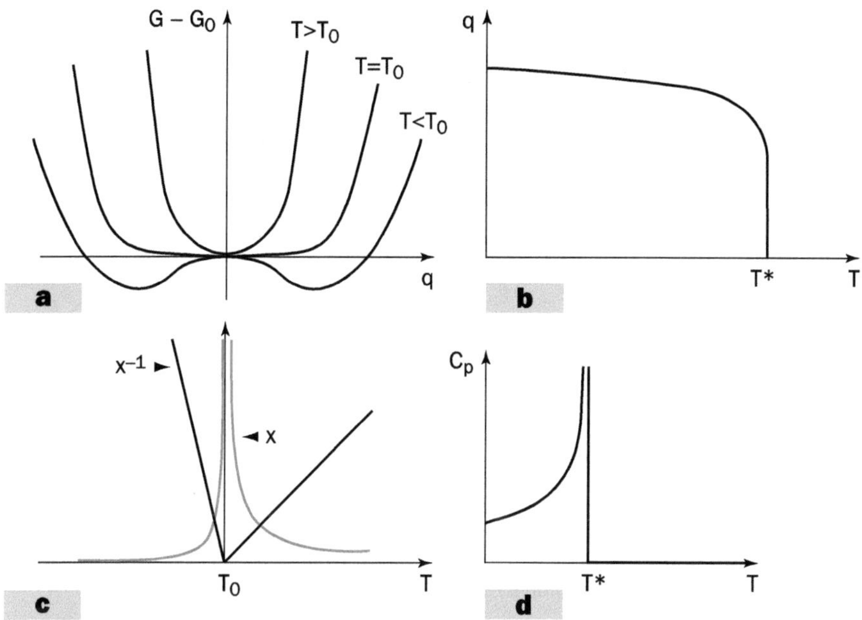

Fig. 1.23 Tri-critical transition: (**a**) Variation of the Gibbs free energy as a function of the order parameter q (e.g. (Ikeda, 1996)); (**b**) variation of the order parameter q with temperature; (**c**) variation of the susceptibility χ and its inverse with T: the slope in the ferro phase of $(\delta q/\delta X)^{-1}$ is in modulus twice as large as that observed for a second-order transition in Fig. 1.16d; (**d**) the variation of the specific heat becomes infinite at the transition temperature

$$G(T,p,q) - G_0 = (a_T/2)(T - T_0)q^2 + (b_T/4)(T - T_b)q^4 + c/6q^6 + \ldots \quad (1.34)$$

With the help of this potential, the value of the spontaneous order parameter, entropy, equilibrium volume change, and susceptibility can be calculated. The graphs in Fig. 1.23 provide an overview of the results obtained.

The representation of phase diagrams allows for a better understanding of the order of transitions and the tri-critical situation depending on experimental conditions. The entire phase diagram is schematically shown in Fig. 1.24a with three parameters: temperature T, pressure p, and X field conjugate to q (which is here E or σ). Also plotted in Fig. 1.24b is the projection of the diagram in the (p, T) plane, thus for a crystal without an applied external conjugate field. The lines represented, with $a = 0$ and $b = 0$, delimit, for the first, the para (a positive) and ferro (a negative) zones, and for the second, zones of first-order (b negative) and second-order (b positive) transitions. The intersection of these two lines occurs at the tri-critical point, first defined by Griffits (1970) with $T = T^*$ and $p = p^*$. Figure 1.25 shows, for crystals of our typical families with para-ferro transitions close to the tri-critical situation, the good agreement between experimental observation and the choice of b linear in T and p in the development G (T, p): the results concern KDP and TMO. There is good agreement over a fairly wide temperature range, unlike simulations with a

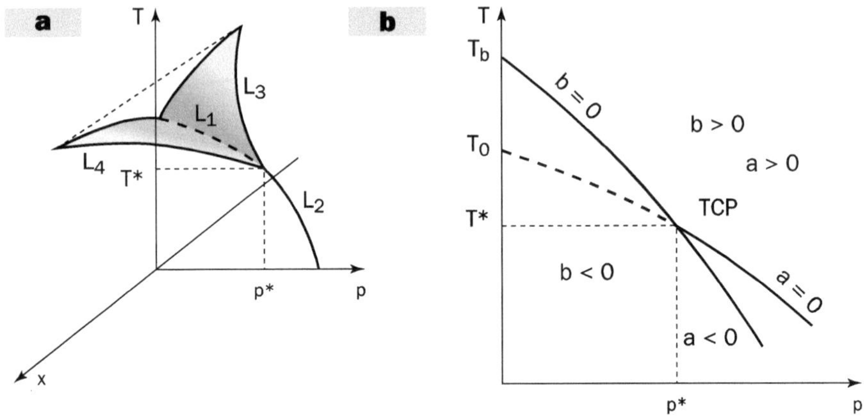

Fig. 1.24 Tri-critical transition: (**a**) Phase diagram near a tri-critical point. X is the field conjugate to the order parameter q, T the temperature, and p the pressure. L_1 is the line that corresponds to $a = 0$ and delimits the para and ferro regions. L_2, L_3, and L_4 correspond to second-order lines that intersect at the tri-critical point (T^*, p^*, $X = 0$); the "wings" correspond to regions of phase coexistence; (**b**) in the $X = 0$ plane, the curves $a = 0$ and $b = 0$ are plotted; the latter delimits the region with positive b (second-order transitions) and that with negative b (first-order transitions)

constant b coefficient in T and p and additional terms in the development. The application of relation (1.9) with $b = 0$ gives a dependence of q_S of the form

$$q_S = \left(A\left(T_0 - T\right)/c\right)^{1/4} \tag{1.35}$$

Figure 1.26 illustrates relations (1.10) and (1.35) with logarithmic scales for various "proper" ferroelectric crystals, the slopes 1/2 and 1/4 for second-order and tri-critical transitions, respectively.

Decades of work on transitions of certain compounds in the KDP family illustrate this tri-critical situation between first-order and second-order transitions. Indeed, depending on the measurement methods or experimental conditions, researchers provided elements that argued either for a second-order transition or for a first-order transition, similarly for RbDP of first or second order. The doubt was lifted by experiments in which hydrostatic pressure and the conjugate field (electric field E in this case) were modified: at zero pressure p and zero applied field E, the transition of RbDP is second order and that of KDP is first order. For KDP, one swings from the first-order transition with zero hydrostatic pressure (Fig. 1.22a) to a second-order transition for a hydrostatic pressure close to 2.3 kbar (Western et al., 1978); or again, at zero pressure p, the critical electric field that swings from first to second order is found by authors between 120 V cm^{-1} (Strukov et al., 1972) and 370 V cm^{-1} (Sidnenko and Gladkii, 1973). However, if the experimental results are unambiguous during a clear first-order transition (e.g. DKDP), it is often difficult to decide experimentally in cases close to the tri-critical (like KDP). Figure 1.27

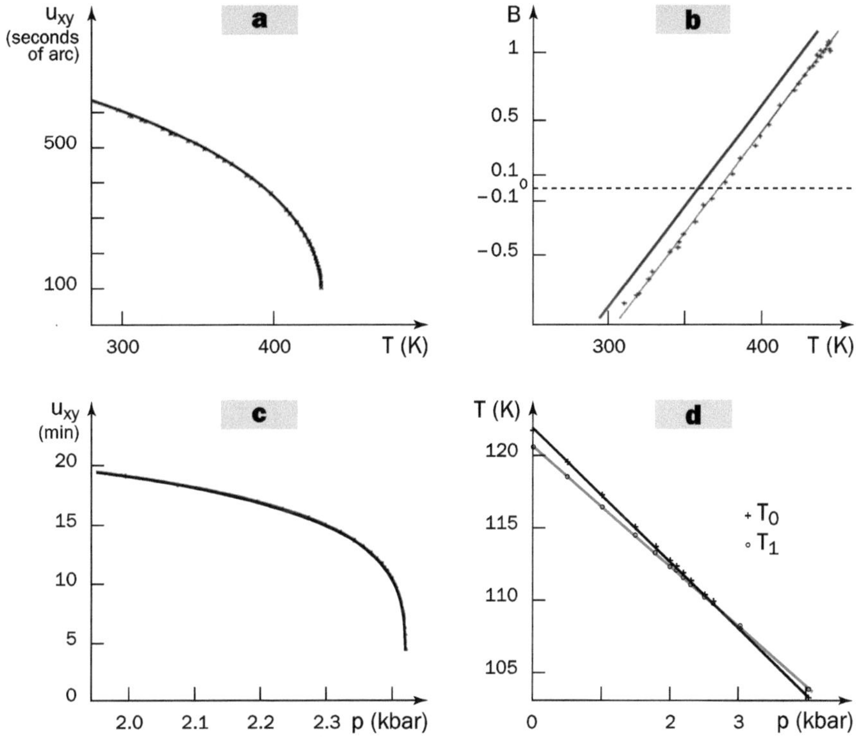

Fig. 1.25 During transitions close to the tri-critical situation, we observe the linear variations with temperature of the parameter b_T (from the Landau development of relation (1.34)): (**a**) and (**b**): In the case of the TMO crystal (improper ferroelectric), the variation of the spontaneous deformation with temperature illustrates a very close first-order transition to the tri-critical (Konak et al., 1975); the B ratio between the coefficient $b(T)$ and its value at the transition shows a linear variation with T (Bastie and Bornarel, 1982); (**c**, **d**): The other example of a transition close to the tri-critical situation is given by the KDP whose variation of the deformation u_{xy} with T was illustrated in Fig. 1.14. The variation with pressure (in **c**) at a temperature of 110.83 K and especially the linear variations of T_0 and T_1 (in **d**) illustrate this proximity to the tri-critical transition (Bastie et al., 1981)

illustrates this by the temperature variation of elastic constants. For crystals near a tri-critical transition, the extremely flat shape of the graph $G(q)$ when q is close to zero (Fig. 1.23a) favours a notable effect of fluctuations and numerous external parameters. The experimenter may encounter difficulties in obtaining stable values of certain physical parameters. This theme of experimental conditions, notably the size of samples (Kobayashi et al., 1971), their history and preparation (Eberhard and Horn, 1975), is addressed in Sect. 1.5. To conclude, the phenomenological description using Landau's development is simple, macroscopic, but static and thus ignores possible dynamic effects and significant fluctuations. However, in many cases, this model accounts with good precision for the observed results, not only near the transition when the value of the order parameter is small but also relatively far from the transition, which may be surprising given the basic assumptions. These good results

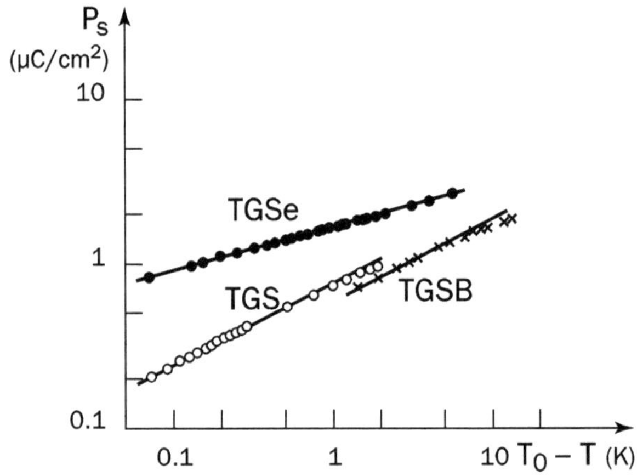

Fig. 1.26 In "proper" ferroelectric crystals with a second-order transition, the spontaneous polarization is given by the relation $P_S = (a/b)\ (T_0 - T)^{1/2}$ hence a slope of ½ in the representation of P_S (T) on logarithmic scales; this is the case for triglycine sulphate TGS (Gonzalo, 1966) and Triglycine fluoberylate TGSB (Stankowska, 1973). When the transition is tri-critical, the relation becomes $P_S = (a/c)\ (T_0 - T)^{1/4}$ hence a slope of 0.25 for triglycine selenate TGSe (Hatano et al., 1977)

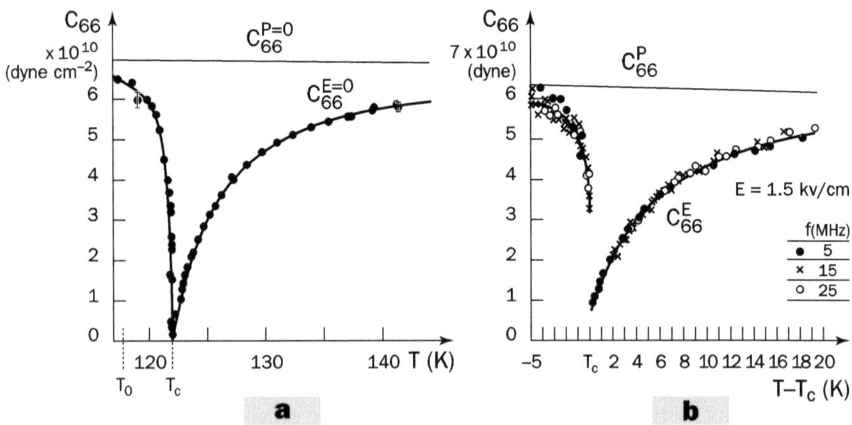

Fig. 1.27 Measurements of the elastic constant c_{66} as a function of temperature (**a**) in the KDP (Brody and Cummins, 1968) and (**b**) in the DKDP (Litov and Uehling, 1968) show almost stability in temperature for c_{66} at constant polarization and a strong variation near the transition for c_{66} measured at constant electric field. The practically zero value of $c_{66}{}^E$ at T_C for the KDP (the soft mode) and the jump for the DKDP illustrate the difference of these transitions: tri-critical for one and clearly first order for the other; the experimental points correspond to various measurement frequencies: 5 MHz (•), 15 MHz (×), 25 MHz (o)

explain the success of this description. Its limits will be seen in Chap. 3 during the reversals (often called "switchings") induced by the applied conjugate field. The properties of the crystal then depend on the more advantageous microscopic or nanoscopic phenomena from an energetic point of view for the reversal.

1.4 Order Parameters and Phase Transitions

Phenomenological approaches to para-ferro transitions often allow predicting the variations of macroscopic quantities as a function of temperature, pressure, applied electric fields, and stresses. Even in these cases, one desires to know what happens at the nanometric level in a volume that transitions from one phase to another. The questions that arise are of various natures:

- Does the detected quantity correspond to the order parameter, or are we in an "improper" situation of a quantity modified by coupling with the true order parameter? For example, in Fig. 1.28, a second-order transition situation is represented with an order parameter q and a deformation that appears by coupling with a certain temperature law, like the elastic constant that will be calculated: the corresponding ferroelastic transition is improper.
- Is the transition of the "displacive" type and the result at the nanometric level of small regular atomic displacements, or of the order-disorder type with sets of regions in which abrupt structural changes have occurred at a determined temperature? The answers to these questions can be obtained if one knows the structural modifications and changes at the atomic level in the cell, which is not necessarily simple from a static and dynamic point of view. In the present work limited to ferro-ferro crystals, an objective is to illustrate the importance of interfaces, that is the walls between domains and the phase fronts at the transition. The field of study is also limited to a single symmetry change (tetragonal $\bar{4}2$ m to orthorhombic mm2) to notably show that it is possible with structurally very different crystals to use identical objects at the microscopic level, such as dislocations or quasi-dislocations (Chaps. 2 and 3). Presented below in a succinct

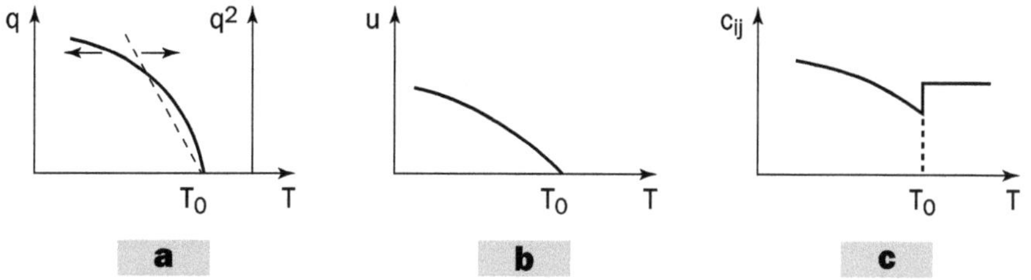

Fig. 1.28 Improper second-order ferroelastic transition: (**a**) the order parameter q follows the classic variation with q^2 linear in T in the ferroelastic phase; (**b**) the deformation u which appears by coupling follows a variation in T that depends on this coupling; (**c**) the same is true for the elastic constant which would be the susceptibility if the order parameter was u

manner are the crystals whose results are partially used, like those of the $Gd_2(MoO_4)^3$ family and Tanane, and in a more detailed manner the crystals of the KDP family (Annexes A4 and A5) for which the link between macroscopic, microscopic, and nanoscopic phenomena is developed. Regarding GMO and its compounds such as TMO, the order parameter is neither the polarization nor the deformation. This crystal has been important for understanding the very existence of improper ferro transitions (Fousek and Kovak, 1972). The crystals obtained by the Czochralski process (Nassau et al., 1965) appeared, from the first crystallographic measurements (Keve et al., 1970; Borchardt and Bierstedt, 1967; Cross et al., 1968) as ferroelectrics or ferroelastics of a new kind. Initially suggested (Pytte, 1970), the explanation was provided from a theoretical point of view (Dvorak, 1971a, 1971b) and then experimentally (Petzelt and Dvorak, 1971; Axe et al., 1971; Dorner et al., 1972): the condensation of a mode actually occurs at the boundary of the Brillouin zone (near the M point) rather than near its centre; the order parameter consists of antiferroelectric displacements that are coupled to the shear deformation u_{xy}. This induces the polarization P_z by piezoelectric coupling. Thus, the spontaneous polarization and deformation of the ferro phase are proportional to the square of the order parameter.

Tanane is a crystal whose para and ferro cells conform to the description in Fig. 1.7. The symmetry variations at the para-ferro transition, similar to those of KDP, should not make one forget the differences between these crystals: for example, KDP and its family compounds are minerals, and Tanane is an organic crystal without hydrogen bonding. For illustration, the modifications of the atomic arrangements of Tanane are presented in Fig. 1.29 (Legrand 1981). The origin of the transition relies on the molecular inversion movement between symmetrical equilibrium positions (average residence time of 10^{-9} s in the para phase) which becomes asymmetrical in the ferro phase. A linear coupling between the Ising parameters and the lattice deformation creates an acoustic instability and imparts a ferroelastic character to the transition with significant spontaneous deformation. The (slight) ferroelectricity occurs due to the piezoelectric coupling (Lajzerowicz and Legrand, 1980; Legrand, 1981). Tanane is now referred to as TEMPO (Tetramethylpiperidine Oxyl). Both names can be found in the bibliography. The KDP crystal, on the other hand, has been known for a long time. It is presented in Fig. 1.30 with the various crystallographic planes that will be consistently used throughout the book. Figure 1.31 shows that the primitive cell contains four $(PO_4)^-$ tetrahedra linked by hydrogen bonds, and K^+ ions located between the tetrahedra along the c-axis. The tetragonal axes indicated as a_1 and a_2 correspond, if we refer to Fig. 1.7a, to the horizontal edges of the cube, while the axes a_1' and a_2' are the horizontal traces of the mirror planes m_1 and m_2. The chemical formula is maintained at the level of each tetrahedron in a region of homogeneous polarization of the crystal and the temperature ranges used here; the position of the hydrogen ions on their bond and the displacement of the potassium and phosphorus ions in the direction of the c-axis are directly correlated. In the para phase, the proton is statistically in the middle of the hydrogen

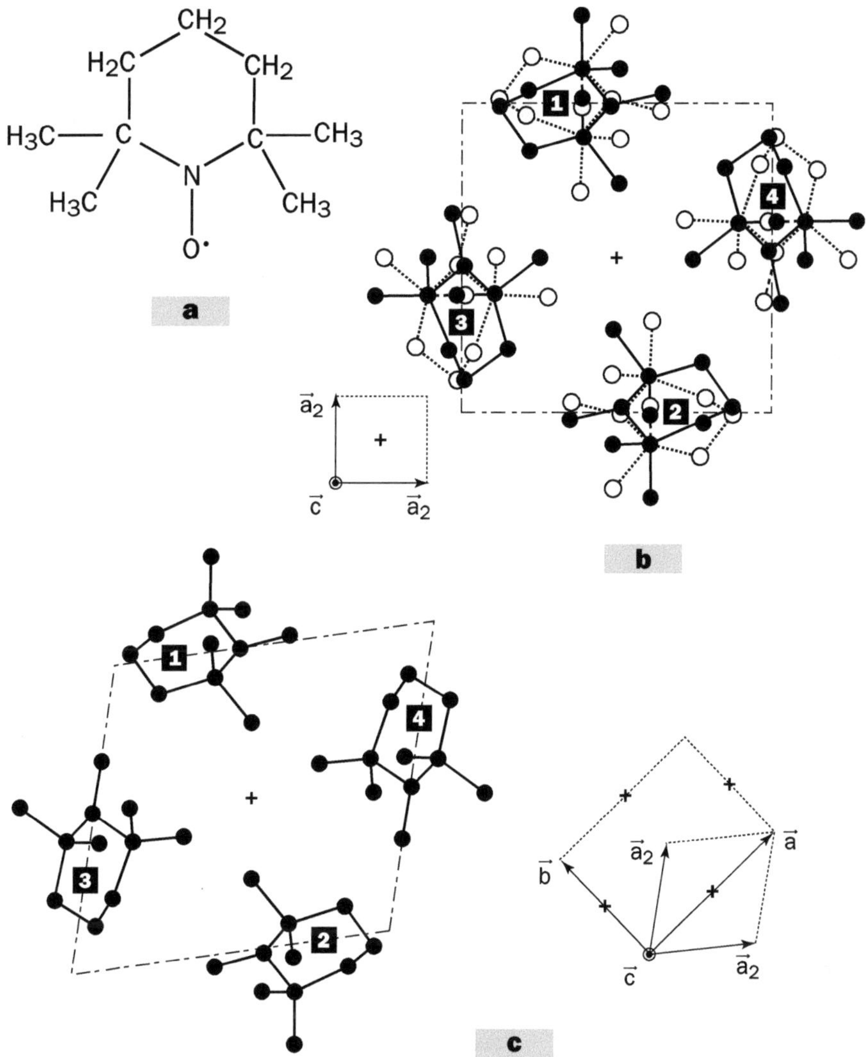

Fig. 1.29 (**a**) Chemical formula of Tanane projected in the plane (001); (**b**) and (**c**) crystal motif and elementary cell in the para phase (**b**) and in the ferro phase (**c**) (Bordeaux et al., 1974)

bond, whereas in the ferro phase, it is closer to one of the two $(PO_4)^-$ tetrahedra. If the hydrogens of the two lower bonds (along the c-axis) are close to a $(PO_4)^{--}$ as shown in Fig. 1.32, the dipole moment contributing to the polarization will be along this c-axis in the negative direction; conversely, if the hydrogens close to the tetrahedron are those of the uppermost bonds, the polarization will be oriented in the positive direction of c. Figure 1.32 contains the minimal information required for the continuation of the book, but more detailed information is provided in Annex A4 regarding the structural parameters of the ferro crystals of the KDP family (density

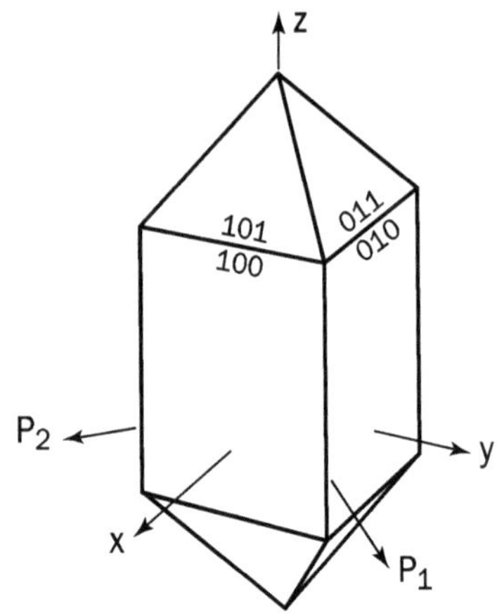

Fig. 1.30 The KDP single crystal has a shape of a double pyramid elongated along the fourfold axis. The twofold axes, *X* and *Y*, are perpendicular to the faces of the prism. The diagonal axes P_1 and P_2 are not equivalent in terms of their piezoelectric response: a compression along P_1 produces a positive polarization along *Z*, while a compression along P_2 creates a polarization in the opposite direction

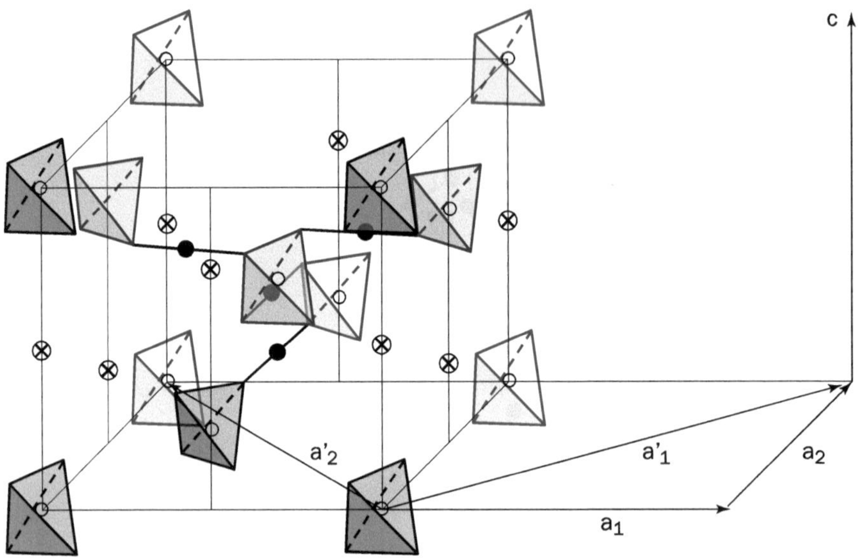

Fig. 1.31 The primitive cell is of tetragonal symmetry $\bar{4}2$ m in the paraelectric phase. The K^+ ions are located along the c axis at equal distance from two PO_4^- tetrahedra. Four hydrogen bonds depart from each tetrahedron to join four others, two of them at the top of the tetrahedron and two of them at the bottom. A hydrogen bond always connects the top of a tetrahedron to the bottom of a neighbouring tetrahedron. There are two types of tetrahedra in terms of orientation. Each tetrahedron of one type is connected to four tetrahedra of the other type. The hydrogen bonds are approximately oriented, either in the direction of the tetragonal axis a_1 or in that of the tetragonal axis a_2

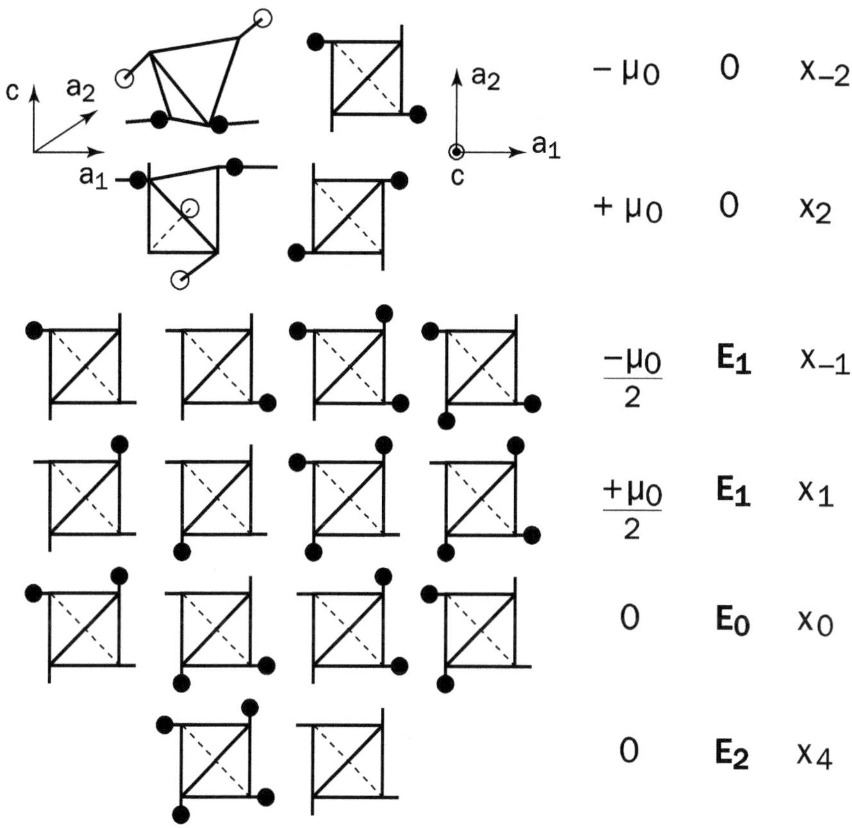

Fig. 1.32 The 3D perspective representations of tetrahedra at the top left continue the presentation of Fig. 1.31, itself inspired by West's initial publication (1930). They remind us that tetrahedra have orientations of two types as also illustrated by Figs. A4.1 and A4.3 of Annex A4 (de Quervain (1944)) with hydrogen bonds approximately oriented either in the tetragonal direction a_1 or in the direction a_2 (the two bonds at the top of a tetrahedron are almost parallel to each other, as are those at the bottom). The simplified representations, used several times in the work, clearly highlight the positions of the protons on their bond but do not account for the type of tetrahedra. For the various situations of tetrahedra with their close protons, the dipole moments associated with the group following the *c* axis, the energy with the stable situations of the ferro crystal as the origin, and the population probability of the concerned tetrahedra are given on the right (the sum of the x_i is equal to 1)

of hydrogen presence on the bond, orientations relative to the tetragonal axes of the lines joining two oxygens and those describing the displacement of H). Similarly, Annex A5 offers an overview of the transition models of these compounds. The various grouping situations presented in Fig. 1.32, their energies, and probabilities of existence will be useful in the following chapters.

1.5 Importance of the Sample and the Experimental Procedure

Many properties of ferroelectric and ferroelastic crystals strongly depend on the experimental conditions. This section focuses on the phenomena that often disrupt measurements and the precautions to take: an effect observed following the variation of an external parameter (temperature, pressure, stress) may originate from the experimental process. The quality of the crystal, its shape, the equipment used, the operating mode, and the way the measurements are conducted, all these elements play a role in the results. Nothing should be overlooked to understand the physical reality, particularly the competition between mechanical energy and electrostatic energy.

1.5.1 Quality of the Crystal and the Sample

Obtaining a high-quality crystal is a prerequisite for conducting relevant experiments, starting from the crystal growth. The crystals in our panel (GMO family, Tanane, KDP family) have been the subject of detailed studies for various reasons:

- The GMO family because its temperature range and hardness made it a candidate for numerous optical applications (Borchardt and Bierstedt, 1965; Kumada, 1972).

- Tanane was the first organic ferroelectric crystal without hydrogen bonding upon its discovery. It requires particular attention because the existence of defects or local stresses in this soft crystal leads to the rapid end of experiments: the orthorhombic phase of the ferroelectric-ferroelastic crystal study is metastable as illustrated in Fig. 1.33 (Legrand, 1981). A too abrupt temperature change, a defect, or a local stress is enough to induce the appearance of the stable monoclinic structure and the transformation of the crystal into powder! This is a case where the experimenter must be attentive!

- The nonlinear optical properties of the KDP family explain the enthusiasm for studying these crystals, which continues with or without doping (Solanski et al., 2021a, b; Bodrov et al., 2023). It is quite easy to grow KDP, RbDP, and DKDP crystals from aqueous solutions (with heavy water for deuterated compounds) but more challenging to obtain large KDA and DKDA crystals. The fine work of the

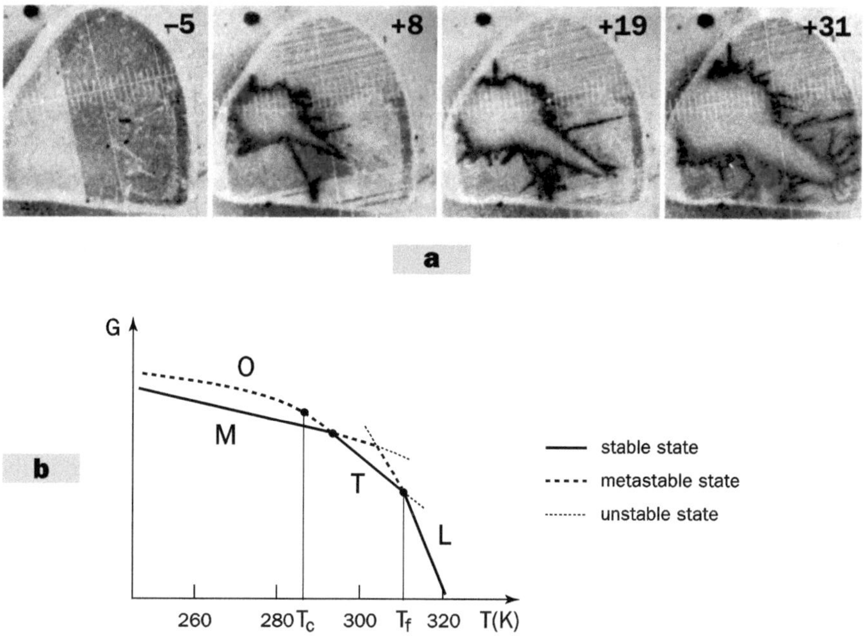

Fig. 1.33 (**a**) In this example of the orthorhombic phase of Tanane, the texture in domains is particularly simple with only one domain wall in a permissible plane. At time $t = 0$, the application of a small constraint causes the disappearance of the metastable orthorhombic phase and the appearance of the monoclinic phase, the progression of which is observed over time, with many polycrystals (Legrand, 1981); (**b**) enthalpy diagram in which the metastable situation of the orthorhombic phase appears

Czech school, then in France by the Dijon group, and now by American industrial groups, has enabled the acquisition of reliable growth methods. The crystal, whose growth is achieved at room temperature from an aqueous solution, appears in the form of an elongated double pyramid along the c-axis, which will become the ferroelectric axis at low temperature (Fig. 1.30). The manner in which the growth process is carried out is very important at each stage (saturated solution, introduction of the seed, supersaturation, withdrawal of the crystal from the solution) to obtain a stress-free, clear, and dislocation-free crystal. The essential parameters are the control of pH, the growth rate, and the agitation speed of the bath (Brezina and Havrankova, 1971; Gavrilova and Kurznjecov, 1964; Remoissemet et al., 1970; Pierre, 1971). For a long time, it has been known how to obtain very large crystals with very good optical properties (Rashkovich, 1991). Industrial constraints have also led to the development of rapid crystal growth technologies (Verma, 2010). Numerous studies aim to understand why ferroelectric crystals grow faster along c than antiferroelectric ones (Ren et al., 2008), the impact of dopants (Solanski et al., 2021a, b; Raj and Perumal, 2022), and mixed crystals. Solid bibliographies on this topic can be found in specialized books and synthesis publications (De Yoreo et al., 2002; Menchon et al., 2018). However, the quality measurement requirements, particularly at the para-ferro transition, are high, and not all parts of a fine crystal are interchangeable. To create a sample of the best possible quality, crystals with zero prismatic part angles are chosen. Furthermore, the sample must come from the pyramidal part free of dislocations and growth defects, and not from the larger prismatic part (Belouet et al., 1974). Cutting with a wire saw moistened with water and alcohol in the tetragonal planes with a precision of a few arc minutes precedes polishing, achieved either with alumina powders of grains smaller than 0.1 μm or on a silk support with a water-alcohol mixture. The quality of a sample is assessed in two ways: microscopic studies in transmitted polarized light and reflected light, with the aid of a compensator and conoscopic observation, should reveal no defects or inhomogeneity. Regarding quality in terms of dislocation density, stress, and mosaicity, small-angle gamma-ray diffraction is effective, as illustrated in Fig. 1.34, with unfortunately few in-depth studies on accidental dislocations (Klapper et al., 1974). The extinction that can be observed in the para phase on the diffraction peaks in Fig. 1.34c is proof of good quality (Bastie and Becker, 1984). To evaluate it, the integrated reflecting power of the experimental peaks R_m is compared with the value that can be calculated by kinematic diffraction theory R_{cin} for a "totally imperfect" crystal. The ratio $y = R_m/R_{cin}$ varies between the value 1, which corresponds to the imperfect crystal, and the value defined by the ratio R_m/R_{dyn}, R_{dyn} being the value calculated in dynamic theory. Figure 1.34c shows how well kinematic diffraction theory suits the ferro phase. As in kinematic diffraction, the integrated reflecting power is proportional to the diffracting volume, and the volume of the domains is easily evaluated by the area of the peaks corresponding to their average deformation. The study in the para phase allows, by comparing the extent of the observed extinctions, the selection of the best samples whose maximum density of accidental dislocations must not exceed 100 per cm^3 (Bastie et al., 1976a, b). Transitions in the ferro phase and the duration of experiments can deteriorate this parameter, but annealing for several months at room temperature allows the recovery of almost the initial values. The mosaic of the crystal, another quality parameter of a sample, can be evaluated using gamma diffraction described in Annex A3 within the experimental divergence limit of the diffraction peak (about 10–20 arc seconds for the results presented in this work); this result is to be compared with high-resolution X-ray diffraction measurements, which allow about arc seconds (Mueller et al., 2002a, b). If the mosaic exceeds this value, it will be detected as illustrated in Fig. 1.34d: within the sample, a peak width at half-height equal to the experimental width of the gamma-ray beam is found, demonstrating the quality of the crystal; however, the peak broadens in the parts of the sample near the edge: the mosaic can be estimated at about one arc minute in a "layer" about 200 μm deep. Thus, between observations with optical microscopy in polarized light, the measurement of extinction, and that of mosaic, one can determine the degree of quality of the samples, assess the impact of this quality on the physical quantities studied, and eliminate samples whose insufficient quality could influence the measurements and their physical interpretation.

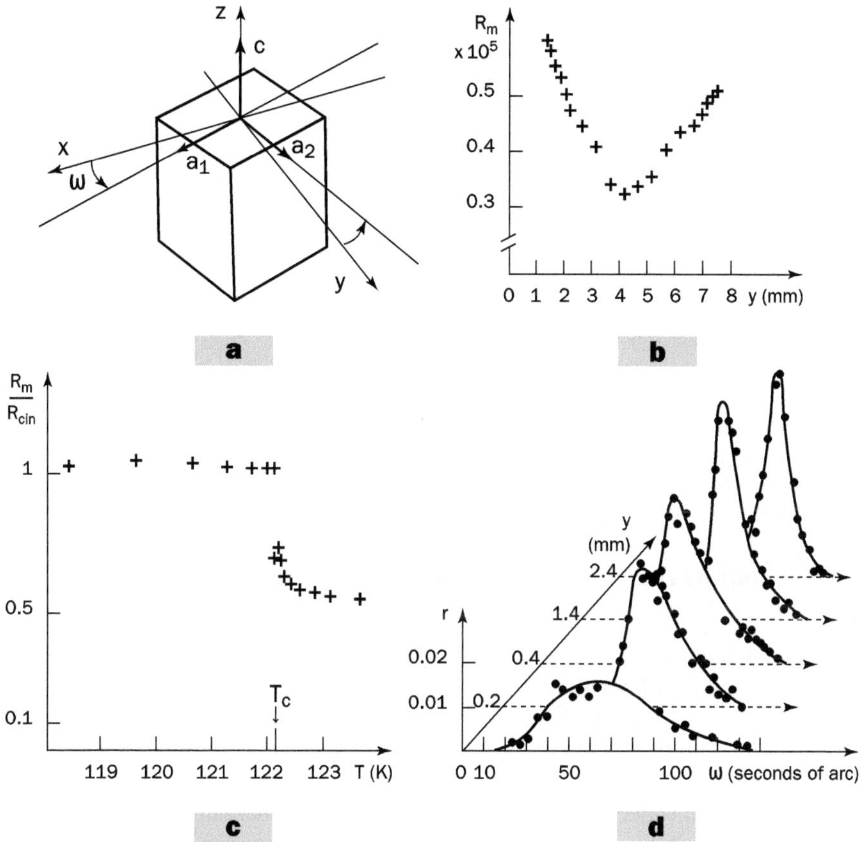

Fig. 1.34 The γ diffraction allows quality measurements for a given sample (Bastie, 1980; Bastie and Becker, 1984; Bastie et al., 1976a, b) (Annex A3): (**a**) Shape of the KDP sample used and definition of the "scan" angle ω. (**b**) Variation of the integrated reflecting power R_m for different positions of the γ radiation beam, concerning the (200) reflection. The best quality is obtained in the centre of the sample with the lowest value of R_m. This variation in quality with depth in the sample can lead to errors in interpreting results. (**c**) Variation of the integrated reflecting power with temperature. The extinction in the para phase demonstrates the quality of the crystal; in the ferro phase, R_m takes the value calculated in kinematic theory R_{cin}, of the "totally imperfect" crystal: the surface of the peaks is proportional to the volume of the domains corresponding to the peak. (**d**) Diffraction curves at 80°K with γ brushes passing through the sample at different depths y. The width of the peak inside the crystal is equal to the experimental limit of the incident beam (10–20 arc seconds), while in a thickness of 200 micrometres near the edge, it can reach about 50 arc seconds, which corresponds to an evaluation of the mosaic

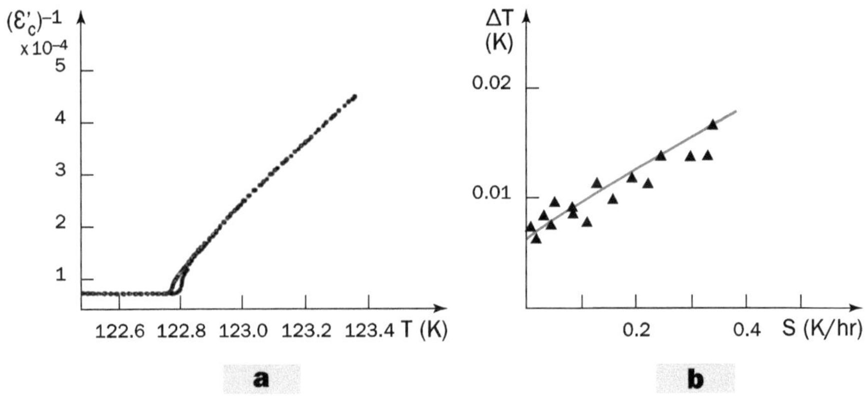

Fig. 1.35 Effect of the rate of temperature change: isothermal or adiabatic situations? (**a**) Inverse of the dielectric constant of KDP as a function of T for a speed >0.4 K/h (Eberhard and Horn, 1975; Western and Schmidt, 1976; Nazario and Gonzalo, 1969), (**b**) amplitude of the thermal hysteresis as a function of the rate of change of T for KDP (Sugié et al., 1972)

1.5.2 *Importance of Temperature Variations Over Time*

A major parameter is obviously the temperature T, but it is necessary to situate the thermodynamic conditions of the experiment conducted: do they correspond to an isothermal or adiabatic situation? The answer is sometimes not obvious: it depends on the thermal properties of the crystal, thermal exchanges, the rate of temperature change, etc. As many characteristics vary drastically around the transition, the greatest attention is necessary. One can thus confuse isothermal susceptibility with adiabatic susceptibility and attribute an observed phenomenon to other causes, as illustrated by the example in Fig. 1.35 (Western and Schmidt, 1976). For this reason, KDP is a particularly critical crystal when measurements are taken near the transition, as the adiabatic application of a conjugate field certainly causes a change in polarization but also in the temperature of the crystal (Strukov et al., 1972). Using the relation between the inverse adiabatic susceptibility $(\chi_{\text{adia}})^{-1} = (\delta E/\delta P)_S$ and the inverse isothermal susceptibility $(\chi_{\text{iso}})^{-1} = (\delta E/\delta P)_T$:

$$(\chi_{adia})^{-1} = (\chi_{iso})^{-1} + (\delta E/\delta T)_P / (\delta T/\delta P)_S = (\chi_{iso})^{-1} + (T/C_P).(\delta E/\delta T)^2 \quad (1.36)$$

with G given by relation (1.9) and $q = E$, it is possible to write, with the approximation $(\delta E/\delta T)^2 = (\text{AP})^2$, thus considering b constant in temperature (i.e., outside the tri-critical situation for which the result would be $(a_T + b_T)^2$):

$$\chi_{adia}^{-1} = A/(T - T_0) + 3bP^2 + (A^2T/C_P)P^2 + 5c\,P^4 \quad (1.37)$$

The adiabatic effect thus amounts to adding a positive contribution to the factor b of the development of G (from the term in P^4): this can transform the order of the transition. For example, if the transition is of the first order and close to the tri-critical regime, the effect can shift it to a second-order transition. This has been shown by electro-calorific (Strukov et al., 1972) and dielectric (Western and Schmidt, 1976) measurements at the transition, conducted with a notable cooling rate (greater than 0.24 K/h in KDP) (Eberhard and Horn, 1975). Also noteworthy is the evidence in KDP of double hysteresis cycles just above the transition temperature when working with slowly varying applied electric fields (typically a few V cm^{-1} per minute and equivalent frequencies of 10^{-4} Hz) unobservable in cycles at 670 Hz (Okada and Sugié, 1971). Similarly, one can use the linear graphs of $\chi_{adia}{}^{-1}(T)$ and $\chi_{iso}{}^{-1}(T)$, results of quasi-static or low-frequency measurements (Okada and Sugié, 1977). Quick information is obtained by comparing in modulus, A^2T/C_P and 3b (approximately 2 in KDP; 2×10^{-3} in $BaTiO_3$, and 6×10^{-2} in TGS (Nakamura et al., 1974)). These examples urge caution regarding the rate of temperature change near transitions. Unfortunately, information on this rate of temperature change is rarely indicated in the literature. Moreover, even when the rate of change of T is known and low, the sample is often considered as a homogeneous entity in temperature. One can genuinely doubt this implicit assumption. To achieve thermal homogeneity with a precision of 10^{-2} K at every point of the sample, it is necessary on the one hand to vary the temperature sufficiently slowly and on the other hand to place the crystal inside a medium with a thermal conductivity better or at least comparable to that of the crystal. It is under these conditions that Meyer and Schneider conducted neutron measurements, allowing for the first time the determination of the natural orientation of the DKDP phase front perpendicular to the c axis (private communication, 1975), with a thermal homogeneity of a thousandth of a degree, thus challenging the certainty of a phase front moment parallel to the ferroelectric axis. This environment is difficult to achieve with dielectric measurements, which involve electrodes on the sample and wires for measurements, or during multi-beam optical observations, especially if one wishes to apply a mechanical constraint. The setups must reconcile the best possible thermal homogeneity by limiting the heat input from the lights associated with observation and control the crystal's environment (exchange gas, copper screens, multiple micro-heaters) to accurately know the thermal gradients. These situations correspond to the experiments presented in this work, particularly in Chaps. 5 and 6. Chapter 5 illustrates how important thermal homogeneity is for the shape of phase fronts at the transition. Moreover, the thermal gradients in the crystal and its vicinity are very unlikely to be identical. It is entirely possible to perform temperature regulations at a specific point with a precision of a thousandth of a degree Kelvin, and to observe in the sample a thermal inhomogeneity exceeding a tenth of a degree. These homogeneity issues are even more critical when the temperature is changed during the experiment.

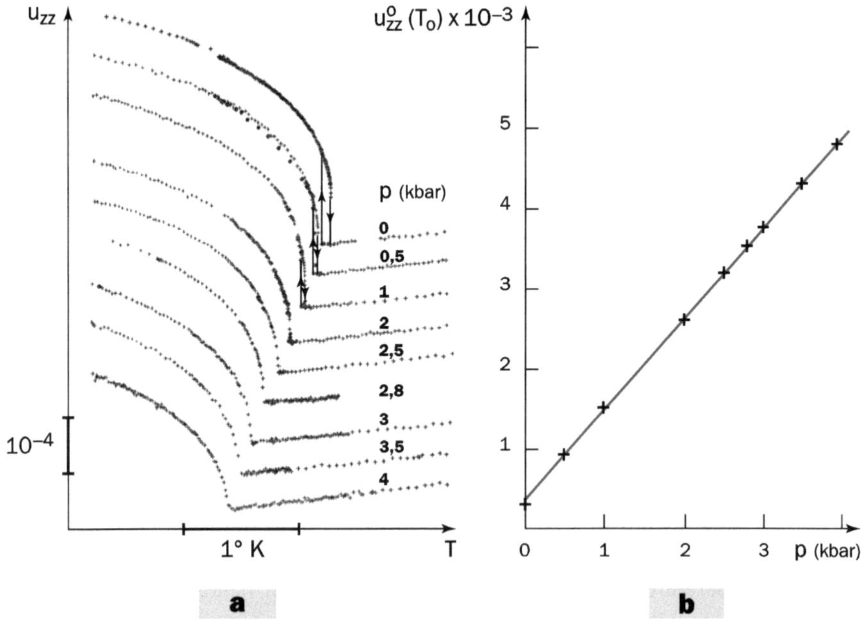

Fig. 1.36 The u_{zz} (T,p) curves of a KDP crystal have been translated in (**a**) along the coordinate axes to highlight their change in shape. These variations can be described by the relation (see (14) and (15) of (Bastie et al., 1981)): $u_{zz}(T,\ p)=f\left(u_{xy}^{\ 2}\ (T,\ p)\right)+\alpha(p)\left(T-T(p)\right)-u_{zz}^{0}\ (T_0,\ p)$ with u_{zz}^{0} (T_0, p) linear function of p, hydrostatic pressure, as illustrated in figure (**b**)

1.5.3 Hydrostatic Pressure and Constraints

As illustrated by the phenomenological approach in Section 1.3, near a transition, the hydrostatic pressure p can play an important role; the sample during measurements is in an exchange gas such as helium; does the pressure of this gas remain constant throughout the experiment when the temperature varies? Figure 1.36 shows how much the measured property, here the deformation as a function of temperature, is modified by a variation of p of a few tenths of a bar, which can occur if precautions to maintain p constant (e.g. with sufficiently large gas expansion vessels) are not taken. The lack of control over this parameter can explain the dispersion of results concerning the order of the KDP transition observed over several decades.

Of course, the effect is even more significant when the studied crystals are subject to transitions close to the tri-critical situation (such as KDP, RbDP). In the context of the work, the parameters conjugated to the order parameters are the shear stress σ_{xy} and the electric field E_z. Regarding the mechanical aspect, one must verify whether the measurement is carried out with a crystal free to deform (the "free" situation) or if the deformation is hindered under the imposed experimental conditions. The inverse of the susceptibility as a function of temperature presented in Fig. 1.37, with the frequency of the alternating measurement field as a parameter, illustrates the phenomenon: at low frequency, the crystal has time to deform; this is

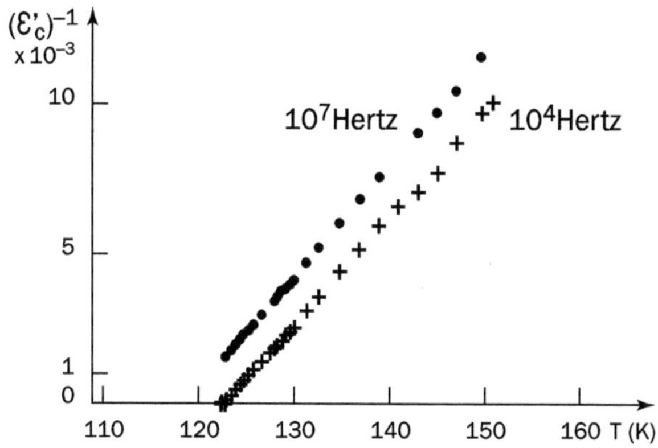

Fig. 1.37 Variation of the inverse dielectric constant in the para phase of KDP for two frequencies that create very different mechanical contexts: at low frequency, the crystal is free to deform whereas this becomes impossible above the piezoelectric resonance frequencies due to its inertia; thus, we obtain a "free" and a "blocked" situation. The temperature translation of Curie's laws by 4K corresponds to the theoretical evaluation produced in the text (Baumgartner, 1951)

no longer possible at excitation frequencies higher than those of piezoelectric resonances. Thus, a temperature shift of the inverse dielectric susceptibility is observed in accordance with theoretical prediction (Baumgartner, 1951; Brody and Cummins, 1968, 1969). The relationship between the dielectric constant of the free crystal $\varepsilon'_{\text{free}}$ and that of the clamped crystal $\varepsilon'_{\text{clamped}}$ is written by considering, in the first approximation, the dielectric constants proportional to the susceptibilities:

$$\begin{aligned} \left(\varepsilon'_{\text{free}}\right)^{-1} &= \varepsilon_0 / \chi'_{\text{free}} = \varepsilon_0 \ A\left(T - T_0\right) \\ &\text{and} \\ \left(\varepsilon'_{\text{clamped}}\right)^{-1} &= \varepsilon_0 / \chi'_{\text{clamped}} = \varepsilon_0 \ A\left(T - T_0 + a_{36}^2 / c_{66}\right) \end{aligned} \tag{1.38}$$

The values of the constants a_{36} and c_{66} from the literature allow for a satisfactory calculation of the 4 K temperature shift observed experimentally. The constraint situation in which a sample may find itself is far from negligible for assessing its response to an excitation. However, "clamping" by inertia does not change the transition temperature: when spontaneous polarization appears, spontaneous deformation also appears due to the piezoelectric effect and consequently the domain texture. Chapter 2 shows that the complexity of domain structures can also prevent the achievement of free deformations in various regions of the sample, even at low frequencies. It is therefore difficult to be certain of the situation, whether free, strained, or partially strained if the domain configuration is unknown. The interpretation of susceptibility measurements can thus become delicate. Any electrical contact with an overly rigid electrode, as illustrated in Fig. 1.38a, can induce a strain and hinder deformation, more or less locally. From a mechanical point of view, if the sample is not subjected to excessive strain, such as that created by a drop of metallic paste, a too constrained electrode attachment, or contact with a support, it

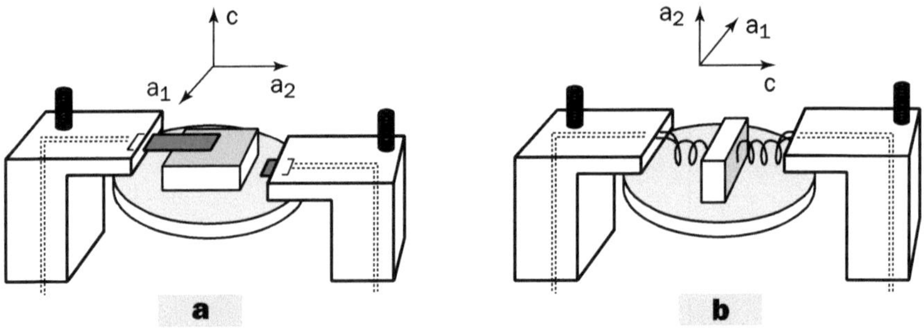

Fig. 1.38 Different modes of electrical contacts on the electrodes: (**a**) with a Teflon stand, a copper strip and a drop of silver paste can create significant stresses; (**b**) with thin accordion copper wires and very small drops of silver paste, we can avoid stresses during thermal cycles

should be considered free. If this is the case, one sample face constitutes a plane of zero strain, and, as demonstrated in Chap. 3, the sample edge can play a role of attraction by image force of certain domains.

1.5.4 Shape, Limits of Samples, and Applied Conjugate Fields

For a ferroelectric, is the experiment conducted at $E = 0$ or $D = 0$, with a short-circuited or isolated crystal? The experiment proposed in Fig. 1.39 demonstrates the importance of the depolarizing field phenomenon on a single-domain sample: a parallelepiped sample of KDP is cut along orthorhombic planes, so that normal stresses applied on these lateral faces create a shear stress σ_{xy} throughout the sample; this is sufficiently large (greater than the critical stress) for the sample to remain single-domain in the ferro phase. The birefringence for light propagating along the ferroelectric axis c allows an easy local measurement from which the polarization P_z and the deformation u_{xy} are deduced. Part of the sample is without an electrode, thus isolated, and the other, with two electrodes connected to each other, is short-circuited. The ratio of 14 between the two measured birefringence values demonstrates the importance of the depolarizing field in an "isolated" situation and the importance of the quality of the electrodes in the short-circuited region! The stable state previously presented in theoretical models corresponds to the short-circuited crystal. If the crystal is isolated, an additional term due to the action of the depolarizing field is added. This term must not be too significant for the appearance of polarization P_S to retain energetic interest, thus for the transition to occur. Of course, as shown in Chap. 2, the appearance of domain textures reduces the importance of the energetic contribution of the depolarizing field and facilitates the occurrence of the transition. The effect of the depolarizing field varies with the shape of the sample and notably with its thickness in the direction of the ferroelectric axis. If the sample is assimilated to an ellipsoid, the homogeneous depolarizing field in a principal direction i will be of the form $E_i = -N_{ii}P_i/\varepsilon_0$ with the importance highlighted of

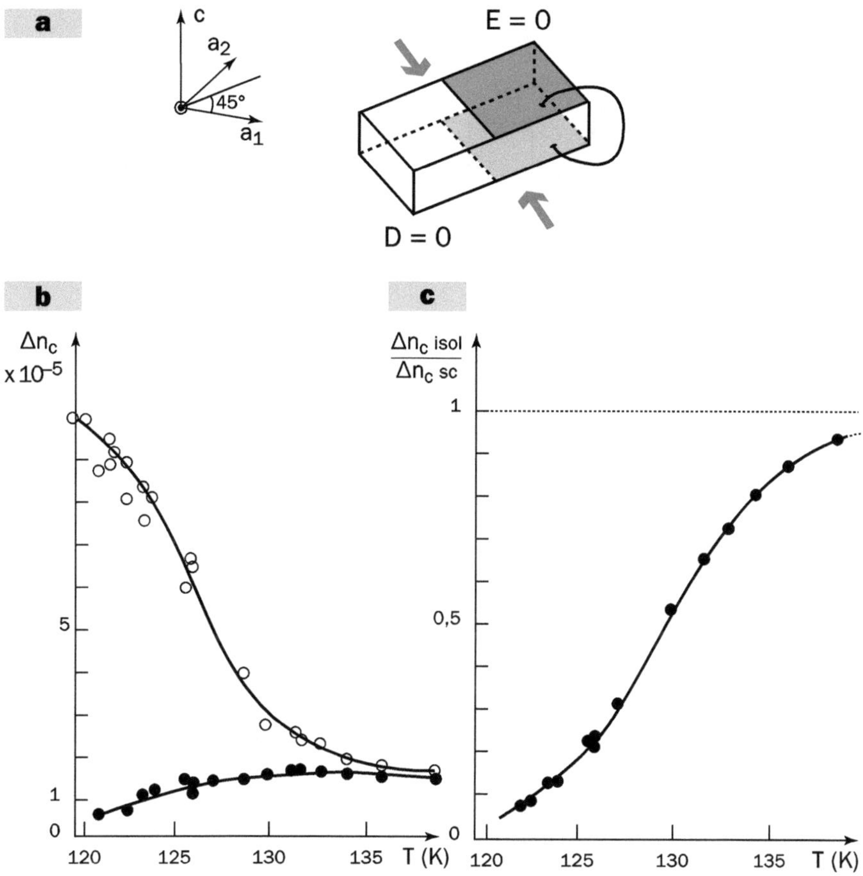

Fig. 1.39 (**a**) A KDP platelet is cut along the orthorhombic planes in such a way that a normal stress on the side faces creates a shear stress. If this is sufficient, we can obtain a single-domain sample in both phases, ferro and para. Part of the sample has short-circuited electrodes ($E = 0$); the other, without an electrode, is electrically isolated ($D = 0$). (**b**) Variation of the birefringence Δn_c for light propagating along the ferroelectric axis as a function of temperature, in each part of the sample: (°) for the short-circuited region and (•) for the isolated region. (**c**) Ratio between the two measured values of the birefringence Δn_c, in the short-circuited region and in the isolated region, obviously equal to 1 in the para phase (Bornarel et al., 1972)

the classical coefficients N_{xx}, N_{yy}, N_{zz}. Figure 1.40 recalls their variation with the shape of the sample in the case of an isotropic sample with a square section. Performing the calculation taking into account dielectric anisotropies does not qualitatively modify the result, and many calculations derived from magnetism of demagnetization factors are transposable to electrostatics (Osborn, 1945, and following). It should be noted that if the effect of the depolarizing field is evident at low temperature since the electrical conduction of crystals is low, it may be much less so at room temperature and at higher temperatures due to the conductivity of the crystal, which will reduce this effect. This is what occurs in the case of GMO, for

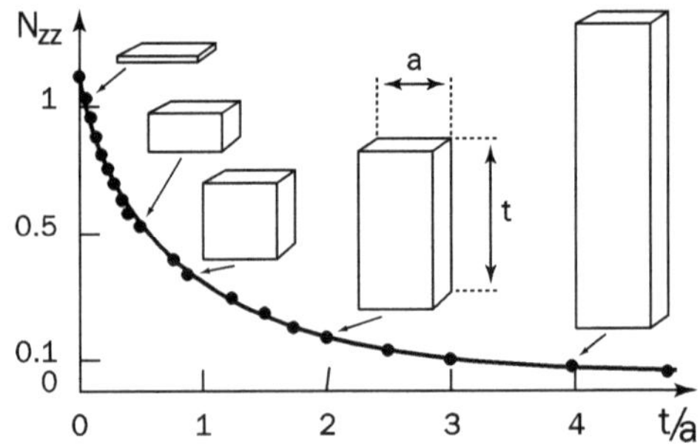

Fig. 1.40 Variation of the N_{zz} coefficient of the depolarizing field as a function of the thickness of the dielectric material wafer. We recall the depolarizing field corresponding to the different situations of sample shape (Durand, 1964):

$$E_d = -N_{zz} P_S, \textit{with}\, N_{xx} + N_{yy} + N_{zz} = 1\, \textit{and here}\, N_{xx} = N_{yy}$$

$$\textit{with for}\, c < a, N_{zz} = \left(a^2 \left(a^2 - c^2\right)^{-1}\right)\left(1 - c\left(a^2 - c^2\right)^{-1/2} .arc\cos\left(c/a\right)\right)$$

$$\textit{and for}\, c > a, N_{zz} = (a^2 \left(a^2 - c^2\right)^{-1} \left(1 - c\left(c^2 - a^2\right)^{-1/2} argch\left(c/a\right)\right)$$

The importance of the depolarizing field is completely different for a platelet like those used for dielectric measurements, for a sample in the form of a cube or for a sample elongated along c like those preferentially used in many diffraction measurements

which no difference was observed between the two regions (short-circuited and isolated) during an experiment similar to that reported in Fig. 1.39 for KDP (Bornarel et al., 1972).

In this overview of some experimental points, electrical phenomena at the edges of the sample must be taken into account, as were the mechanical effects with the necessity of electrodes for dielectric measurements. The search for the optimal electrode is important, especially in a ferroelectric and ferroelastic crystal: rigid electrodes with strong adhesion to the crystal surface could constrain it; however, an insufficiently adherent electrode would make it difficult to short-circuit the sample with significant effects at the transition. The optimal solution for the crystals used corresponds to gold electrodes vaporized under vacuum with very low chrome vapours for the crystals used. However, whatever precautions are taken, the phenomenon known as "surface layers" cannot be completely eliminated, modelled in Fig. 1.41a by a layer of thickness d/2 between the electrode metal and the crystal, a layer with a dielectric constant ε_2 much smaller than that of the bulk crystal ε_1 (Chincholkar and Unruh, 1968; Mansingh and Prasad, 1976). The inverse of the measured capacitance C of the sample with surface S and thickness t is easily written as:

$$C^{-1} = \left(1/\varepsilon_0 S\right)\left(t/\varepsilon_1 + d\left(1/\varepsilon_2 - 1/\varepsilon_1\right)\right) \tag{1.39}$$

Figure 1.41b demonstrates the effect of these "surface layers" in the para phase of KDP (Bornarel and Torche, 1992). Figure 1.42 illustrates the significance this

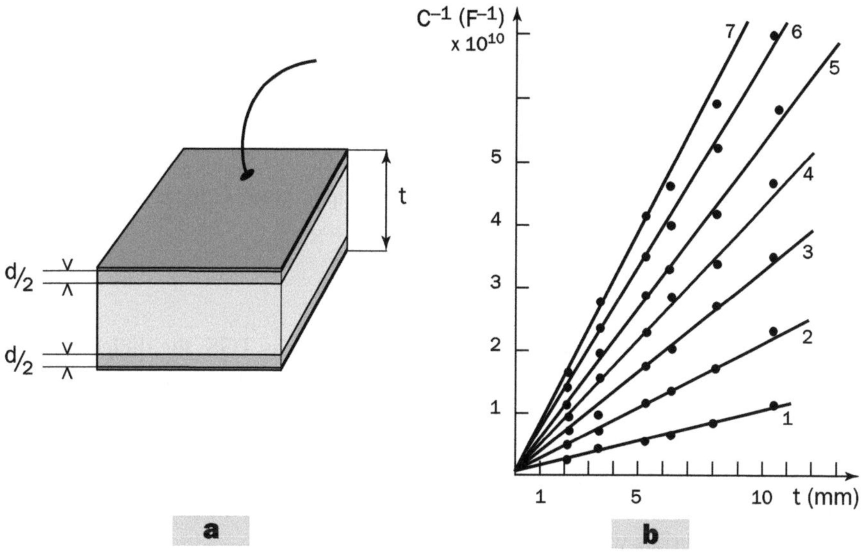

Fig. 1.41 (**a**) Schematic representation of a sample with electrodes and contact wires glued with silver paste. The surface layers of a thickness of $d/2$ are represented (d remains very small compared to the thickness t of the sample). (**b**) Variation of the measured inverse capacity as a function of the thickness t of the KDP sample with the difference $(T - T_0)$, T measurement temperature and T_0 transition temperature (Bornarel and Torche, 1992). The ordinate at $t = 0$ allows to evaluate the ratio d/ε_2

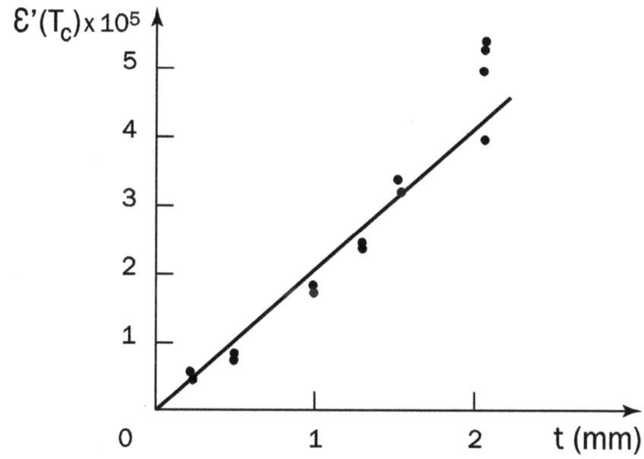

Fig. 1.42 Variation of the maximum dielectric constant at the second-order transition of a TGS crystal therefore at $T = T_C$. Using the diagram from Fig. 1.41a, we call ε_1 the dielectric constant of the bulk crystal and ε_2 that of the surface layers. As ε_1 becomes infinite at $T = T_C$ (second-order transition), $\varepsilon'(T_C) = \varepsilon_2\, t/d$, as shown in the figure (Sekido and Mitsui, 1967)

effect can have for crystals with a second-order para-ferro transition such as TGS (Sekido and Mitsui, 1967) without the mechanical issues of our typical crystals since this crystal is not piezoelectric in the para phase. In this case, one can write:

$$1/\varepsilon_2 >> (T - T_0)/A$$

hence the dielectric constant deduced from the capacitance C measurement at the transition:

$$\varepsilon'(T_C) \approx t\varepsilon_2/d \tag{1.40}$$

Thus, for a crystal with a second-order transition such as TGS, the dielectric constant measured at the transition T_C varies linearly with the thickness of the sample. As the bulk crystal's capacitance is in series with the capacitances created by the "surface layers," these play an increasingly important role as the intrinsic dielectric constant of the crystal is large, especially in second-order transitions where it is theoretically infinite. Particular attention is paid to this phenomenon when applying ferroelectric materials with large dielectric constants.

In the context of this book, attention will be paid to the phenomenon when studying the equilibrium domain texture in Chap. 2, polarization reversals in Chap. 3, and dielectric measurements in Chap. 4. For reference, the dielectric constant of the "surface layer" region is lower than that of the bulk crystal and often comparable to that of a crystal under strong stress or a strong applied electric field. The proposed models assume that this region is either polar or non-polar with various depolarizing field effects (Hadni et al., 1983). Studies conducted on crystals such as perovskites or compounds of the TGS family show significant effects in the para phase or in the mono-domain ferro phase for samples thinner than a micrometre. The results in this book concern samples with thicknesses between 200 μm and several millimetres. Of course, the importance of the depolarizing field in a mono-domain sample exists, as the experiment reported in Fig. 1.39 demonstrates. And, as the electrodes never allow for a perfectly short-circuited sample, the role of the surface layers cannot be overlooked in the exploitation of experimental results.

1.5.5 Comments on Some Real Situations for Ferro-Ferro

A theoretical memento that accumulates edge effects when the samples are of a size comparable to those of the disturbed "layers" (Fousek, 1970) and using the usual Landau-type developments can be proposed. If the importance of the mechanical edge energy and the effects of the depolarizing field become notable in a crystal, the undisturbed part of the crystal would become almost non-existent. Let us mention the experiment by Jaccard and Känzig (1953) with colloidal particles which shows a near disappearance of the transition. In this case of very small samples, the

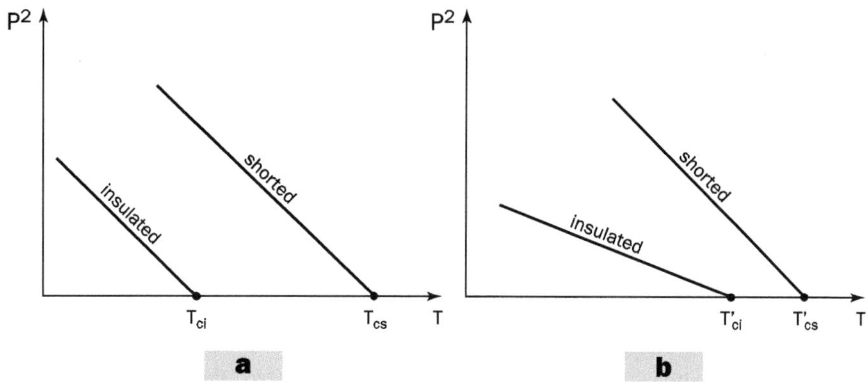

Fig. 1.43 Theoretical variation of the square of the polarization P^2 as a function of temperature T (Fousek, 1970) in a "proper" second-order transition, and (b) in an "improper" transition. (**a**) *If* $G - G_0 = (A/2)(T - T_0)P^2 + (b/4)P^4 + (c/6)P^6 + duP + (f/2)u^2 - EP$, the transition temperatures are for the short-circuited crystal $T_{CS} = T_0 + d^2/Af$ and for the isolated crystal $T_i = T_{CS} - 4\pi/A$. The difference between these two transition temperatures would be so great that the transition of the isolated crystal could not occur without the creation of domains. (**b**) In an "improper" transition, the differences in transition temperature between isolated and short-circuited are small. For example, if we write $G - G_O = (C_0/2)(T - T_0)q^2 + B_0q^4 + DqP + (A/2)P^2 - EP$, we will find for the temperature of the short-circuited crystal $T_{CS} = T_0 + D^2/CA$ and for that of the isolated crystal $T_i = T_0 + D^2/C(A + 4\pi)$ which differs greatly from the variations in transition temperature of a "proper" crystal

mechanical energy at the sample's surface due to its own deformation (upon bending) when domains are created is estimated in the form

$$U_S \text{ as } K u_{xy}^2 / n \tag{1.41}$$

with n being the number of domains and K a function of the material's elastic constants. This additional term in the development of G induces a shift in the free crystal's transition

$$\Delta T_C = a_{36} C\left(\left(c_{66}^P + 2K/n\right)^{-1} - \left(c_{66}^P\right)^{-1}\right) \tag{1.42}$$

a_{36} being the elastic coefficient and C the Curie constant. In Jaccard and Känzig's (1953) evaluation, this transition temperature variation would be of the order of a Kelvin degree for KDP.

The reasoning for each ferro-ferro crystal is as follows: if the crystals are very small in size, the appearance of domains is made difficult due to the surface mechanical energy and the wall energy it generates; but, if the crystal remains mono-domain and is electrically isolated, the para-ferro transition can only occur at a temperature lower than that expected for the short-circuited crystal. This is illustrated in Fig. 1.43 with the variation of the square of the polarization as a function of temperature for a "proper" ferroelectric. One can even predict the impossibility of the transition appearing in certain electrically isolated crystals. Fousek (1970) notes that this effect will not have the same importance if the polarization only appears through coupling with the true order parameter as in the case of crystals "improper". The result obtained by Jaccard and Känzig corresponded to an almost disappearance of the transition due to a very small sample size, in a somewhat extreme situation (colloidal particles); but, even when the size of the crystals is larger, effects can exist, for example, during X-ray diffraction measurements often carried out with samples of very small size, very different from those used for mechanical or dielectric experiments. Therefore, great caution is required in the interpretation of the results; the size and shape of the sample are major parameters. The situation of thin films, much studied over the past fifteen years, is not that of these "bulk" crystals.

In the following chapters, the precautions summarized above concerning the quality of the crystals, their preparation, and the experimental conditions have been taken to ensure, as far as possible, the relevance of the physical interpretations of experimental results. We will be careful to recall the importance of experimental conditions in contentious cases.

Memo Chapter 1

A ferroic transition, with a non-zero order parameter in the "ferro" phase and zero in the "para" phase, corresponds to symmetry change with symmetry group of the ferro phase subgroup of that of the para phase. This change allows to know the nature and number of possible primitive cells in the "ferro" phase. The modifications of the tensors of physical properties in the ferro phase compared to those in the para phase can be predicted.

The existence of para-ferro transitions can be modelled with the development of free energy (known as "Landau") in even powers of the order parameter for virtual crystals of infinite dimensions, homogeneous temperature T, under constant hydrostatic pressure p, in stable situations. Even the simplistic assumption of constant coefficients in T and p in the development, except for a linear variation in T of the square power term, is sufficient to account for second-order and weakly first-order transitions. In more delicate situations, such as for tri-critical transitions, linear variations in T and p for the coefficients of the development are supposed. The success of predicting physical properties, with such models, would even tend to make us forget the limits of the initial assumptions. However, the example of real coercive fields a hundred times smaller than those given by these theoretical models brings us back to reality: The previous simple development assumes a mono-domain ferro crystal unless we consider the order parameter as an average value in a continuous medium; but as this simple development does not contain space and time gradient terms, it does not allow to study the domains and the dynamics of reversal ("switching").

Limiting ourselves to the case of "ferro-ferro" crystals with a tetragonal-orthorhombic transition with proportional polarization and deformation, the importance of the specificities of the sample (quality, shape, nature of the edges: with or without electrodes, constrained or relaxed), thermodynamic conditions (isothermal, adiabatic), and time (history and mode of measurement) is experimentally demonstrated. Many questions remain unanswered, including:

- Why is the ferro phase not mono-domain in the real crystal?
- How can possible domains, allowed by symmetry, coexist and with what boundaries between them? What walls?
- How does the order parameter change in the walls?
- Is it possible to correlate the phenomena at the microscopic level with atomic arrangements at the nanoscopic level concerning the walls between domains?
- How are electrostatic and mechanical energies competing in "ferro-ferro" crystals while one or the other is practically negligible in other ferroics?
- What are the domains (shapes, number, arrangements) in reality, depending on the situations (samples, history) and how to obtain the simplest arrangements? The more stable?

Chapter 2 will answer these questions for ferro-ferro crystals in "natural structure", i.e. under the sole actions of temperature and time, for real samples with their specificities (quality, shape, edges). The answers with the action of conjugate fields (electric field, mechanical stresses) will be addressed later, in Chaps. 3 and 4.

Chapter 2
Textures in Natural Domains

2.1 Existence of Domains

In the first chapter, basic elements on ferroelectrics and ferroelastics were presented, as well as the necessary experimental precautions for their study. The interfaces (domains and their walls, phase fronts during first-order transitions) were not addressed. Phenomenological models, for example with the Landau development, did not take them into account since they implicitly assumed the crystal to be mono-domain and of infinite dimensions.

The importance of the depolarizing field demonstrated in the experiments of Figs. 1.39 and 1.40 highlighted the impact of the electrostatic context, isolated or short-circuited from the sample. The situation is evident for a non-piezoelectric ferroelectric crystal in the paraelectric phase: domains appear in the ferro phase to minimize the electrostatic energy. Figure 2.1 illustrates this with barium titanate at room temperature. The mechanical energy is considered in this example, as almost negligible compared to the electrostatic energy (Fousek and Safrankova, 1965): if the sample is immersed in a liquid with high conductivity, the existence of domains is unlikely; on the contrary, if the conductivity of the liquid is low, the domains are numerous: we find the short-circuited and isolated situations. Comparable studies have been conducted with TGS (Gilletta, 1972) and many other crystals.

Similarly, for a ferroelastic crystal, the nature of the sample edges, free or constrained, changes the deformations and textures in domains. Thus, the texture of twins can be modified, and a virtually null macroscopic deformation can be obtained in the ferro phase if relaxations of the sample edges are possible. The "ferro" domains in both cases allow to decrease the polarization and/or the macroscopic deformation of the ferroic crystal sample compared to a mono-domain situation. How does this happen for a ferroelectric-ferroelastic crystal such as KDP, whose first observations of textures in domains were quite late (Fomitchev, 1965; Oettel, 1965; Bornarel et al., 1966) after their demonstration by the frost method (Toshev,

J. Bornarel, *Ferroic Crystal Domains, Walls, and Phase Fronts*, Springer Series in Solid-State Sciences 209, https://doi.org/10.1007/978-3-031-98099-2_2

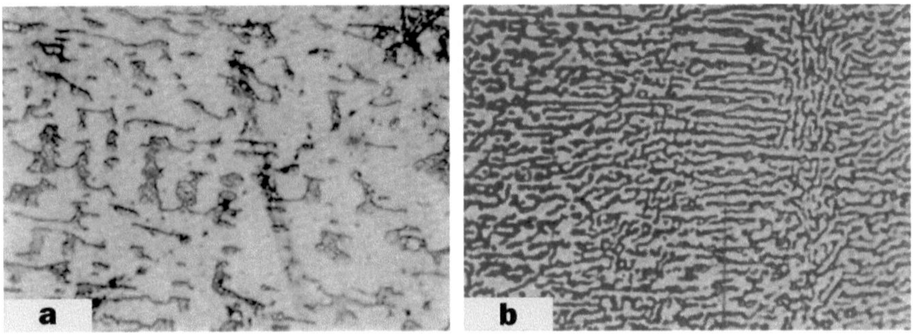

Fig. 2.1 Textures in domains of a non-ferroelastic, non-piezoelectric ferroelectric crystal in its high temperature phase, the barium titanate ($BaTiO_3$): (**a**) the sample is immersed in a liquid with a conductivity $10^{-3}\ \Omega^{-1}\ cm^{-1}$ at room temperature, and the existence of domains is energetically not very interesting, and (**b**) the external liquid is not very conductive (conductivity of $10^{-6}\ \Omega^{-1}\ cm^{-1}$); the existence of domains reduces the overall energy of the crystal in this situation of almost isolated sample (Fousek and Safrankova, 1965)

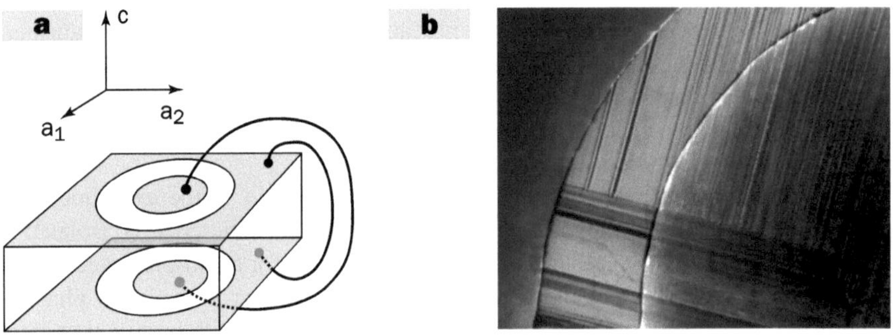

Fig. 2.2 (**a**) Representation of a KDP sample with partial electrodes on the surfaces perpendicular to the ferroelectric axis: a ring-shaped region is without electrodes therefore electrically isolated; the electrodes being connected by a conductive wire, the corresponding regions of the sample are short-circuited. (**b**) Regardless of the temperature history between the para and ferro phases, no difference is observed for the texture in domains between the short-circuited and isolated regions

1963)? Figure 2.2 presents a KDP sample with specially shaped electrodes evaporated on the faces perpendicular to the c axis (gold electrodes under chrome vapour with ring-shaped interruptions): the majority of the sample can be short-circuited electrically, and a part remains constantly isolated. During the observations made in para-ferro, ferro-para temperature cycles, etc., no notable variation of the texture in domains is observed between the isolated part and the short-circuited part (Bornarel, 1971; Nakamura et al., 1984). Some authors have proposed, to explain these identical textures in an isolated or electrically short-circuited crystal, the existence of surface layers with a low dielectric constant, for example equal to five for a surface layer thickness of 1 μm (Bjorkstam and Oettel, 1967).

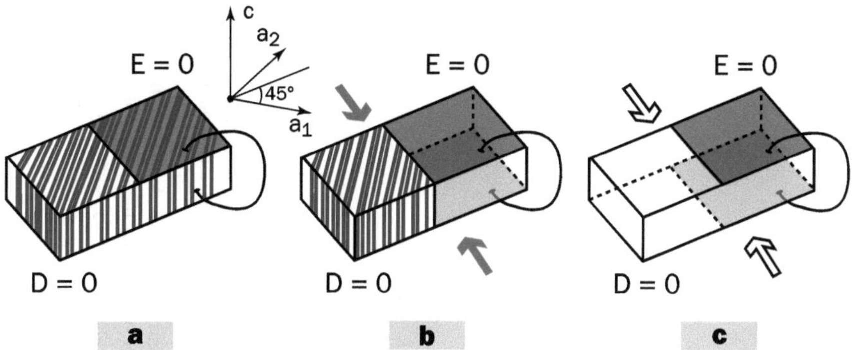

Fig. 2.3 The KDP sample has a parallelepiped shape with the large faces perpendicular to the ferroelectric axis *c*. The side faces are cut at 45° from the tetragonal planes (100) and (010), therefore in the planes of the orthorhombic lattice. Electrodes connected to each other by a conductive wire delimit a short-circuited half of the sample; the other half of the sample is electrically isolated (Bornarel et al., 1972). A normal stress on the side faces creates a shear stress σ_{12} throughout the crystal. The experiment is carried out a few degrees below the transition temperature. (**a**) Without applied stress, the domain texture is established in both regions with walls in the tetragonal planes (100) or (010), therefore at 45° from the faces of sample. No difference in texture is observed between the two regions (isolated and short-circuited). (**b**) A moderate applied stress proves sufficient to make the short-circuited part of the sample mono-domain, the isolated part retaining a finely layered domain texture. (**c**) Under the effect of a higher applied stress, the domain texture also disappears in the isolated region

But the experiment in Fig. 2.3 highlights the competition sometimes balanced in the ferro-ferro between mechanical energy and electrical energy (Bornarel et al., 1972): in this example, the texture that appears without applied mechanical constraint is the same in the short-circuited or isolated regions. This is not the case under the action of an applied constraint in coherence with the result of Fig. 1.39.

Thus, a notable incidence on the texture in domains of the sample shape is predicted, which will be studied in detail in Chap. 4 with susceptibility measurements. Figure 2.4 summarizes a general qualitative result: when the dimensions of the crystal are of the order of the micrometre, such as colloidal particles, domains may not be observed (Jaccard et al., 1953); if the samples are platelets perpendicular to the ferroelectric c axis, of thickness equal to one tenth of the other dimensions and between one tenth of a micrometre and a few micrometres, the textures in domains are fine and regular; they remain dense for samples of shape between the plate and the cube; on the other hand, if the samples are elongated along the ferroelectric axis as in Fig. 2.4d, some poly-domain volumes still exist, but the sample is increasingly mono-domain as its length along c grows. These observations are consistent with the evolution of the depolarizing field, but this is not the only reason: mechanical energies also play their role.

In this chapter, the domain textures of a crystal subjected only to temperature variations (so-called natural domain textures), without application of an external

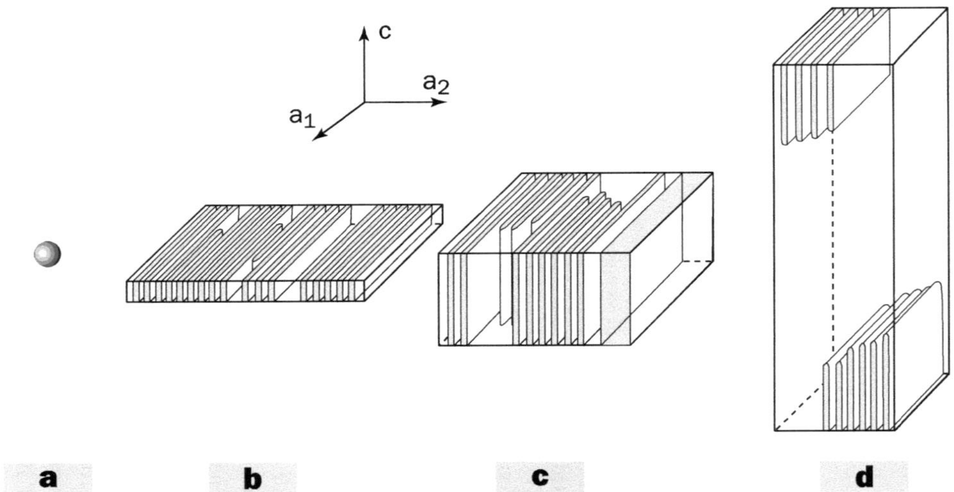

Fig. 2.4 Effect of the sample shape on the domain texture for electrically isolated samples: (**a**) A very small sample (colloidal sphere) remains mono-domain and the transition becomes difficult (Jaccard et al., 1953; Känzig and Sommerhalder, 1953); these results are not inconsistent with the size-dependent studies conducted since (Zhong et al., 1994) nor with those obtained in thin layers (Salje et al., 2012). (**b**) A texture of fine domains is observed in a platelet of thickness following *c* of 0.2 μm with, often ends of blocked domains, without omitting the influence of shape on the depolarizing field (Fedosov, 1980). (**c**) In a platelet of thickness close to half of the other dimensions, textures with the different orientations of walls allowed, or only one of them, are observed; the domains are very mobile. (**d**) Samples elongated in the direction of the ferroelectric axis c have vast mono-domain regions, especially towards the middle of the crystal following *c*

field conjugated to the order parameter (electric field or mechanical stress), will be studied. We wish to answer the following questions:

- Are there optimal orientations of the walls, boundaries between two domains of different order parameters?
- Are all the wall elements observed experimentally located in these optimal wall orientation planes?
- What are the structures of the walls and what impact do they have on the physical properties of the crystal?
- What shapes and dimensions of domains exist? With what organizations? What is the impact of experimental conditions?

We attempt to answer these questions for ferro-ferro crystals and particularly for those whose transition occurs between a para tetragonal phase $\overline{4}\,2\,\mathrm{m}$ and a ferro orthorhombic phase mm2, but we can draw from this more widely applicable teachings.

2.2 "Permissible" Walls

2.2.1 *Definition of a Permissible Wall*

In an electrically isolated ferroelectric crystal, the existence of domains reduces the overall electrostatic energy, but the existence of walls brings an additional energy contribution. How is the total energy minimized? The electrostatic energy created by the wall will obviously be noticeable if it is electrically charged. If the mutual orientation of the polarizations in two contiguous domains is different from 180°, walls will be observed sometimes with compensation of electric charges, the positive ends of the dipoles of one domain joining the wall with the negative ends of the dipoles of the adjacent domain (Merz, 1954). In the case of domains with 180° polarisations, the walls will preferably be parallel to the ferroelectric axis to avoid being charged. We then expect in equilibrium situations, domains crossing the entire sample in this direction. If this is not the case, with an angle φ between the wall and the ferroelectric axis, the energy of this wall can be high. The models use the continuity conditions of the electric field, the displacement vector on the wall, and Maxwell's equations and show that φ remains less than a critical value φ_1 function of the ratio between the dielectric constants along the ferroelectric axis ε_z and those perpendicular to this axis ε_x:

$$tg\,\phi_1 \text{ varies as} \left(\varepsilon_x / \varepsilon_z\right)^{1/2} \tag{2.1}$$

the values of φ_1 vary according to the crystals between 1 and 3 degrees of arc. When this condition is not met, the appearance of ferroelectricity may prove impossible. Sometimes, with this limit value, zigzag walls may appear (Chervonobrodov and Roytburd, 1988).

The concept of "permissible walls" was introduced by Fousek and Janovec (1969) for ferroelectric crystals. These crystals being always piezoelectric in the ferro phase, the optimal orientations from a mechanical point of view are sought, and therefore the surfaces in which the deformation with respect to the prototype phase is identical in the two domains that the wall delimits. For each pair of symmetry (ferro and para phase), the resolution of the equations obtained gives three types of permitted walls schematized in Fig. 2.5, in addition to the cases for which no wall can exist:

- Walls called W_∞, parallel to the polarization, therefore not electrically charged which can take any orientation in the plane perpendicular to it. The concerned crystals are non-piezoelectric ferroelectrics in the para phase and non-ferroelastic in the ferro phase. Any wall orientation that contains the ferroelectric axis is suitable from the point of view of mechanical compatibility. Figure 2.5a illustrates this as well as Fig. 2.1 for the domain structures of $BaTiO_3$.
- Walls called W_f which retain the memory of the symmetry operations lost at the para-ferro transition. These are thus 90° planes for a lost order 4 symmetry axis as in Fig. 2.5b. These permissible wall planes therefore correspond to elements of symmetry of the para phase.
- Walls called S (for "strange") with a permissible orientation function of the value of the polarization and the electromechanical coefficients, therefore of the temperature, as illustrated in Fig. 2.5c.

The same reasoning was used subsequently for ferroelastic crystals in response to the same question: in which surfaces is the deformation with respect to the prototype phase the same in the two domains that the wall delimits? The equations of the planes of permitted walls have been simply expressed in Cartesian coordinates in the reference system of the prototype phase (Sapriel, 1975) or under other presentations (Erhart, 2004). The number of possible domains in the ferro phase is given by the ratio between the number of elements of symmetry of the para phase and that of the elements of symmetry of the ferro phase as indicated in Chap. 1. We can also use

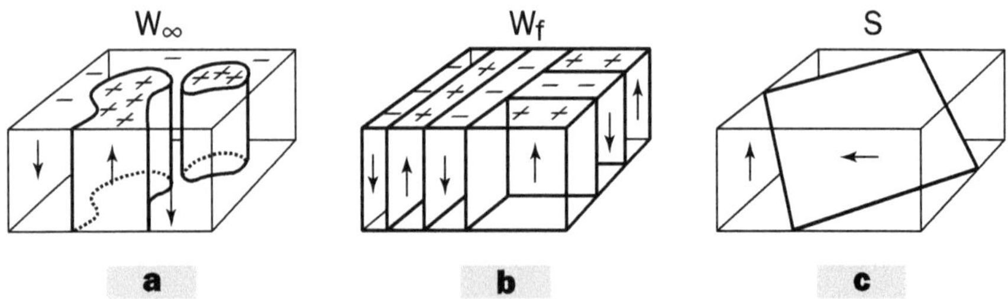

Fig. 2.5 Representations of the walls allowed in the model of Fousek and Janovec (1969) (electric polarization indicated by arrows): (**a**) W_∞ walls in the form of cylinders with sections of arbitrary shapes in the plane perpendicular to the polarization, (**b**) plane W_f walls in crystallographic planes corresponding to elements of symmetry lost at the para-ferro transition, and (**c**) walls S whose orientation varies with polarization and electromechanical coefficients and therefore with temperature

this criterion if we know the symmetry of the ferroelastic crystal and its domain texture, to deduce what would be the symmetry of the high temperature phase, useful reasoning for example, in the case of fusion of the crystal in ferro phase (Dvorak, 1978).

To illustrate this concept of permissible walls, Fig. 2.6 presents domain textures of the ferroelastic crystal $LiKSO_4$ with a transition from $P6_3mc$ to $Cmc2_1$: the angles between permissible walls are 120° in memory of the hexagonal symmetry of the prototype phase lost at the transition (Cach et al., 1985). Furthermore, Fig. 2.7 proposes textures of some of our favourite crystals for which the planes of permitted walls are the planes (100) and (010) of the prototype para phase because of the lost order 4 symmetry axis at the transition.

2.2.2 *Experimental Realities of Permissible Walls*

The first step is to optically observe such "permissible" walls. When the optical axes differ by any angle between one domain and the neighbouring domain, observations are easily made between crossed polarizers and analysers as illustrated by certain photographs in this book of GMO, TMO, and Tanane textures. If the optical axes rotate exactly 90° between two adjacent domains, contrast figures due to reflections on the walls can still allow the observation of the walls as explained in Fig. 2.8.

This observability disappears if the angle of incidence on the crystal becomes too large, about 10 degrees of arc for the KDP (Oettel, 1965). For larger angles, the walls accumulate the phenomena of light transmission and reflection in accordance

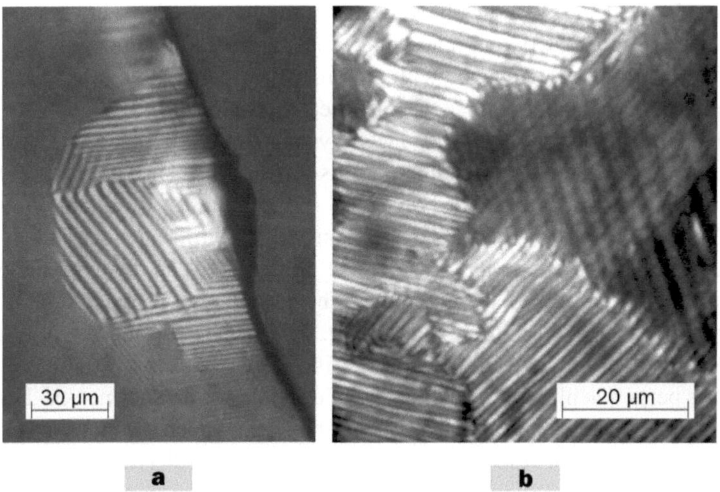

Fig. 2.6 Textures in domains of the $LiKSO_4$ crystal in the ferroelastic phase (transition at 190K): (**a**) at the appearance of the ferro phase on one sample edge at the transition and (**b**) in the ferroelastic phase with a very dense blocked texture (Cach et al., 1985)

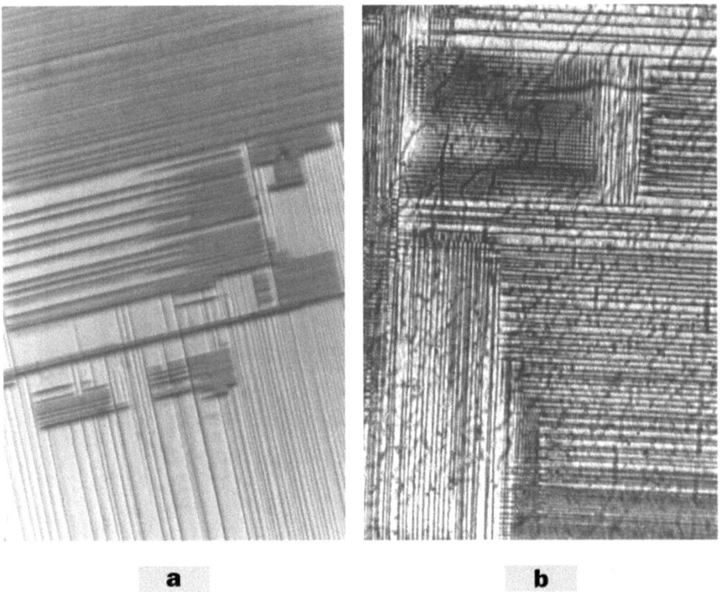

Fig. 2.7 Textures in domains of reference crystals (with the symmetry change $\bar{4}$ 2 m–mm2) of a KDP (**a**) crystal and a TMO (**b**). The orientation of the domain walls corresponds in both cases to the (100) and (010) tetragonal planes due to the loss of fourth-order symmetry at the transition

with the variations in light index from one domain to another for the polarization of the light used (Takagi and Shigenari, 1973). Beautiful works of light dispersion have thus been carried out and are not developed here (Hill et al., 1965; Kitaeva et al., 1995). Deflection studies have been numerous, especially in crystals whose optical axes are not exactly perpendicular (Tsukamoto and Futama, 1993).

The variation of birefringence with the order parameter can be exploited during precise experiments as in Fig. 2.9 (Bastie et al., 1976a, b): a laser beam passes through a fairly thick sample of the GMO crystal, parallel to the ferroelectric axis c and the unique domain wall it contains in its ferro phase. The sample is free from any external excitation and therefore from induced light diffraction (Shandarov et al., 2019). The variation of the deformation and the birefringence in the wall creates a classic mirage effect and interference figures between deviated light beams and non-deviated beams, observed on the light exit face. A change in light polarization (following an orthorhombic axis in Fig. 2.9c then following the other axis in Fig. 2.9d) changes the interference zone. This demonstrates the variation of optical properties perpendicular to the wall over a width of several micrometres. This result is in agreement with others obtained in visible light under similar conditions, by studying Fraunhofer diffraction figures of an isolated wall (1.3 μm for a sample thickness of 0.78 mm) (Suzuki, 1972).

Other measurements in visible light give "wall widths" of the order of a micrometre without being in the same situation of an isolated wall, at rest: either the wall is

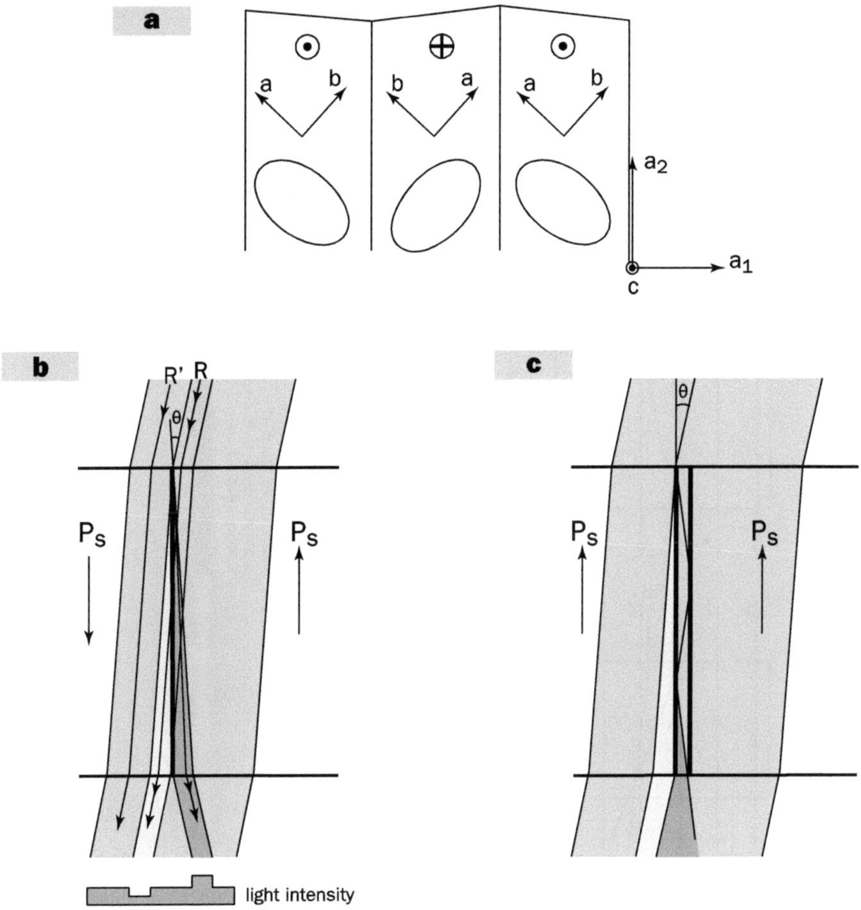

Fig. 2.8 (**a**) Representation of domains in projection along the c-axis of the type crystals; the direction of polarization, the orthorhombic axes and sections of ellipsoids of indices are indicated. (**b**) Path of a light beam which, outside the sample, makes an angle of incidence ϑ with the wall plane: after having penetrated into the crystal, depending on the polarization of the light and therefore the direction of index variation between the two domains, refraction or total reflection occurs. A small diagram (below) illustrates the contrast in light intensity observed on the exit face. The difference in intensity on the exit face is of $t\vartheta(n_a - n_b)/(n_a n_b)$ (Oettel, 1965), with t being the thickness of the sample. (**c**) Path of light beams when a domain channels some of the light through total reflections on the walls

in fact moving in the Raman scattering experiment of a wall excited at 400 Hz (Sheperd and Barkley, 1972), or the situations correspond to textures with numerous walls. The boundary conditions differ, and moreover, in the GMO, deflection problems with low intensities can complicate interpretations (Staniorowski and Bornarel, 2000).

Other means of observing the walls do not give the same orders of magnitude for the "widths" of the wall, that is to say for the regions with properties not conforming

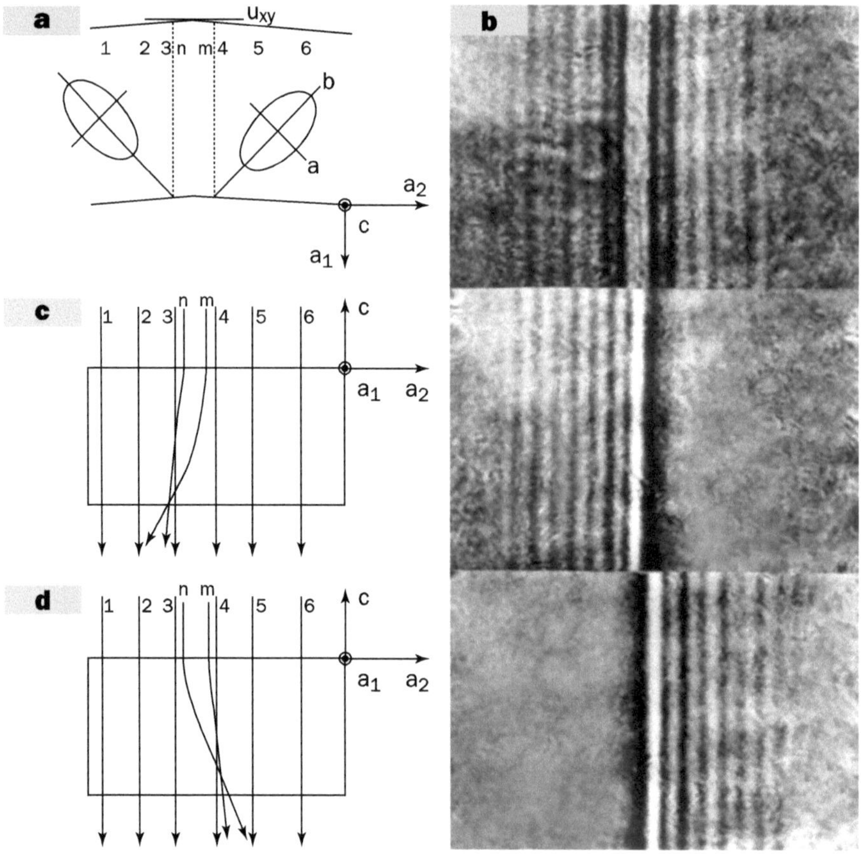

Fig. 2.9 Deviation of a laser beam propagating along the ferroelectric *c*-axis, mirage effect in GMO (Bastie et al., 1976a, b): the modification of the light index perpendicular to the wall causes a curvature of the light path as in mirage effects; one can observe on the exit face of a sample, the interferences between the parts of the laser beam deviated, those that are less so and others that are not. The experiment is carried out in a static situation, at regulated ambient temperature, on a single wall in the middle of a massive sample 8 mm thick. (**a**) Schematic representation of the domain wall in section (001), of the optical indicatrices and of the shear deformation u_{xy}; (**b**) interference figure on the exit face of the sample with polarization of the light at 45 degree of arc of extinction positions, therefore parallel to the wall; (**c**) represented on the left diagram are laser beams deviated by the change in light index in the wall for a light polarized in the direction of the orthorhombic *a*-axis; on the right, photograph of the interference figures observed on the exit face; (**d**) schematic representation (on the left) and photograph of the interference figures (on the right) for a polarization of the light in the direction of the orthorhombic *b*-axis. On both photographs of (**c**) and (**d**) the vertical black limit corresponds for the incident light to the most deviated beams; the gap between the two black traces of about 10 μm gives an estimate of the region in which the existence of the wall at rest sufficiently modifies the birefringence to create a mirage effect, a curvature of the laser beam

to those observed within the domains that a wall delimits. Reconstructed images of electron microscopy, on the GMO (Yamamoto et al., 1977a, 1977b), lead to estimates of wall width between 30 and 70 Å, but the variations in intensity of scattered light cover regions of the order of a micrometre. These results are obtained with thin crystals (0.2 mm before chemical attacks), and the method provides surface information. The wavelength corresponding to the diffracted energies is about 10 Å. For their part, results obtained in neutron diffractions, X-rays (using wavelengths of about 10 Å to 1 Å), lead to estimates of wall width in the KDP, between one and a few tens of elementary cells. These results are usually obtained with textures in domains whose characteristics are often unknown.

To conclude, the estimated wall widths depend on the experimental technique (Tagantsev et al., 2010) and the wavelength used in ferroelectric or ferroelastic crystals: of the order of a micrometre with optical measurements or X-ray topographies, of the order of the Å or a few tens of Å in electron microscopy, X-ray, or neutron diffraction. It is clear that the physical phenomena detected are different. This theme, already pointed out by Kittel (1972), will be revisited in Chap. 4 with electric fields (or stresses) weak enough to create only variations around average positions, and in Chap. 3 with regularly moving walls.

The long-distance interactions between walls in Chap. 3 will also allow to question the concept of wall width. A good illustration is given by the works of diffuse X-ray dispersion in high resolution in the ferroelastic crystal WO_3: if with the usual profile calculations, the width of the wall is estimated at four elementary cells, the complete intensity profile shows intensities that vary inside the domains at all values located between the extreme peaks (Locherer et al., 1998). What is the important information? Is it the distance from the centre of the wall at which a modification of such physical magnitude occurs? Is it the maximum interaction distance with another object and especially with another wall? Is it a length chosen by definition, within the framework of a theoretical model with its assumptions, linked to the form of law obtained for such a physical parameter which generally accounts for its variation as a function of distance?

2.2.3 Wall Models

Various types of theoretical models are proposed for a permitted plane wall: simulations by atomic-level calculations taking into account a number of neighbouring atoms and a number of more or less important elementary cells, and others based on a continuum hypothesis of the medium with an extension of the Landau type energy development. Ordinarily, the authors assume the medium to be infinite, others less numerous take into account the boundary conditions imposed by the edges of the sample, or by the presence of defects, inclusions. Some models summarized below are adapted to the tetragonal-orthorhombic symmetry change of the work.

2.2.4 Atomic Level Models

Domain walls are boundaries between domains at the macroscopic or microscopic level, and the site of atomic arrangement modifications in the cells, at the nanoscopic level. Knowing these modifications, linking them to results with various measurement methods is not simple. In addition, models with assumptions of short-distance atomic interactions, within the elementary cell, may not account for its deformations for long-distance effects. Few models are validated by all known experimental results.

Let's first consider what the wall could be in the simplest case, without fluctuations, at absolute zero temperature. Regarding the crystals of the KDP family, an arrangement of hydrogens on their bond was proposed by Barkla and Finlayson (1953) (Fig. 2.10): two organizations account for a permissible wall while respecting the conservation of the chemical formula at the level of each tetrahedron (rule known as Slater's (1941)). Because of the orientations of the dipoles represented at the level of the tetrahedra within the wall, in projection in the (001) plane, one of the walls was called "neutral" (Fig. 2.10a), and the other "polarized" (Fig. 2.10b). Energy calculations per unit of wall surface γ, with assumptions about the position of protons (see Appendices A4 and A5), limiting to the interaction of the first 25 neighbouring tetrahedra (Bjorkstam and Oettel, 1966), give different results for the neutral wall and for the polarized wall:

$$\gamma_{\text{polarized}} = -30 \times 10^{36}\ \mu^2\ \text{erg}\,\text{cm}^{-2} = -3 \times 10^{-2}\ \text{J}\,\text{m}^{-2}$$
$$\gamma_{\text{neutral}} = -20 \times 10^{36}\ \mu^2\ \text{erg}\,\text{cm}^{-2} = -2 \times 10^{-2}\ \text{J}\,\text{m}^{-2}$$

with μ dipole moment linked to each tetrahedron ($\mu = 10^{-18}$ ucgs).

This result would argue for the preferential existence of a single type of wall, and one could quickly conclude that obtaining permitted walls at rest (at least at absolute zero) in KDP would not pose a problem of junction at the atomic level. The upper diagrams of Fig. 2.10a, b, however, remind us that proton displacements occur in correlation with tetrahedron rotations (and modifications of $P-K$ distance not illustrated by these diagrams) (see Annex A4); the problem remains more open.

Some models are based on simulations of atomic interactions with the nearest neighbours in the walls of ferroelastic crystals with simple symmetries going from cubic to tetragonal or orthorhombic, with few atoms in the cell, and use Van der Waals and Lennard-Jones potentials. The aim is often to link atomic interactions to macroscopic parameters such as the c_{ijkl}. Ordinarily, harmonic relations are chosen between first and third neighbours and anharmonic with second neighbours (to obtain a possibility of phase transition). The ferroelastic perovskites are thus well studied such as WO_3 (two-dimensional model (Novak and Salje, 1998a) and three-dimensional (Lee et al., 2003a, b, c)). These simulations involve at most a few hundred atoms or cells. They are motivated by possible applications that would use the difference in properties between the walls and the interior of the domains in terms of ion diffusion, oxygen vacancy diffusion (Wang et al., 2001; Harrison and Redfern, 2002; Calleja et al., 2003), and more generally, conductivity including

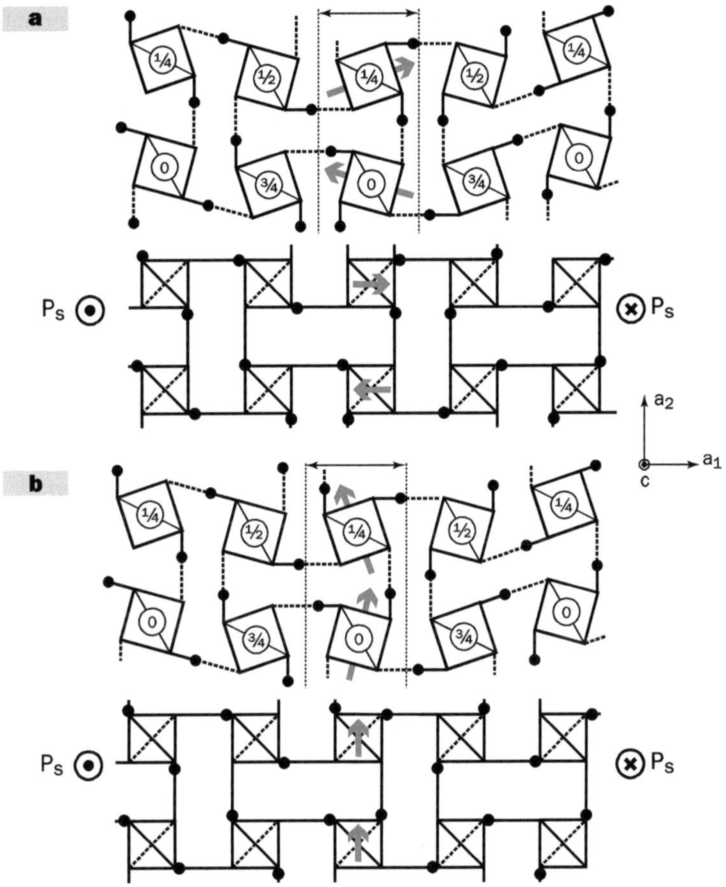

Fig. 2.10 Only the protons and the bonds close to a PO_4^- tetrahedron are represented in projection in a (001) plane. In (**a**) and (**b**), the upper representations were proposed by Barkla and Finlayson (1953) with the coordinates of the centres of tetrahedra along (001) indicated in fraction of primitive cell dimension *c* in the circles. The lower, more schematic representations will be used in the rest of the work when only the positions of protons are of interest. However, we must not forget that there are two types of tetrahedra in different orders along *c*. Also indicated are the polarizations P_S inside the domains by traces of arrows perpendicular to the sheet (up and down). The arrows in the plane of the sheet account for the dipoles linked to the tetrahedra in the wall: (**a**) the dipoles linked to the tetrahedra of the wall are oriented approximately in the direction a_1 and in opposite directions; the wall is said to be "neutral"; (**b**) the dipoles linked to the tetrahedra of the wall are this time approximately in the same direction a_2; the wall is said to be "polarized"

superconductivity (Viehland and Salje, 2014). Diffusion is often greater in the wall than in the domain (Na^+ ions in WO_3 (Aird and Salje, 1998, 2000)), but the opposite is also found (alkali ions in the Dauphiné twins of α quartz (Calleja et al., 2001)) or reduced conductivity (in Pb, Cu, PO_4) (Camara et al., 2000). Let's also mention the case of non-polar ferroelastic $CaTiO_3$ whose walls have a strong anti-ferroelectric component with dipoles perpendicular to the wall planes (Goncalves-Ferreira et al.,

2008), which recalls the case of Fig. 2.10a with coupling between deformation and dipole moment.

Generally, atomic-level simulations suggest certain explanations regarding phenomena, for example, different interstitial diffusion between the wall and the domains, with:

- A widening of the mesh that "opens" the passage in the wall
- Weaker atomic interactions in the wall, in another phase than inside the domains
- More numerous atomic vacancies
- Accumulated electric charges

These hypotheses are taken up in various forms in the continuous medium treatments presented below.

2.2.5 *Simple Continuum Models Based on an Extension of Landau's Development*

By hypothesis, in such phenomenological models, it is considered that there are no singularities for the physical parameters considered and that their variations are continuous. The difference compared to the relation (1.9) of Chap. 1 is to consider the medium inhomogeneous. The order parameter will be defined at a meso-scopic level, with enough atoms to have a thermodynamic meaning, and possibilities of modification at dimensions small compared to those of the crystal to be able to see the effect of interactions, for example elastic. The exercise involves often elegant presentations with partition functions, Hamiltonian, etc. strictly applicable for small fluctuations and this, sufficiently close to transitions. The arguments used should exclude situations with order-disorder transitions. But in fact, these restrictions often seem excessive if we are satisfied with an agreement with experimental results, as is often the case with predictions from Landau-Ginzburg-Zhirnov developments.

Let's therefore consider the case of a single wall between two domains of opposite order parameter values, and a profile assumed to be identical when the wall moves perpendicularly to itself at constant speed v. Various forces are in competition, one tends to make the order parameter be in a situation of global minimum energy, another is created by the imposed energy gradient, and, a friction force is added during this movement. Progress in this approach over the years is observed in the literature (Zhirnov, 1959; Ginzburg, 1960; Collins et al., 1979; Gordon, 1983; Barsch and Krumhansl, 1984; Jacobs, 1985; Gordon, 1986; Salje, 1988), synthesized from the 1990s (Salje, 1990, 1991; Huang et al., 1995). For a single flat wall, the laws are continuous functions of the variable x, distance to the centre of the wall. The limiting hypotheses compared to the complete initial development of Zhirnov (1959) allow the use of a simple form of the Gibbs type potential as a function of the single order parameter q and a single gradient term with respect to the x direction:

$$G(T,q) = G_0 + (a/2)q^2 + (b/4)q^4 + (c/6)q^6 + (\delta/2)(dq/dx)^2 \tag{2.2}$$

δ positive is called the correlation coefficient (Smolenskii et al., 1984). Only the square gradient term is retained, considering by hypothesis that the variations are not too abrupt. Using the Euler-Lagrange relation to optimize the search for equilibrium, we get

$$\delta\left(d^2q/dx^2\right) = aq + bq^3 + cq^5 \tag{2.3}$$

If the transition is of the second order (with b positive and c negligible), and the values of q varying from $+q_S$ for x tending towards $+\infty$, to $-q_S$ for x tending towards $-\infty$, we find a variation of q perpendicular to the wall plane (Zhirnov, 1958, 1959; Salje, 1990, 1991) of the form:

$$q(x) = q_S \ \tanh\left(x/t_0\right) \ \text{with} \ q_S = |a|^{1/2} b^{-1/2} \ \text{ and } \ t_0 = (2\delta)^{1/2} |a|^{-1/2} \tag{2.4}$$

the coefficient δ provides a major indication concerning t_0 usually called half-width of the wall. In this hypothesis of second-order transition, with $|a| = A\ (T_0 - T)$, q_S varies as $(T_0 - T)^{1/2}$ therefore increases as we move away from the transition (relation (1.10)); in this model, the wall width $2t_0$ varies with temperature as $(T_0 - T)^{-1/2}$ and therefore becomes infinitely large at the transition as illustrated in Fig. 2.11a by the slope of the curve α in $x/t_0 = 0$. It is also possible to calculate the energy per unit area of this flat wall γ_w:

$$\gamma_w = \int_{-\infty}^{+\infty} \left(G(q,x) - G(q_s)\right) \mathrm{d}x \text{ from which } \gamma_w = \left(2\delta a^3\right)^{1/2} b^{-1}.$$

This energy per unit area of the wall varies like $(T_0 - T)^{3/2}$ and therefore becomes very small as we approach the transition; conversely, it grows as the temperature decreases and faster than other parameters, firstly the order parameter which varies like $(T_0 - T)^{1/2}$. In the simple assumption of an activation process for the creation of walls, we could thus expect a variation in the number of walls as:

$$n \approx \exp\left(-f\left(T_0 - T\right)^{3/2} / T\right),$$

hence a very large number of walls at the transition. Given the initial assumptions of the model, we can only expect this kind of variation in the few degrees below the transition, since other parameters intervene at lower temperatures, as demonstrated in the following section.

If the transition is of the first order, that is to say if b is negative and c positive, q(x) takes the form (Lajzerowicz, 1981; Falk, 1983; Jacobs, 1985):

$$q = q_S\, sh\left(x/t_0\right)\left(D + ch^2\left(x/t_0\right)\right)^{-1/2} \tag{2.5}$$

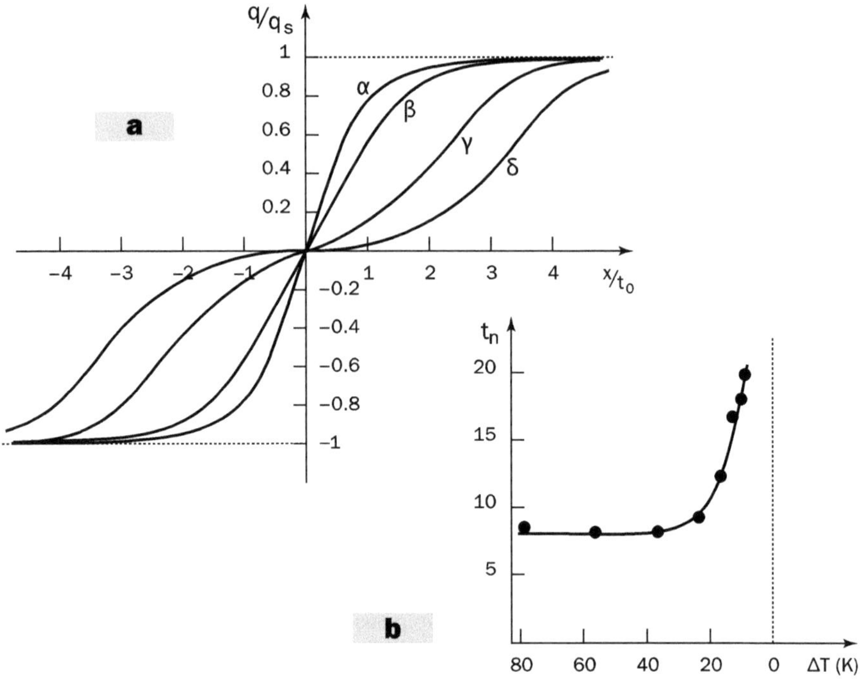

Fig. 2.11 (**a**) Relationship between the order parameter and its saturation value as a function of the distance to the centre of the wall (in half-width units). It is the result of a model of a flat wall in a continuous medium using a Landau-type development with a gradient term of the position × perpendicular to the wall (Zhirnov, 1958) (relations (2.4)–(2.6) of the text). (α) corresponds to a second-order transition, (β) to the tri-critical transition, (γ) and (δ) to a first-order transition with respectively $D = 50$ and $D = 500$ (D defined in the text). In these latter cases, a significant part of the wall can be assimilated to a para region (Ishibashi and Suzuki, 1984). (**b**) Wall width calculated following X-ray diffraction experiments for KDP (Andrews and Cowley, 1986) (t_n wall width in elementary cell dimensions and $\Delta T = T_C - T$). For KDP, this wall width remains almost constant between 10 K and 90 K, then grows with $T_C - T$, reaching about ten cells at $T_C - 2$ K and then diverges; the values close to T_C have not been estimated because of noise. The same measurements in DKDP are qualitatively similar: the width is estimated at two cells around 100 K (Bruce, 1981), 13 cells at $T_C - 1.7$ K; the values closer to T_C, not estimated because of noise, are higher. Such variation laws are also observed, sometimes more brutal, in pure ferroelastics (Sun Wenyan et al. 1985)

$$\text{with } t_0 = (2\delta)^{1/2}\left(q_S^2\left(b + 2cq_S^2\right)\right)^{-1/2}$$
$$\text{and } D = c\left(2c + \left(3b/2q_S^2\right)\right)^{-1} \tag{2.6}$$

The above relations remain valid both when the transition is of the second order (with $c = 0$) and when b tends towards zero in the case of a tri-critical transition. Figure 2.11a presents in various transition situations, the variations of $q(x/t_0)$ predicted by this phenomenological model derived from the relation (2.2) (the middle of the wall is at $x = 0$). In fact, the variation $q(x/t_0)$ is not radically different between a second-order transition and a first-order transition close to the tri-critical situation in which b cancels out like $(T - T_b)$ with T_b defined by the relation (1.34). On the other hand, the shape of $q(x/t_0)$ changes noticeably in a clear first-order transition if the parameter D takes large values. In such cases, we could find a wall of notable width with a value of the order parameter close to zero (Fig. 2.11a curve δ), therefore comparable to that of the para phase (Lajzerowicz, 1981).

By setting as before $a = A(T - T_0)$ and q_S increasing in the ferro phase as the temperature decreases, the above relations give in the tri-critical case (Gordon, 1983):

$$q = q_S\left(3/2\coth^2\left(x/t_0\right) - 1/2\right)^{-1/2}$$
$$\text{with } q_S = \left(-a/c\right)^{1/4} \quad \text{and} \quad t_0 = \left(\delta/-a\right)^{1/2} \tag{2.7}$$

The energy per unit wall area varies like $(T_0 - T)$ with a very large wall width when approaching T_0. X-ray diffraction measurements concerning the KDP crystal with its first-order transition near the tri-critical, carried out on a texture with many domains, give a result that goes in this direction, as illustrated in Fig. 2.11b (Andrews and Cowley, 1986). These results are to be compared to Gordon's estimate with the previous model: $t_0 = 30$ elementary cells for $(T_0 - T) = 6.02\times10^{-2}$ K (Gordon, 1983). At this temperature, using the relation $\delta = (\pi/15)\, a_1^2$ (Ginzburg, 1960) with $a_1 =$ 7.453 Å, and average values of a and c (Western et al., 1978), we find for the wall width $2t_0 = 442$ Å, and for its energy density $\gamma_w = 0.029$ erg/cm^2. These results are very different from those obtained at lower temperatures by various methods: wall widths of a few lattices and γ_w between 10 and 40 erg/cm^2 (Känzig and Sommerhalder, 1953). But the experimental situation is quite different: the proximity of the transition widens the walls and significantly reduces their surface energy which results from the initial assumptions of the model (continuum, etc.).

We cannot forget that the previous theoretical model assumes an infinite crystal, both in the x direction perpendicular to the plane domain wall and in the plane of the wall. Generally, theoretical calculations carried out with these implicit assumptions of infinite crystals conclude for temperatures far from the transition, to wall widths of the order of the elementary cell or a few ones. This result for the wall width is usually obtained without taking into account all the deformation components correlated to the order parameter (Tagantsev et al., 2016). The assumption of infinite crystals does not correspond obviously to the real situation. Similarly, it will sometimes be difficult to find experimental situations conforming to the assumption of the isolated wall (with x varying from $-\infty$ to $+\infty$), except in a few cases like those described during reversals in Chap. 3 with GMO or Tanane crystals. However, if the isolated wall situation is defined as "without proven interaction with other objects", it could also exist in some forms in dense domain textures, at temperatures lower than the freezing of domains as demonstrated by dielectric measurements with simultaneous observations of domains (Chap. 4).

2.2.6 Models with Several Order Parameters

In the assumption where only the distance x to the centre of the wall is considered, some theoretical continuum models introduce, in addition to the main order parameter, a second order parameter (or several) and a suitable Gibbs potential (Pöttker and Salje, 2014). Some crystals are symbolic of these situations with interesting implications in terms of physical properties. We will not develop these situations here despite their richness with the concept of chirality (Houchmandzadeh et al., 1991; Salje, 2002; Salje and Zhang, 2009) and wall widths often different for the various order parameters. We usually find variations in thx/t_0 for an order parameter (in the case of a second-order transition) and for the other order parameters, variations in x either of the same form, or different. This is the case for GMO: the domain system is more complex than that of KDP and its family, with antiphase walls that could be a theme for book development in itself. In the context of this book, only the aspects common to our favourite ferro-ferro crystals due to losses of point group symmetry and translation symmetry are retained.

2.2.7 Models with Associated Deformations

It is possible to take into account other contributions to the free energy, implicitly considered negligible when we limit ourselves to the relation (2.2) (see Annex A2). Elastic contributions have thus been introduced in very diverse situations, for example for twinned domain walls or antiphase (Cao and Barsch, 1990).

If we consider that the chosen order parameter q is for KDP, the polarization P_3, considering only the variations perpendicular to the wall, and implicitly an infinite medium, the additional energy density contributions can be written with the usual notations (Annexes A1 and A2):

Piezoelectricity $-a_{36}u_6P_3$,
Elasticity $½\Sigma\ c_{ij}u_iu_j - \Sigma u_ip_i$,
Thermal expansion $-\Sigma\ \delta_iu_i\Delta T$,
Electrostriction $\Sigma\ (q_iu_iP_3^2 + r_iu_iP_3^4)$

and gradient contributions of the form $\delta_i(\delta u_i/\delta x)^2$ in this simple case where only the variable x intervenes (Troussaut, 1987).

Considering the crystals of the KDP family, the deformations $u_1 = u_{xx}$ and $u_3 = u_{zz}$ with the same approach as before, relations such as the following are obtained:

$$u_{xx} = u_{xx}^0 + Q_{13}P_3^2 + R_{13}P_3^4 \quad (2.8)$$

$$u_{zz} = u_{xx}^0 + Q_{33}P_3^2 + R_{33}P_3^4 \quad (2.9)$$

in which u_{xx}^0 and u_{xx}^0, not coupled to the order parameter, are only functions of the temperature T and the hydrostatic pressure p; Q_{13} and Q_{33} are the coefficients of electrostriction, and R_{13} and R_{33} are the coefficients of higher order electrostriction. The volume change of the cell is obvious:

$$\Delta V / V = u_{zz} + 2u_{xx} \tag{2.10}$$

The variations $u_{xx}(T)$ of Fig. 2.12 are the result of simultaneous measurements of neutron diffraction and gamma diffraction (Annex A3) (Troussaut, 1987). They highlight, at constant pressure, the linear variation of u_{xx}^0 (T) in the para-electric phase. In the ferro phase, the results are interpreted as the sum of a linear temperature contribution different from that observed in para phase, and a contribution correlated to the polarization. The use of electrostriction coefficients accounts for the experimental results.

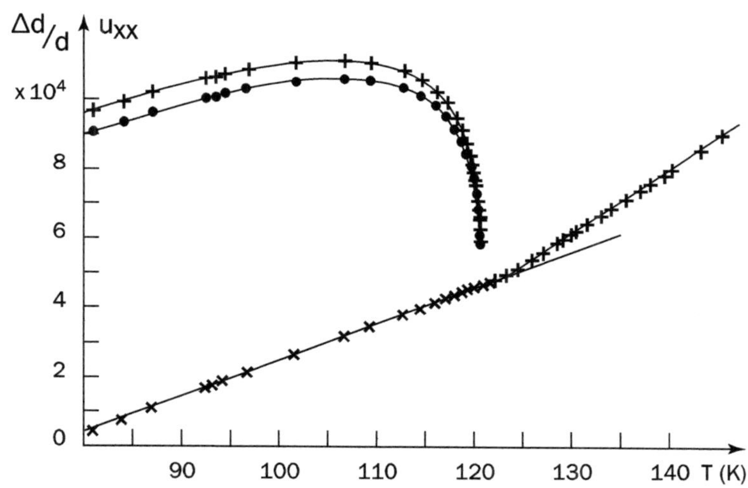

Fig. 2.12 Relative variation $\Delta d/d$ of the distances between lattice planes (•) of KDP along the tetragonal axis a_1 as a function of temperature, obtained by neutron diffraction. Simultaneous measurements of neutron diffraction and γ diffraction allow to obtain $u_{xx}(T)$ (+) by eliminating the effect of possible rotations of lattice planes (Troussaut, 1987; Bastie et al., 1988) and the possible surface effects encountered with X-rays. The variations due to simple dilation (x) are deduced by calculation. The results are in agreement with the relation:

$u_{xx}(T) - u_{xx}(T_0) = \Delta u_{xx}^0(T) + Q_{13}P_3^2 + R_{13}P_3^4$

with $\Delta u_{xx}^0(T) = \alpha_T{}^i(T - T_0)$. As the thin lines of the figure demonstrate, the dilation slopes are different in the para phase ($\alpha_T{}^P = (2 \pm 0.1) \times 10^{-5}$ K^{-1}) and in the ferro phase ($\alpha_T{}^F = (0.6 \pm 0.05)$ $\times 10^{-5}$ K^{-1})

Assuming temperature-constant electrostrictive coefficients, we find the values $Q_{13} = (2.8 \pm 0.1) \times 10^{-12}$ esu = $(2.52 \pm 0.09) \times 10^{-1}$ m^4 C^{-2} and $R_{13} = (3.1 \pm 0.1) \times 10^{-21}$esu = (25.11 ± 8.1) m^8 C^{-4}, results in agreement with published values (Von Arx and Bantle, 1943; De Quervain, 1944; William and Cook, 1967; Boiko and Liu Tan Fat, 1970; Zeyen and Meister, 1976, b; Bastie et al., 1975, 1978a, b, 1981; Torstweit, 1979)

Comparable analyses concerning u_{zz} (T), carried out by neutron diffraction (Deguchi et al., 1980; Bastie et al., 1981), give the results presented in Fig. 1.36. The value of Q_{33} obtained is $(3.1 \pm 0.3) \times 10^{12}$ esu; that of R_{33} is more imprecise: $(0 \pm 1) \times 10^{-21}$ esu; the expansion coefficients from Deguchi's work would have values $\alpha_T{}^P = 2.9 \times 10^{-5}$ K^{-1} and $\alpha_T{}^F = 1.9 \times 10^{-5}$ K^{-1} with more uncertainty on this last parameter than that evaluated for $u_{xx}(T)$

By the approach of the continuous model on the basis of a development of Landau type with gradient terms in x, we find the variation of $P_3(x)$ (hence of $u_{xy}(x) = u_6$) of the form $\mathrm{th}(x/t_0)$ represented in Fig. 2.13a. The deformations $u_{xx}(x)$ and $u_{zz}(x)$ instead of cancelling out in the middle of the wall (at $x = 0$) pass through a minimum as illustrated by Fig. 2.13b, c. The relations (2.8) and (2.9) with terms in $\mathrm{th}^2(x/t_0)$,

a

b

c

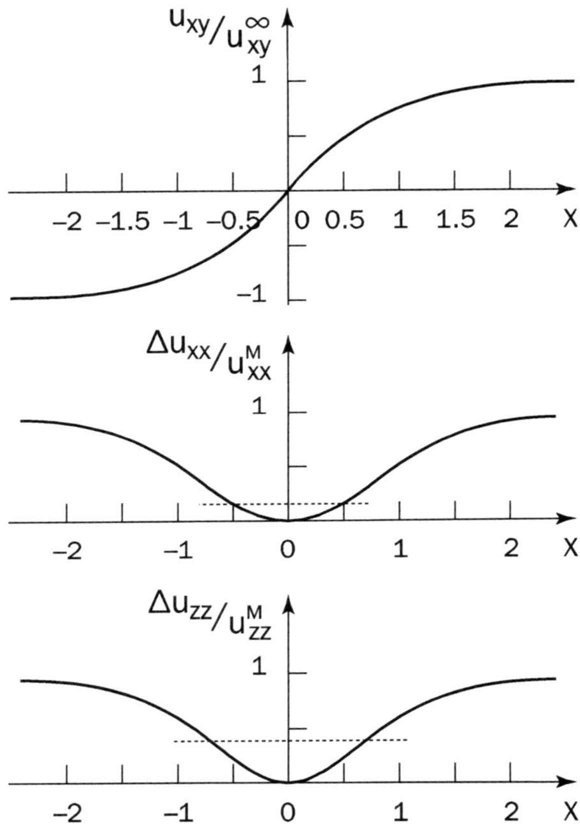

Fig. 2.13 For the KDP crystal at a temperature of 95 K: (**a**) Variation of the ratio u_{xy} (95 K, X)/u_{xy}^{∞}(95 K, ∞) with a temperature T_0 of 120.324 K (Troussaut, 1987) as a function of $X = x/t_0$, t_0 defined by the relation (2.4). (**b**) Ratio between Δu_{xx} $(T, X) = u_{xx}(T, X) - u_{xx}^{0}(T, 0)$ and the maximum value $\Delta u_{xx}^{M}(T, \infty)$ for X infinity. The dotted line reminds the value of u_{xx} $(T_0, 0)$ and allows to evaluate at $X = 0$, the part due to the expansion and due to the polarization P_z (null at the centre of the wall). (**c**) Ratio between Δu_{zz} $(T, X) = u_{zz}(T, X) - u_{zz}^{0}(T, 0)$ and $\Delta u_{zz}^{M}(T, \infty)$, maximum value for X infinity. The dotted line reminds the value of $u_{zz}(T_0, 0)$ and allows to evaluate at $X = 0$, the part due to the expansion and that due to the polarization P_z (null at the centre of the wall). The curves (**b**) and (**c**) show that the variations following x due to the wall have a greater range for u_{xx} and u_{zz} than for u_{xy}. The relative variations between u_{xx} and u_{zz} are not obvious because of the importance, in both cases of the term in $\mathrm{th}^2(x/t_0)$. The effect of temperature is double: the values of u_{xy}, Δu_{xx} and Δu_{zz} decrease with $T - T_0$ while the wall width increases by a factor of 10 between 95 K and a temperature a few tenths of degrees below the transition

and in $th^4(x/t_0)$ if we take into account the higher order electrostrictive effect, lead to an increase in the effective wall width greater than that observed for $P_3(x)$ (or $u_{xy}(x)$) (Lee et al., 2003b).

This phenomenon, well studied in certain ferroelastics, explains the changes in wall properties compared to the domains it delimits. For example, a softening that allows a greater density of vacancies, or on the contrary, a higher concentration of defects, thus different mobilities (Aird and Salje, 1998, 2000; Lee et al., 2003a, c; Calleja et al. 2003), a growth or a decrease by compression of the distance between the atoms (Lee et al., 2003b), different correlation energies, charged walls (Tagantsev et al., 2016), different chemical properties (Camara et al., 2000), greater conductivities (Aird and Salje, 2000; Lee et al., 2004), weaker ones (Bartels et al., 2003), etc.

Regarding the deformation u_{zz} under the assumption of a simple model with only the variable x and an infinite crystal, a gradient term in the form of $g_z(du_{zz}/dx)^2$ in the development of G leads to the same consequences as those observed with u_{xx}: $u_{zz}(T)$ which is qualitatively comparable with a notable growth in the ferro phase compared to the para phase as demonstrated in Fig. 1.36. The essential difference compared to $u_{xx}(T)$, in KDP, is an electrostrictive effect which essentially varies in $Q_{33}P^2$ given the low value of the coefficient R_{33}. Figure 2.13c illustrates this modification along the ferroelectric axis (in a very exaggerated way) under the assumption that only the variable x would play a role.

Figure 2.14 schematizes in the case of our favourite crystals with the simplistic assumption of negligible edge effects of the samples, how the presence of a wall between two domains induces compression or dilation stresses. The differences in situation between KDP, GMO, and Tanane (completely opposite to KDP regarding normal deformations along x and along z) are illustrated. In certain ferroelastics such as APFA, the stresses induce such variations in birefringence that the walls appear as bright between crossed polarizers (Bornarel et al., 1997). These deformations within the wall also shed light on some properties of poly-domain crystals, such as at very low temperature, the channelling of slow transverse phonons (Weibert et al., 1993). Taking into account normal stresses changes the estimates of wall widths as demonstrated notably with GMO (Haas and Jaep, 1974).

2.2.8 *Taking into Account the Limits of the Sample*

Models that take into account all the limits of the sample are rare despite the obvious importance of these. If we consider all directions, it becomes necessary to write the gradient terms in the most general forms, with the possible variations of quantities along x, but also along y and along z.

For example, if we take into account the role of the edges of the sample bent by the existence of the wall, the deformation can change according to the distance to the sample edge parallel to the wall. As illustrated in Fig. 2.15b, the deformation created by a ferroelastic wall obviously varies depending on the distance to the centre of the wall perpendicular to its plane, but also depending on the depth in the sample. The models made, either by numerical simulation considering the atoms

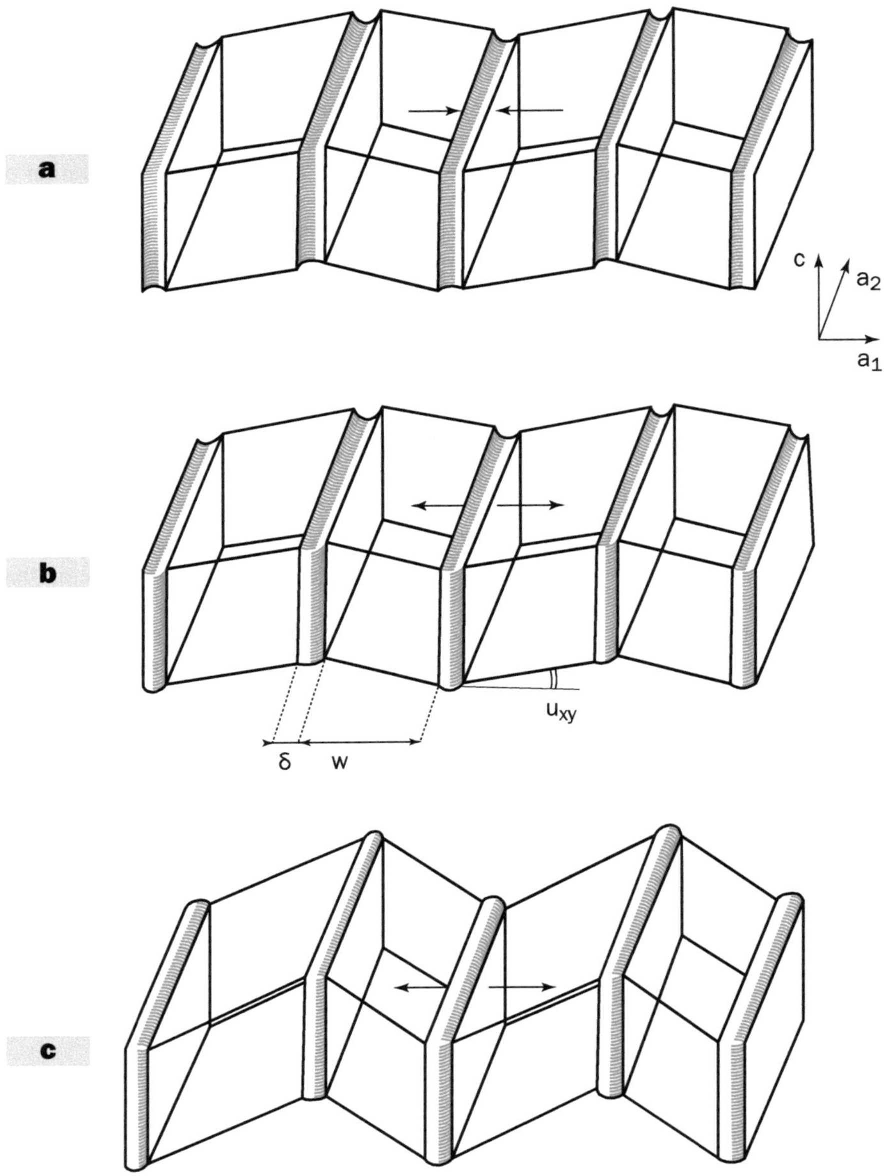

Fig. 2.14 "Artist's representation" made using dozens of measurements collected in the bibliography of lattice parameters in para and ferro phases. Only the average values of the lattice parameters are considered; their possible variations depending on the distances to the surface are neglected. Adjacent domains are of opposite polarities and deformations, and the walls that separate them are in tetragonal a_1 planes. The walls are considered wide enough for the realization of this small drawing with lattice values close to those of the para tetragonal phase. For example, if according to c, the length of the primitive cell is smaller in the para phase than in the ferro phase, the wall appears as a "groove", and if it is the opposite, as a "bulge". The same exercise is carried out in the tetragonal a_2 direction. The differences between the crystals are illustrated: (**a**) KDP, (**b**) GMO, (**c**) Tanane. The horizontal arrows indicate whether the domains exert pressure on the wall as in the KDP or, on the contrary, facilitate the expansion of the wall (especially the Tanane)

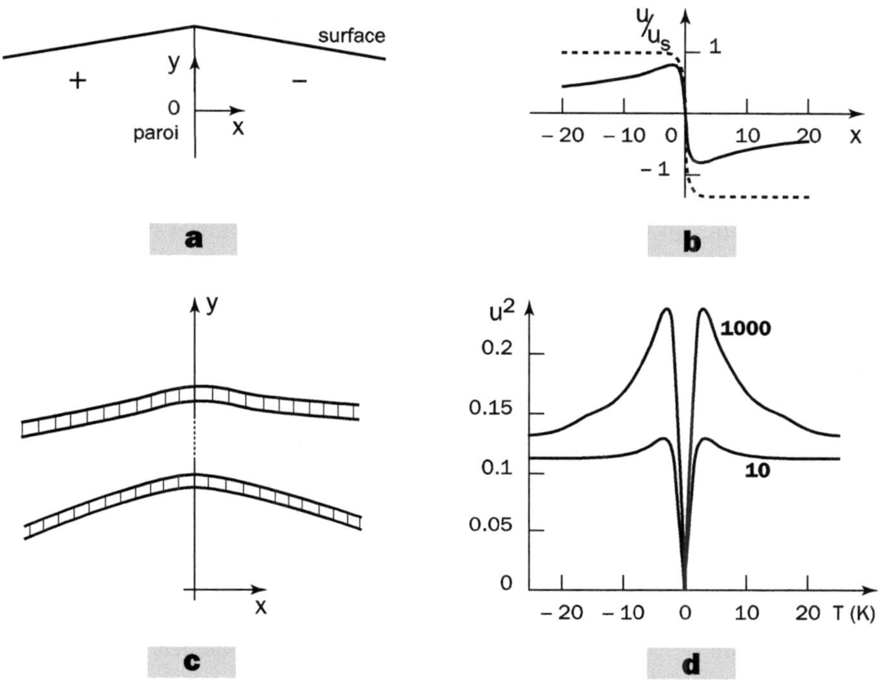

Fig. 2.15 Results from numerical simulations using a two-dimensional atomic model (NS) (Novak and Salje, 1998a, b) and other energy minimization models in the hypothesis of an elastic continuum (CS) (Conti and Salje, 2001). These models are adapted to textures close to isotropic structures of ferroelastic crystals. (**a**) A wall in the $x = 0$ plane delimits two domains with opposite deformations; this wall cuts the sample edge perpendicularly; the surface profile represented is theoretical. (**b**) In the NS models, the wall width does not depend on the depth relative to the surface; however, the surface deformation (continuous line) is less than that existing in the medium (interrupted lines) with a saturation value at infinity u_S; x is the distance from the centre of the wall, in units of its width. (**c**) Schematic representation of the elementary cells near the twin wall (centred at $x = 0$) at the surface (top) and in the crystal (bottom) in the case where the shear at the edge of the sample is not as marked as the shear in the middle and where a deformation in the direction parallel to the wall causes a growth. (**d**) In the context of the CS model, the square of the deformation u^2 as a function of the distance to the centre of the wall varies according to the value of the ratio c_{11}/c_{44} indicated (at $y = 0$, with $c_{12} = 0$ and $\gamma = 2$) (Conti and Salje, 2001). This representation illustrates a double relaxation at the edge of the sample, with maximum chemical properties of the wall, not at the centre of the wall, but at its edges, at distances of two or three times its width (Salje et al., 1999)

and their nearest neighbours, or by calculations in continuous elastic media, lead to the same conclusions (Houchmandzadeh et al., 1992; Novak and Salje, 1998a, b; Conti and Salje, 2001; Lee and Salje, 1999): the deformation at the surface is different from that inside the crystal. Even in the simple case of a cubic symmetry, the assumptions introduced in the models play a major role regarding relaxation at the surface and in the volume. Thus, we can expect surface profiles with bumps or hollows at the intersection between the wall and the sample edge. The more or less great correlation between the variation of the order parameter perpendicular to the wall and that depending on the depth in the crystal, the coupling between the order parameter and the other stresses, the ratios between the elastic properties are all determining parameters.

Furthermore, as demonstrated by the variation of extinction in γ diffraction (Fig. 1.34c, d), the change in mosaic and quality between the edge of a crystal and the interior is effective. It is therefore necessary to take into account all the experimental characteristics of a given crystal to propose a realistic model concerning the walls and the domains of ferroelastic-ferroelectric crystals.

Studies on thin layers as a function of thickness match the experimental observations made in ferro-ferro crystals with deformations induced by a wall at distances of the order of the thickness of the sample (Shapovalov et al., 2014). The order of magnitude of the dimensions is different, but the principle is the same.

Therefore, for a ferro-ferro domain wall, isolated, in equilibrium in a permissible plane, theoretical models predict walls of narrow widths (of one or a few unit cells) if the hypotheses are infinite crystals in all dimensions without taking into account induced deformations, only considering the direction perpendicular to the wall. Models that impose more realistic boundary conditions lead to wall widths of a few tens of cells, and experimental results prompt questions about the meaning of this wall width in relation to observed interactions. Questions remain:

- Can the wall be considered without fluctuations? Many models are implicitly assumed at absolute zero?
- Are the effective boundary conditions realistically explained (presence of other walls, distances from the sample boundaries perpendicularly, parallel to the wall, etc.)?
- Is the wall stationary or is it moving? The properties of one and the other have no reason to be identical; this point will be the subject of Chaps. 3 and 4.

2.3 Textures in Domains of a Set of Permitted Parallel Walls

2.3.1 Domain Textures of Permissible Parallel Walls

In many crystals, the permissible walls are parallel to crystallographic planes (W_f walls), and a natural texture includes sets of walls parallel to each other as presented in Figs. 2.6 and 2.7. In other crystals with usually W_∞ walls, therefore of any shape in section perpendicular to the ferro axis, the presence of impurities often creates an anisotropy that favours the orientation of the walls along a crystallographic plane; the texture is then comparable to those of crystals with W_f walls. If the domains are lens-shaped, the presence in the walls of mechanical defects (disinclinations, screw dislocations, etc.) is all the more likely that the curvature of the wall is greater; over time, the rearrangement of these defects usually favours the elongation of domains. Ferroelectric crystals TGS or $BaTiO_3$ provide examples, as illustrated in Fig. 2.1b.

In the simple case of plane walls parallel to a single direction and crossing the entire sample, several situations exist depending on the experimental conditions:

- Textures whose distances between walls differ as illustrated in Fig. 2.16 for a Tanane sample. Large domains (few walls) exist in this electrically short-circuited sample, whose edges give little grip to the walls since Tanane sublimes easily at room temperature with "soft" and rounded sample edges (Legrand, 1973).
- Finely layered natural textures, whose domain walls seem to be equidistant from each other. The theoretical proposals of fractals between distances have not been experimentally confirmed (Ozaki et al., 1995). In the first situations, the distributions of distances between walls are spread out as in Fig. 2.16b in Tanane, and Fig. 2.17 in TGS (Jaśkiewicz et al., 1977). This is also true in regular textures

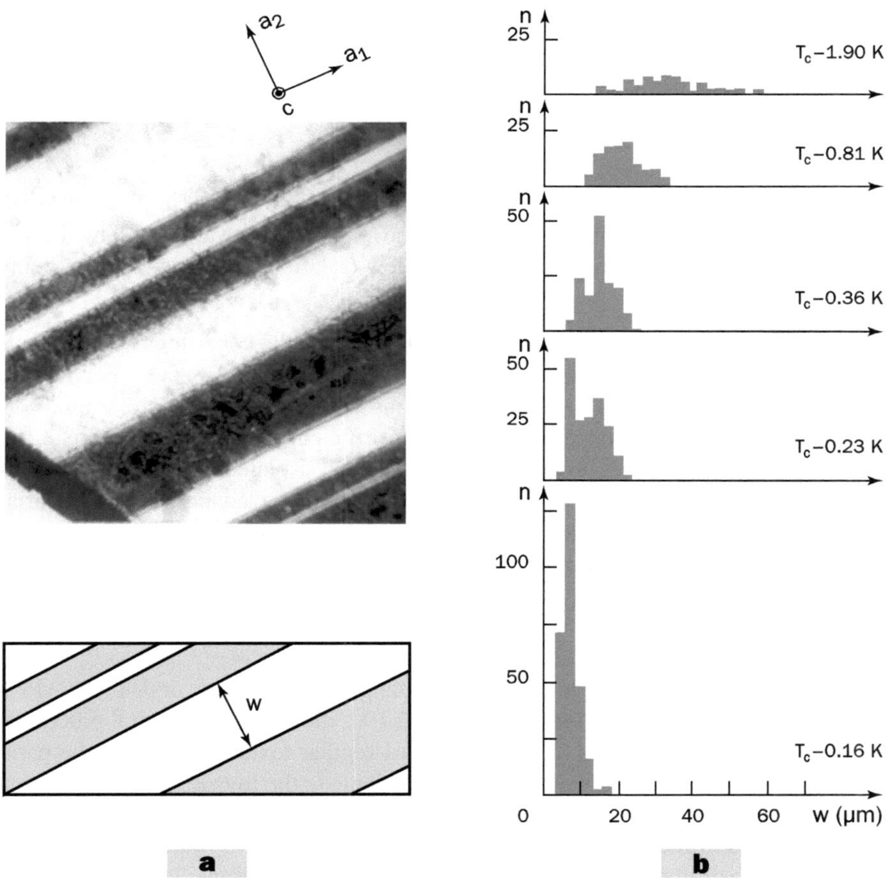

Fig. 2.16 (**a**) Simple domain texture (photograph in section (001) and schematic representation) of the Tanane crystal at a temperature of 5 °C (T_C – 8.5 K) and (**b**) distribution of domain widths at different temperatures in the ferro phase of Tanane (Legrand, 1973)

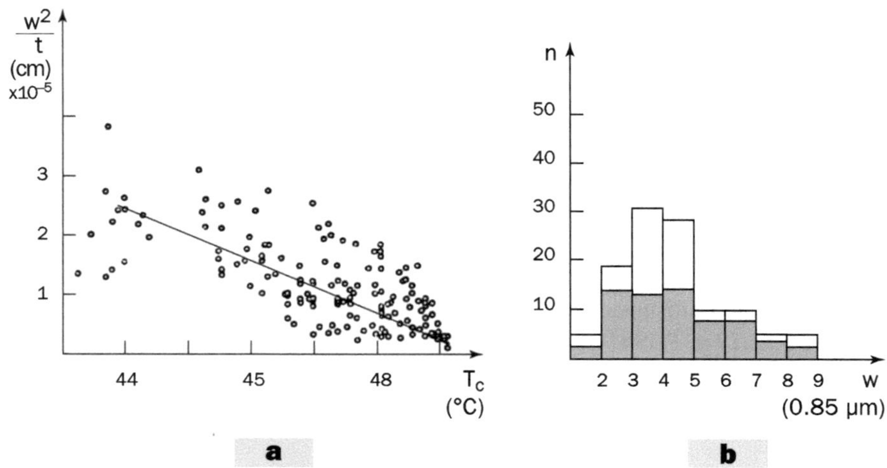

Fig. 2.17 (**a**) Correlation between the square of the domain width and the thickness of the samples at various temperatures in the TGS (Strukov et al., 1973) and (**b**) distribution of domain width in the TGS at 48 °C (Jaskiewicz et al., 1977)

depending on the thickness of the samples as illustrated in Rochelle salt (Fig. 2.18a) and KDP (Fig. 2.18b). The square root variation law of the Kittel-Mitsui-Furuiki model described below (relation (2.13)) is far from being followed:

For Rochelle salt, one sample out of eight gives results consistent with this law, and the observed domain widths are smaller than might be expected. For KDP, a collection of experimental results from various works shows, at a minimum, how the displayed experimental parameters are insufficient to conclude. Figure 2.18b illustrates the difference observed for w between isolated samples and short-circuited samples. In isolated samples, the observation of a law $w(t)$ close to a variation in $t^{1/2}$ requires a very slow cooling allowing domain rearrangements and stress releases, obtaining plane domain walls in one direction with identical dimensions along their tetragonal axis a_1 or a_2, and similar histories during measurements (Ha and Kim, 1985). Even in these cases, the values do not tend towards zero with the thickness of the sample (Fig. 2.18b) (Abe, 1987).

The evolution of the texture during the cooling of a ferroelastic crystal, the APFA (ammonium pentafluoro antimonate) in Fig. 2.19d, illustrates the general process of establishing natural textures: a very fine quasi-regular texture appears at the para-ferro transition with a periodicity that some attribute to the freezing of the mode that condenses at the transition. When the temperature decreases, the polarization and spontaneous deformation increase and also the widths of the walls decrease. The equilibrium texture obtained at a temperature T close to the transition, which would correspond to the optimization of energy, is obviously no longer the optimal texture at a temperature $T - \Delta T$; as the average distance between the walls increases, walls must disappear in the sample. A quasi-continuous variation of the number of walls

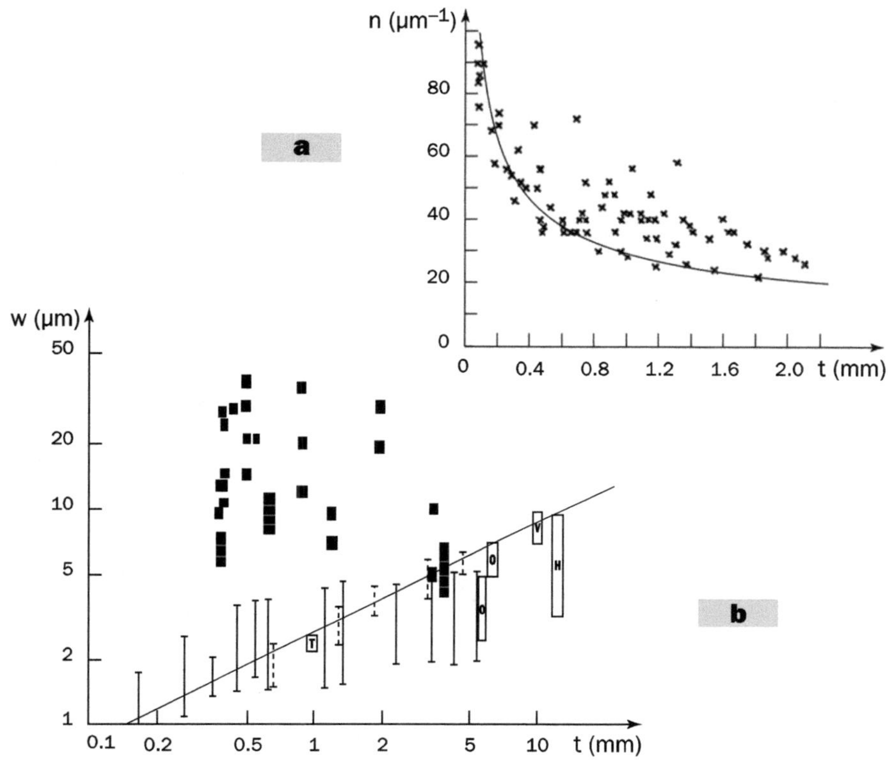

Fig. 2.18 (**a**) Variation of the number of domains per unit length as a function of the thickness of the Rochelle salt samples (Mitsui and Furuichi, 1953). The continuous curve represents a variation of w of the form $t^{1/2}$: the observed domain widths are statistically smaller than those predicted by this so-called Känzig model. (**b**) Compilation of domain width results as a function of the thickness of the KDP crystal samples (textures in dense and regular domains). The black rectangles correspond to samples with a single domain texture during dielectric measurements at 100 K; these samples are therefore short-circuited; domain widths can grow after several cycles of electric field and temperature. The other results are from the bibliography on isolated crystals: the continuous vertical bars (Abe, 1987), those in discontinuous traits (Ha and Kim, 1985), and the rectangles marked H (Hill and Ichiki, 1964) correspond to measurements by light diffraction. The results O (Oettel, 1965), T (Toshev, 1963), and V (Velichkhina et al., 1969) are obtained by direct microscopic observation. The drawn line would correspond to a variation of w as $t^{1/2}$. The results of Ha and Kim, which are the closest, are imprecise regarding the temperature. The thickness of the sample is evidently not the only parameter of variation of the width w of the domains; w tends towards a non-zero value when t tends towards zero as predicted by dislocation models (Torres et al., 1982)

can occur if the walls translate perpendicularly to their permitted plane and disappear on the edge of the sample parallel to them; even in this case, more or less abrupt rearrangements are observed. The most frequent rearrangement corresponds rather to a disappearance of domains by withdrawal of a number of walls multiple of two, followed by more or less significant lateral displacements of the remaining walls.

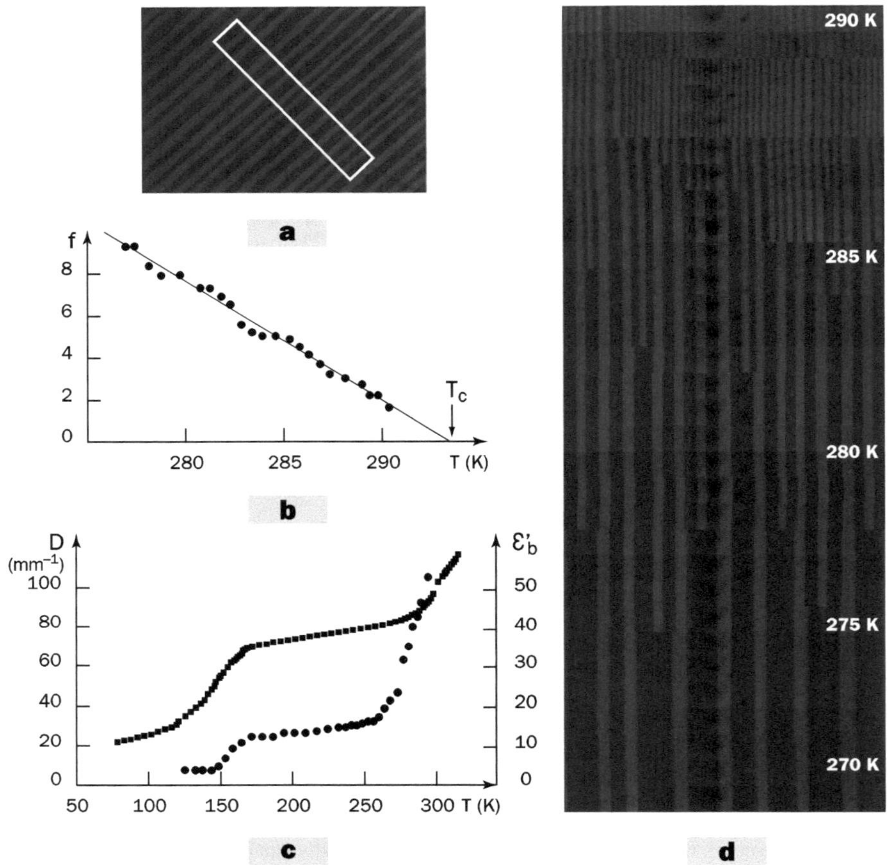

Fig. 2.19 (**a**) In the ferroelastic crystal of pentafluoro antimonate of ammonium $(NH_4)_2$ SbF_5 called APFA, the texture in domains of a region of the sample defined by the white rectangle, during cooling in the ferro phase (Bornarel et al. 1997), is studied. (**b**) The logarithm *f* of the number of domains *N* measured in the rectangle defined in (**a**) plotted as a function of temperature shows various modifications of *N*: a classic law of thermal activation process would give for the evolution of the number of domains with *T* (*a* and *b* are coefficients) (Salje, 1990):

$N = a\ e(-b(T_C - T)^{3/2})/T)$ with T_C Curie temperature, law generally satisfied with, however, alternations between stable situations and rapid rearrangements of the texture. (**c**) Evolution as a function of decreasing temperature, of the domain density *D* (number of domains per unit of distance perpendicular to their wall plane allowed) (black circles), and of the dielectric constant (black squares). (**d**) Photographs taken in the rectangle defined in (**a**) during the decrease of T. The disappearance of domains at certain temperatures is observed by rapid longitudinal withdrawals corresponding to the changes observed in (**b**)

We thus observe the existence of different characteristic times in the rearrangements of textures: brief times during a longitudinal withdrawal of domains and longer times of evolution towards a more regular texture by small lateral displacements of walls. The rate of change of the crystal temperature is decisive: a sufficiently slow variation allows successive rearrangements and the obtaining of regular textures with stress relief; conversely, a too rapid variation creates well-known quenching phenomenon in metallurgy, during which local stresses block the evolution of textures. This is confirmed by the plotting of the first polarization curves of Chaps. 3 and 4.

Even when all precautions are taken during cooling, the textures are metastable, and rearrangements bring them closer to the optimal equilibrium situation. Figure 2.19b shows such an evolution concerning the ferroelastic APFA with successive plateaus of variation in the number of domains as a function of temperature (Bornarel et al., 1997). The continuous line of the graph corresponds to the variation predicted by a classic model of thermal activation (Salje, 1990). The series of photographs in Fig. 2.19d demonstrates the link between notable modifications and withdrawal of needle domains. To understand natural domain textures, we are therefore interested in the processes of creation and disappearance of domains (following section and Chap. 3). The cooling process described with the APFA allows us to understand the importance of thermal history (cooling rate, time to obtain a stable texture at a given temperature). During warming in temperature, the texture often retains the memory of the lowest temperature, and for some crystals (and too high heating rates), the texture can only change at temperatures close to the transition. This explains the wide distributions of Figs. 2.17a and 2.18a, b. The problem is even more complex if defects interact with the walls, migrate into the walls, or modify their properties. The phenomenon is general: memory effects have been well studied, in quartz (Frondel, 1945; Heaney and Veblen, 1991; Xu and Heaney, 1997), in $YBa_2Cu_3O_7$ (Voronkova and Wolf, 1993), etc. The same is true of typical cases of sodium ion migration in WO_3 (Calleja et al., 2001) and in quartz (Redfern and Sartbaeva, 2002). The density of defects in the walls thus leads to a more or less large distribution of wall widths (Lee et al., 2005). The greatest caution is therefore necessary in ferro-ferro to interpret results related to domain texture: it is necessary to know this texture and ensure its stability during measurements.

2.3.2 Models Proposed to Explain Simple Periodic Textures

The scheme of Fig. 2.20a corresponds to the oldest model proposed to explain a regular domain texture in a ferroelectric, following a model concerning ferromagnetic domains (Landau and Lifshitz, 1935; Kittel, 1949a, b). Adapted to ferroelectricity (Annex A6) (Mitsui and Furuichi, 1953), only two types of energy are considered: the electrostatic energy of a set of equidistant domains with the existence of a depolarizing field W_{el}, and the energy due to the walls W_w, assuming a constant energy per unit of wall area. Then, the total energy per unit of surface perpendicular to the ferroelectric axis c is written:

$$W_t = Kw + \gamma_w t w^{-1} \tag{2.11}$$

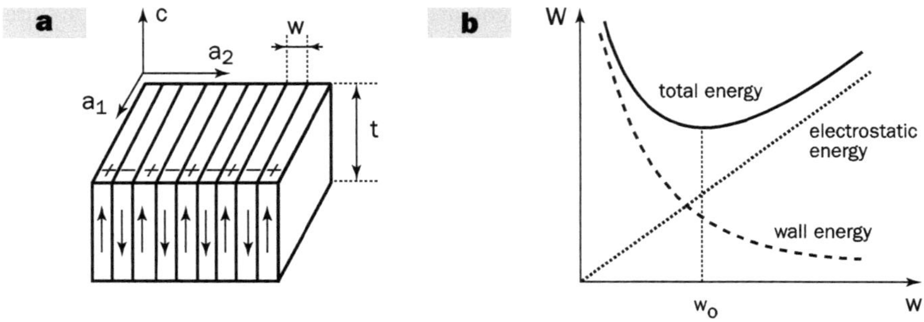

Fig. 2.20 (**a**) Representation of a crystal of the so-called Känzig model with a single wall orientation allowed and equidistant walls between them and (**b**) variation of energies for a unit surface sample perpendicular to the ferroelectric axis: electrostatic energy (dotted line), wall energy (discontinuous traits), and total energy (continuous line)

with γ_w energy per unit of wall surface, t thickness of the sample following c, w distance between two walls, therefore average width of the domains, with K of the form:

$$K = 0.26\, P_S^{\,2} \,/\, (\varepsilon_0 \left(1 + \left(\varepsilon_a \varepsilon_c\right)^{1/2}\right) \tag{2.12}$$

ε_a and ε_c are the dielectric constants in the directions respectively perpendicular and parallel to the ferroelectric axis of values much >1. By optimizing the total energy, we find the variation of w proportional to the square root of t:

$$w_0 = \gamma^{1/2} K^{-1/2} \; t^{1/2} \tag{2.13}$$

A detailed presentation of these types of calculations from the publication of Mitsui and Furuichi can be consulted in the case where the widths of domains of opposite polarities differ (Kopal et al., 1997) in the work of Tagantsev et al. (2010); a summary is proposed in Annex A6. The assumptions made implicitly suppose an infinitely sized sample except in the direction of the ferroelectric axis, walls without interaction between them, without interaction with the sample edges, and moreover, of zero width therefore at absolute zero. These assumptions give fairly good results for simply ferroelectric crystals like those presented in Fig. 2.17 for TGS even if there is some dispersion of experimental points (Strukov et al., 1973; Jaskiewicz et al., 1977). However, these assumptions require that the thickness of the sample be greater than a critical value t_{cri}

$$t_{cri} = 4\varepsilon_a^{-1/2} \varepsilon_c^{1/2} K^{-1} \gamma_{\mathrm{w}} \tag{2.14}$$

which increases as $(T_0 - T)$ decreases. The limits of the model are clearly apparent: near the transition, the walls would be infinite in number, nevertheless, without interaction between them and of zero energy. Despite the problems of critical thickness for the sample, this model has been used for thin layers.

In the case of a ferroelectric crystal, the existence of surface layers on the faces perpendicular to the ferroelectric axis has also been considered; a model still predicts a variation in the width of the domains as the square root of the thickness of the sample (Kopal et al., 1999):

$$w_0 = \left(B\left(\sqrt{\varepsilon_a^{\,1}\varepsilon_c^{\,1}} + \sqrt{\varepsilon_a\varepsilon_c}\right)\gamma_w\right)^{1/2} t^{1/2} / P_S \tag{2.15}$$

the index 1 corresponds to the dielectric constants inside the surface layers, perpendicularly (ε_a^1) and parallel to the ferroelectric axis (ε_c^1); B is a coefficient that depends on the chosen units and the ratio between the thickness of the surface layers and the sample.

Similarly, if the conductivity of the crystal is taken into account, the variation of w in $t^{1/2}$ can be theoretically obtained in the case where the dielectric constant in the direction of the ferroelectric axis is much higher than that of the perpendicular directions, which is usually the case (Mitsui and Furuichi, 1953):

$$w_0 = \gamma_w^{\,1/2} t^{1/2} \; f\left(\varepsilon_a, \sigma_a, \varepsilon_c, \sigma_c, \tau\right) K^{-1/2} \tag{2.16}$$

σ_a and σ_c are the conductivities along the a and c axes; τ is the delay between the creation of the domain texture and their width measurement; f is a function increasing with τ: the width of the domains would therefore increase over time, usually over tens of minutes, sometimes hours or days (Dabrowska et al., 1977) as demonstrated (Nakatani, 1985). These times vary considerably from one crystal to another and, of course, depending on the observation temperature. In addition to the value of this temperature, its difference from the transition temperature plays a role.

Keeping the same assumptions as before (infinite dimensions in the directions perpendicular to the ferroelectric axis, isolated walls and of zero thickness), models have been proposed to explain the existence of textures with non-equidistant walls: some authors consider them as transitional situations before obtaining a perfectly regular texture; others postulate the existence of fractals (Ozaki et al., 1993, 1995) and even demonstrate that for certain ranks (the third for pentad Cantor sets) the energy would be lower than with equidistant textures. These models only take into account the electrostatic energy and the wall energy, neglecting the mechanical energies of domain ends (Ozaki et al., 2002). Experimental observations in support of these models remain rare.

The model of Sect. 2.2 of isolated walls, plausible at very low temperatures, is much less so when we observe interactions between walls like those described in Chaps. 3 and 4, for example in the 20° under the transitions of KDP or DKDP; also, to explain the observed domain widths, an interaction force between domain walls inversely proportional to their distance was introduced (Abe, 1987) based on experimental observations which actually concerned moving walls (Bornarel and Legrand, 1981). If we took into account such interactions between walls, then in the relation (2.11) an interaction energy of the form would be added:

$$W_{ww} = C\,\mathrm{Ln}(\mathrm{H}/\mathrm{w})\ (H\ \text{and}\ C\ \text{positive}) \tag{2.17}$$

The minimization of energy with this contribution gives for the domain width:

$$w_0 = C\left(1+\left(1+4AP_0^2\gamma_w t/C^2\right)^{1/2}\right)\left(2AP_0^2\right)^{-1} \tag{2.18}$$

The interaction energy between walls would therefore tend to increase the optimal width of the domains as illustrated in Fig. 2.21, summary of theoretical curves with various assumptions: with or without "surface layers" for a ferroelectric, with or without interaction between the walls, with or without conductivity σ. These factors can explain the wide distributions of observed wall widths. Note that the addition of a law of interaction between the walls leads to a value of w non-zero C/AP_0^2 for an extremely thin sample, which is more in line with observations, even when these are close to the transition. All these calculation assumptions implicitly assume a thickness t much greater than w. These models should therefore not be suitable for very thin samples.

A big difference exists between ferroelectricity and ferroelasticity for the periodic textures of parallel walls: even in the simple model (Kittel-Mitsui-Furuichi), the necessity of domain wall existence is not in doubt for a ferroelectric; indeed, in an electrically isolated sample the charges on the surface perpendicular to the polarization create a depolarizing field, with a significant energy if the sample remained

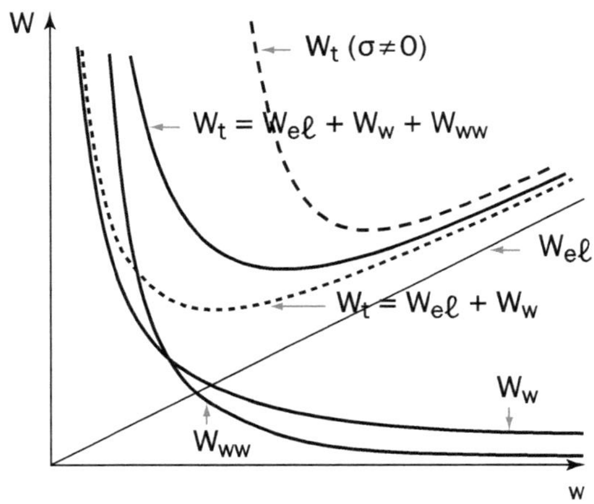

Fig. 2.21 Energies per unit of crystal area in a section perpendicular to the ferroelectric axis for a texture in domains of allowed plane equidistant walls (Fig. 2.20a) as a function of their distance w (width of the domains)

Various contributions to this energy are represented: the electrostatic energy W_{el}, the intrinsic energy of the walls W_w, and interaction energy between the walls W_{ww}. The graphs of the total energy $W_t(w)$ depend on the considered contributions: simple model called Känzig (dotted curve), with the interaction between walls (curve in wide continuous line) and with a non-zero conductivity (curve in wide discontinuous line). The effect of the surface layers is not taken into account mono-domain. Therefore, it is necessary to create domains so that this type of energy decreases, in a way still limited by that of the walls. But, for a ferroelastic, with a model as simplistic (infinitely thin walls parallel to the plane of permissible walls), the need to create domain walls does not appear. However, a parallel can be drawn concerning the importance of the sample edges, of its finite dimension: the depolarizing field has allowed to obtain a sample, with the domains, of macroscopic polarization close to zero; the existence of domains in a ferroelastic, similarly, allows to obtain a sample of macroscopic deformation close to zero. But the forms of energies are not comparable; for the ferroelectric aspect in the relation (2.11), the energy per unit of surface introduces a term in P^2w. Regarding a ferroelastic, a comparable term does not easily arise when writing, for example that the stresses on sample surface must cancel out. If we except the models explaining the necessity of domains when we are in the presence of a phase front for first-order transitions (Roytburd, 1998a, b), the need for domains is not of the same nature for the ferroelastics. To obtain such a term, in u^2w, it is necessary to take into account the dislocations (Torres et al., 1982), therefore of walls not perfectly in the permissible plane as the track was given by the interactions at long distance demonstrated with the tips of domains (Bornarel and Lajzerowicz, 1968). Or, it is necessary to admit that the crystal edge will not conform to the theoretical design with sharp edges like that of Fig. 2.8a and will present shapes that induce long distance energies. This aspect is implicitly addressed in Sect. 2.4 ("realistic" domains).

2.3.3 *Intersection of the Permissible Wall with the Sample Edge*

Regarding the effect of the sample edges, when a permissible wall cuts a sample edge perpendicularly, the sample deformation law near the surface of a "groove" is different to that of a "bulge" as shown in Fig. 2.22c.

The elastic continuum models that use as a basis the development of Landau-Ginzburg with gradient terms and as the only variable the order parameter do not account for these differences (Novak and Salje, 1998a, b; Conti and Salje, 2001). On the other hand, if coupling terms between the order parameter (here proportional to the shear u_{12}) and normal deformations (u_{11} and u_{22}) are introduced, it is demonstrated that when the edge-wall intersection occurs, the wall can widen or narrow. Figure 2.22a, b accounts for these results (Lee et al., 2002), transposable to ferro-ferro crystals. Figure 2.22c proposes a very exaggerated illustration of the process. Taking into account the edges of the sample usually omitted in the models is a lead explanation of many underutilized results (Darinskii and Sidorkin, 2016).

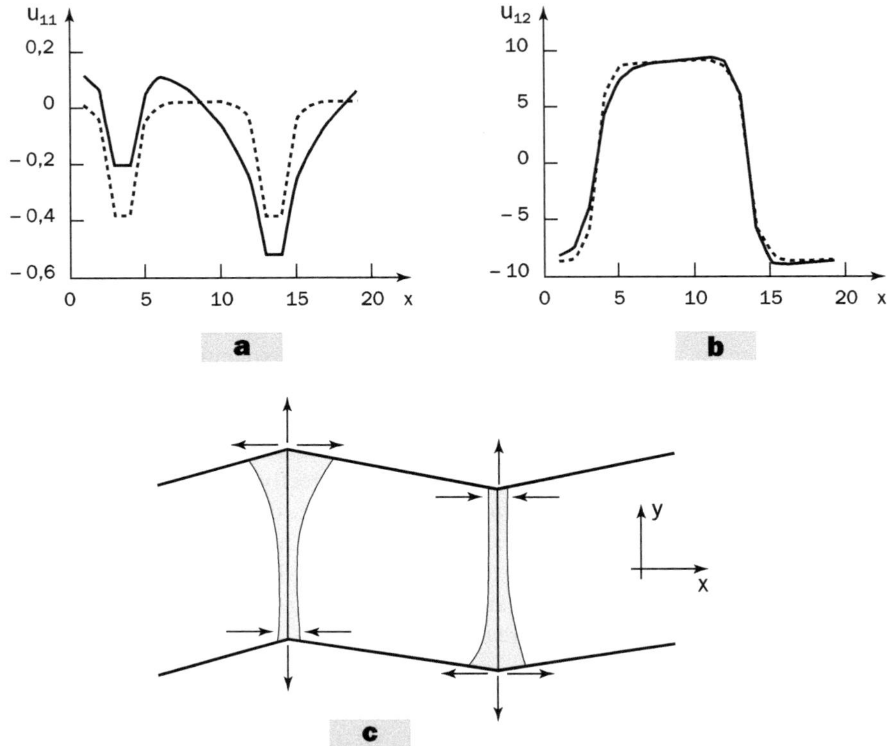

Fig. 2.22 The model (adapted to the ferroelastic WO_3), extrapolable to the crystals of the work, takes into account to evaluate the relaxation in the wall, the coupling between the order parameter and the volume deformations (Lee et al., 2002). The relaxation in the wall, for the various components of the deformation, is not the same inside the crystal and near its surface. The variations $u_{11}(x)$ in (**a**) and $u_{12}(x)$ in (**b**) illustrate for the width of the wall at the surface, increases in some cases and decreases in others as schematized, very amplified, in (**c**). These modifications, if they exist in a ferro-ferro, could, through the charged wall parts, act in a probably marginal way on the dielectric properties (Darinskii and Sidorkin, 2016), induce interactions between walls, and facilitate the creation of ends (Lu et al., 2019). In these regions, the linearity of the variations of u_{11}, u_{33}, and u_{12} can be questioned (Pöttler and Salje, 2014)

2.4 Usual Shapes of Domains and Textures

2.4.1 Simple Domain Textures and Tips

Even within the framework of equilibrium domain textures, almost immobile, it is not always easy to obtain domain walls strictly in the permissible planes. For domain textures with walls approximately parallel to each other, two situations can be differentiated:

That of lenticular-shaped domains as illustrated with TMO in Fig. 2.23. TMO, GMO crystals also offer at ambient, examples of walls whose orientation around

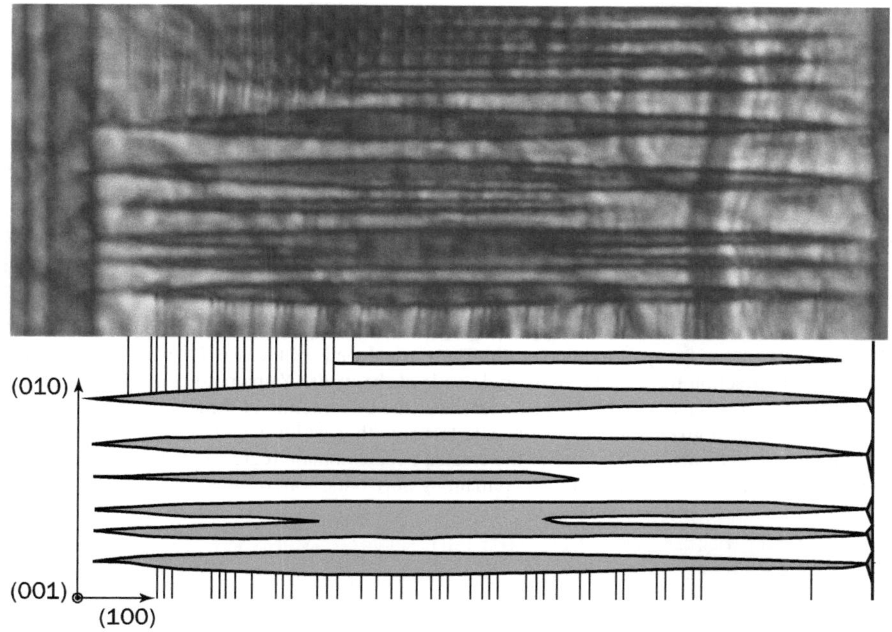

Fig. 2.23 Lenticular-shaped domains in a TMO crystal observed in section (001) (width of the following photograph $a_1 = 1$ mm of the sample)

that of a permissible plane can change continuously, without jolting, depending on the temperature and the environment. In addition, electron microscopy experiments have shown how much the wall curvature can be continuously modified, and Fig. 2.24 presents an example of creation of a needle-like domain end (Yamamoto et al., 1977a, 1977b): parallel branches of two right-angled walls approach, the corners deform until the creation of a needle; the wall that remains, perpendicular to the formed needle, is locally deformed along the axis of the needle, then by relaxation, becomes flat again when the needle is far enough away. These short-distance interactions (from 0.1 to 1 μm) of walls with curvature deformation, in line with various models (Indenbom and Chamrov, 1980), are frequently observed with very slight modifications of atomic positions in these crystal families.

Another situation is illustrated in Figs. 2.25, 2.26, and 2.7a, which concern KDP. The situation is quite different from previous easy curvatures: the needles essentially have wall surfaces in the permissible wall planes and only one domain tip connects two walls −+ and +− with a more or less elongated shape depending on the crystal and the experimental conditions. In such cases, the creations and disappearances of domain tips usually occur on the edges of the samples in a brutal way, as opposed to the smooth and continuous variations of Fig. 2.24.

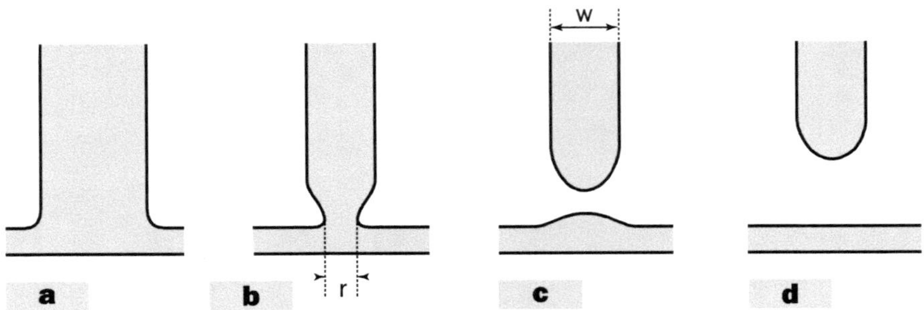

Fig. 2.24 In this GMO crystal, the white and grey areas are of opposite deformations: (**a**) Two 90° domain walls are too far apart to interact. (**b**) The vertical walls of the two 90° walls interact with a horizontal attraction force at the right-angle ends of the walls (Yamamoto et al., 1977a; Torres et al., 1982; Roucau et al., 1979). These forces cause the deformation of these ends and their rapid approach. (**c**) For a given value of r, a needle forms with a width w and, at the same time, the remaining wall deforms. (**d**) The end of the created needle domain regresses in the vertical direction of the figure; after a certain time, the remaining horizontal wall resumes a flat shape

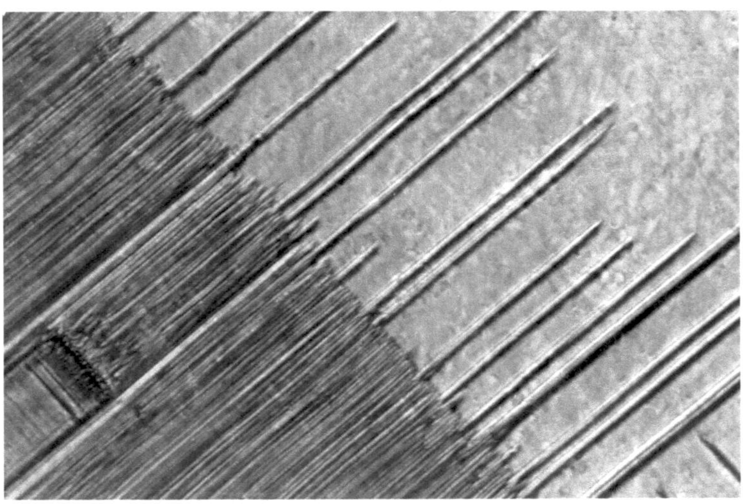

Fig. 2.25 Domain texture in KDP at 95 K in section (001). Domain ends (called "tips") stabilize following a front perpendicular to their longitudinal direction

2.4.2 Models of Walls Outside the Permissible Plans and Tip Shapes

Many models have been proposed in ferroelastic crystals after being in ferromagnetic crystals (Kleman and Schlenker, 1972; Miltat and Kleman, 1973). They often rely on a description using densities of fictitious dislocations (Mura, 1968) and fictitious disinclinations modelled using elastic theory in an anisotropic medium (Torrès

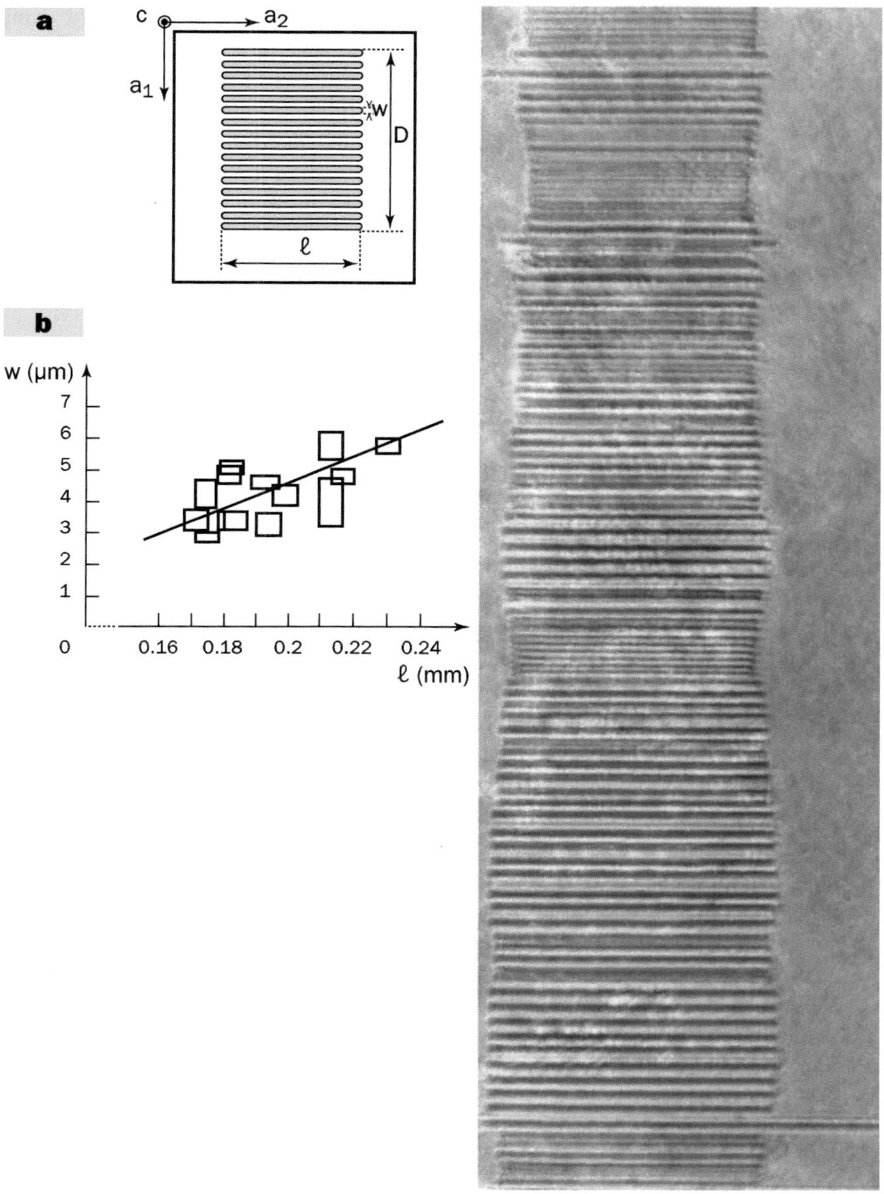

Fig. 2.26 Right-hand photograph in section (001): set of needle domains located within a sample without other domains (KDP at a temperature of 119 K (Bornarel, 1990)). This arrangement recalls that of ferroelastic bubbles (Meeks and Auld, 1985, 1988). (**a**) Schematic representation of a set of identical domains. (**b**) Correlation between the width of the domains of the photograph and their length; the continuous line illustrates the variation resulting from the model developed in Annex A7 (shape law with lengths in μm: $w = 0.0374\,l - 2.86$)

et al., 1982; Salje and Ishibashi, 1996). Various situations are accounted for using screw dislocations, edge dislocations, and rotations: a plane wall inclined relative to the permissible wall plane is modelled by a distribution of fictitious dislocations, and a curved wall requires a distribution of disinclinations; all induce long-distance stress fields (Pertsev et al., 2000). Thus are explained electron microscopy observations of walls like that described in Fig. 2.24 of creating a needle using two right-angled walls. For example, the optimal curvature radius R of the wall that links the two perpendicular permissible orientations (Pertsev et al., 2000) has been calculated:

$$R = 3.7\, l_0 \tag{2.19}$$

with the characteristic length l_0 defined by

$$l_0 = (1-\nu)\Upsilon\left(\mu u_0^{\,2}\right)^{-1} \tag{2.20}$$

u_0 is the spontaneous deformation, ν the Poisson's ratio, μ the shear modulus, and Υ the energy per unit wall area. This relationship between R and l_0 remains true far enough from a transition within the assumptions retained (infinite crystal, continuous medium without singular point, infinitely thin wall). The quantities involved in l_0, therefore in R, vary with temperature differently depending on the crystals (proper or improper ferroelastic, ferroelectric) and depending on the order of the transitions.

Knowing precisely what happens at the atomic level when the wall deviates from the permissible wall plane is not easy, especially in crystals whose atomic structure is not simple. However, for crystals of the KDP family, proposals for wall steps have been made based on the models of Barkla and Finlayson wall (Fig. 2.10) (Bornarel, 1972). Figure 2.27 thus presents lateral wall steps in the case of polar and neutral walls (at absolute zero) respecting the principle of conservation of the chemical formula at the level of each tetrahedron and cell. We can evaluate the energy per unit length of these quasi-dislocations for a lateral translation of the wall, for example of an elementary cell, a few tens of degrees below the transition for our favourite crystals, using relation A7.3 (Annex A7).

We find approximately for GMO 4×10^{-13} J m^{-1}, for KDP 8×10^{-13} J m^{-1}, and for Tanane 8×10^{-11} J m^{-1}. Without exaggerating this comparison since other elements play a role, such as the state of the edges of the non-comparable crystals, temperatures, etc., it can be affirmed that creating walls out of the permissible plane poses more problems in the Tanane than in the other two crystals.

Figure 2.28 proposes in the same theoretical exercise a diagram at the atomic level of square-shaped domain ends. In these simplistic diagrams of square end, a "neutral" wall end corresponds to largely "polarized" walls (Fig. 2.28a), and on the contrary, a "polarized" wall end to "neutral" walls (Fig. 2.28b). As the energy densities per unit of wall surface calculated (Bjorkstal and Oettel, 1966) are not the same (Sect. 2.2), a quick reasoning would argue for the more probable existence of large polarized walls connected by a neutral perpendicular wall, the domain width acting through this difference of energy per unit of surface, in addition to the other

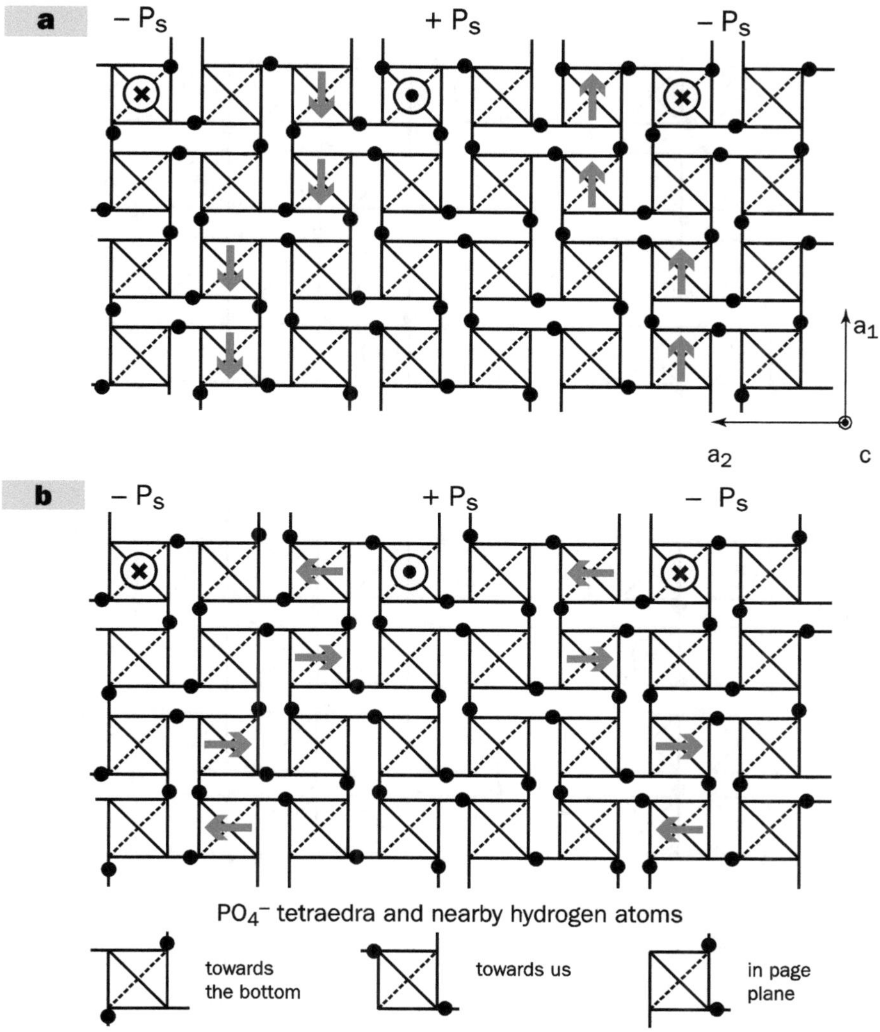

Fig. 2.27 Representation of quasi-dislocations allowing a lateral wall displacement in the KDP family respecting the chemical formula at the level of each PO_4^- tetrahedron in the mode of representation introduced in Fig. 2.10 (the tetrahedra are represented in projection in the (001) plane and are not in the same plane along this axis): (**a**) so-called polarized wall and (**b**) so-called neutral wall

variations already mentioned. However, it is difficult to substantiate such hypotheses at the atomic level even if these diagrams can be a source of inspiration for simulating modifications at the lattice level.

If a model of continuous elastic medium is retained, various types of energy are taken into account at the meso-scopic level (Salje and Ishibashi, 1996): the elastic energy of anisotropy which grows with the angle between the wall and the

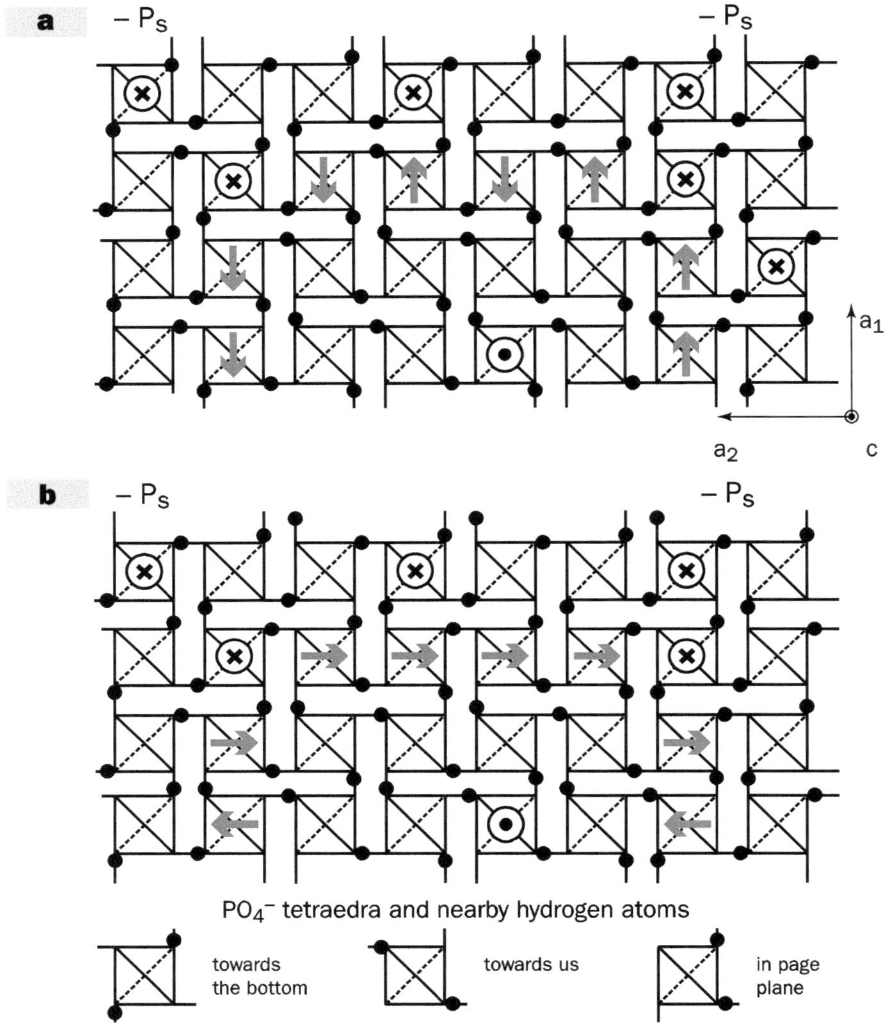

Fig. 2.28 With the same elements of representation as in Fig. 2.27, that is, by only showing the displacements of protons on the bonds and not, the rotations of the tetrahedra, representations (at absolute zero) of 90° walls in (**a**) a square end of a needle with "polarized" side walls and (**b**) a square end of a needle with "neutral" side walls. In both cases, if the Slater rule is maintained for each tetrahedron, the nature of the wall that closes the end of the domain is not that of the longitudinal walls of the needle

permissible wall plane (as $(\mathrm{d}y/\mathrm{d}x)^2$ if x is the direction of the permissible wall plane and y the perpendicular direction as illustrated by Fig. 2.29), the wall curvature energy (in $(\mathrm{d}^2y/\mathrm{d}x^2)^2$), that is induced by possible external forces, not to mention the Peierls forces. The examples of Fig. 2.29 show the variety of possible solutions. These models based on the hypothesis of an elastic continuum assume regular

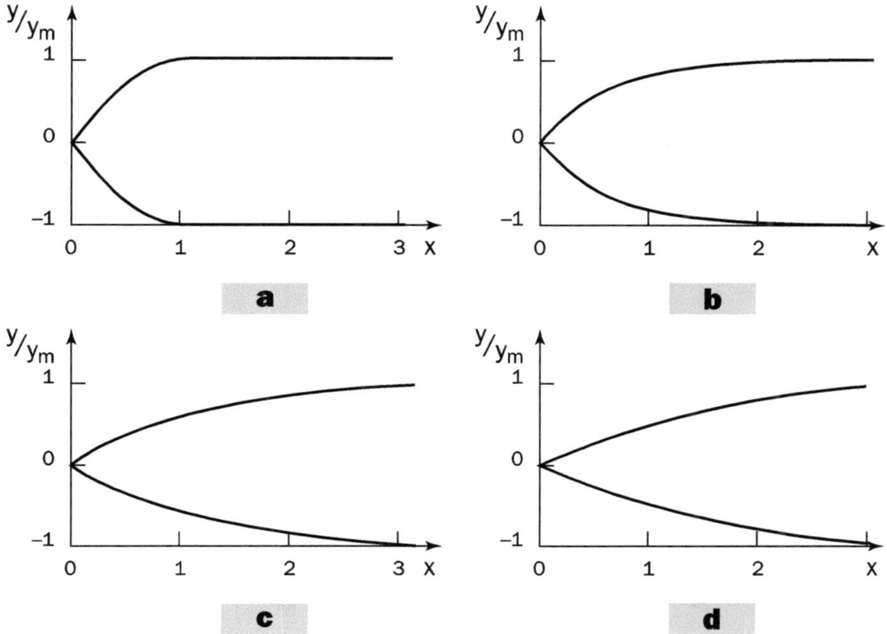

Fig. 2.29 Shapes of twin ends calculated with a continuum assumption using elasticity theory and various reports on anisotropy, curvature, and Peierls energies. The needle is elongated in the x direction with its end at $x = 0$, its maximum width for large x is equal to 2 y_m (Salje and Ishibashi, 1996). Results of profiles in various scenarios: (**a**) large curvature energy without relaxation with the network, (**b**) strong anisotropy and relaxation, (**c**) periodic Peierls force with a strong anisotropy energy, (**d**) very large Peierls force and large anisotropy energy

deformation variations in the walls. By abstracting from deformations non-linearly coupled to the order parameter, this hypothesis can be retained in crystals such as TMO and GMO, given the observations of electron microscopy and optics. The same can be true for crystals such as KDP in the few degrees below the transition. But this hypothesis cannot be maintained as the temperature decreases, especially as it approaches the domain freezing. A more realistic modelling of a permissible wall translation perpendicular to its plane at the level of the elementary cells uses (Friedel, 1964) (Annex A7) as in Fig. 2.30a a quasi-dislocation of wedge type with Burgers vector

$$b_e = 2a\, tgu_{xy} \tag{2.21}$$

Such elementary quasi-dislocations are compatible, at equilibrium, at low temperature, for the KDP crystal, with the conservation of the chemical formula on each tetrahedron as demonstrated in Fig. 2.27 for the two permissible walls of Barkla and Finlayson. The exact atomic arrangement at the core of this quasi-dislocation could differ without modifying the Burgers vector (Bornarel, 1972). At the end of a

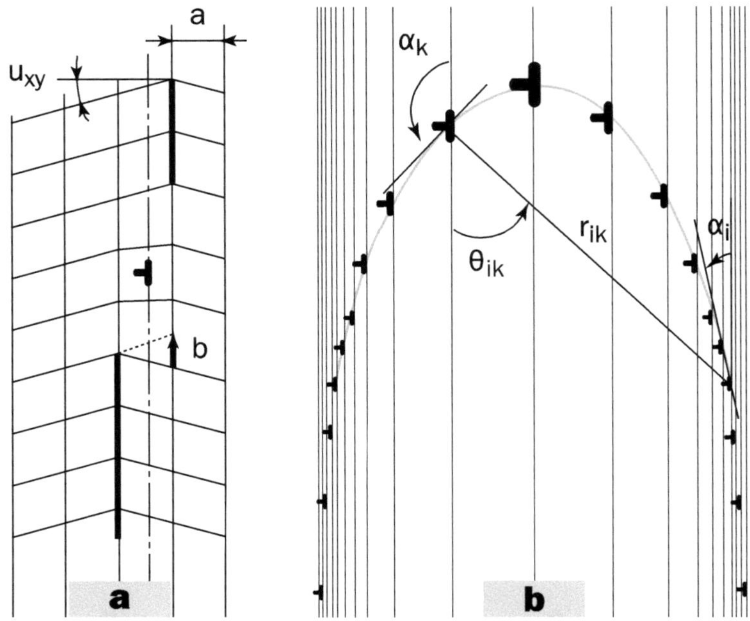

Fig. 2.30 (**a**) Representation in section (001), of a wedge-type quasi-dislocation of Burgers vector $b = 2a\ tgu_{xy}$ creating a lateral displacement of the permissible wall of value a, length of the unit cell, or $a/2$ in the case of Fig. 2.27 (see in Annex A7 the difference between dislocation and quasi-dislocation) (Friedel, 1964). (**b**) Assembly of quasi-dislocations at the end of domains. Their core radius r_{0k} increases with the value of the angle α_k

domain, assemblies of such dissociable quasi-dislocations can model the profile as represented in Fig. 2.30b, with core radii r_{0k} and different Burgers vectors (Annex A7). Their Burgers vector will be zero if α_k is equal to 0 or π and, on the contrary, maximal if it is equal to $\pi/2$. The relationship between the angle α_k and the wall translation na (n integer) will be $\sin\alpha_k = na/2r_{0k}$ and the corresponding Burgers vector $b_k = nb_e$. (The choice of the core radius r_{0k} is a delicate problem but studied for a long time (Gilman, 1969).) The balance of the domain tip and its profile are the result of the competition between two contributions:

- One corresponds to the sum of the self-energies of the dislocations or quasi-dislocations which tends to dissociate the elementary dislocations (nb_e^2 is smaller than $(nb_e)^2$ to achieve a lateral dislocation of the wall of na).
- The other comes from the long-distance interaction of the dissociable dislocations between them; these wedge-type dislocations having parallel glide planes will have a minimum interaction energy when they are located in the same plane perpendicular to the glide plane.

Depending on the relative importance of these energies, the domains will be either in the form of a wedge or elongated lens shape, with wall surfaces slightly inclined relative to the permissible wall plane, on the contrary, in the form of needles with parallel lateral planes to the permissible plane, with very localized ends (tips) as in Fig. 2.31. This problem of the shape of the domain has been dealt with in detail by many authors (Friedel, 1964; Kosevich and Boiko, 1971; Torres et al., 1982; Salje and Ishibashi, 1996). The stresses induced by these assemblies of dislocations are long-range, and the corresponding deformation energy can concern the entire sample. However, to preserve the simple formulas of dislocation theory (of infinite length), it is usual to choose for a dislocation or an assembly of dislocations, a cut-off radius r_1, for example, of the order of the thickness of the sample in the direction of the dislocation. With this assumption, it is demonstrated in the case of Fig. 2.31, for a sufficiently large distance r (a few domain widths between the point M centre of the domain end and an outer wedge dislocation) to the domain end and sufficiently distant, that the assembly of quasi-dislocations of the end can be summarized by a single dislocation in M with a Burgers vector $B = 2w\ tg\ u_{xy}$. A calculation of the mechanical energy of a domain tip assumed to be semi-circular in shape and the variation of this energy as a function of the domain width is found in Annex A7.

Thus, an intermediate presentation is obtained between atomic-level simulations and those of continuous models poorly suited to low temperatures. The assumptions made are limited to taking into account the deformation correlated linearly with the order parameter (in the KDP). But we cannot forget the other dilation deformations discussed in Sect. 2.2. For example, if a domain tip of width w detaches from the edge of the crystal, it will have to accommodate the disappearance of this staircase of (maximum) value $2w\ \sin u_{xy}$ therefore of a hundred Angstroms for a domain width of about one micrometre. During such events, compressions-dilations will occur in the direction of the needles (u_{yy} and u_{zz}) and perpendicular to the walls (u_{xx}).

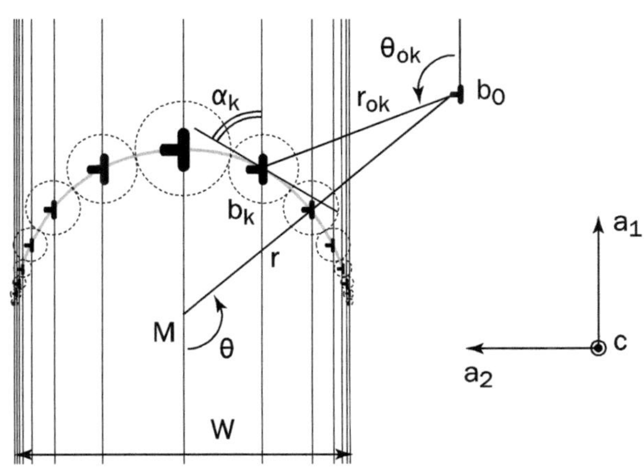

Fig. 2.31 Schematic representation of an assembly of quasi-dislocations at the chosen domain end by hypothesis of semi-circular shape, interacting with a wedge-type dislocation of Burgers vector b_0. The core radii of the dissociated dislocations are qualitatively represented by dotted circles

2.4.3 Equilibrium of a Domain Texture with Needle Tips

In Annex A7, the interaction energy between two needle domain tips and that of a set of similar needle domain tips are calculated. The energy of this set is minimal when all the ends are in the same plane perpendicular to their slip plane (Friedel, 1964). The existence of tips leads to introducing an additional contribution W_n in the energy of the twinned region, which corresponds to the long-distance deformations created (Torres et al., 1982). As demonstrated in Annex A7, it can be put in the form:

$$W_n = a + bw \tag{2.22}$$

with w width of the needle domains of the twinned region, b function of the mechanical properties of the crystal and the dimension D of the texture; the same applies to a which also depends on r_0 and r_1. Figure 2.26a takes up the schematic representation of Annex A7 and models the texture presented in the photograph on the right. In this set of domains located in the middle of a sample without other domains, the correlation between the width w and the length l of the domains (Bornarel, 1990) is clearly observed. The corresponding experimental results and those of the calculations of Annex A7 are summarized in graph 2.26b.

To model more realistically a texture in equilibrium domains for the KDP crystal (at low temperatures, electrical conduction can be neglected), four contributions to the total energy W_t can thus be considered: the electrostatic energy with the depolarizing field, the intrinsic wall energy, the interaction energy between walls, and the energy of the domain tips (see relations (2.11), (2.17) and (2.19)). By minimizing W_t, the equilibrium width of the domains w is obtained:

$$w = \left(lc + \left((lc)^2 + 2l\gamma_w t(2lK + bt) \right)^{1/2} \right) / (2lK + bt) \tag{2.23}$$

This formula allows for the various experimental situations to be considered:

- Those in which the mechanical energy is minor, situations which lead to a variation of w as $t^{1/2}$ (t thickness of the sample).
- Those in which the electrostatic energy is in turn minor and which correspond to variations of w as $l^{1/2}$ (l length of the needle domains) when domain tips are present with their assemblies of quasi-dislocations.
- Those where the two energies co-exist as in the example of Fig. 2.26b, with ends of needle domains. If in this case, we neglect the interaction between adjacent walls (assumption relevant only at low temperatures, far from the transition):

$$w = (2\gamma_w lt)^{1/2} (2lK + bt)^{-1/2} \tag{2.24}$$

This relationship applied to the average variation observed in Fig. 2.26b gives a good agreement with the experimental values for an energy value per unit of wall area of about 10 erg/cm^2 (10^{-2} J/m^2). The agreement is better if we consider that interactions between nearby walls modify the wall surface energy. The values then vary between 6 erg/cm^2 for small domain lengths and 16 erg/cm^2 for the largest, which covers all the experimental values from the literature.

2.4.4 The Limits of Theoretical Models

The limits of the previous models are easily seen: at low temperatures, domain walls can be considered as isolated as demonstrated in Chap. 4 for temperatures lower than those of the freezing of domains with immobile domain ends and domains of maximum width. Similarly, in this context of low variation with temperature of deformation and spontaneous polarization, the hypothesis of a constant value of energy per unit area of the wall can be considered robust. The situation is very different when interactions between domains are detected with collective phenomena (including with low amplitude excitations as in Chap. 4) and rapid variations of physical parameters as near the transition.

The existence of assemblies of quasi-dislocations explains the results schematized in Fig. 2.4 at temperatures low enough for electrical conductivity not to play a short-circuit role below a certain size of a massive crystal, the domains cannot appear because the shear energy of the surface and that of the walls are too great, which adds to the effect of the depolarizing field (Jaccard et al., 1953): the crystal must be mono-domain and in some cases, the para-ferro transition may even become unlikely (2.1.5a). If the crystal is in the form of a thin plate perpendicular to the ferro direction, the quasi-dislocations are quickly blocked by the "pinning" on both faces and the domain tips are too. There is an optimal ratio between the thickness of the crystal and its other dimensions for which the domains are most mobile, as demonstrated by the dielectric constant measurements depending on the shape of the samples in Chap. 4 (Fig. 4.2.3). Finally, if the sample is very elongated along the ferro axis, the influence of depolarizing fields will be weaker, but the interaction of the assemblies of quasi-dislocations with each of the side faces (image forces of Fig. 3.5.6) will also decrease the probability of obtaining domains. The crystals, even short-circuited, will usually show significant mono-domain regions and this, despite the permanent electric charges and the assemblies of dislocation knees, at the limits of interrupted domains along the ferro axis in the crystal.

It is therefore not systematic, in ferro-ferro crystals, to obtain domain textures with all the wall surfaces in the permissible wall planes. When parts of these walls deviate from these planes, they systematically contain quasi-dislocations (wedge type in our case), dis-inclinations related to curvatures, and compression-dilation deformations. The context varies depending on the temperature: the continuum assumptions remain solid at temperatures close to transitions. They become unrealistic when singularities in the walls impose themselves and especially at the ends of needle domains. If the domain ends are very localized (domain tips), they can, in the textures in equilibrium, be summarized by large dissociated dislocations. The existence of these assemblies of quasi-dislocations explains the dispersion of the experimental results observed concerning the domain widths as a function of the thickness of the samples since it leads to a variation of the width with the length of these domains (in a plane perpendicular to the ferroelectric axis). In this section, only textures with domain walls generally parallel to a single permissible wall plane were studied in order to simplify the models. The experimental conditions and especially the history of the crystal do not always allow these situations. Without compiling a complete register, the following section will illustrate various possible natural textures in the simple example of the chosen crystals.

2.5 Ordinary Domain Textures

2.5.1 Thermal History: An Important Parameter

Figure 2.19 illustrates the creation of a natural domain texture and its evolution with a single permissible wall direction, obtained in the APFA crystal. During a slow decrease in temperature, the texture evolved quite regularly by lateral displacements of walls and disappearance of domains of width too small for the reached temperature, by rapid longitudinal displacement of needle domains. The variation of average domain width could not be perfectly regular: for needles to form and disappear, there must be a notable deviation from the optimal equilibrium situation. Thus, constraints are eliminated and become too significant as in any quenching system. In the case of Fig. 2.19, no blockage is observed. The phenomena occur easily.

There are comparable situations in the retained crystals (in GMO, for example at a temperature higher than room temperature), but for others, especially at low temperatures, it can be very difficult to easily obtain lateral wall displacements. The crystals then go through quenching situations with textures in more or less complicated domains during a temperature drop, to abrupt rearrangements that lead to the disappearance of certain domains, sometimes in avalanches (see Annex A8). These quenched phenomena without applied external constraints essentially depend on the average cooling rate $\delta T/\delta t$. The examples proposed in Fig. 2.32 (Bastie and Bornarel, 1982) concerning domain textures of the TMO crystal illustrate a general rule: the faster the cooling takes place, the greater the number of different domain sets, with many incoherent twin boundaries (see Fig. 2.33). On the contrary, a slow cooling rate (in the concerned crystals, <0.01 K min^{1}) allows to obtain a single wall orientation permitted throughout the sample if the thermal homogeneity, accidental constraints, shape, and state of the sample edges do not prevent it.

Similarly, a thermal inhomogeneity (of a tenth of a degree, for example under the transition inside the sample) or the existence of mechanical defects (of growth, due to cutting or polishing) can induce blocked domain textures, thus a sample with particular physical properties.

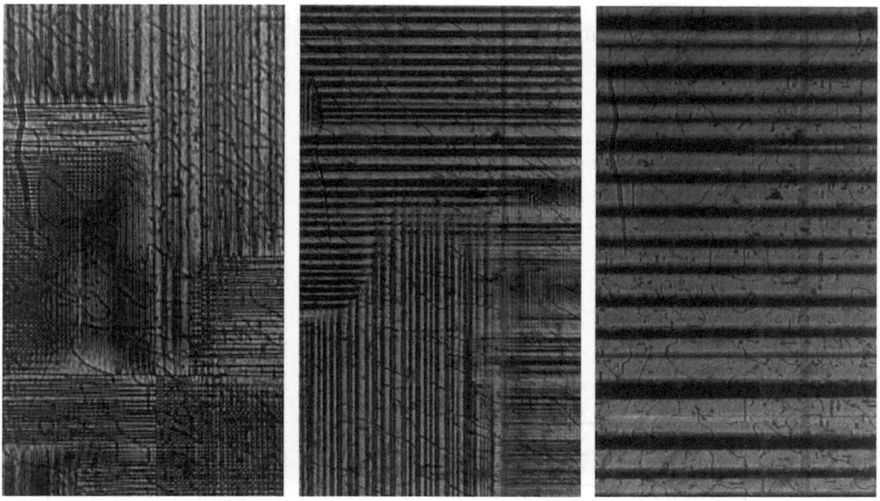

Fig. 2.32 Equilibrium domain textures in the TMO (6 mm × 2 mm × 7 mm) crystal at room temperature in (001) projection obtained with different cooling rates during the para-ferro transition: (a) a few degrees per minute, (**b**) 0.1 K min^{-1}, (**c**) <0.01 K min^{-1} (the height of the photographs corresponds to 3 mm of the crystal) (Bastie and Bornarel, 1982)

2.5.2 Sample Shape and Texture

The shape of the sample also plays a notable role on the texture illustrated in Fig. 2.34 by some correlations between sample shape, orientation of faces with respect to tetragonal planes, and domain texture. Even in the hypothesis of a sample cut and prepared in the best way with its faces in the tetragonal planes and an optimal thermal history, it is possible to obtain regions with different permissible wall orientations as illustrated in Fig. 2.33, thus with incoherent twin boundaries between domain sets. Two possibilities then arise:

- The domain sets have permissible wall planes at 90° arc from each other as in 2.5.2c with a well-defined incoherent twin boundary, thus an obvious limit (the EF line in the figure) estimated by X-ray diffraction of a width in a (001) plane of about 1 μm (Afonikova et al., 1987).
- The texture is not dense, when an external electric field or an external constraint is applied, or when the crystal is short-circuited, then the angle α of Fig. 2.33d may not be equal to 90° arc. The examples of Fig. 2.35 correspond to these situations with α equal to 90°+u_{xy} or 90°−°u_{xy}. Figure 2.36 shows how in a sample that includes the two orientations of wall planes, one can go during the realization of a hysteresis cycle, from an angle α of 90°+°u_{xy} to an angle α of 90°(Bornarel, 1991). In such a case, a significant region of the sample deforms, and not only at the incoherent twin boundary between two domain sets (species): the small angle γ diffraction peaks deform and move. X-ray diffraction experiments (Shekhtman

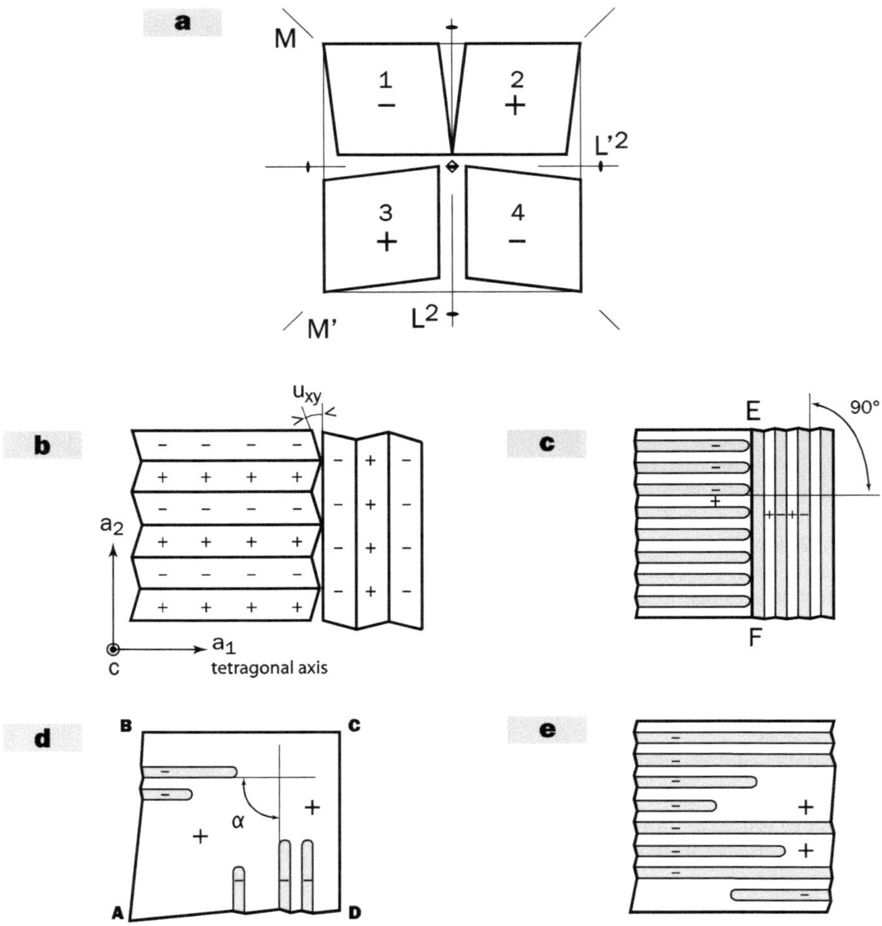

Fig. 2.33 (**a**) Theoretical representation of possible cells in the studied crystals with indication of spontaneous deformation. (**b**) Theoretical texture diagram: we notice in the polarizations and deformations the role of the inversion operation lost at the transition. (**c**) More realistic representation: EF is an incoherent twin boundary. (**d**) Polarized sample with two permissible wall directions making an angle α between them which in the representation is considered approximately equal to 90°. (**e**) Simple texture with only two types of elementary cells, one permissible wall orientation, and therefore only coherent twin boundaries in the walls

et al., 1973; Aknazarov et al., 1975), synchrotron topographies (Aleshko-Ozhevskij et al., 1985), and neutron diffraction experiments (Balagurov et al., 1980, 1983) confirm the phenomenon. This is complicated when an incoherent twin boundary ends on the surface, as illustrated by the X-ray diffraction results in Fig. 2.37: one can go continuously from cells of two adjacent domains of opposite directions to cells of domains at 90° near the surface with a mean shear deformation decreased compared to that observed within the crystal (Afonikova et al., 1987).

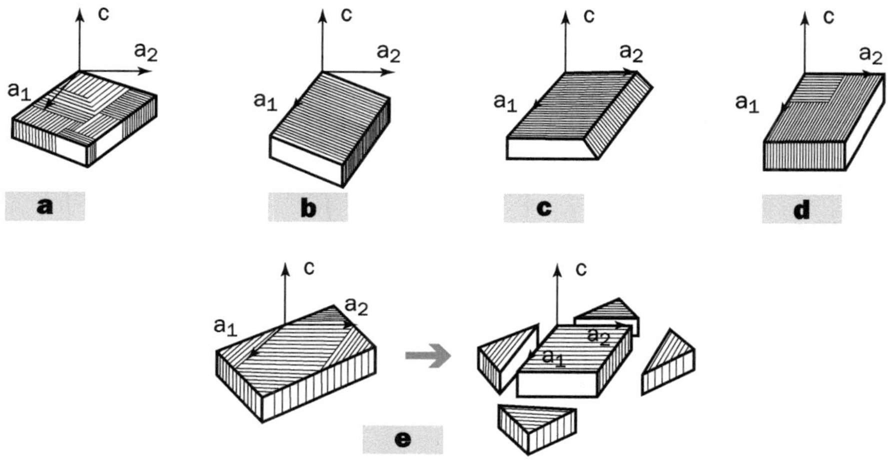

Fig. 2.34 Domain textures of a ferro-ferro sample in various situations: (**a**) The misorientation of the platelet with respect to the tetragonal planes induces complicated and quenched domain textures. (**b**) With this misorientation, one permissible wall orientation is favorized: the domain walls are in the plane of the misorientation angle; constraints exist on the edges of the sample. (**c**) Cut of a lateral face out of the tetragonal plane: the domains are perpendicular to this face; there are constraints on this face. (**d**) Lateral faces perfectly located in the tetragonal planes: it is possible to obtain several species, then only one if the cooling rate is <0.01 K min^{-1}. (**e**) Lateral faces cut according to the orthorhombic planes: we observe blocked domain textures in each corner of the sample; the application of a very brief electric field pulse allows to break the sample along the incoherent twin planes and to obtain a smaller sample with lateral faces in perfect tetragonal planes (typically for KDP, a pulse of a maximum duration of 10 ns and amplitude 2 $kVcm^{-1}$ at 110 K). After a slow rise in temperature and an annealing of a few days, we have a sample with edges of such quality that no domain nucleus can happen there. This method allows to obtain lateral displacement walls in crystals whose textures are usually in needles

2.5.3 Particular Arrangements of Domains

In the case where two quasi-perpendicular orientations of needle domains coexist in the sample in natural structure, the boundaries between the domain species provide good illustrations of the theory of twins treated either by elastic models, or with the help of dislocations as with Fig. 2.38: the domain tips are comparable to edge dislocations on perpendicular slip planes. Thus, the interweaving of domains is observed to optimize the global energy of this incoherent twin region which constitutes a texture blocking zone, even in a very good crystal. Very different phenomena exist from one crystal to another: in Tanane, these constraints can create the transition from the orthorhombic phase to the monoclinic phase (see Fig. 1.33a); in GMO and TMO, they occur easily in a somewhat quenched texture and disappear without leaving traces when the temperature decreases if the cooling rate is slow enough (Fig. 2.32); in crystals of the KDP family, with twins closer to needles than to corner domains, they can be sources of constraints difficult to eliminate if the temperature cooling and thermal homogeneity are not optimized. Figure 2.38 presents the

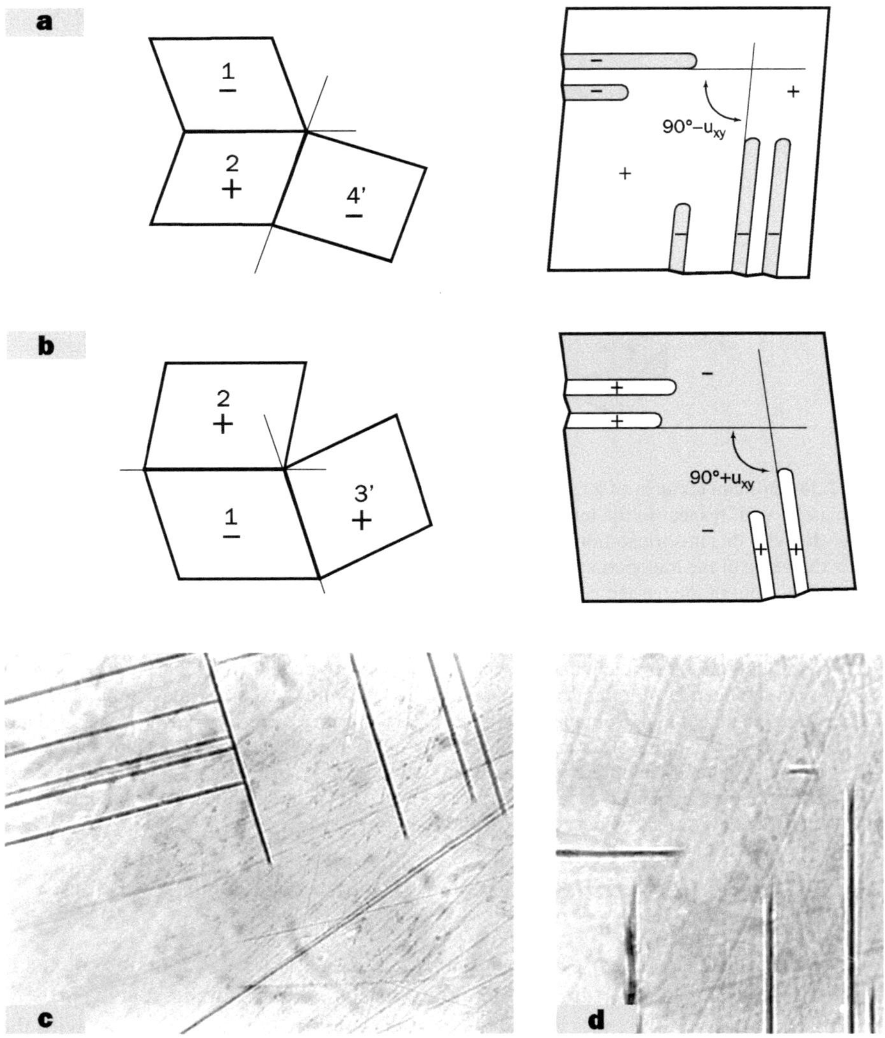

Fig. 2.35 Schematic representations (**a** and **b**) and corresponding photographs (**c** and **d**) of a polarized sample with two wall orientations and only three types of cells: in the sample (**a**) only the 2+ cell exists for the positive domain; in the sample (**b**) only the 1-cell exists for the negative domain. In (**c**) and (**d**), the angle between the walls is $90° \pm u_{xy}$ (KDP type crystal with u_{xy} of about 30 arc seconds)

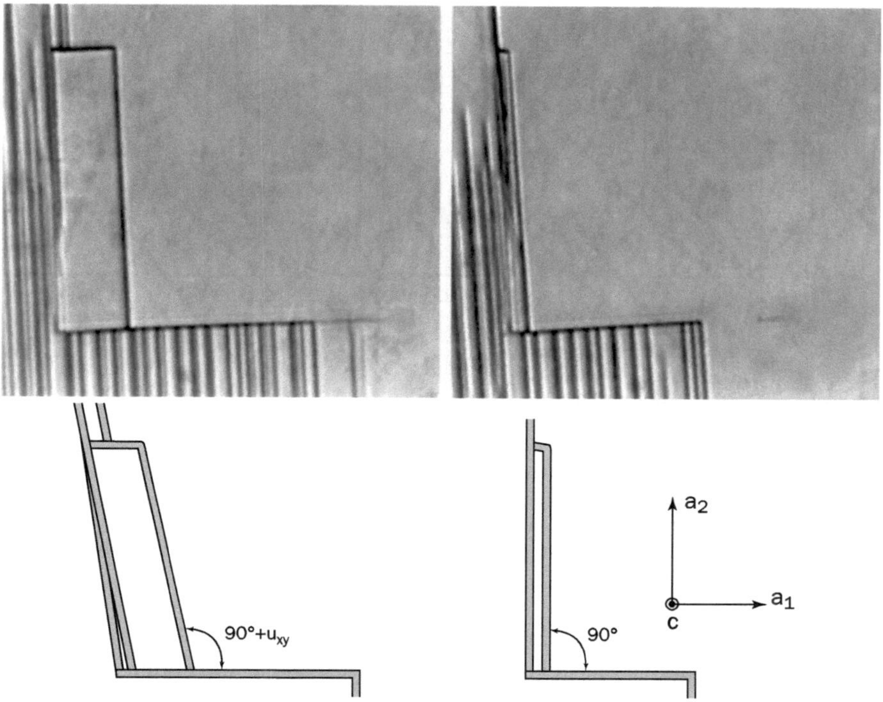

Fig. 2.36 Modification of the relative orientation of the walls of two species within the sample during a hysteresis cycle: the angle of $90°+u_{xy}$ becomes equal to 90° (KDP with t = 0.3 mm, T = 110 K, E = 1 kV cm^{-1}) (Bornarel, 1991)

situations encountered: if the boundaries between two species of domains with domains at about 90° are practically at 45° from these wall planes, situations of twin intersections (Fig. 2.38d) or interconnections with creation of micro-twins (Fig. 2.38f) are observed, as in most ferroelastics, for example $YBa_2Cu_3O_7$ (Rosova et al., 1993). These sets, results of the crystal's history, allow for local strain accommodation but greatly complicate the interpretation of the macroscopic responses of the sample. These situations can be avoided by optimizing crystal quality, sample cut, and thermal history. This will be the case for the results produced in Chaps. 3 and 4.

Rarer situations can be observed in ferro-ferro crystals as in other twinned textures. Figures 2.39 (Bornarel, 1991) and 2.40 provide examples. These situations are classic in ferroelastics with significant deformation angles, often greater than the arc degree. In Fig. 2.41, the S-shaped junction between species of domain walls approximately perpendicular is explained using the model of Fig. 2.42. All these examples occur in good crystals, with correctly oriented, cut, and polished samples. They are more frequent in the GMO family (Flippen and Haas, 1973) than in the KDP family.

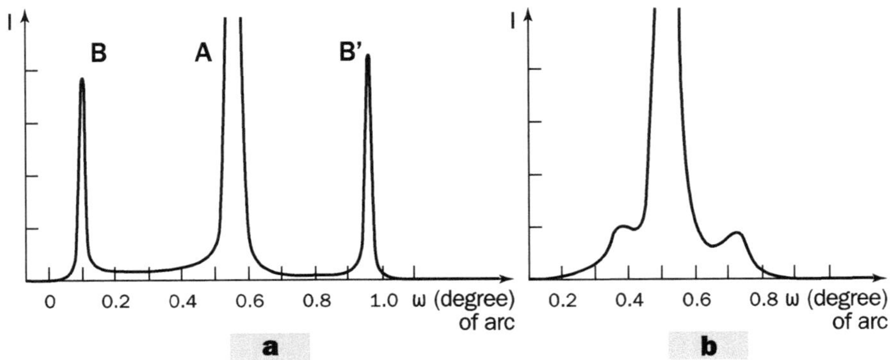

Fig. 2.37 X-ray diffraction spectrum of KDP (3 × 3 × 6 mm sample) (Afonikova et al., 1987). (**a**) At 20 K below the transition of a region of the crystal near the surface, in which there are two species of domains: peak *A* corresponds to the twin planes of the permissible walls of one set and the peaks *B* and *B′* to the lattice planes of the other, of two twins of opposite deformations (shear angle of about half a degree of arc). (**b**) A few tenths of a degree Kelvin below the transition, the region concerned by the diffraction includes an incoherent twin plane perpendicular to the surface of the crystal. The low value of spontaneous deformation and the continuum between the three diffraction peaks are observed

In conclusion, optimal heat treatment and sufficiently slow cooling rates are necessary to obtain simple and regular textures with a single wall orientation and a minimization of strains. It is on the basis of such textures that the results presented in Chaps. 3 and 4 were obtained. Chapter 3 deals with texture modifications in domains during the transition from a mono-domain state to the opposite mono-domain state. In Chap. 4, the textures are often quite close to those of natural textures since the electrical or mechanical excitations, to perform "zero field" susceptibility measurements, are of amplitudes sufficiently weak not to modify the average position of the domains.

Memo Chapter 2

The finite dimensions of samples impose boundary conditions: sample shape, orientations, and natures of the edges (with or without connected electrodes for faces perpendicular to the ferroelectric axis, constraints on the sample edges more or less relaxed). The existence of the domains depends on it and often leads to global polarization and deformation of the sample almost null.

The orientation of the domain walls is defined by the symmetry elements lost at the transition: in a wall the deformation relative to the para phase is the same in the two domains it delimits. Three types of domain walls exist, S, W_∞, and W_f, the latter corresponding to the case of ferro-ferro crystals retained in the work (permissible walls are the tetragonal planes (100) and (010)). These "permissible" walls are not zero-thickness. Their impact can be noticed at distances that depend on the measurement mode used: from a few cells to a few tens of micrometres, or limited by

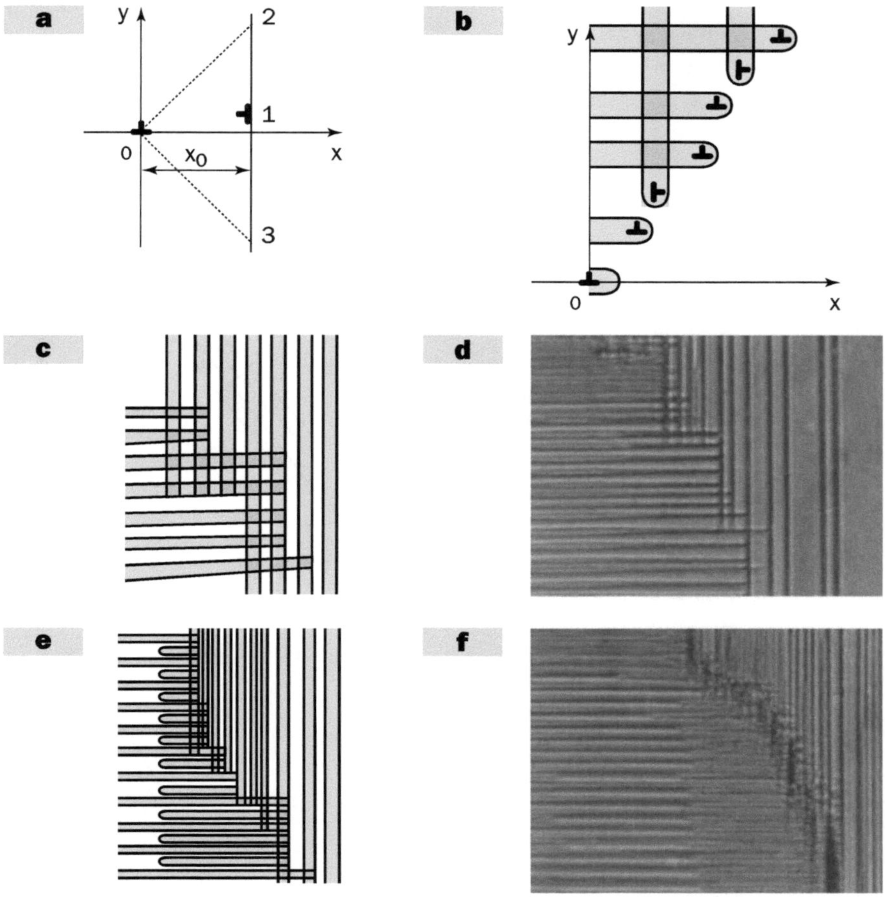

Fig. 2.38 (**a**) Two edge dislocations of the same Burgers vector magnitude have perpendicular slip planes and a minimum interaction energy in positions 2 and 3: between these two positions, the force F_y along the y-axis exerted on the dislocation 1 by the dislocation located at O tends to bring the dislocation 1 into positions 2 or 3, where this force is nullified (when y is equal to $\pm x_0$). (**b**) This explains the typical arrangements of dislocations summarized from the ends of domains (edge type on perpendicular slip planes). (**c**) and (**d**): The descriptive diagram (**c**) and the photograph (**d**) illustrate a situation of intersections of domains in perpendicular needles. The vertical needles are of almost constant width, while the horizontal domains are wedge-shaped in this (001) section. (**e**) and (**f**): The descriptive diagram (**e**) and the photograph (**f**) illustrate situations called "interconnection" (Rosova et al., 1993) with the creation of micro-domains linked to the meeting region between the families of perpendicular domains: a better accommodation of constraints is created. When sets of perpendicular domains meet, depending on the type of junction, it is possible to encounter situations with twin planes at 90°or at 90°± u_{xy}

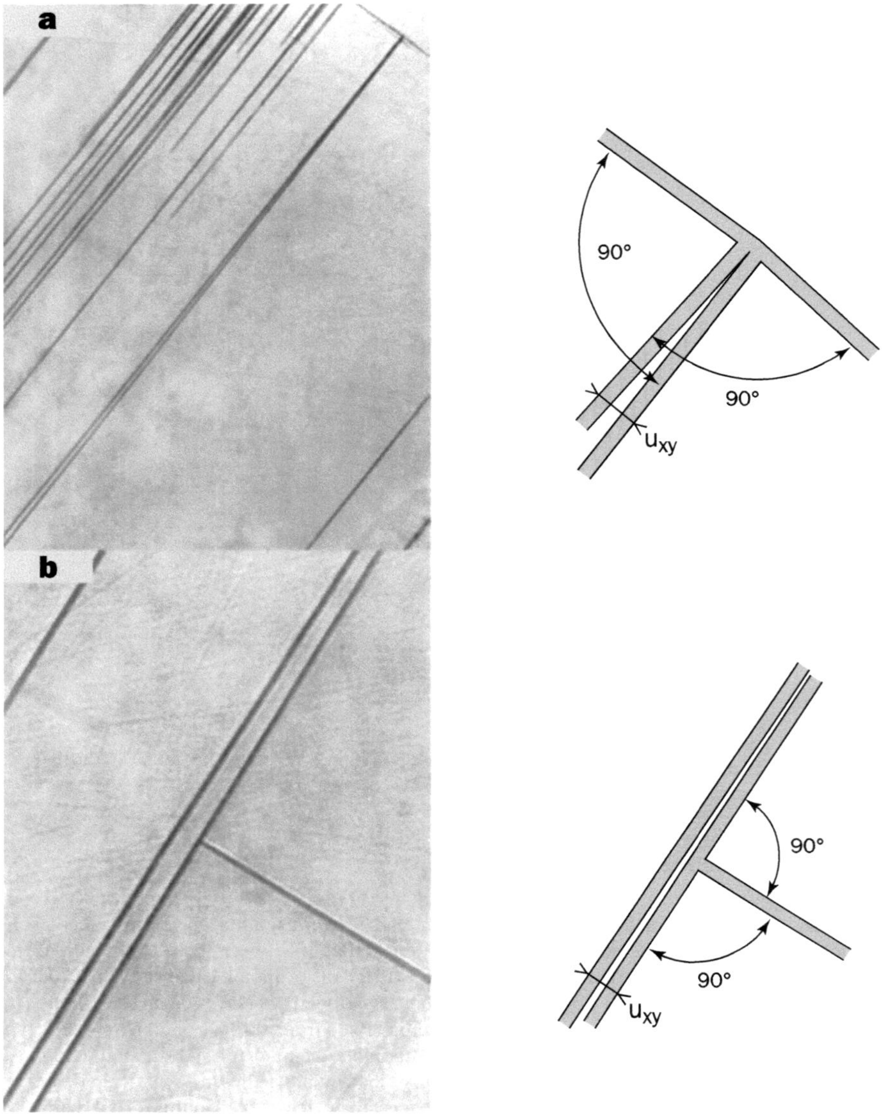

Fig. 2.39 Unusual situation of two domains whose walls make an angle equal to the spontaneous deformation u_{xy} (Bornarel, 1991) which can be modelled with, in its plane of symmetry, a set of disclinations

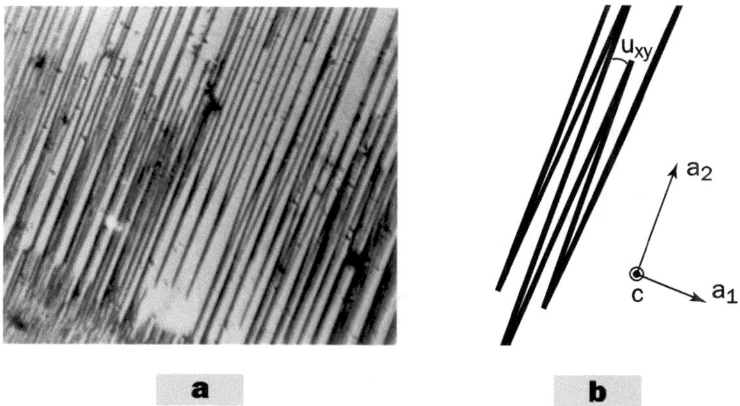

Fig. 2.40 Photograph (**a**) and schematic representation (**b**) of a zig-zag domain arrangement in a KDP crystal

Fig. 2.41 S junction between two sets of domains, due to the minimization of the interaction energy of the domain tips for irregular textures (texture photograph in a (001) section of KDP). The boundary between the two regions avoids the intersection of the needle domains; the needles make an angle of 90° ± u_{xy}

the smallest dimension of the sample. This experimental "wall width" can change depending on the depth in the sample. It grows near the transition and decreases with temperature. It does not only concern the order parameter, such as the shear deformation u_{xy} in our context, but also other quantities, such as the normal deformations u_{xx} and u_{zz}, identical in domains of opposite polarities but different between a domain and a wall.

The wall number varies with temperature, increases towards the transition, and decreases when it decreases. The domain shape (width and length of needle domains in a section (001), for example) varies depending on the sample, notably its thickness. Even in the case of a single wall orientation, the modification in temperature of the domain number is usually only possible if parts of walls outside the

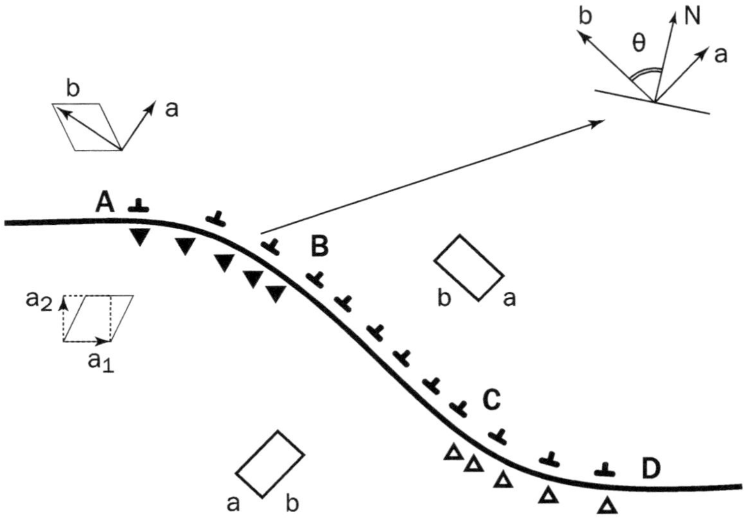

Fig. 2.42 In section (001), a wall located in a permissible wall plane (horizontal in the diagram) performs an S shift to ultimately find the same direction. The tetragonal (**a**$_1$, **a**$_2$) and orthorhombic (**a**, **b**) cells are represented. Also schematized are densities of fictitious dislocations to account for the wall's misorientation with respect to the permissible plane. These densities are all the more important as the spontaneous deformation and cos2θ are large (θ angle between the normal to the wall and the vector **b** of the orthorhombic cell): this dislocation density is therefore maximum when the wall is 45° from the permissible wall plane. Triangles indicate densities of disclinations that account for the curvature of the boundary between the two species; the whites and the blacks correspond to opposite curvatures. Dislocations and disinclinations disappear beyond A and D since the wall is in the permissible wall plane (Pertsev et al., 2000)

permissible planes (needle tips, for example) can be created. Thus, the domain textures are often metastable.

If a part of a wall is not in a permissible plane, imperfections are created. The most obvious here correspond to lines of quasi-dislocations wedges parallel to the ferroelectric axis with Burgers vectors in the permissible wall planes, proportional to u_{xy} with slip planes in these planes; the other deformations also induce other screw or wedge dislocations. Given the deformation values and the preponderance of shear, almost all the quasi-static phenomena observed at the domain level are explained by limiting ourselves to the wedge quasi-dislocations of the walls and their assemblies at the domain tips: tip shapes, interaction between tips, domain species, number of domains in a set, particular arrangements of domains in real crystals.

Many questions remain after studying natural domain textures:

- Do the walls interact with each other?
- Is there a minimum critical distance between two permissible walls (−+) and (+−)?
- Is the order parameter within a domain a function of the domain size?

- How do the domains, their sets change when a conjugate field is applied to the sample?
- How does the dynamics of a wall occur? Of a domain? Of a set of domains?
- Do the hysteresis cycles, symbols of ferro crystals, change shape with the domain texture?
- What macroscopic properties of ferro-ferro crystals can be attributed to the domains?

Answers are given in Chap. 3, with applied fields sufficient to switch the sample from a single domain to the opposite polarity single domain, and in Chap. 4, with sufficiently weak periodic fields which allow understanding the dynamics of a wall itself without changing the texture.

Chapter 3
Switching Process and Hysteresis Cycles

3.1 Macroscopic Properties and Microscopic Phenomena

3.1.1 Characteristics Quantities of the Switching Process

The hysteresis cycles $P(E)$, $u(\sigma)$ characterize the ferroic properties of a crystal, and their plots require the application of periodic excitations (electric fields or stresses), amplitude, and frequency of which are imposed. The numerous improvements to the initial setup (Sawyer and Tower, 1930) take into account conduction problems (Diamant et al., 1951) and compensate for effects not directly related to the switching phenomenon (Roetschi, 1962; Sinha, 1965; Hatano et al., 1992). A review of these methods can be found in the work of Tagantsev et al. (2010), particularly those adapted to thin films (Glazer et al., 1984). In this context of forced oscillations, the maximum measured values $P_{\max}$, $u_{\max}$, the remanent values P_r, u_r and the coercive fields E_c (or σ_c) are determined. The coercive field for the KDP crystal varies with the amplitude $E_{\max}$ as illustrated in Fig. 3.1b (Guyon and Lajzerowicz, 1966; Bornarel, 1995). It usually increases with the frequency f of the applied external electric field (or applied stress) up to a critical value beyond which the switching process does not have time to occur (Jung et al., 2002).

For applications, crystals are sought that can switch as quickly as possible from a + situation to a—situation, with the lowest possible excitation ($E_{\max}$, $\alpha_{\max}$) to consume as little energy as possible (with a minimum surface area of the hysteresis cycle). Particular attention is paid to materials that can be used at room temperature for obvious practical reasons. These include perovskites and inexpensive ceramics, capable of withstanding polarization reversal without deteriorating over many cycles. Over the past 20 years, interest in bulk crystals has declined in favour of thin films, not studied here; however, some of the results from thin films will be used when they enrich the discussion and remain relevant in the context of the book.

J. Bornarel, *Ferroic Crystal Domains, Walls, and Phase Fronts*, Springer Series in Solid-State Sciences 209, https://doi.org/10.1007/978-3-031-98099-2_3

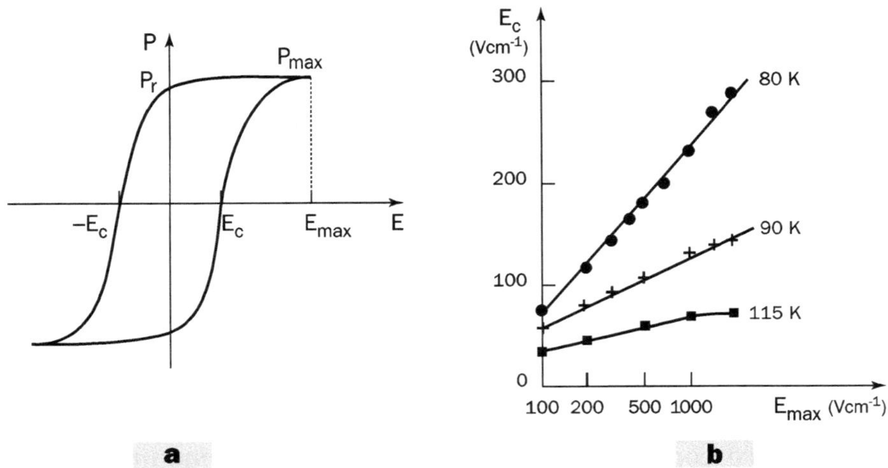

Fig. 3.1 (**a**) Representation of a low-frequency hysteresis cycle (50 Hz); E_c is the coercive field, and E_{max} is the maximum applied electric field; (**b**) for a KDP crystal, the coercive field increases logarithmically with the maximum applied field; each curve is plotted at a constant temperature indicated (Guyon and Lajzerowicz, 1966)

In this search for rapid and frequent reversals, non-piezoelectric ferroelectric crystals in the para phase involve low mechanical energies and rarely break. This is not the case for ferro-ferro crystals due to dislocations and dis-inclinations, the movement of which creates a fatigue phenomenon. The aim of the book is not to seek the best reversal performance of a crystal, but to understand the phenomena at the microscopic level, or even at the nano level, and to relate them to macroscopic properties.

As the hysteresis cycle is the result of a process imposed by the experimenter, in frequency and amplitude of excitation, the plot of a switching current $i(t)$ illustrated in Fig. 3.2 can provide more intrinsic responses from the crystal (Merz, 1956). The values i_m, t_m, t_s by definition characterize the switching. In these experiments, the crystal is initially mono-domain. An electric field (or a mechanical stress) is imposed in the form of a step, of sufficient amplitude and appropriate polarity to achieve switching. The switching time is not imposed, but observed. The study of switching currents complements that of hysteresis cycles to detect the various phenomena of the process and to understand them.

Measurement methods using a succession of applied pulses have also been developed: the minimum pulse duration required to set the domains in motion and the periodicity between pulses are major parameters (Fatuzzo and Merz, 1959, 1966; Taylor, 1965, 1966). Other measurements with linear variations of the excitation are between hysteresis cycles with sinusoidal excitation and switching by application of step excitation. Beyond the analysis of hysteresis cycle shapes (Gan and Cao, 2020), the aim of this chapter is essentially to relate as concretely as possible the macroscopic responses to microscopic phenomena.

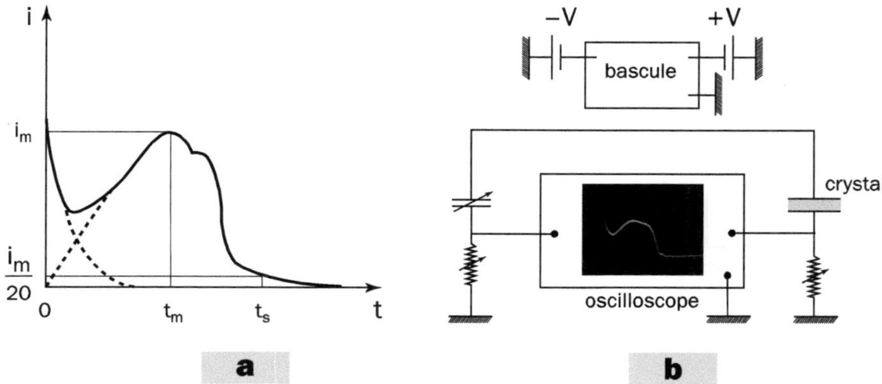

Fig. 3.2 (**a**) Typical variation of the switching current $i(t)$ of a ferroelectric crystal obtained with the setup schematically represented in (**b**); the response peak at $t = 0$ corresponds to the linear capacity of the sample unrelated to the reversal phenomenon; i_m is the maximum value of the $i(t)$ curve and t_m corresponds to it. The duration of the switching phenomenon t_s is subject to various definitions; here, the time interval between the application of the excitation field and the fall of the current $i(t)$ to one twentieth of its maximum value is considered

3.1.2 Switching of Polarization and Deformation: A Non-linear Phenomenon

The experiment summarized in Fig. 3.3 (Bornarel, 1971) illustrates the non-linearity of the phenomena: On a sample of the KDP crystal in the ferro phase, with a domain texture at equilibrium, electric field pulses of the same polarity are applied, sufficiently large to make the crystal mono-domain, and sufficiently spaced in time so that after each pulse application, the domain texture can be recreated. From pulse to pulse, the progressive saturation of the crystal is observed, with the remanent polarization increasing with a slower return of the domains and a growth of the region mono-domain in the sample. Are these results due to changes in polarization (deformation) within the domains? To changes in the shapes of the domains? To surface layer phenomena? Some answers to these questions are proposed below.

3.1.3 Different Arrangements of Domains and Walls

In the considered crystals, the domains quickly cross the entire sample in the direction of the ferroelectric axis due to mobility differences as in practically all crystals, parallel and perpendicular to this axis (Fang et al., 2024); three modes of switching are observed (Bornarel et al., 1974):

- Switching with a major role of lateral wall displacements (Fig. 3.4a)
- Switching with domains in the shape of corners whose apex angles can remain constant or, on the contrary, vary during the reversal. This process corresponds to

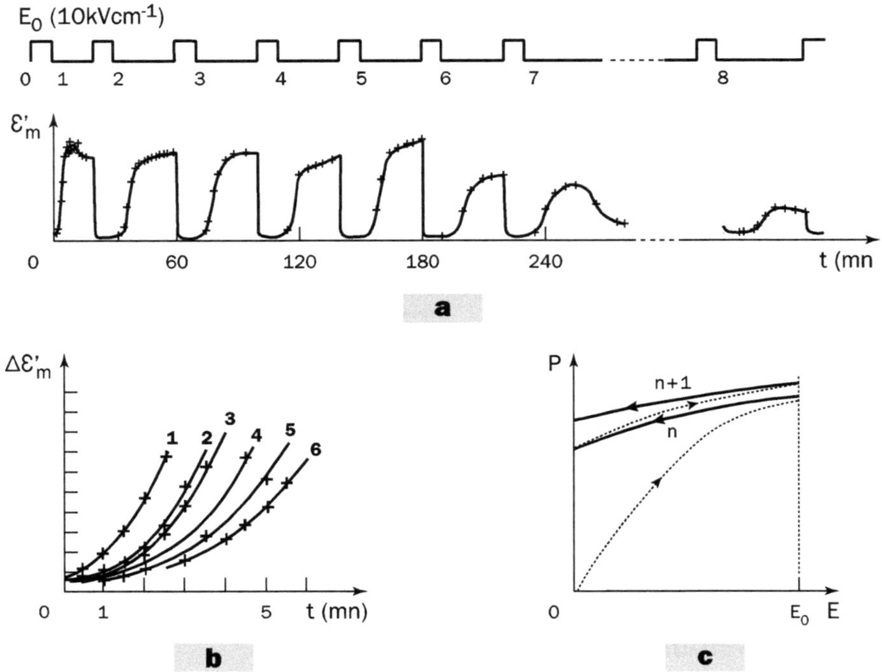

Fig. 3.3 (**a**) Identical electric field pulses of the same polarity (+10 kV cm^{-1}, 10 min), exceeding the coercive field are periodically applied (period = 40 min) to a KDP crystal with a laminar domain texture at a temperature of 98 K; the dielectric constant following c is calculated using capacity measurements (Bornarel, 1971); (**b**) variation of the dielectric constant at the beginning of the reappearance of domains ($t = 0$ corresponds to the end of the applied electric field pulse) for various ranks n of pulses: the response is slower when n increases; the characteristic time varies like the logarithm of n; it is several hundred seconds in the example presented; (**c**) the polarization of the crystal $P(E)$ under the effect of successive pulses gradually increases and the return of domains after a pulse is more difficult when n increases

This type of experiment is comparable to those carried out in ferroelasticity in which, to maintain contact between the sample and the stress application piece, a continuous stress greater than the amplitudes of the variable stress variation is permanently applied: there is no real switching, but only changes in domain size – (for example) in a polarity + matrix (deformation or polarization)

the models studied from the 1950s with nucleation, wall displacement, and disappearances ("collapses") (Fig. 3.4b)

- Switching with needle-shaped domains, of almost constant widths, and a major role of their tips (Fig. 3.4c)

In the following of the chapter, the lateral displacements of walls and the movements in needles will be successively studied, starting from the simplest situations to end with the most complex ones. The quasi-static studies will allow to decompose and understand the isolated phenomena; the interactions between domains will bring us closer to the complex situations that often correspond to macroscopic measurements, especially at the highest frequencies.

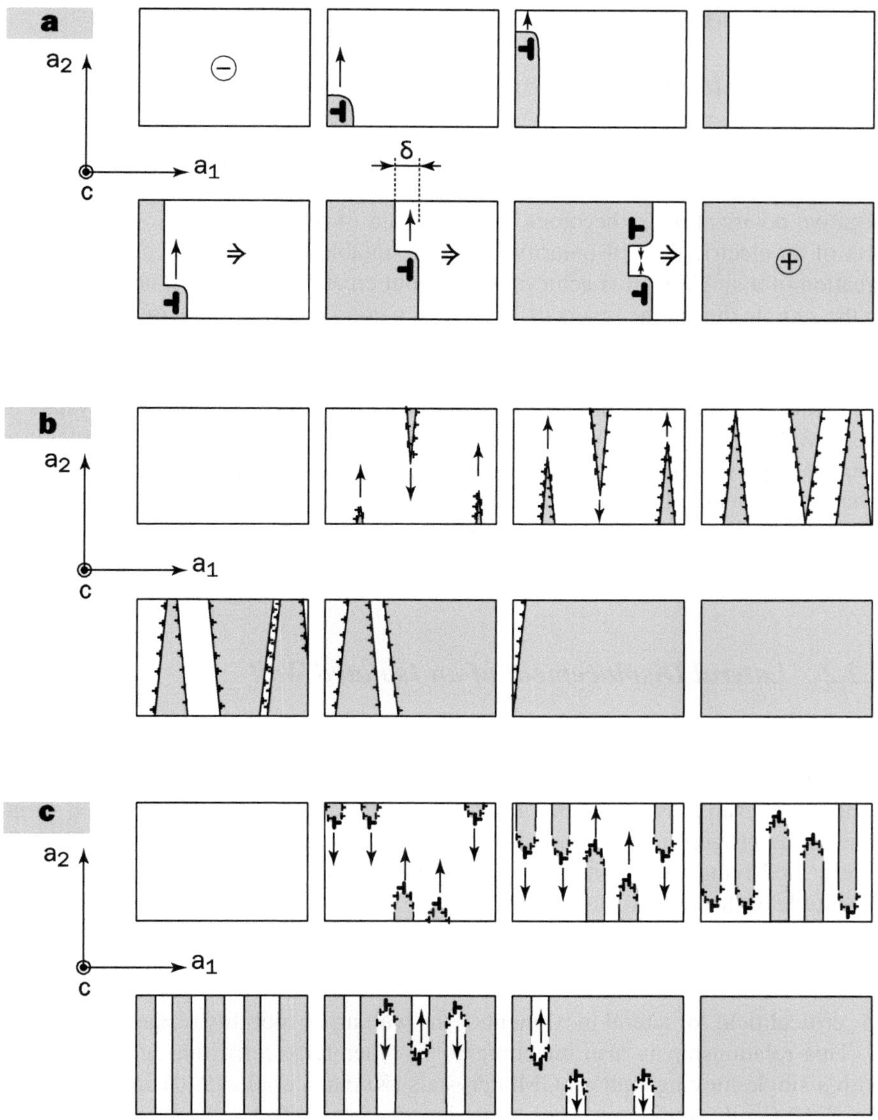

Fig. 3.4 Different modes of polarization-deformation reversal (Bornarel et al., 1974) are presented with symbols of dislocations or wedge-type quasi-dislocations: (**a**) reversal by lateral wall movement, (**b**) reversal using wedge-shaped domains, (**c**) reversal using needle-shaped domains of almost constant widths

3.2 Switching by Lateral Displacement of a Domain Wall

3.2.1 How to Obtain a Single Domain Wall in a Sample?

The simplest switching by lateral movement of a domain wall involves a single wall that sweeps the entire crystal. Thus, a sample initially mono-domain, for example of negative polarization −, becomes mono-domain of positive polarity + by application of an electric field of polarity + and of suitable value. This is possible if the creation of a single wall is achievable, without creating a corner or needle domain in the sample during the reversal. To do this, a good quality crystal must have sample edges with particular characteristics. This is the case of Tanane which sublimates somewhat at room temperature, hence sample edges without possible grip for the creation of a domain nucleus. A single wall can also be obtained in crystals of the raw GMO family (Kumada, 1969; Kumada et al., 1970). It is more difficult to obtain samples with perfect lateral faces for crystals of the KDP family, but a somewhat violent method allows it, as explained in Fig. 2.34e. The process will be used in Chap. 4 to correlate macroscopic dielectric measurements, and total lateral surface of domain walls.

3.2.2 Lateral Displacement of an Isolated Wall

Figure 3.5a shows a Tanane sample with a single wall moving under the action of an applied electric field; in Fig. 3.5b, for a sufficient value of the continuous field applied, the wall moves laterally in a regular way (Legrand, 1973); the small variations in its progression are due to imperfections of the edges without this significantly modifying the displacement law. The speed v of the wall varies with the amplitude of the applied electric field E as:

$$v = k\left(E - E_S\right) \tag{3.1}$$

(E_S critical field for lateral movement of the wall and k mobility of the wall).

This relationship is also encountered in other ferro-ferro during experiments with a single moving wall on GMO crystals (Kumada et al., 1970) and crystals of the KDP family. It demonstrates that friction forces effectively oppose the forces induced by the applied electric field (or mechanical stress). For their origin, the proposal of dislocations created in the wall during its movement is old for both KDP (Barkla and Finlayson 1953) and ferroelectrics such as $BaTiO_3$ (Nakamura, 1954). In the context of "tetragonal-orthorhombic" ferro-ferro crystals, wall-level models as in Fig. 3.6 have been proposed. The practical shear deformations of the domains separated by the wall define the Burgers vector of the quasi-dislocation, hence the dynamics of the system. The lateral shift of the wall created by the quasi-dislocation and the existence of knees depend on the experimental contexts. It is not always

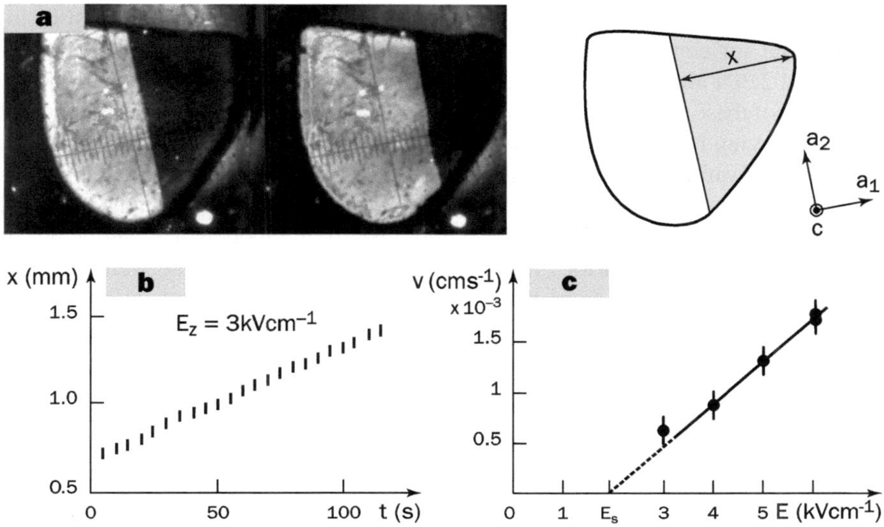

Fig. 3.5 (**a**) Lateral displacement of an isolated wall in the Tanane crystal at 12 °C: photograph and schematic representation; (**b**) lateral displacement of the wall over time with an applied electric field $E = 3$ kV cm^{-1}; (**c**) wall speed as a function of the amplitude of the continuous field E applied: $v = k\ (E - E_S)$. The coefficients k and E_S vary with temperature and sample thickness and do not change with the size of the wall following a_2 (Legrand, 1973; Bornarel and Legrand, 1981)

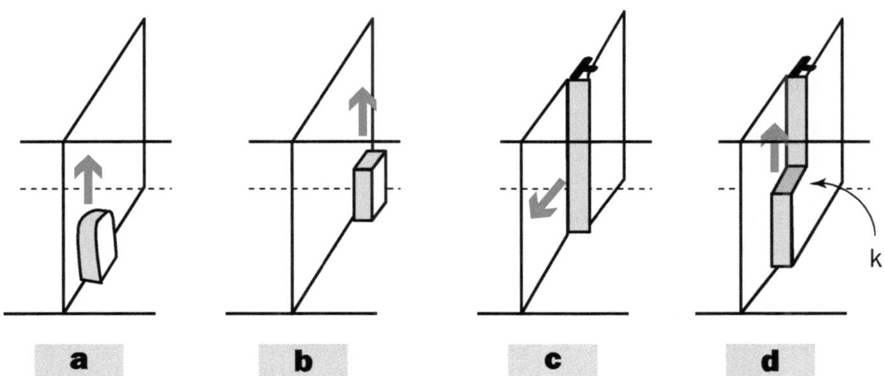

Fig. 3.6 Representations of nucleation on a domain wall: (**a**) and (**b**): a small nucleus of low thickness is created on a (100) wall and crosses the sample in the direction of the c axis (here vertical), (**c**) lateral displacement in the plane of the wall using a wedge-type quasi-dislocation that would move as a single block in its slip plane, (**d**) a knee moves in the c direction along the quasi-dislocation and thus contributes to its displacement parallel to the wall

simple to propose coherent atomic arrangements for the core of such a quasi-dislocation. But for compounds of the KDP family, the representations of Figs. 2.10, 2.27, and 2.30a allow to keep the chemical formula. This is no longer possible for the knees of these quasi-dislocations, which create (Fig. 3.7) Takagi ions HPO_4^- and $H_3PO_4^+$ moving by successive jumps of protons on their bonds as indicated for the two types of walls (polarized or neutral). These jumps account for a global displacement of the ions in the direction of the ferroelectric axis. One can imagine, in the logic of Annex A5 and the work of Schmidt (2020), the models that can be developed with these tools. Various models of modifications to moving walls are proposed at the end of the chapter, but for the smallest lateral wall translation in KDP, the previous model is sufficient.

3.2.3 Switching of Polarization (Deformation) with a Single Wall

The displacement laws illustrated by relation (3.1) concern a wall under well-defined conditions of temperature and constant applied field. In this context of switching with a single wall, what can we expect at the macroscopic level, such as a complete hysteresis cycle, achieved in sinusoidal alternating field or triangular signals?

Two very different situations are presented: the cycle of Fig. 3.8 for a GMO sample with a single wall is described correctly by relation (3.1) with a quasi-constant wall speed and sufficient for a reversal time small compared to the period. The value of the critical field E_S and that of the mobility μ are easily estimated.

$$b' = a' tg\left(|u_+| + |u_-|\right) \tag{3.2}$$

The situation described for Tanane in Fig. 3.9 is quite different: during the lateral displacement of the wall, the deformation within the two domains changes according to the applied electric field and this change also depends on the orientation of this field relative to the local polarization. This results in a particular form of hysteresis cycle which illustrates the need to increase the amplitude of the field to complete the wall's journey and reach the single-domain situation. The crystal presented is particularly soft with significant deformation, but this phenomenon can exist in other crystals even if it is less obvious to detect.

The hypothesis of domains with constant spontaneous deformation modules during switching is obviously to reconsider. Then, Fig. 3.10 gives some theoretical forms of hysteresis cycles for various deformation laws within a domain as a function of the applied electric field (or mechanical stress). These cycles, and that of

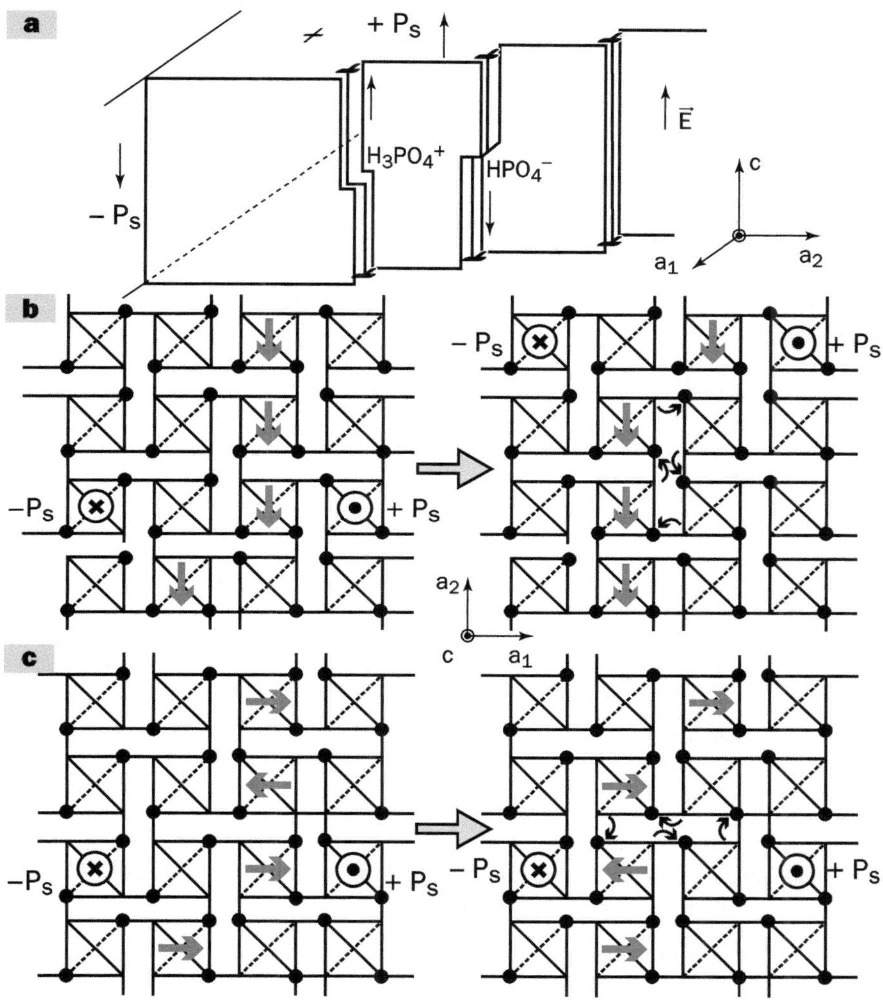

Fig. 3.7 (**a**) Models of lateral displacements of walls using wedge-type quasi-dislocations in the KDP crystal. The indicated knees correspond to HPO_4^- and $H_3PO_4^+$ ions that move along the *c* axis respectively in the opposite direction and in the direction of the applied electric field (Bornarel, 1975). (**b**) and (**c**) Schematic representation of the knees in the standards of Figs. 2.10 and 2.27, with tetrahedra and hydrogen bonds in projection in a (001) plane. The successive jumps of protons allowing the displacement of the knees along the c axis are indicated (these tetrahedra are not in the same plane perpendicular to the ferro axis and two types of tetrahedra exist; (**b**) corresponds to a polarized wall and (**c**) to a neutral wall

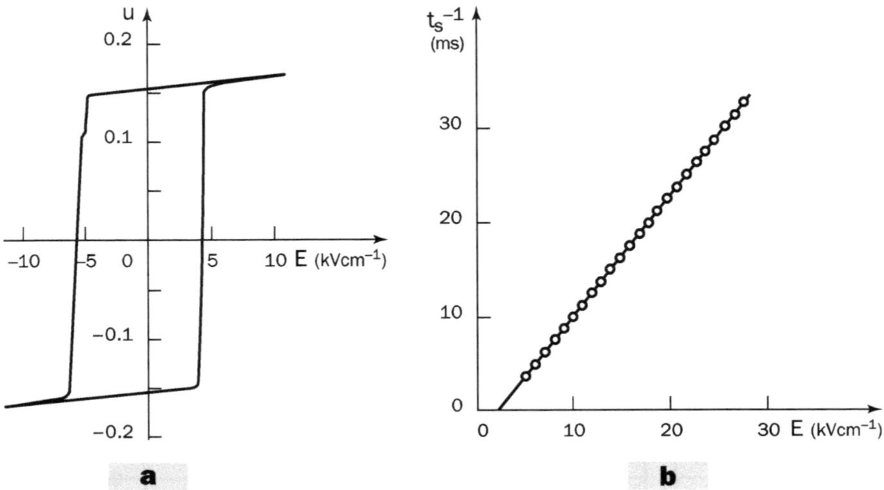

Fig. 3.8 In a GMO sample at temperatures below 80 °C, only one domain wall has a lateral displacement speed proportional to the applied electric field (Kumada, 1969). (**a**) A hysteresis cycle (5×10^{-3} Hz, 20°C) is almost square with a moving wall as soon as the minimum field of movement E_S of the relation (3.1) is applied, and a speed which, here, allows to cross the sample in a time much less than the period of the applied electric field. These results can be modified according to the experimental conditions (at room temperature) because of the conductivity and space charges by phenomena of evolution over time: "ageing, deageing" (Shur et al., 2005). (**b**) In this example, the inverse of the time t_S necessary for the wall to cross the sample is proportional to the amplitude of the electric field applied to the crystal (the dimension of the sample perpendicular to the wall is 1.55 mm, and its thickness is 0.114 mm)

Fig. 3.9, in the case of a switching with a single wall in lateral displacement propose some links between macroscopic response and microscopic phenomena.

We will remember for reversals with a single wall:

- The wall can follow simple lateral displacement laws as illustrated by relation (3.1) when the external excitation applied is constant and the polarizations and deformations within the + and – domains can be considered as constant.
- When the movement of the wall under the action of an applied field occurs at the same time as changes in deformations and polarizations in the domains it limits, its speed can depend on this practical deformation change (other reasons may contribute).
- In a hysteresis cycle with a single wall, the situation with equal volumes of + domain and - domain (situation X of Fig. 3.10) does not necessarily correspond to a polarization and a macroscopic deformation null.
- The shape of a hysteresis cycle in the simple case of a single wall can suggest phenomena at the level of the domains and the wall that limits them.

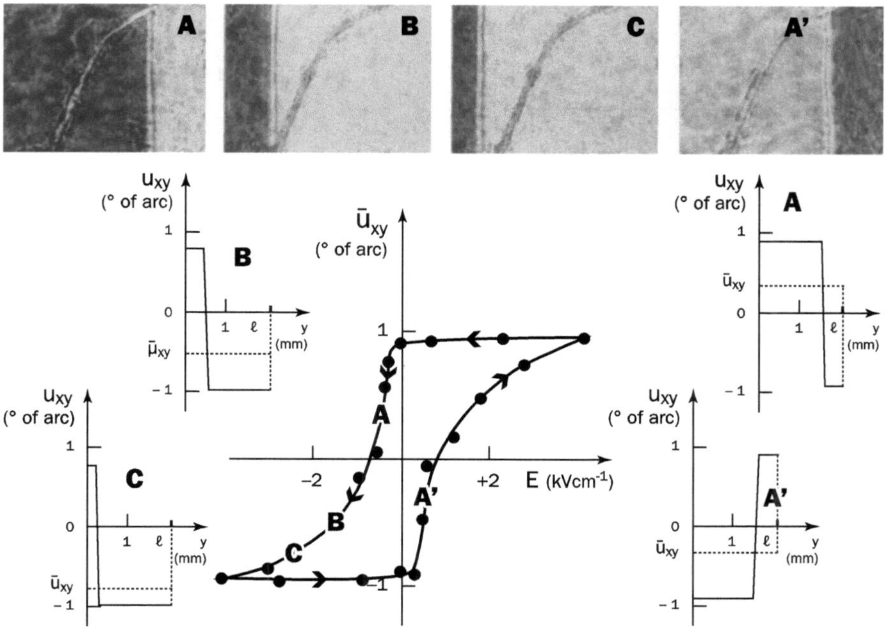

Fig. 3.9 Macroscopic deformation and local deformations during the reversal of a Tanane crystal ($a_1 = 2$ mm, $a_2 = 3$ mm, c = 0.5 mm; $T = 13$ °C). The applied electric field E varies linearly between values of maximum amplitudes $E_{max} = 4$ kV cm^{-1} with a period of 2 h. The measurement of the birefringence Δn_c for a light propagating along the c axis allows to calculate the local deformation u_{xy} and to estimate with the surface measurements of the domains, the macroscopic deformation by integration over the whole sample (Legrand, 1973; Bastie et al., 1976a, b). This deformation is plotted as a function of E; the photographs presented are marked with letters on the cycle

The experiment begins with a single-domain sample of maximum shear deformation u_m (1 degree of arc) positive in our representation; the decrease in E causes that of the deformation in the single domain from 1% to 2% compared to its maximum value u_m. The field E changes direction, and a domain wall appears when $E = -0.3$ kV cm^{-1}. It moves laterally (A) under the action of the electric field of opposite direction to the existing global deformation. The deformation in the regressing domain is about 0.9 u_m while the modulus of that of the favoured domain (which still occupies the smallest volume) is 0.92 u_m; these values do not change significantly between the creation of the wall and the situation of zero global deformation of the sample. The verticality of this part of the cycle shows that the wall, set in motion for a field $E = -0.3$ kV cm^{-1}, continues its displacement quickly: the critical field E_S of the wall is therefore lower than the value observed at the creation of the wall

The situation changes in the part of the cycle between the zero global deformation and the complete disappearance of the wall (**b** and **c**). The average deformation of the positive polarity region is now about 0.8 u_m while that of the domain of negative polarity favoured by the applied electric field is in modulus about 0.97 u_m. The shape of the cycle demonstrates the difficulty encountered by the wall to progress: its displacement requires an increasing amplitude of applied electric field. The practical deformation ($|u_+| + |u_-|$) between the two domains changes between situation A (equal to 1.82 u_m) and situations B or C (1.77 u_m). These significant effects are characteristic of Tanane, a soft crystal, with a spontaneous deformation of 2 degrees of arc at 0 °C. This changes the magnitude of the Burgers vector of an edge quasi-dislocation that would produce a lateral translation of the wall by a distance a'. Is this modification of deformations also at the origin of the changes in E_S and k of the relation (3.1), which make it more difficult, in the case of Tanane, the end of the switching of the polarization-deformation?

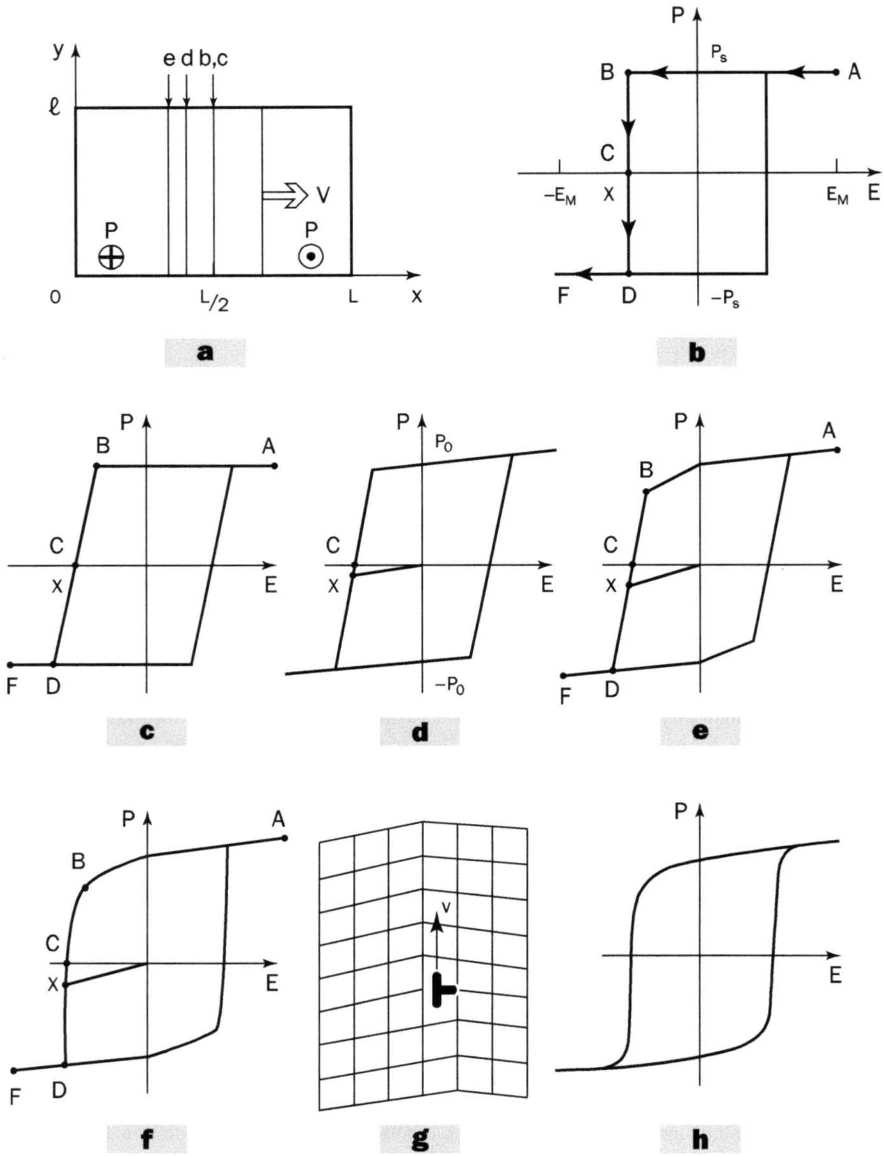

Fig. 3.10 Theoretical shapes of hysteresis cycles for various laws of variations of the local polarization (deformation) of domains as a function of the applied external field E and various wall speed laws, in the simple case of a single wall moving laterally. The ferro-ferro crystal is assumed to be subjected to an external excitation (E or σ) in a triangular signal of maximum amplitude E_M and of period T. The polarization P and the deformation u are assumed to be proportional at all points. The electrical conductivity is assumed negligible. (**a**) A single wall assumed very thin, parallel to a permitted plane, translates laterally in the x direction with speed V. (**b**), (**c**) We move from a mono-domain situation in A to the opposite mono-domain situation in F (see the arrows

(continued)

from A to F) by creating in B the wall, passing in C to the coercive field ($P = 0$) and equality of volumes of opposite polarities in X; the wall disappears in D. From D to F, the crystal is again mono-domain, with polarization opposite to that from A to B. In (**b**) and (**c**), the polarization (deformation) in the domains is assumed of constant modulus (P_S) whatever the value of E (or σ) and the points C and X are coincident: $x_C = x_X = L/2$. In (**b**), the wall, as soon as it starts moving, moves with a notable limit speed, such that the time to go from B to D is small compared to the period T of the excitation. In (**c**), the limit speed is lower and does not change with E: the time to go from B to D is not negligible compared to T. (**d**) The local polarization P (or local deformation u) varies linearly with E: $P = \pm P_0 + \alpha E$. At the coercive field (point C), the wall is at the abscissa $x_C = L/2\ (1 + \alpha E/P_0)$ smaller than L/2; the equality of volumes of opposite polarity at point X corresponds to the value $P_X = \alpha E$ of slightly negative value. (**e**) If the local deformation varies linearly with E but in a different way when E and P are of the same sign (α_1) or of opposite sign (α_2), the hysteresis cycle takes the form shown (in which $\alpha_2 > \alpha_1$). At point C, the position of the wall is defined by $x_C = L/2\ (1 + \alpha E/P_0)\ (1 + (\alpha_2 - \alpha_1)\ E/2P_0)^{-1}$; the polarization in the situation X is $P_X = (\alpha_{2+}\ \alpha_1)E/2$ of slightly negative value. (**f**) If the speed V of the wall varies with E (or σ), for example with a simple law such as $V = \gamma\ (E_B - E)$, the $P(E)$ curve between B and D is no longer linear but quadratic as illustrated (between B and D, $P =-P_0 + \alpha E - P_0\ \gamma T/4L\ E_M\ (E_B - E)^2$). The phenomena on the wall during its displacement may not be simple. (**g**), (**h**) We assume a simple translation in the wall plane of a quasi-dislocation: the elementary Burgers vector b for a lateral wall translation a would be proportional to the sum of the deformations of the domains it separates, so if these vary with E, variations of the $P(E)$ cycle will be observed as in (**h**) in which, in addition, a saturation law of $P(E)$ (or $u(\sigma)$) is introduced. As the force on the wall that conditions v (dislocation displacement speed in its slip plane) is proportional to b, this is also true in a simple case for V; it is thus possible to construct models by observing the slopes around the equilibrium points of the free energy as a function of the polarization (or deformation): the slopes of $P(E)$ are not the same if we increase or if we decrease P with E applied, therefore with the values of α_2 and α_1 and the sign of their difference (assumed positive in these figures)

3.3 Lateral Interactions Between Moving Domain Walls

The existence of planar walls in lateral displacement is possible if the state of the sample edges excludes the creation of domain nuclei in wedge or needle form. In this context, under the action of an electric field or a constraint, the entry and exit of planar walls in the sample can lead to the existence of one or several walls. Knowing the dynamics of an isolated wall, what new phenomena appear when several walls are in lateral displacement?

3.3.1 Interactions Between Two Moving Plane Walls

A first answer is given in Fig. 3.11 by two walls moving laterally towards each other, in a KDP sample (Bornarel, 1987). The experiment demonstrates the interaction of these two planar walls at distances of several hundred microns. A model of either dipoles or dislocations in the wall explains this interaction of walls, perpendicular to their permissible plane at distances of about the thickness of the sample (direction of the ferroelectric axis c) and a greater complexity of the phenomenon beyond.

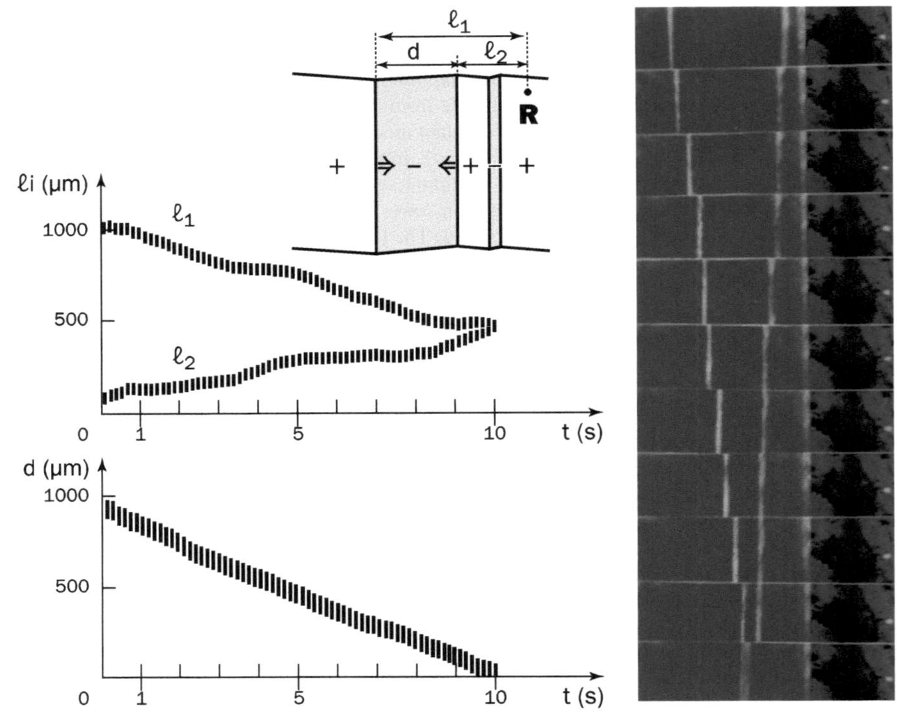

Fig. 3.11 Long-distance interaction between two domain walls in lateral movement towards each other. This ferro-ferro crystal includes a stable domain preserved throughout the experiment and two walls that move towards each other under the action of the applied electric field (KDP crystal, a_1 =10.2 mm, a_2 = 5.06 mm, c = 2.8 mm, T = 100 K, E = 1.7 kV cm^{-1}, relative wall speed 10^{-4} ms^{-1}) (Bornarel, 1987). A moving wall can be slowed down by defects, and its movement over time undergoes some blockages. However, the variation in distance between the two walls as a function of time remains regular, demonstrating their constant interaction at distances of several hundred micrometres. This result illustrates the reversal (switching) experiments with acoustic waves carried out on ferro-ferro such as KDP (Meeks et al., 1983; Navirian et al., 2010)

When two planar walls approach laterally under the action of an electric field or a stress, at distances less than the previous interaction distance, the displacement laws of each wall remain unchanged with regular displacements, as long as their relative distance is greater than another critical dimension, that of the stable domain width they will be able to constitute. If their distance is of this order, the walls repel each other with a force inverse to their distance, as illustrated in Fig. 3.12. The displacement of the walls becomes more and more difficult as observed in Tanane (Bornarel and Legrand, 1981), KDP and also Rochelle salt (Wieder and Collins, 1960; Styrkowiec, 1985). Models of moving walls that rely on the existence, either of parallel current lines following (001) or of dislocation lines parallel to this axis, allow to explain this interaction law. Following such interactions, thin and stable planar domains are created in sometimes significant characteristic times which will require strong external actions to be dissociated or removed.

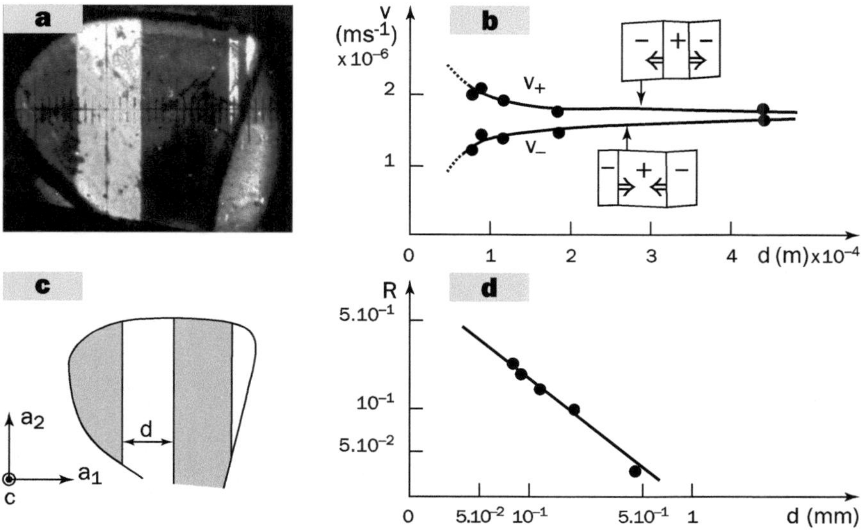

Fig. 3.12 Photograph (**a**) and schematic representation (**c**) of a texture in Tanane crystal. Two domain walls, depending on the direction of the applied continuous electric field, see their relative distance d, increase (speed v_+) or decrease (speed v_-). (**b**) Variation of the relative speed of the two walls as a function of their distance d in the two situations v_+ and v_-. (**d**) If by hypothesis $v_\pm = k(E - E_S) \pm F_{rep}$ with F_{rep} repulsion force between the walls, $F_{rep}/k(E - E_S) = R = (v_+ - v_-)/(v_+ + v_-)$ varies as the inverse of d (slope of the line in logarithmic scales equal to −1) ($T = 9$ °C, $E = 4.3$ kV cm^{-1}) (Bornarel and Legrand, 1981)

The two very different previous examples, one with a soft, deformable crystal at room temperature (Tanane) and the other with a rigid crystal at low temperature (KDP at 100°K), demonstrate the generality of the phenomena, with obviously critical distances functions, for the maximum interaction distance, the dimensions of the sample, and for the repulsion between the walls, the nature of the crystal and the experimental conditions.

3.3.2 Interactions Between Plane Domain Walls and Stable Domains

If plane domain walls can enter and leave the sample under the action of an electric field or a constraint, there can exist walls distant from each other and others close together which together constitute a stable domain. The action of an electric field or a constraint will first cause the lateral displacement of the isolated walls before observably modifying the stable domains. The interaction between moving walls and stable domains is a major process in a texture of parallel walls.

This process is illustrated in KDP in Fig. 3.13 with a moving wall attracted by the stable domain, destabilizing this domain at its two walls, one creating a new

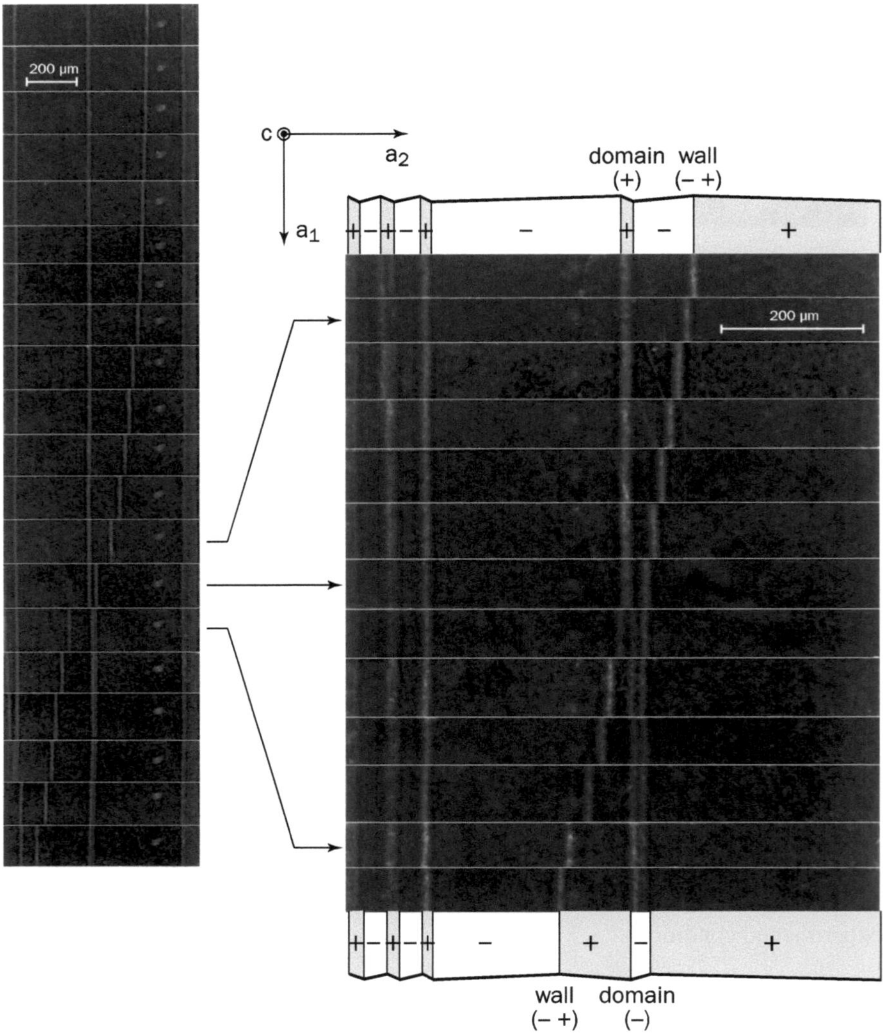

Fig. 3.13 Domain wall relay phenomenon during a reversal of the KDP crystal ($T = 90$ K, $t = 0.3$ mm, $E = +3$ kVcm^{-1}, photographs every 1/72 of a second): a domain wall moves from right to left, increasing the volume of the crystal of positive polarity. The speed of this −+ wall is almost constant as long as it is more than 200 μm away from the stable domain of + polarity. When it is closer, it is attracted, and its speed increases; the separation of the walls of the stable domain of + polarity, the constitution of a new stable domain of – polarity with the ejection of a new −+ wall, occur quickly; the new wall then resumes a speed comparable to that of the incident wall before the interaction, when the distance between this wall and the new stable domain is greater than a hundred micrometres (Bornarel, 1987)

stable domain with the incident wall, and the other ejected, resuming its dynamics (Bornarel, 1987). Figure 3.14 illustrates a succession of such relay phenomena, and Fig. 3.15 demonstrates how the dissociated walls of a stable domain can then move as isolated walls or contribute to relay phenomena. In all cases, the lateral displacements of the wall obviously change the macroscopic polarization (deformation) of the sample under the action of the applied electric field. This relay phenomenon between wall and stable domains in KDP, at low temperature, in this rather rigid crystal, can occur quite abruptly, a source of Barkhausen peaks and other avalanche manifestations (see Annex A8).

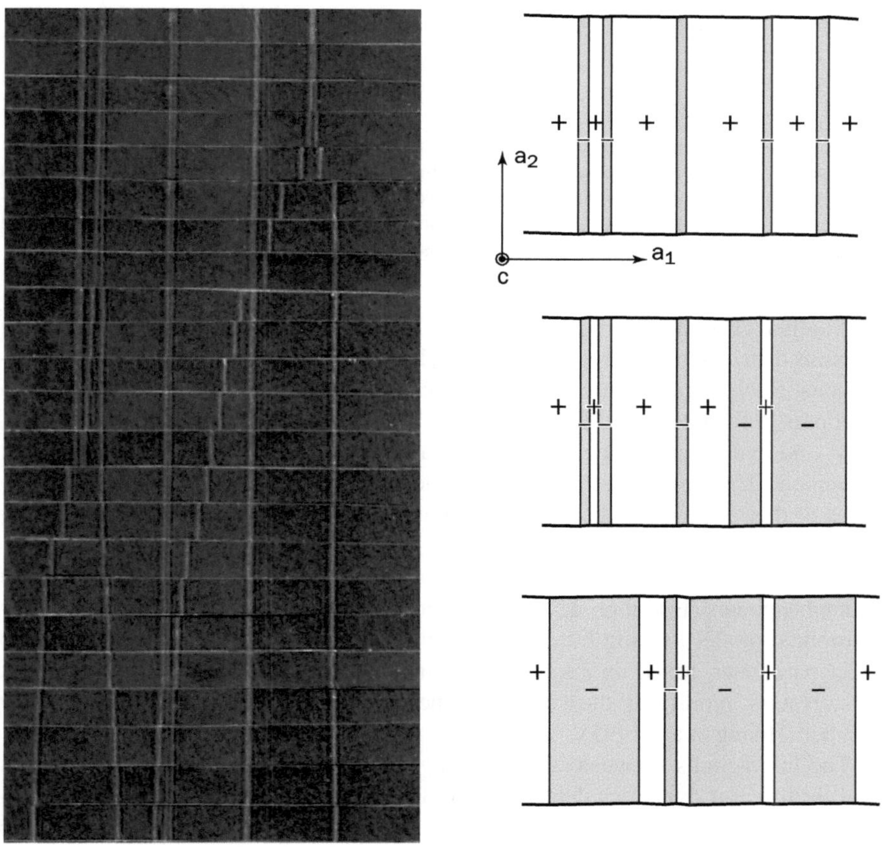

Fig. 3.14 The montage on the left of photographs taken every 1/72 of a second and the diagram on the right illustrate the phenomena observed during this polarization reversal of KDP (at $T = 90$ K): a stable domain to the right separates into its two walls, which then move by lateral movement. The lateral movement of the walls, a relay phenomenon and then interactions with the reconstitution of a stable domain are observed. The polarization switching is not complete, but this region of the crystal, which was globally positive at the start, has become globally negative in the last photographs

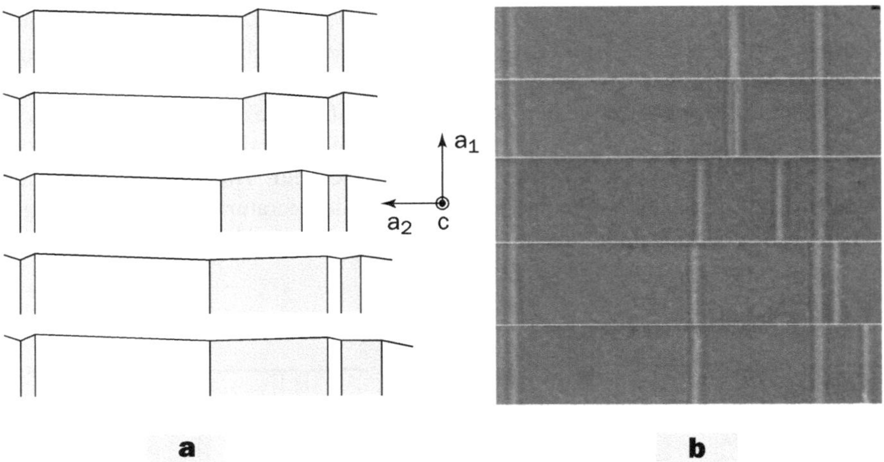

Fig. 3.15 (**a**) Diagram and (**b**) photographs of lateral displacements of walls in KDP. Initially, three stable needle domains exist. Under an applied electric field that favours their polarization (deformation), one of these needles widens; the two walls suddenly separate and then move laterally. The one on the left moves slowly; the other on the right interacts with an existing needle and creates a relay phenomenon with it, as already illustrated in Figs. 3.13 and 3.14

Another example, in Fig. 3.16 with a Tanane crystal, presents a set of stable domains distributed quite regularly and wall displacements with the same relay phenomena. Such situations are good examples for models of displacements between "stoppers" (Tagantsev and Fousek, 1999). The dynamics is different from that of KDP, and everything happens as if a polarity front swept the crystal in a regular movement. The regular distribution of stable domains and the different elastic properties of this "soft" crystal explain this change in dynamics compared to the situation of KDP.

This kind of situation is encountered in many ferroelastics such as rare earth pentaphosphates (Budin et al., 1975; Hu et al., 1994; Huang et al., 1995). Simple theoretical models, taking into account only the shear deformations related to the order parameter, based on a system of spins under constraint, are consistent with these results: modifying the volume of the domains is easier than changing their number (Huang et al., 1995).

The interactions between walls moving laterally and stable domains can thus lead to different dynamics depending on the crystals and the experimental conditions; these relay phenomena are at the basis of the change in macroscopic deformation-polarization.

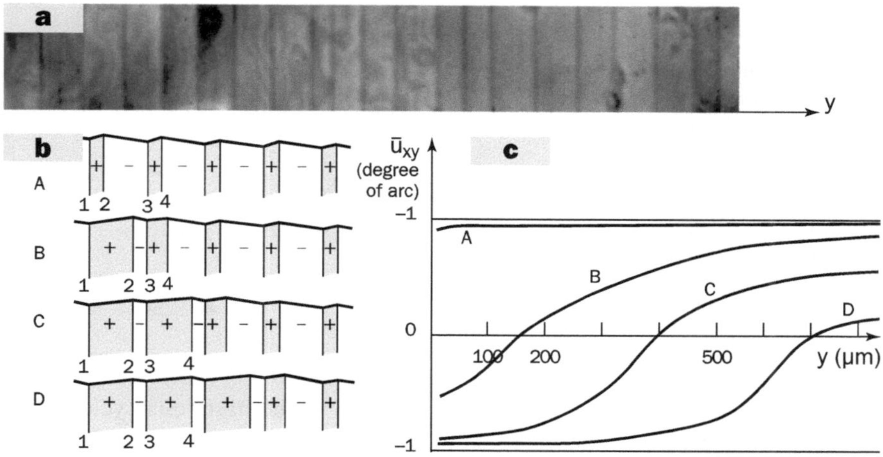

Fig. 3.16 (**a**) Photograph of a Tanane crystal with stable domains of width w_m distributed quite regularly at distances w_M of a few micrometres (T = 12 °C). (**b**) Schematic representation of the wall relay process under the action of a positive electric field: a wall in this regular set moves towards a stable domain at a nearly constant speed, creates the dissociation of the domain and the departure of the wall that is similar to it (for example + – for walls 2 and 4 in A and B), then, the new wall moves in turn to dissociate the next domain, and so on. We thus move from a sample of overall negative polarity – with small positive stable domains + to a sample of overall + polarity with small stable domains–. If the stable domains are regularly distributed in the sample, the moving wall is like in a medium with homogeneous average deformation. The moving walls are different over time but, everything happens as if a front of polarity change moved almost regularly in the sample as illustrated in Fig. **c**). (**c**) Value of the local deformation calculated on volumes of dimension 100 μm in the y direction (perpendicular to the wall planes) as a function of this coordinate, for the various situations of the reversal schematized in (**b**) (Bornarel and Legrand, 1981)

3.3.3 Macroscopic Responses and Microscopic Phenomena

If during macroscopic measurements, therefore of change in global deformation and polarization, we retain the hypothesis of simple lateral displacements of plane walls, how to detect these microscopic phenomena?

Relation (3.1) summarizes the friction process that limits the speed of a wall. The mobility and the critical field of this speed vary from one crystal to another, and for the same crystal depending on the experimental context. But, in the case of walls moving laterally, associating a friction with the displacement of the wall at the microscopic level is a solid hypothesis, experimentally verified. Thus, when the changes are not too rapid compared to the measurement methods, we can use the losses (the loss tangent during dielectric measurements) to detect the number of moving walls, as illustrated in Fig. 3.17.

If the measurement is more global as during a hysteresis cycle, we can detect the contributions of the number of walls in action, their speed, and variations in local properties (deformation, polarization, etc.) in the following Figs. 3.9 and 3.10.

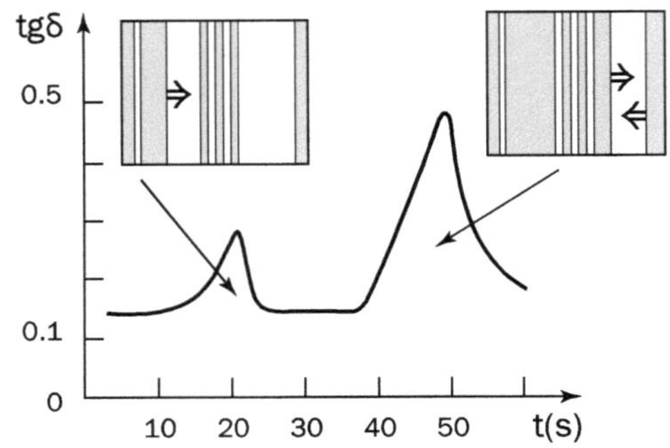

Fig. 3.17 The loss factor tgδ measured macroscopically increases when a wall moves laterally within the sample illustrating the friction process. In this schematic representation, the first peak corresponds to a moving wall and the second to two walls simultaneously in motion

Figure 3.18 allows to deepen the approach by a correlation between the local measurements at the microscopic level and the values of the hysteresis cycle: in this example of Tanane, the very square local cycles, integrated over the whole crystal, allow the drawing of a macroscopic cycle revealing the collective difficulty of the walls and deformations in the domains to complete the cycle towards the mono-domain.

These analyses are easy when the deformation modifications in a domain are limited which is the case in many measurements on ferro-elastics under stress. The problem is different during more complete cycles with creation of stable domains. Figure 3.19 underlines in a very amplified way the different local property (deformation, polarization, etc.) within the stable domain and in a domain of the same orientation of larger size. Figure 2.22 illustrated this phenomenon difficult to neglect when the number of walls moving laterally is notable in a sample. The following sections concerning the domains in needles will provide the complementary light on the creation and disappearance of these stable domains which will become domain needles.

3.4 Longitudinal Movements of Needle Domains

3.4.1 Creation of Needle Domains

The studies carried out during quasi-static modifications of polarization-deformation allow to measure the local properties in situations of quasi-equilibrium. When the dynamics is faster, the phenomena can become more complex, but the following results remain qualitatively relevant.

Needle domains appear in a crystal, either by creation of a nucleus on the edge of the mono-domain sample, or by detachment from the edge of the sample, from one end(tip) of a stable domain previously created by lateral displacement of two plane walls.

For the first case, Fig. 3.20 provides a schematic representation. The characteristic times of the various situations are different: time of creation and crossing of the

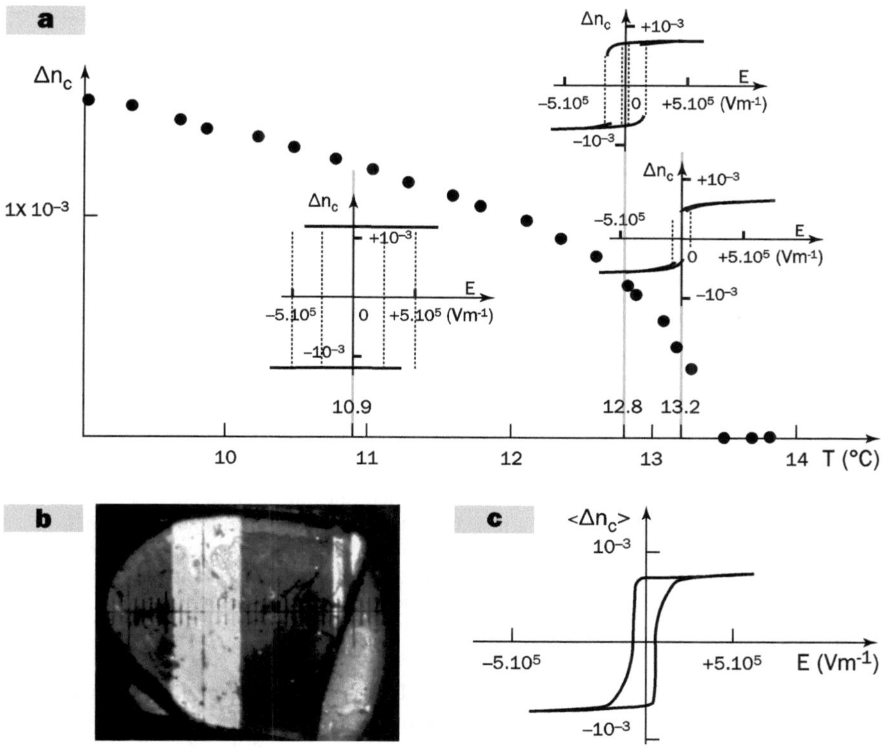

Fig. 3.18 (**a**) The birefringence Δn_c in Tanane is proportional to the deformation u_{xy} according to the relation $\Delta n_c = (4.9 \pm 0.1)\times 10^{-2}\, u_{xy}$. The measurement of birefringence is carried out in each domain of Fig. b), at each temperature, as a function of the field E. We can thus plot local hysteresis cycles and the curve of the values of Δn_c at saturation (Bastie et al., 1976a, b). (**b**) Typical domain texture used for these birefringence measurements Δn_c (horizontal dimension of the sample: 3 mm). (**c**) Local birefringence measurements allow us to plot hysteresis cycles such as those presented in (**a**). In addition, knowing the volumes of domains, it is possible to calculate a value proportional to a global deformation of the crystal: $\langle \Delta n_c \rangle = 1/S_t\, (\int_{S+} \Delta n_c\, dS - \int_{S-} \Delta n_c\, dS)$ and to plot a "deformation" macroscopic hysteresis cycle at each temperature: result at 12.8 °C (Legrand, 1981)

sample along the ferro axis, release of the end of the needle domain from the sample edge, movement in the longitudinal direction parallel to the permissible wall plane (axis a_2 in the example).

The second mode of creation of needle domains, illustrated in Fig. 3.21, occurs during the second part of a switching of polarization (deformation) and notably towards the end of the hysteresis cycle, before obtaining the mono-domain state: two walls, by approaching each other, have created a stable domain whose width can no longer decrease. The detachment of one sample edge of the domain suddenly creates a tip, which then moves in the longitudinal direction of the needle. The disappearance on the other sample edge usually occurs quite abruptly. As described in Chap. 2 and in Annexe A7, the tip of a needle domain is an assembly of

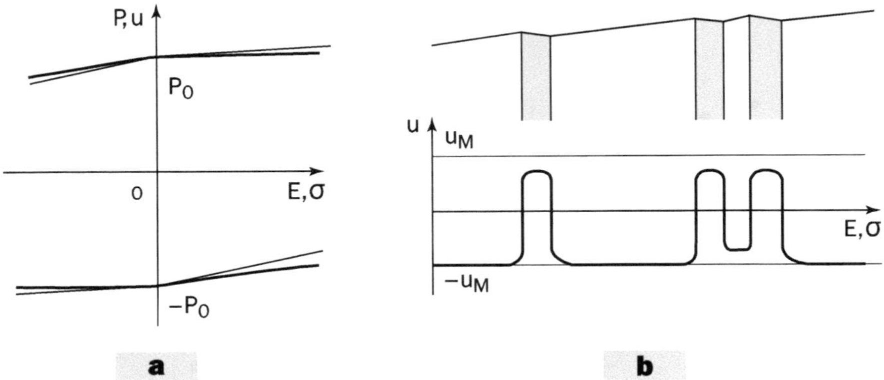

Fig. 3.19 Local deformation. (**a**) If the application of an electric field E (or a mechanical stress σ) changes the local polarization and deformation, we can account for it in first approximation by relations $\pm P_0 + \alpha E$ as the thin lines illustrate, with different values of α depending on the direction of the polarization and the field (greater if they are in opposite directions in our case: $\alpha_2 > \alpha_1$). The curves in thicker lines correspond to a possible saturation of the phenomenon when E (or σ) increases. (**b**) In this diagram, the maximum value of the deformation in a large domain is $\pm u_M$. It is hypothesized that this value is not reached if the dimension of the domain is too limited. The 20% drop represented is obviously exaggerated in most crystals, but barely in the case of Tanane (see Fig. 3.9)

quasi-dislocations of Burgers vectors parallel to the permissible wall planes. It can be considered at a distance of a few domain widths w, as equivalent to a wedge dislocation along the c axis, with the symmetry plane of the domain as the slip plane, and a Burgers vector parallel to this plane of module $B = w\, tg(u_+ + u_-)$, u_+ and u_- modules of the deformation in the domain and outside it around the tip.

3.4.2 *Law of Displacement of an Isolated Domain Tip*

We call an "isolated" or "free" domain tip an end of a needle domain whose movement is not modified by interactions with other tips, with a sample edge or a defect. The proximity of stable domains without a domain tip does not usually disturb its dynamics. Figure 3.22 presents simple displacement laws, linear with the duration of the pulses and the amplitude of the applied electric field.

This is not entirely accurate since in the presented case, the tip has a start-up time of a few nanoseconds, not detectable in the figure. These simple laws can be verified over significant distances, but if the pulse application continues long enough and over significant distances, a phenomenon of displacement saturation is observed for the majority of needle tips (Fig. 3.23). The tip's movement will become effective again with the application of an external field of higher amplitude. Thus, at the level of a domain tip, the same phenomenon is observed as that seen at the macroscopic level in Fig. 3.3. As the width of the domain has not varied in an observable way, we

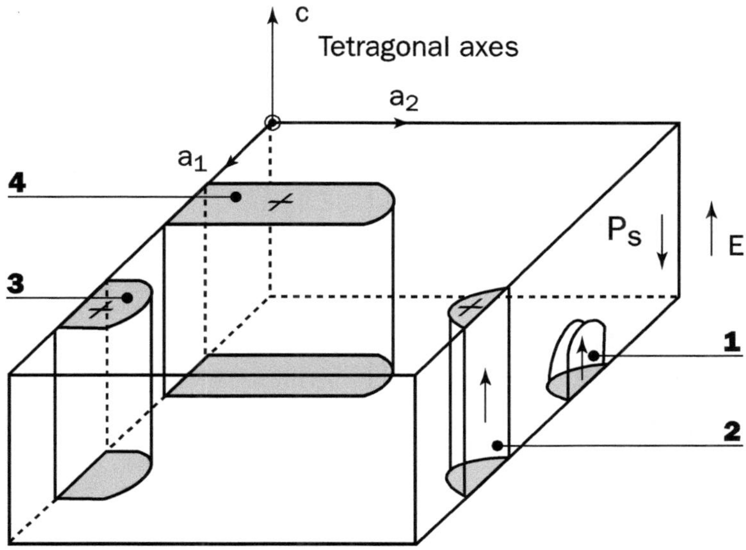

Fig. 3.20 Schematic creation of a needle domain: (1) creation of a nucleus on an sample edge; (2) set of knees and electric charges on the part perpendicular to the *c*-axis creates an excess of energy which tends to disappear by rapid crossing of the sample along the ferroelectric axis; (3) arrangement of the assembly of quasi-dislocations along the *c*-axis which defines the width of the domain; (4) longitudinal growth of the needle domain (observed in section (001)) along a_2

can suppose a modification of the arrangement of the tip's quasi-dislocations following a change, even minimal, of the local deformations. Thus, an electric field (or a shear stress) which was sufficient to start the movement before is no longer sufficient when the local deformations have changed; even if the KDP does not have the same elastic constants as the Tanane, we cannot forget the results of Fig. 3.9 concerning local deformations in the domains during the switching.

3.4.3 Modification of the Dislocation Assembly of a Domain Tip and Displacement Law

The modifications of the quasi-dislocation assembly according to the stresses, the situations, with relaxation times that are specific to it, have been calculated (see Fig. 2.29 (Salje and Ishibashi, 1996)). The photographs in Fig. 3.24 confirm these rearrangements of a needle domain tip with an electric field applied to the crystal which, either favours the growth of the needle, or tends to make it regress. The deformation field, consistent with those of wedge-type dislocations (Bullough, 1958), differs depending on the orientation of the applied field.

Thus during a tip movement, the arrangement of the quasi-dislocations is modified, due to a change, even small, of the matrix average deformations in which the

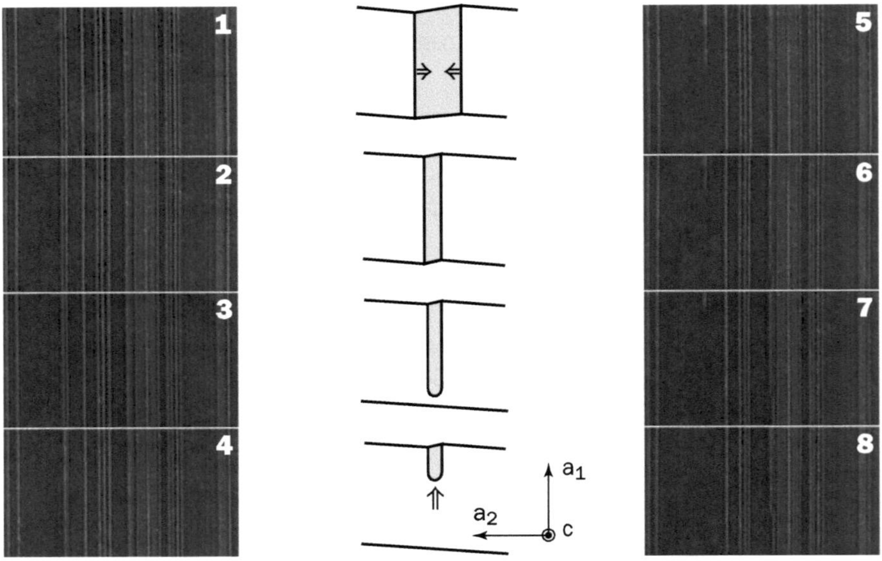

Fig. 3.21 Domains whose walls cross the entire sample in the directions c and a_1 (KDP sample: $a_1 = 10$ mm; $a_2 = 8$ mm; $c = 0.3$ mm; $T = 90$ K). The applied electric field $E = 3 \times 10^3$ V cm^{-1} tends to decrease the size of existing domains by lateral wall displacement along a_2. On the left side near a metastable domain, two walls approach between photographs 1 and 4, then this domain of minimum width detaches from the edge in 5, and then regresses along a_1 (schematic illustration in the centre) until it disappears in photograph 8

needle domain evolves with deformations inside it. These modifications obviously affect the critical field of start-up movement E_S of the domain tip and on its mobility.

To better understand the phenomena of start-up movement of a tip and the speed laws, results of moving tips in large regions of the crystal are presented in Figs. 3.25 and 3.26. They correspond to domain tips whose distances to other tips are greater than the thickness of the sample. The presence of nearby stable domain walls has no effect. The set of these observations can be summarized as follows:

- The speed law of a domain tip is of the form $v = k(E - E_S)$ over significant distances; the coefficient k does not depend on the value of the applied electric field and seems rather to depend on the width of the domain (Fig. 3.25).
- Each domain tip has its own characteristics, and, in the absence of interaction, describes its own hysteresis cycle under the action of a periodic excitation (as observed for another ferro-ferro, Rochelle Salt from 1952 by Marutake) (Mitsui and Furuichi, 1953, 1954).
- The critical field to move the domain tip E_S, changes after each interaction it undergoes, for example with a defect that modifies the arrangement of the assembly of quasi-dislocations (Fig. 3.26).

But what happens if several tips are closer, for example at distances less than the thickness of the sample? the answers are given in the next section.

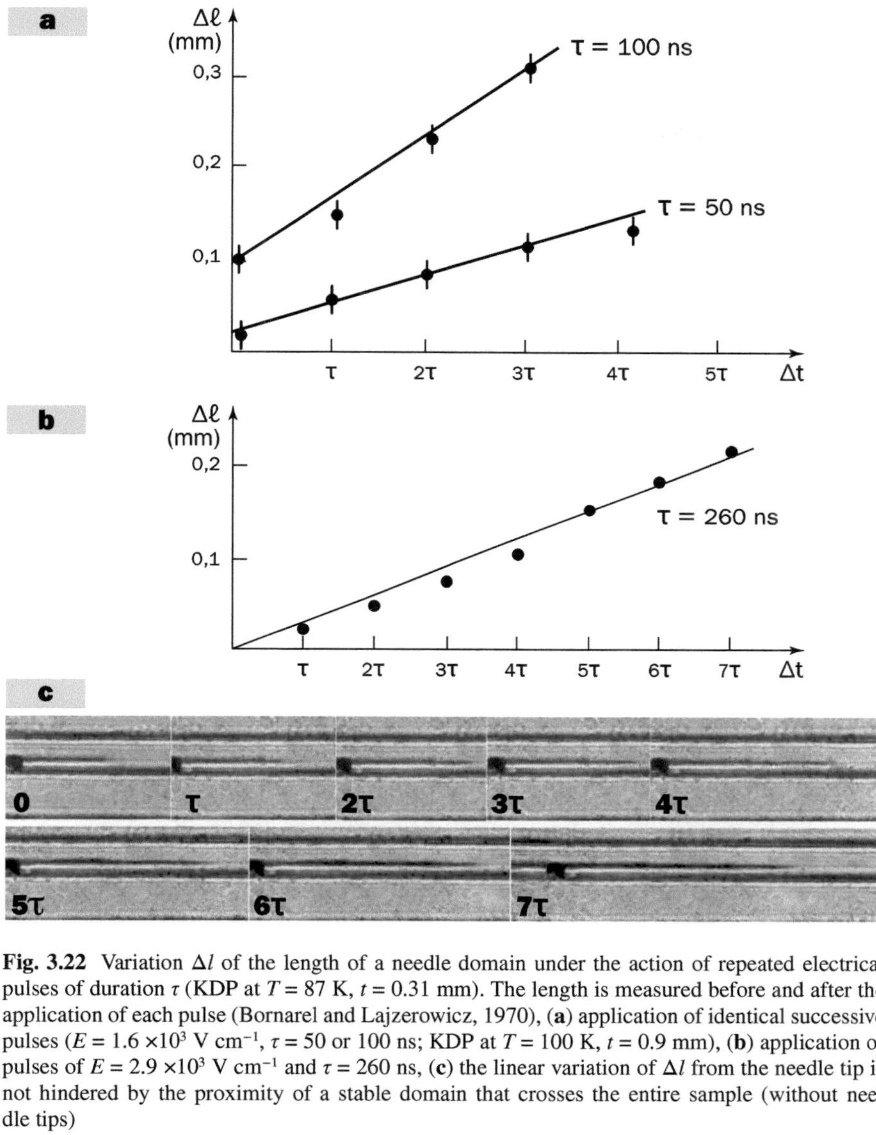

Fig. 3.22 Variation Δl of the length of a needle domain under the action of repeated electrical pulses of duration τ (KDP at $T = 87$ K, $t = 0.31$ mm). The length is measured before and after the application of each pulse (Bornarel and Lajzerowicz, 1970), (**a**) application of identical successive pulses ($E = 1.6 \times 10^3$ V cm^{-1}, $\tau = 50$ or 100 ns; KDP at $T = 100$ K, $t = 0.9$ mm), (**b**) application of pulses of $E = 2.9 \times 10^3$ V cm^{-1} and $\tau = 260$ ns, (**c**) the linear variation of Δl from the needle tip is not hindered by the proximity of a stable domain that crosses the entire sample (without needle tips)

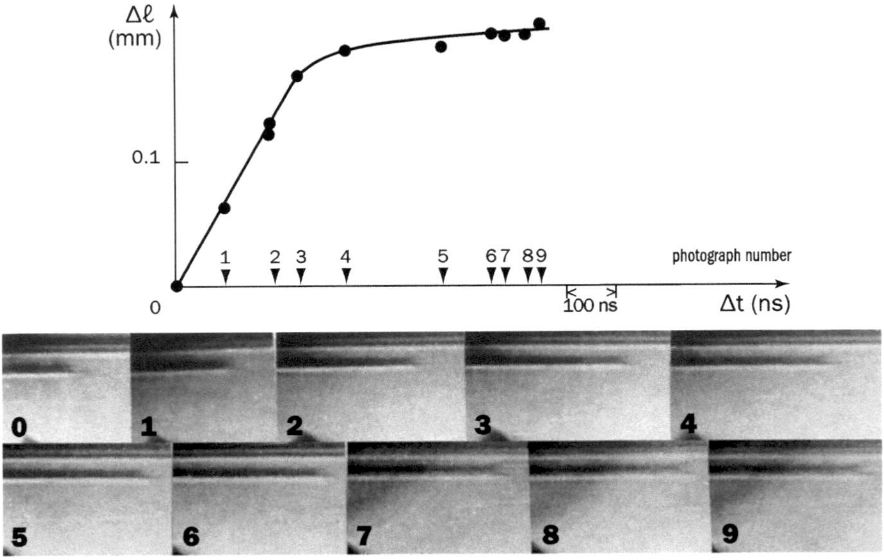

Fig. 3.23 Variation of the length of a needle domain under the action of electric pulses (the amplitude of the field remains the same $E = 2.7 \times 10^3$ V cm^{-1} but the duration of the pulses is modified): the linear variation is maintained over a certain distance (with a speed of 680 ms^{-1}); then the displacement decreases until it stops (KDP, $T = 89$ K and $t = 0.37$ mm) (Bornarel and Lajzerowicz, 1972). Only an increase in the field amplitude would allow a resumption of movement

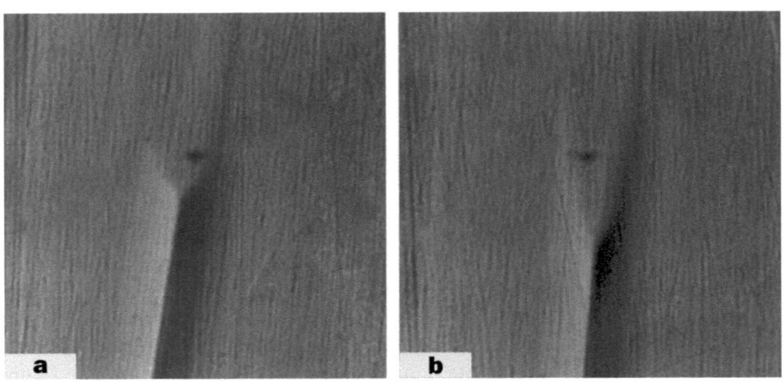

Fig. 3.24 Demonstration of the constraints at a needle domain tip in KDP ($T = 90$ K) in polarized light with crossed polarisers, located at 45° from the extinction positions, for light propagating along the c-axis (Bastie et al., 1976a, b): (**a**) the applied electric field favours the growth of the needle domain; (**b**) the applied electric field tends to make the needle domain regress

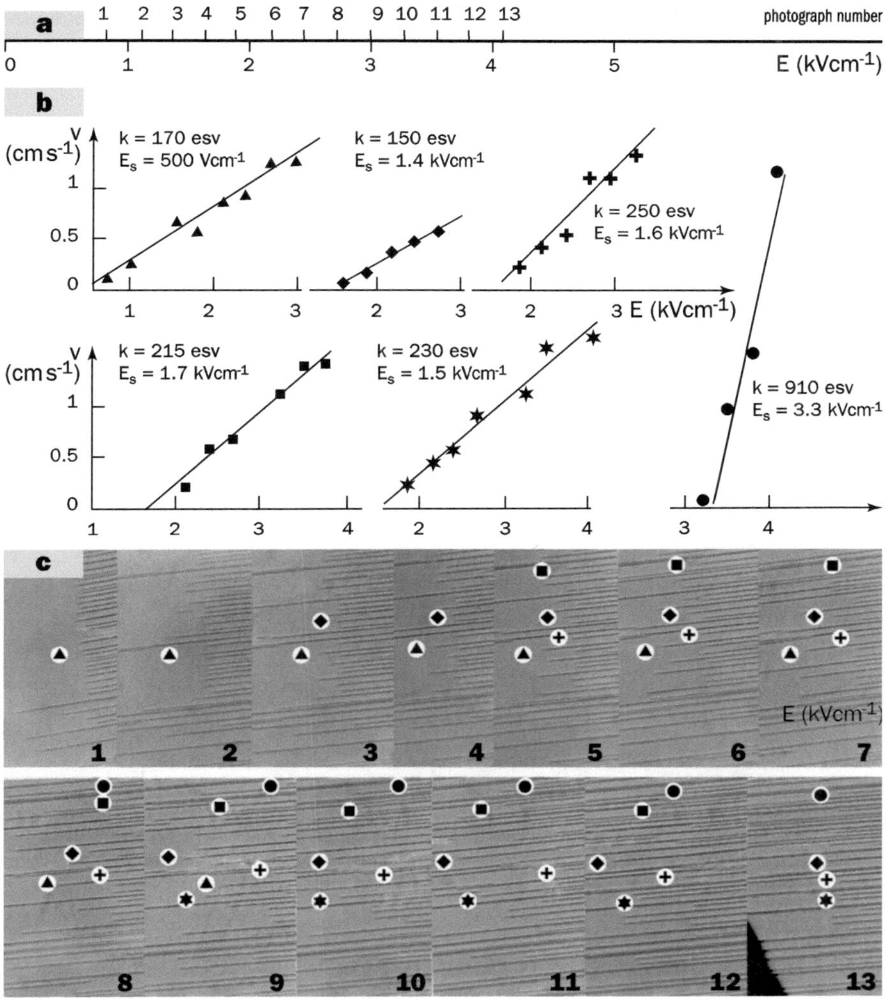

Fig. 3.25 The speed of the "free" domain tips is measured by the technique of applying successive pulses (KDP sample, $T = 89$ K, $t = 0.37$ mm, $\tau = 1.3$ ms and increasing E). Each tip follows a speed law of the form $v = k(E - E_S)$ with a certain distribution of k and E_S values (Bornarel, 1975)

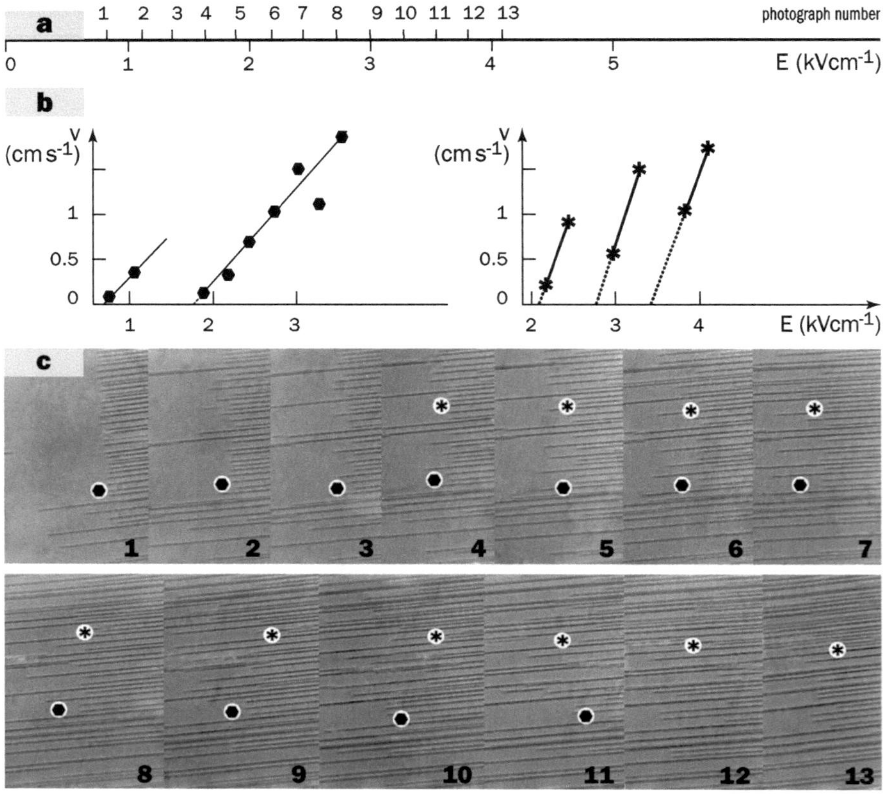

Fig. 3.26 During the same experiment as that illustrated in Fig. 3.25, the interaction of certain domain tips with defects changes the value of the critical field for setting in motion E_S but does not change the mobility k in the speed law $v = k(E - E_S)$ (Bornarel, 1975)

3.5 Interactions Between Domain Tips

3.5.1 Interactions Between Similar Domain Tips

The tip of a needle domain of width w, at a distance of a few domain widths, can be likened to a dislocation coin summarized by Burges vector $B = w\, tg\, (|u_+| + |u_-|)$. The most obvious illustration is given by the textures represented in Figs. 2.25, 2.26, and 3.27a: the energy of this set of domain tips is minimal when the assemblies of dislocations of the tips are all in the same plane perpendicular to the slip planes, therefore to the wall planes (Annexe A7).

The interaction energy between a summarized tip dislocation and another tip A is recalled as a function of their relative position in x (Fig. 3.27c): when a domain tip leaves the front, it attracts its two nearest neighbours, then repels them by

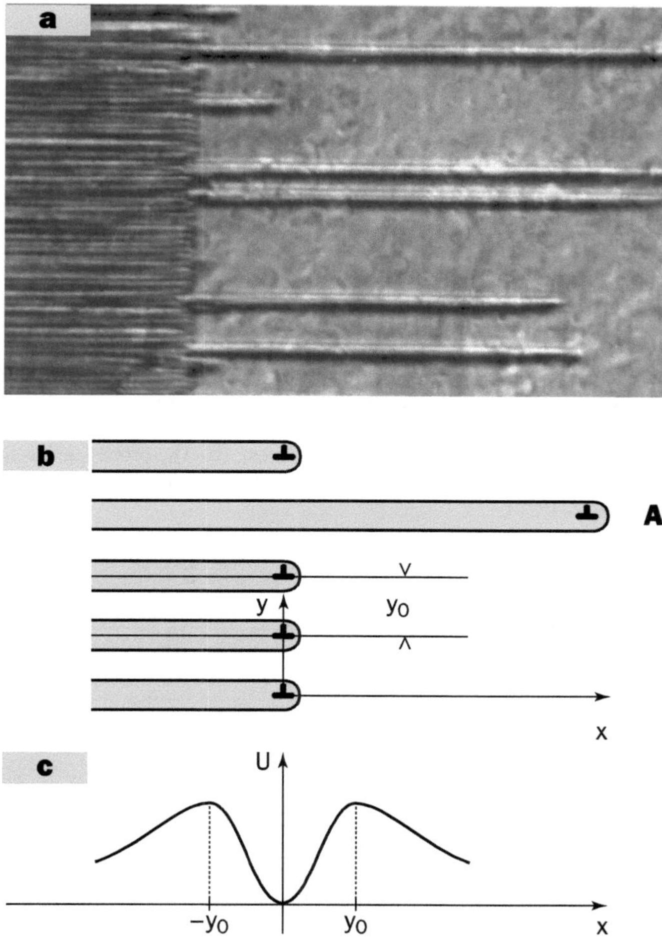

Fig. 3.27 As shown in Chap. 2, the tips of needle-like domains in quasi-static switchings can be summarized by edge dislocations with a Burgers vector B (proportional to the width w of the domain) with slip planes in the plane of symmetry of the domain. Its interaction energy U with another domain tip is represented as a function of the abscissa of the latter in c). This classic result of dislocation theory explains the observed end fronts: the stresses created by the tip front on one of them are null at $x = 0$, and equilibrium is obtained with a front perpendicular to the slip plane (Annexe A7)

attracting the next ones as illustrated in Fig. 3.28a (Bornarel and Lajzerowicz, 1972). Quite regular textures like that of Fig. 3.28b are thus created. One can observe "waves" of tips during the reversals of the polarization (deformation) when enough needles are created on the same edge with good simultaneity, which occurs in good crystals after several hysteresis cycles. This type of interaction between identical needles with parallel slip planes is observed even through other stable

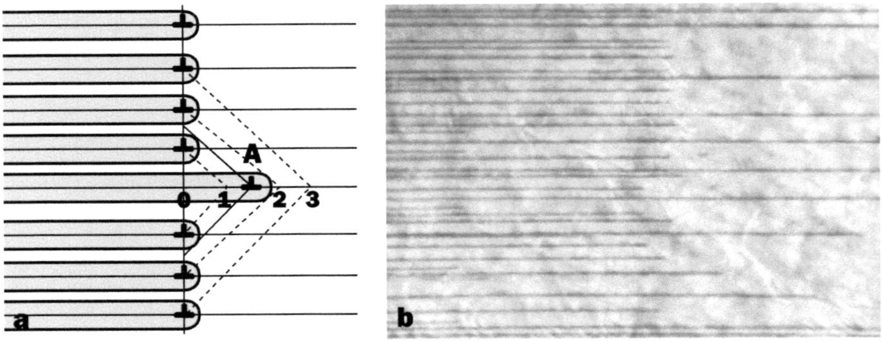

Fig. 3.28 (**a**) The interaction energy between domain tips recalled in Fig. 3.27c explains that a domain tip that comes out of the front attracts the two nearest tips if it is between 0 and 1. If the domain tip is between positions 1 and 2, it repels the nearby tips but still attracts the other ends, then, between positions 2 and 3, repels the 4 nearest tips, etc. (**b**) The process gives simplified structures with successive waves of domain tip fronts. Stable domains without tip play no role (Bornarel and Lajzerowicz, 1972)

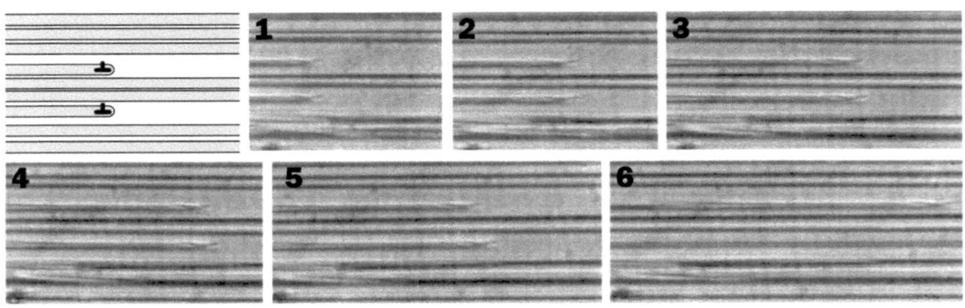

Fig. 3.29 Interaction through the walls of stable domains, of two moving domain tips: the two domain tips remain during their movement in the plane perpendicular to their slip plane (Bornarel and Lajzerowicz, 1972)

walls and domains as illustrated by the series of photographs in Fig. 3.29. In the same way, two domain tips (summarized by similar Burgers vectors) whose displacement speeds are not exactly identical will, when they are in the same plane perpendicular, interact for a certain period of time following the same displacement law (Fig. 3.30).

It can be noted that "domain waves" are observed in other contexts for ferroelastics, for example, in the formation of domain structures by laser irradiation in $LiNbO^3$ crystals (Mingaliev et al., 2010).

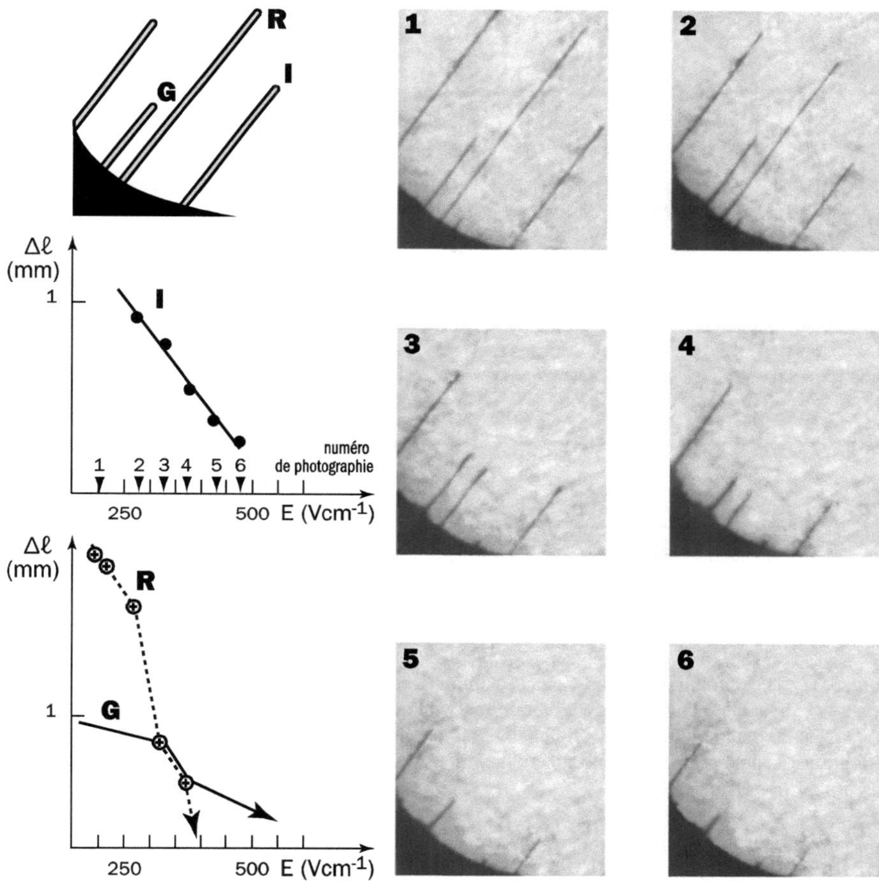

Fig. 3.30 During a quasi-static hysteresis cycle, needle-like domains decrease in length at different speeds. A domain tip I, far from any other end, follows a regular displacement. The tips *G* and *R* distant by <500 μm move at a speed that is their own, but when they are in the same plane perpendicular to their wall plane, they take the same speed (photographs 3 and 4), and after this interaction, regain their initial speeds (Bornarel and Lajzerowicz, 1972)

3.5.2 Interactions Between Domain Tips During Switching

When two tips of opposite Burgers vectors intersect, the law of wedge dislocations works: the tips will still tend to block in the situations of minimum interaction energy which correspond this time to situations 3 and 4 of Fig. 3.31, i.e. with tips located in planes at 45° from the slip planes (Friedel, 1964).

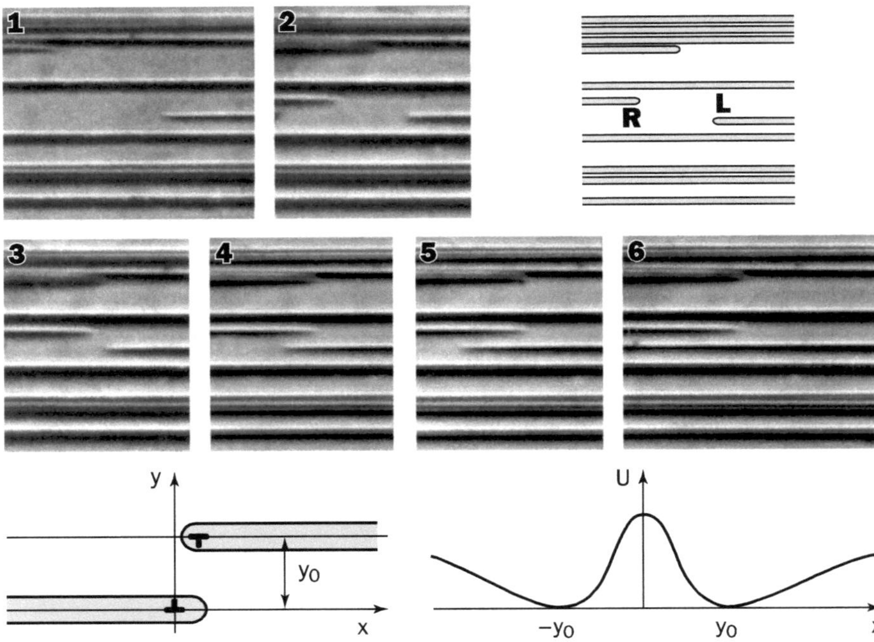

Fig. 3.31 The domain tips *R* and *L* can be summarized by edge dislocations with opposite Burgers vectors. Their interaction energy *U* is minimal when the plane containing them is at 45° (Annex A7) from their slip planes (the wall planes) as illustrated by the blocking during their movement in situations 3 and 4 (Bornarel and Lajzerowicz, 1972) and that they create a dislocation dipole with its properties (Krupa, 1966)

Interactions between all the tips of needle domains exist when they are close enough, and the phenomenon is observable at distances of the order of the thickness of the sample following c. But, in addition, a sample is necessarily limited by its edges. Each sample edge is a plane of zero constraint. Thus, a domain tip whose slip plane is perpendicular to the sample edge will be attracted by its image (mirror dislocation in symmetrical position with respect to the edge plane and with opposite Burgers vector). The action of these image forces leads to brutal attractions of the moving tips towards the sample edge as illustrated in Fig. 3.32.

The set of these domain tip interactions occurs systematically in every ferro-ferro crystal. Of course, defects play a role. If a domain nucleus remains blocked on

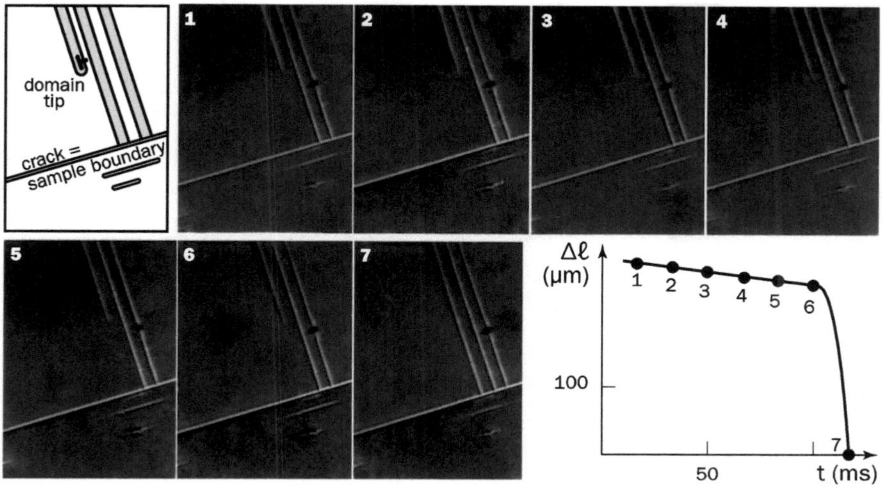

Fig. 3.32 A sample edge in KDP is perfectly in a tetragonal plane following the treatment explained in Fig. 2.34. This clean break plane is zero-stress plane. We observe in photographs from 1 to 5 a domain tip that moves very regularly towards the edge in the perpendicular tetragonal plane ($v = 0.13$ cm s^{-1}; $T = 90$ K, $t = 0.3$ mm). When the domain tip is 250 μm from the break, it is suddenly attracted. The domain tip can be summarized by a line of edge dislocation (perpendicular to the photograph) attracted by its image dislocation (relative to the zero stress plane). The range of the interaction is comparable to the thickness of the sample (Bornarel and Lajzerowicz, 1972)

a localized defect, it will interact with the moving domain tips. It can play a role of relay that accelerates the movement of certain tips, as in Fig. 3.33, or, on the contrary, induce a delay phenomenon, as in Figs. 3.34 and 3.35. Such a nucleus can be the origin of complex interactions between several domain tips, as in Fig. 3.36.

The results presented in this section correspond to quasi-static experiments. The dynamics of reversals can be faster and the dynamic phenomena more complex; however, the stroboscopic observations confirm the existence of these phenomena in all experimental situations up to frequencies of about 10 kHz, therefore for all classic dielectric measurements.

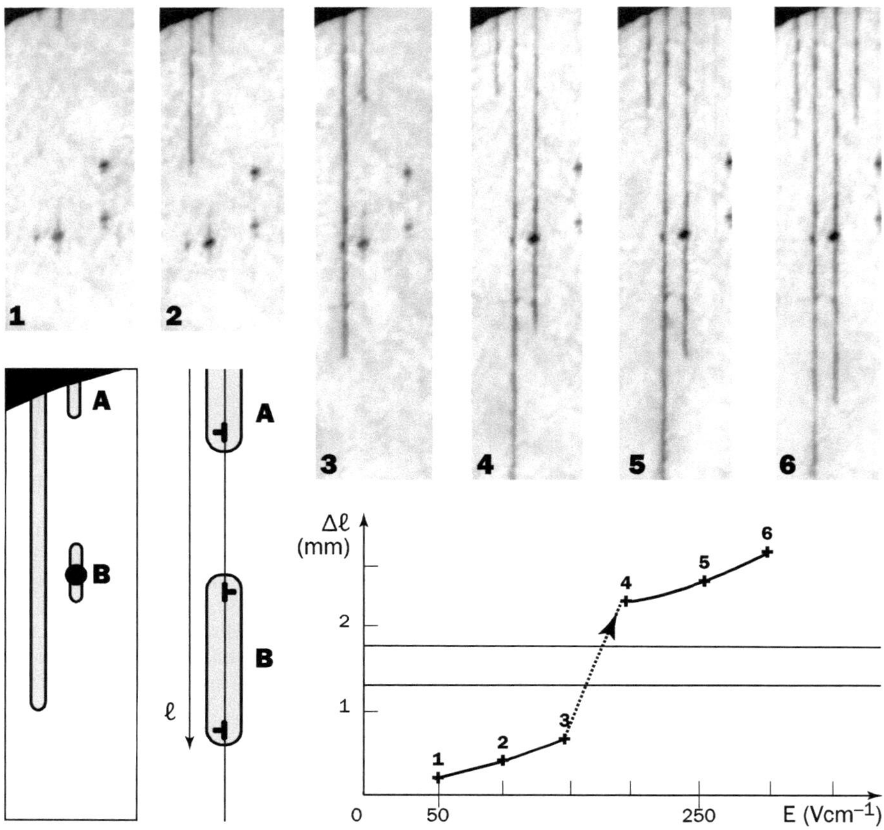

Fig. 3.33 The speed of the domain tip *A* during this quasi-static switching is regular (photographs 1, 2, 3) except when it interacts with the micro-domain *B*, which serves as a relay in its progression. It then resumes a speed comparable to the initial speed (KDP, $T = 90$ K)

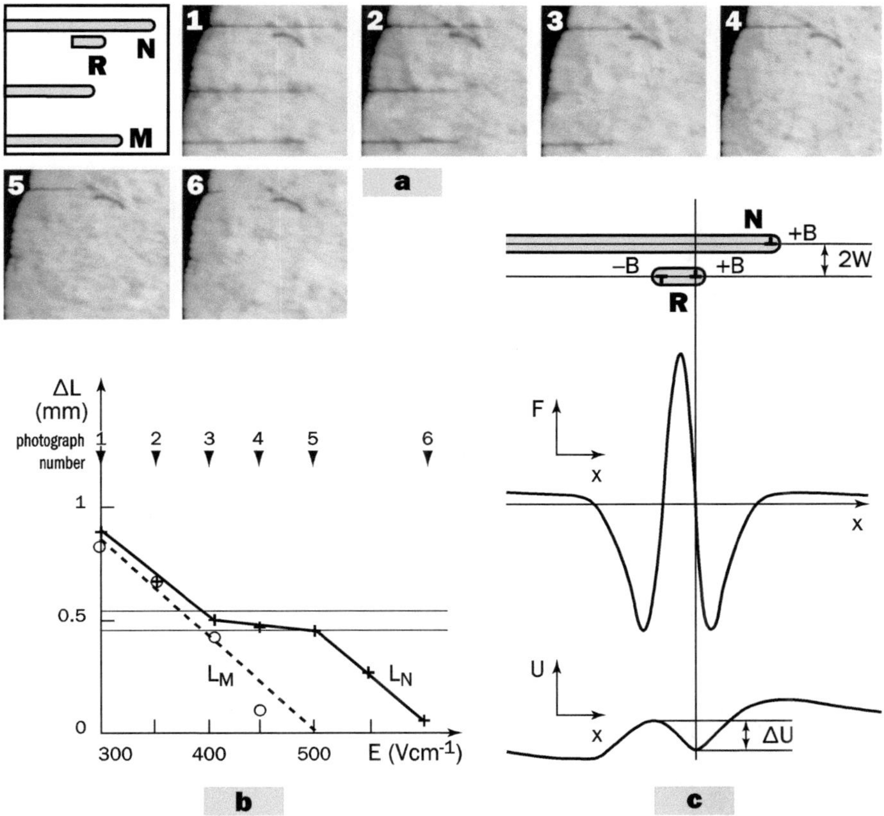

Fig. 3.34 During a hysteresis cycle in KDP, (**a**) a needle domain tip M follows the same displacement law as another domain tip *N*. But the latter interacts with a domain nucleus *R*; this interaction creates a delay for *N* as can be seen in graph (**b**). (**c**) Shear stress created by *R* on *N* and interaction energy

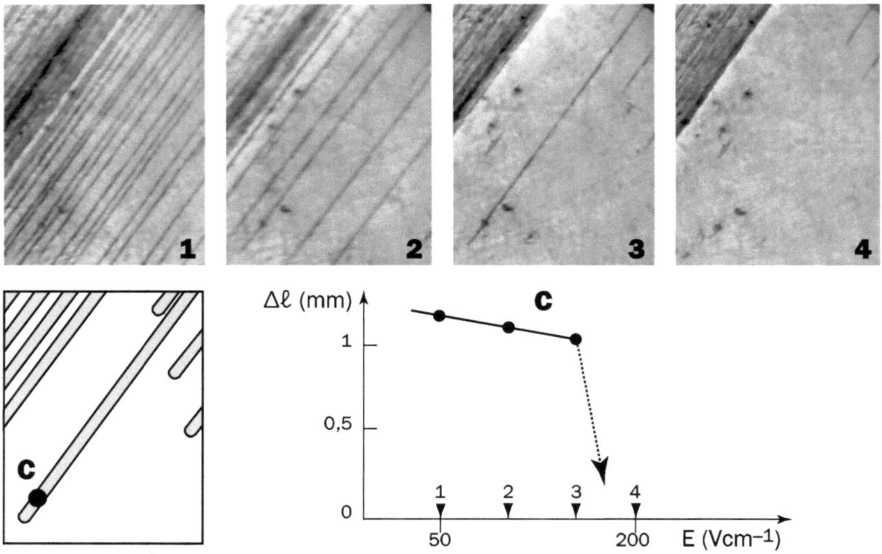

Fig. 3.35 During the withdrawal of needle domains, a defect blocks one domain tip which then detaches, leaving a small residual domain (Bornarel and Lajzerowicz, 1972)

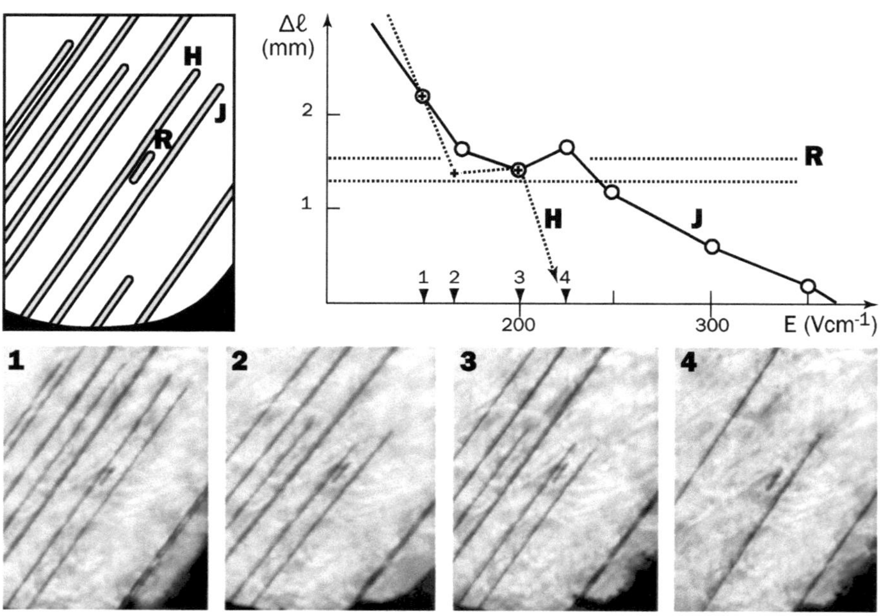

Fig. 3.36 Different and almost simultaneous interactions during a polarization reversal: interaction between the domain tip *H* and the latent nucleus *R* on the one hand, and interaction between the tips *H* and *J* on the other. The breakdown of interaction between *H* and *R* leads to a repulsion for *J*

3.6 Reversal of Polarization (Deformation) with Needle Domains

The previous sections have shown laws of lateral displacements of domain walls and longitudinal variations of needle domains in a ferro-ferro sample. In both cases, the proposed examples involved, one or several walls, one or several domains. The often quasi-static experimental conditions facilitated observation and stress relaxation after a change. But the usual macroscopic measurements integrate more complex phenomena. The objective of this section is to rely on the phenomena already understood at the microscopic level to make the link with the usual macroscopic results (switching currents, hysteresis cycles). KDP will still be the reference crystal as it encompasses the majority of situations encountered in ferroelectrics and ferro-ferro.

3.6.1 General Phenomena

Some phenomena are found in all these materials. For example, the effect of temperature compared to the transition temperature determine intervals with different properties as illustrated by the model in Fig. 3.37: wide walls near the transition favourable to statistical models and thermodynamic descriptions, narrow walls at low temperature which to move require the creation of singularities, such as in KDP at the atomic level the "Slater and Takagi defects" (Schmidt et al., 2000), and at the microscopic level dislocations.

Similarly, in all situations, moving certain singularities requires specific times: to create a Takagi ion, to create a domain nucleus, to set in motion a needle domain tip, and to modify a set of domains. Figure 3.38 illustrates how the minimum application time τ^* of an external field for the movement of domain tips varies, obviously with temperature, field value, but also by a factor of 10 between the various domain tips of the crystal. Considering the previous results (Fig. 3.24), we can correlate this time τ^* between 10ns and the microsecond to the rearrangement of assemblies of quasi-dislocations of tips depending on domain widths and tip interactions. These property distributions are usually overlooked in the macroscopic result analysis.

Speed laws offer another example of choice: the inverses of ferroic switching times vary linearly under certain experimental conditions with the excitation (as in this section in Fig. 3.46c for KDP). As the speeds of domain tip movements also follow linear laws, one could conclude the similarity of a demonstrated friction phenomenon. Figure 3.39 shows how much the reality of the phenomena would be neglected: the domain tips that move at a given moment (and which may be immobile the next moment) interact and drive each other; the maximum of their speed distribution V_M at time t grows exponentially with the amplitude of the excitation.

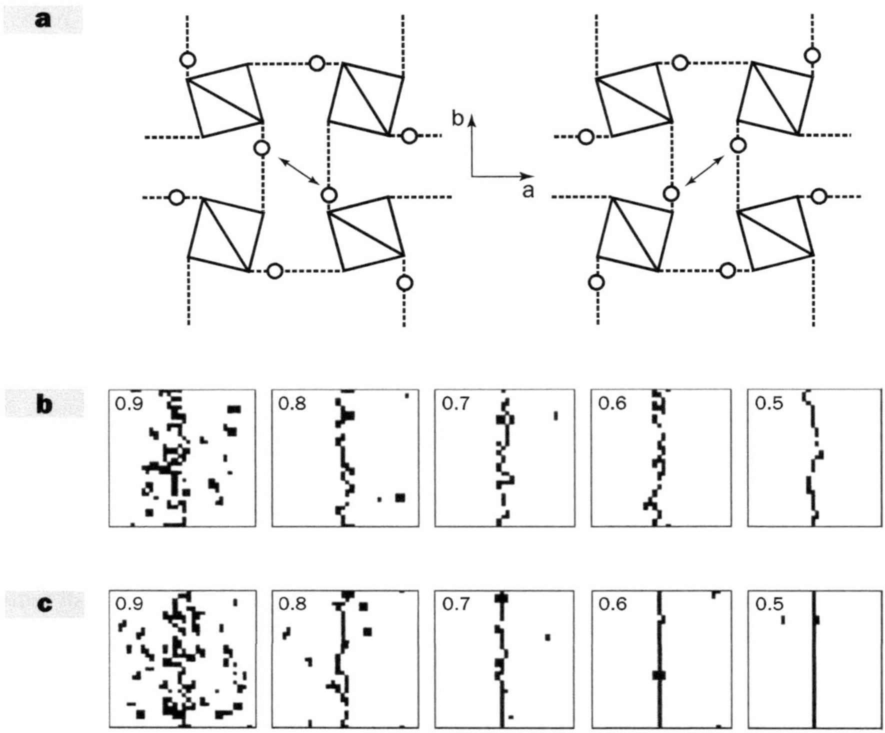

Fig. 3.37 (**a**) Projection representation in a (001) plane of the PO_4^- tetrahedra in the KDP crystal (the *P* are spaced *c*/4 along *c*) with two close hydrogens either in the upper position or in the lower position. The arrows indicate the interactions between close hydrogen bonds, approximately parallel, located at different ordinates along the c axis (their distance is less than the length of a unit cell as can be seen in Fig. A4.3 of Annex A4) (Ishibashi et al., 1972; Fukami et al., 1985). (**b**, **c**) Using a Monte Carlo simulation and following the approach of Sinitski and Schmidt (1996), possible distribution of Slater and Takagi groups (Fig. 1.32) in a domain wall without an applied electric field on the crystal (**b**), and with a small electric field E applied along the ferroelectric axis (**c**). The interaction energy of Ishibashi introduced in (**a**) is taken into account. The indicated temperature scale takes the transition temperature as a unit

In the proximity of the transition, the defects of Slater and Takagi form like clouds of a rather wide wall (Fig. 2.11) whereas at low temperature, and especially if the anisotropy is increased by a small applied electric field, their singularity argues for the quasi-dislocation model

These results concern tips in the middle of the sample and demonstrate their collective interactions, a phenomenon in proximity with those of avalanches. But the macroscopic results that take into account all the domain tips, isolated, in front, interacting with the sample edges, give results proportional to the excitation field: therefore, other phenomena will create enough losses so that the whole presents itself in fine as a simple friction process.

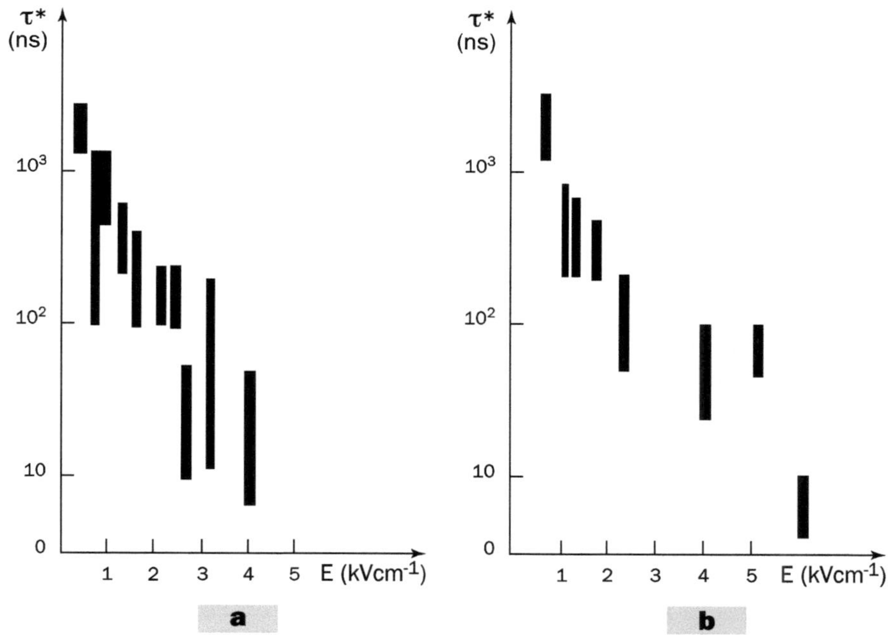

Fig. 3.38 The application of electric field pulses (E, τ) on a KDP sample (**a**): $T = 89$ K, $t = 0.37$ mm and (**b**): $T = 95$ K, $t = 0.29$ mm creates a displacement of domain tips when the duration of the pulses is greater than a critical value τ^*. The distribution law of τ^* on all the domain tips of a sample demonstrates an exponential decrease as a function of E (results obtained in the experimental contexts described Figs. 3.25 and 3.26), (Bornarel, 1975)

3.6.2 Detecting Microscopic Phenomena

The phenomena at the microscopic level that integrate the macroscopic results cannot be guessed if they are unknown. In this sense, the theoretical hysteresis cycle of Fig. 3.40 only highlights phenomena that occur during the tracing of such a polarization-deformation switching. This simplistic model is experimentally extended by Fig. 3.41 by linking the hysteresis cycle under sinusoidal field and the domain texture in the case of a single needle domain direction. Starting with a mono-domain sample of positive polarity, the applied electric field E is decreased; a few nuclei of negative polarity are created on the faces of the sample, sometimes facilitated by a collective phenomenon (Bratkovsky and Levanyuk, 2000); these domains grow longitudinally by freeing themselves from image forces due to the existence of the sample edge (1 in Fig. 3.41a). The movements of domain tips may seem disordered to the observer as the field values for moving the tips, their own speeds, and interactions between tips come into play. In the rest of the cycle, the needle texture densifies and the domains cross longitudinally the entire sample (2 then 3 in Fig. 3.41a); it is in 3, in situation X of Fig. 3.40, close to the coercive field value on the cycle of Fig. 3.41b, that the phenomenon of lateral wall displacement

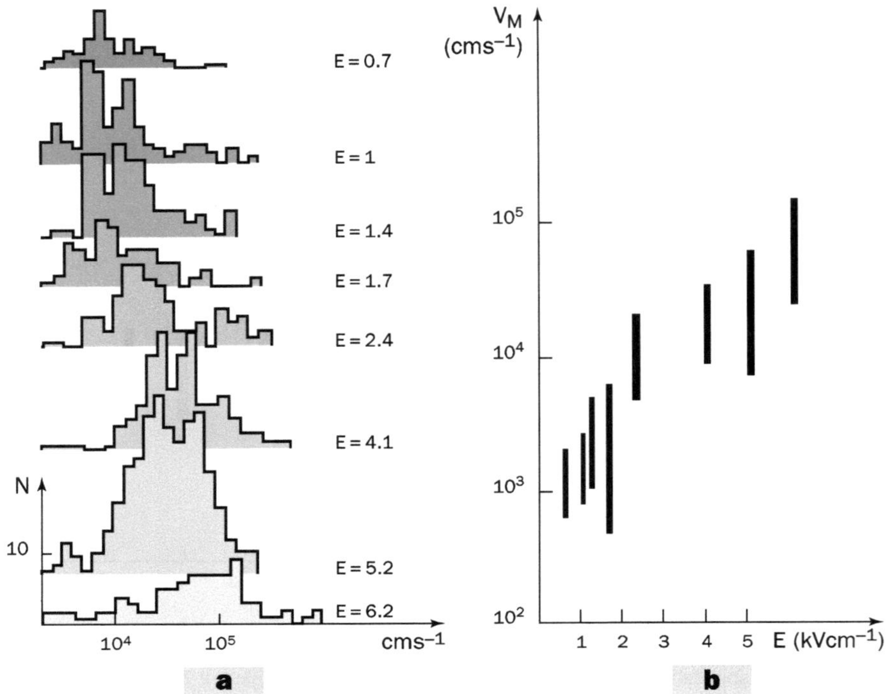

Fig. 3.39 Measurement of speeds by the pulse (square wave) technique carried out in a KDP sample; all domain tip motions were considered (sample at $T = 95$K, $t = 0.29$ mm, with $\tau = 1300$ ns) (Bornarel, 1975), (**a**) distribution law of speeds (in kV cm^{-1}) as a function of the applied electric field, (**b**) V_M is the maximum of the speed distribution; the bars correspond to the widths of the distributions at half-height: V_M grows almost exponentially with E

occurs which transforms the regular and finely laminated texture of a set of needle domains of polarity− in a matrix of polarity+, into a set of needle domains of polarity+ in a matrix of polarity− (such exchanges between domains of opposite polarities were observed for the first time in Rochelle Salt and correlated to the production of Barkhausen pulses (Annexe A8).

The rearrangements correspond to extremely fast and very small amplitude lateral displacements: we are in the situation of a wall attracted by a significant set of domains as illustrated by the interactions of lateral wall movements in Figs. 3.14 and 3.15. In the rest of the cycle, domain tips are again created, detach from the sample edges by freeing themselves from image forces, progress by interacting with each other with their own laws, then disappear, attracted by the opposite sample edge. After a few field cycles, nucleations on the sample edges reproduce at the same place, and it is possible to spot as on the photographs of Fig. 3.41 conservations of walls (slightly translated laterally) between the needles of polarity—before the passage to the coercive field and the needles of polarity + after. These memory effects have been observed in ferroelastic crystals. For $PbTiO_3$, a model with defect

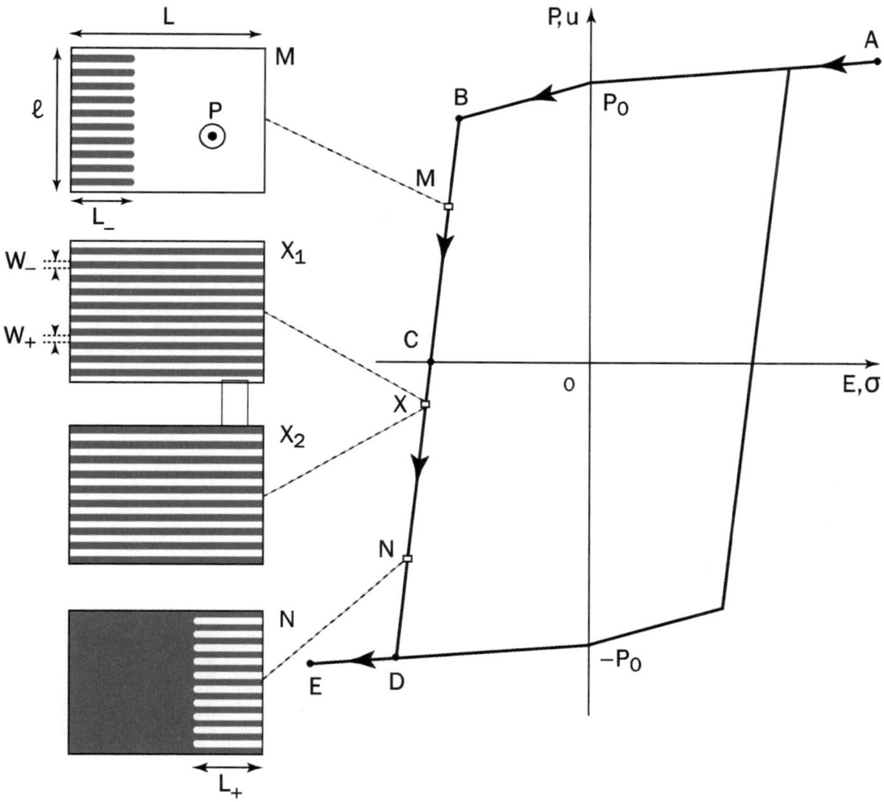

Fig. 3.40 Theoretical hysteresis cycle for a simple context of needle texture, with domain tips in front; the local polarization is assumed to be of the form $\pm P_0 + \alpha_i E$, with α_i larger if P and E are of opposite sense (α_2) than of the same sense (α_1), and a speed V of displacement of the domain tip front assumed constant. The situations of a set of needles of one polarity within the predominantly globally polarized sample (in M and in N) are schematized. The distance between two walls in the positive matrix between points B and X of the cycle is $w_+ = (1 - N\, w_-)\,(N + 1)^{-1}$, N is the number of needles

In this switching from A to E with domains present in the sample only between B and D, it is assumed that at X, two walls eject laterally on the edges of the sample. The global polarization of the crystal can be written between B and X

$P = P_0 + \alpha_2\ E - (L_-/L).\ (2P_0 + E\,(\alpha_2 - \alpha_1)).\ (Nw_-/l)$ (3.3)

Depending on the values of $\alpha_i\, E$, of P_0, and of w_j, situation X can occur above C (P positive), at C, or as in the figure, below C (P negative). In all situations, except in the hypothetical case with $E\,(\alpha_2 - \alpha_1) = -2P_0$, a lateral displacement of the walls Δw occurs at X:

$\Delta w = -\, w_+\ (P_0 + \alpha_2\ E).\ ((2N - 1)\, P_0 + E\,((N - 1)\,\alpha_2 - \alpha_1))^{-1}$ (3.4)

In a purely geometric context, Δw takes values of about 10 Å. These lateral displacements may seem to have little effect on the plotting of hysteresis cycles, but they will become major for susceptibility measurements.

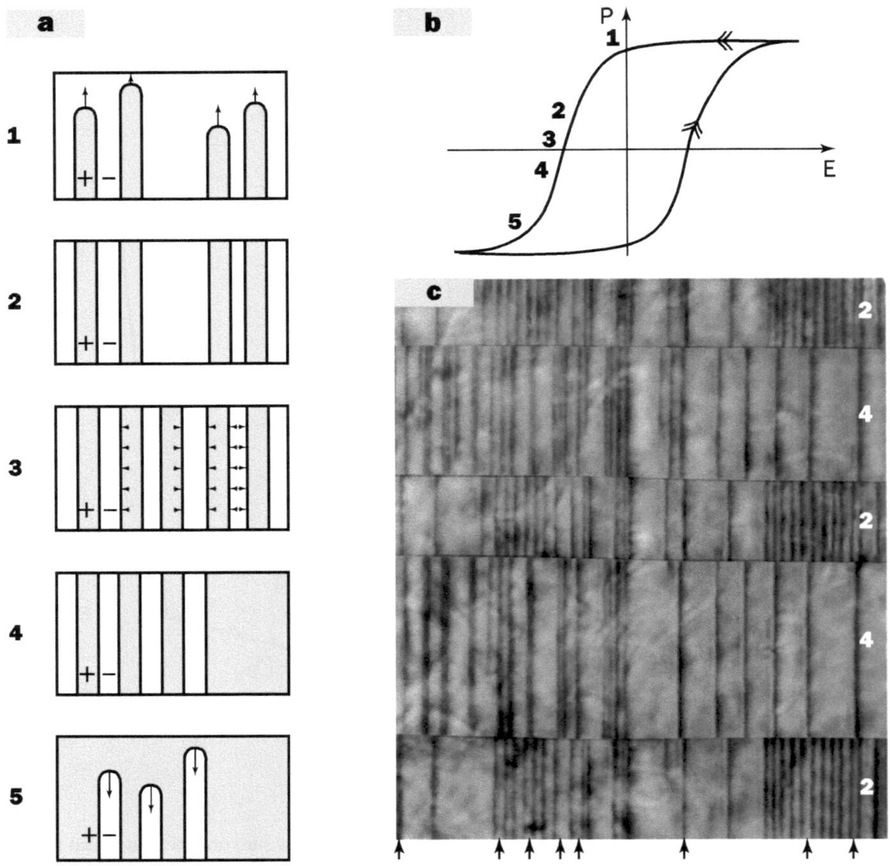

Fig. 3.41 (**a**) Schematic representation of a switching cycle of polarization-deformation with needle domains following a single wall direction. Domain tips are created in situations 1 and 5. (**b**) As illustrated in 3 of (**a**), very small lateral displacements occur abruptly in a situation close to the coercive field and this, the more the $P(E)$ curve approaches the vertical (close to the X situations of Fig. 3.40). (**c**) Assembly of texture photographs in domain sections (001) in situations 2 and 4 of a): −+ needle domain walls before the passage to the coercive field become +− needle domain walls after the passage

migration, interactions and cluster creation in the wall has been proposed to explain the distributions detected by atomic force microscope and the short-range repulsions between the walls themselves (Lee et al., 2005).

The variations in macroscopic properties with the applied field (or stress) result from the accumulation of the previously described phenomena. For example, Fig. 3.42 qualitatively shows the difference for image forces between an undeformed edge (in X_1) or deformed in the other three cases. It is always in the case of X_1 that the attraction of the tips by the sample edge is most evident: it corresponds

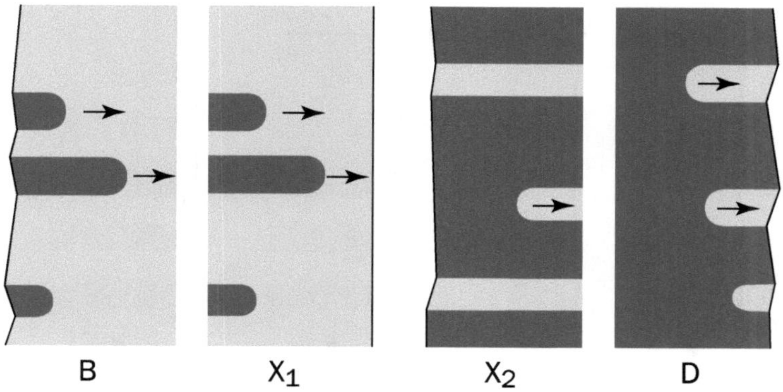

Fig. 3.42 Image forces. The photograph in Fig. 3.32 illustrates the attraction of a needle domain tip by the sample edge perpendicular to the domain axis. This interaction occurs four times during a polarization switching: when domain tips are created towards *B* in the theoretical cycle of Fig. 3.40; when tips are attracted by the edge where they disappear in *X* (X_1); when new tips of needle domains of opposite polarization are created beyond *X* (X_2); when these needles disappear with their tips in the sample edge towards *D*. In all cases, the dislocations that summarize domain tips (Fig. 3.27 and following) are attracted by their symmetrical images with respect to the surface of the sample edge at zero stress. But, only one situation corresponds to tips attracted by an undeformed edge (in X_1); this is what is experimentally observed with strong acceleration and significant $P(E)$ slopes. In the other situations, the deformation of the edge induces relaxations that can attenuate the effect of the image forces variably depending on the period of the excitation

to the increase in the slope of ($\mathrm{d}P/\mathrm{d}E$) towards the end of the BX path (Fig. 3.40 did not take this into account).

Similarly, the existence of a single front of domain tips, if it can occur near the transition with the phase front as described in Chap. 6, is rare in the ferro phase during a switching: the tips are created and propagate less regularly. However, two types of collective phenomena with different characteristic times are observed and described by Figs. 3.43 and 3.44:

- The interactions between domain tips during their longitudinal movements at the beginning and during the switching.
- The small lateral movements of the walls when the texture is at its maximum leafiness. This phenomenon can be more or less abrupt depending on the texture and temperature; the slope of the hysteresis cycle at $P = 0$ provides information: if this slope is relatively low, we may be dealing with a wave propagation phenomenon as illustrated in Fig. 3.16. On the contrary, if the slope is very large, with a curve close to the vertical when cutting the horizontal axis, the texture corresponds to the leafing in needles of maximum density (situation X of Fig. 3.40), in a situation close to a violent polarity reversal (around zero for *P* at the macroscopic level, situation *C*) with collective phenomena.

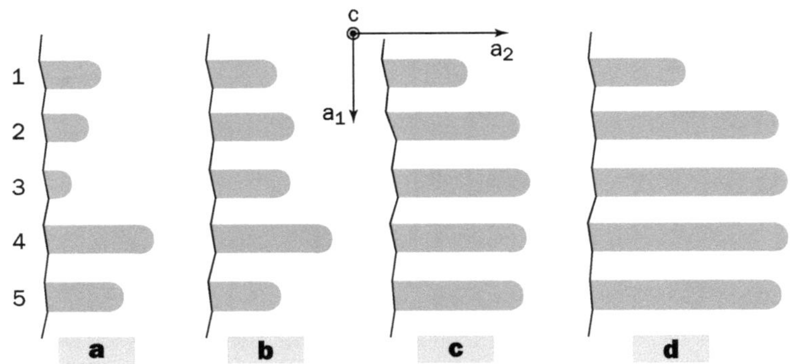

Fig. 3.43 Example of longitudinal avalanches by interactions between domain tips: domain tip 1 moves without any real change in speed. On the other hand, the significant dynamics of domain tip 3 leads to a change in dynamics from 2 in (**b**), then from 4 in (**c**), which then leads to 5. These avalanches that accelerate the switching have been observed. But a domain tip blocked by a small defect or one slower than its neighbours, can also initiate an avalanche of switching braking. In addition, the domain tips moving at a given moment of the reversal will not necessarily move at another moment

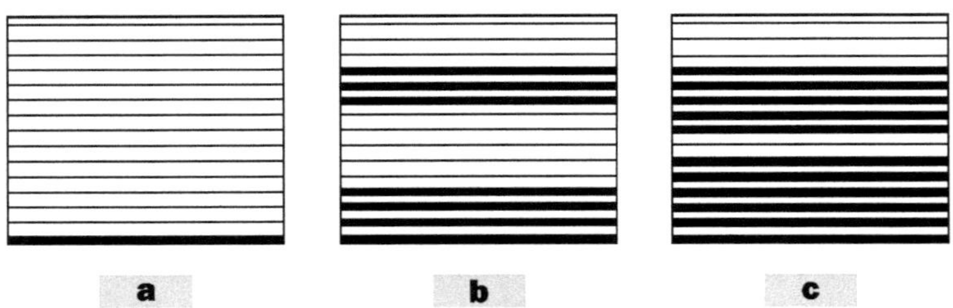

Fig. 3.44 Lateral avalanches: During a switching with needle domains, the walls cross the entire sample in situation X (Fig. 3.40) where small lateral displacements of walls occur. There is rarely a wave of wall displacements starting from one sample edge (as in **a**) to end at the other. Rather, lateral rearrangements propagating in both directions are observed, triggering sequences of successive rearrangements as (**b**) and (**c**) illustrate; the result of meetings between two waves of displacements from (**c**) can be abrupt and unpredictable

3.6.3 *Macroscopic Indicators of the Switching Phenomenon*

Consider the common macroscopic measurements as a function of the usual external parameters: temperature T, frequency f, and maximum amplitude E_m of the applied external electric field E. Figure 3.45 illustrates the measurements by hysteresis cycle at 50 Hz and by switching current.

Regarding the hysteresis cycle, Fig. 3.45a presents the variation over time of the applied electric field in the sinusoidal case and of the induced polarization, at a fixed temperature, for maximum amplitudes E_m in a 2:1 ratio: an increase in E_m creates a faster switching (t_{C2} smaller than t_{C1}) and a growth of the coercive field (E_{C2} larger

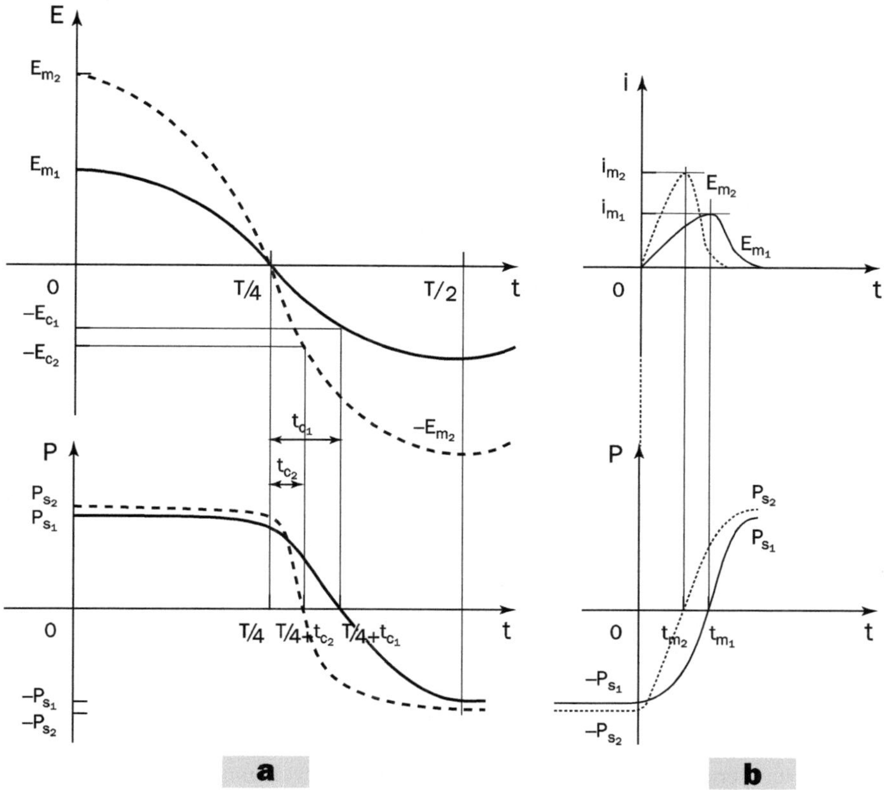

Fig. 3.45 (**a**) Sinusoidal variation as a function of time t of an electric field applied to the crystal during the plotting of a hysteresis cycle (top), and measured polarization (bottom), for E_{m1} and E_{m2} values of the field amplitude in a 2 ratio; the transition from P to the null value defines the coercive field (E_{c1} and E_{c2}). (**b**) Switching currents as a function of time after application of an electric field in the form of a square wave for values E_{m1} and E_{m2} in a 2 ratio (top); variation of P during the switching (bottom). The rise time of the step of E of a few nanoseconds can be considered negligible in comparison with t_m. The maximum of the switching current i_m at time t_m corresponds to the inflection point of the curve $P(t)$

than E_{C1} as illustrated by the result of Fig. 3.1b). Therefore, during a hysteresis cycle, the experimenter imposes both the switching frequency and the amplitude of the applied electric field, which changes continuously during the measurement.

To reduce the influence of these external parameters, the switching current is plotted with square steps and a duration of application of E_m chosen judiciously: it must be significantly greater than the time required for a complete switching; one must also perform enough switchings with opposite polarities to obtain reproducible results for determined values of E_m and T and, at the same time, have many different values without breaking the sample during the experiment! The results were obtained with application durations of E_m of the order of one second (while the switching times vary from a few microseconds to a hundred microseconds for these crystals). Figure 3.45b presents typical curves of reversal current $i(t)$ with a maximum i_m, for values of E_m in a 2:1 ratio. t_m is the time interval between the start of application of E_m and the maximum of $i(t)$). An increase in E_m results in a faster switching (t_{m2} smaller than t_{m1}) and a larger current maximum (i_{m2} larger than i_{m1}).

What does the analysis of these indicators allow?

The variation of the coercive field E_c (Fig. 3.46a) and the simultaneous observation of the domains allow to dissociate in the KDP crystal, various temperature intervals:

- The two to three degrees below the transition: the domain texture easily disappears, in block, the coercive field is low, the creation and disappearance of needle domain tips are carried out globally; the domain texture is a set that is difficult to dissociate. Therefore, in this temperature range, continuum models are relatively satisfactory.
- Within the range of twenty to thirty degrees Kelvin below the transition temperature: the domains appear in optical observation increasingly contrasted (the birefringence and deformation are larger) and the hysteresis cycles facilitate rearrangements between the domains, more regular textures and larger domain widths; the role of the domain tips and the moment of maximum leafiness are observed as illustrated in Fig. 3.40 with often brutal collective rearrangements.
- For temperatures below 95°K or 90°K, domain tip blockages occur at various times in the cycle: domains may remain for some time in the sample while adjacent domains have disappeared; collective phenomena are no longer observed as for higher temperatures; on the other hand, interactions between tips, between one tip and the edge of the sample, are easily observable.
- For a temperature below 70 °K, the plotting of the cycle may become impossible if E_m is not sufficient: it is still possible to plot hysteresis cycles at 50 Hz with an E_m of 2 kV cm^{-1}up to a temperature of 40°K using very good crystals and samples; but, increasing the value of E_m to obtain results at lower temperatures may cause the samples to fracture.

The variations, as a function of E_m and T, of the current i_m and of $(t_m)^{-1}$ presented in Figs. 3.46b, 3.46c, and 3.611a (Guyon and Lajzerowicz, 1966) illustrate the existence of two regimes:

- For large values of E_m and T, linear variations are obtained as in many ferro crystals:

$$i_m = k_i\left(E_m - E_i\right) \tag{3.5}$$

$$\left(t_m\right)^{-1} = k_t\left(E_m - E_t\right) \tag{3.6}$$

It is noted that, whatever the macroscopic method used, we are faced with a forced process with a global friction phenomenon integrating a set of microscopic phenomena. It appears for the lowest values of E_c and t_m with the slope 1 of their reciprocal dependence in Fig. 3.46d. We find almost linear variations with T of the coefficients k and the critical fields in the temperature region called the "plateau"

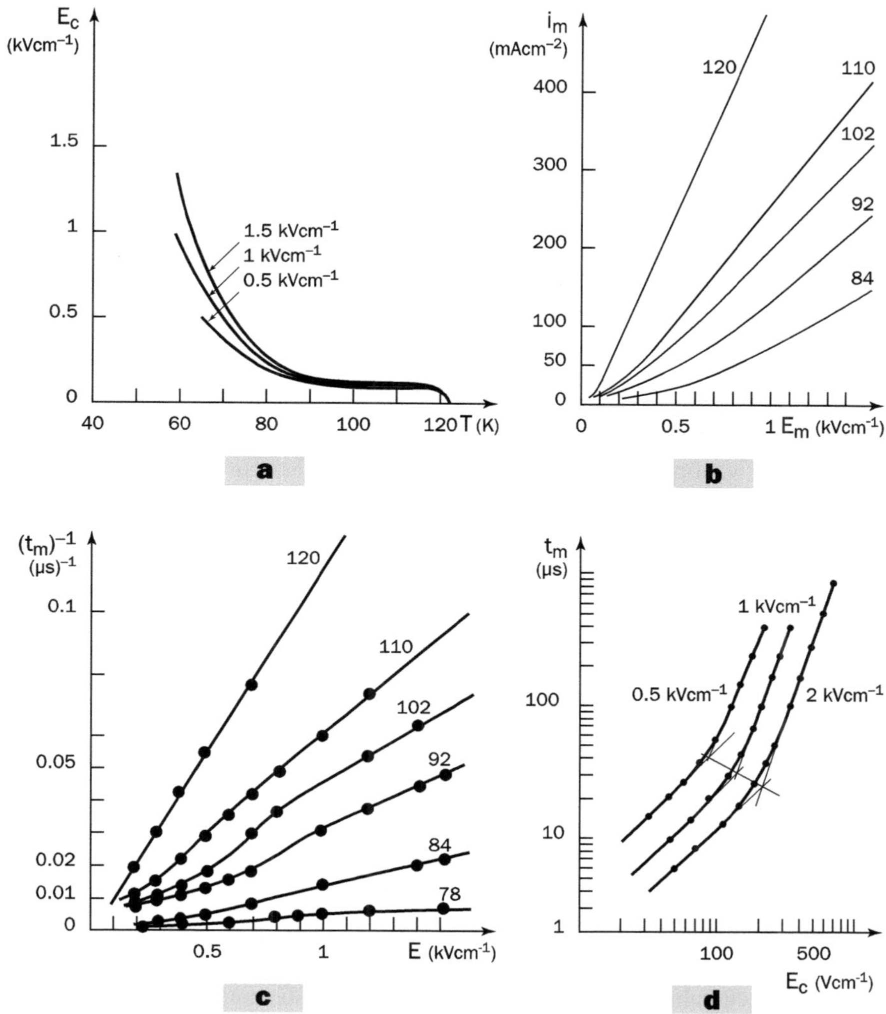

Fig. 3.46 (Macroscopic results of a KDP crystal). (**a**) Variation of the coercive field E_c as a function of the temperature at 50 Hz for the maximum amplitudes E_m of E indicated (Guyon and Lajzerowicz, 1966). (**b**) Maximum value of the switching current i_m as a function of the amplitude of the electric field pulses at the indicated temperatures (in degrees Kelvin). (**c**) Inverse of the reversal time corresponding to the maximum i_m of Fig. 3.45b(top) as a function of the electric field E applied at the indicated temperatures (in degrees Kelvin). (**d**) This representation (in logarithmic scales) of the switching time t_m as a function of the coercive field E_c for all temperatures and amplitudes of applied electric field shows in the ferro phase, a proportionality relationship for the highest temperatures (thus with low values of E_c) and a variation of t_m like E_c^3, therefore very non-linear at low temperatures

(because of the fairly constant value of the dielectric constant) consistent with those of i_m and t_m (as T increases, the first increases and the second decreases). This context of global friction allows a remark: t_m corresponds to the maximum value i_m of the current as illustrated in Fig. 3.45b and therefore to the relationship:

$$\delta^2 P / \delta t^2 = 0 \tag{3.7}$$

Similarly, if we consider Fig. 3.45a for the hysteresis cycle, t_c corresponds to the same relationship (3.7), in this other measurement context, with this time, E variable over time. As

$$t_c = \left(1/100\pi\right)\arcsin\left(E_c / E_m\right) \tag{3.8}$$

is close in this case to $(1/100\pi)$ (E_c/E_m), the linear variation between t_m and E_c means that, despite the difference in experimental context, t_c and t_m vary linearly. We do not really know what phenomena contribute to the overall friction result if we do not observe the microscopic textures, but these results and performances for applications are reproducible.

- The situation is completely different at lower temperatures where to reach the maximum reversal current, longer application times of E_m are necessary and the complete hysteresis cycles require larger coercive fields. These times t_m is more important than those obtained by extrapolating the linear dependence with E_m observed at higher temperatures. The phenomena are different and very nonlinear: t_m no longer varies like E_c but like E_c^3 (Fig. 3.46d). In addition, the product i_m t_m is not really constant as often assumed for "proper" ferroelectric crystals. $(t_m)^{-1}$ varies like $e^{-\alpha/E}$ with a nearly constant α coefficient (Fig. 3.47b), unlike what is found in these ferroelectrics. These forms of law are revisited in the next section and usually explained by modes of wall displacement by two-dimensional nuclei on the walls.
- The experiment summarized in Fig. 3.48 completes the correlations between proven microscopic phenomena and macroscopic measurements: in addition to Barkhausen pulses (Kipenko et al., 2023) and acoustic emission pulses, one can observe light emission pulses during a polarization switching cycle (Bornarel and Kvitek, 1993), a phenomenon known for a long time (Sujar and Biedrzycki, 1985). The reversals near the transition happen as in a quasi-continuum, with only more significant emission peaks during the rearrangements of domain assemblies. At lower temperatures, other phenomena appear; the significant peaks correspond to notable modifications of needle domains, but it is difficult to dissociate in the light emission intensity, what is due to domain tip movements or micro-lateral wall displacements. However, it is observed that emission peaks appear in increasingly large E intervals, evidence of less and less regular displacements with more frequent "stop and go" during almost all the part of the switching in presence of domains. These results are consistent with the work on Barkhausen peaks conducted in the ferroelectrics (Shur et al., 2001; Udalov et al., 2015; Akhmatkhanov et al., 2020), light emissions from the surfaces of

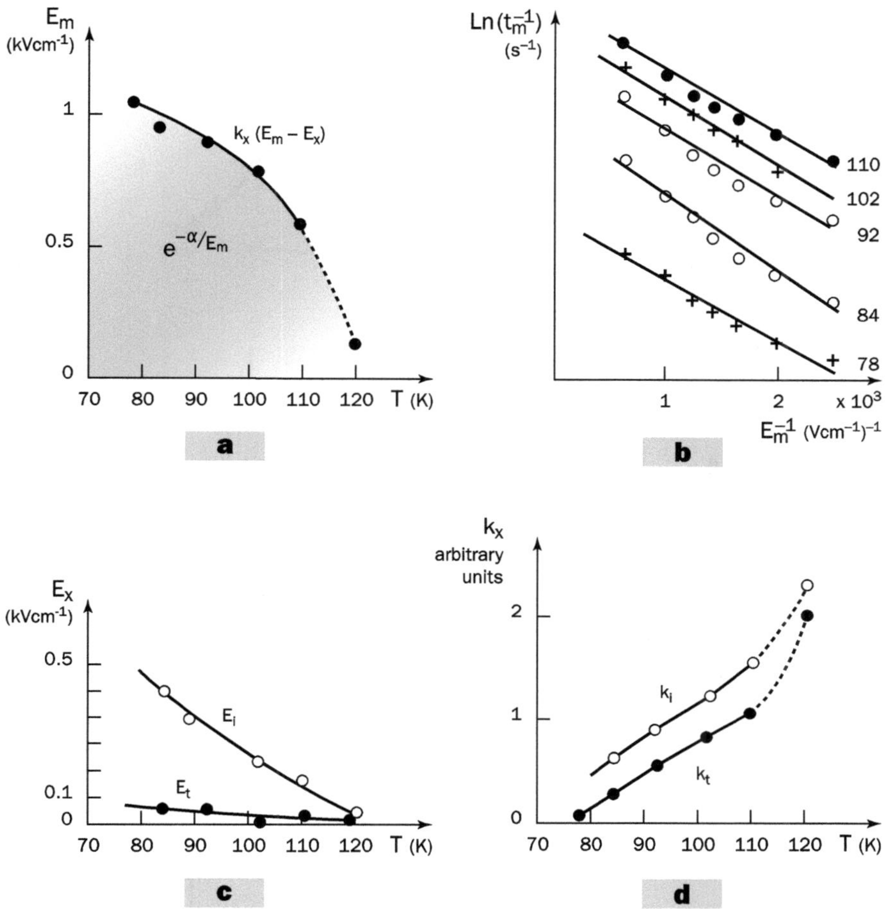

Fig. 3.47 (**a**) The inverse of the switching time $(t_m)^{-1}$ varies as a function of the applied electric field following laws $e^{-\alpha/E}$ or $k_x\,(E_m - E_x)$ as illustrated in Fig. 3.46c. The representation in (T, E_m) plane delimits the conditions of existence of these laws for KDP. (**b**) In the existence zone in $e^{-\alpha/E}$, the representation in semi-logarithmic coordinates of $(t_m)^{-1}$ as a function of E_m^{-1} shows a quasi-constant temperature value of the α coefficient unlike the results obtained in ferroelectrics such as $BaTiO_3$ (Savage and Miller, 1960). (**c**, **d**) In the existence zone in $k_x\,(E_m - E_x)$, the variation of E_x with temperature is negligible for t_m^{-1} and notable for i_m (**c**); the values of k_x are comparable for the two quantities with difficult measurements near the transition

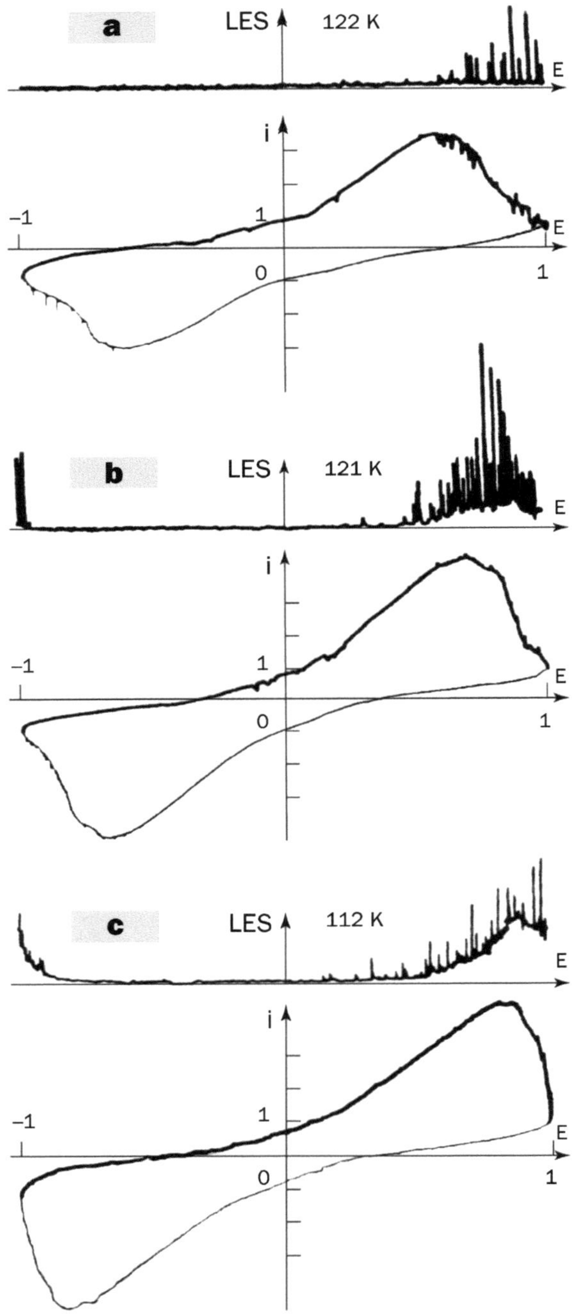

Fig. 3.48 During the plotting of the switching current $i(E)$ with a field E in the form of a sinusoidal wave ($f = 2 \times 10^{-3}$ Hz), in the KDP crystal, light emission signals are emitted, correlated with

(continued)

sudden changes in wall and texture into domains. These light emission phenomena (LES in arbitrary units) occur at the transition from the polarization (deformation) negative of the sample with $E = -1$ kV cm^{-1} to the positive polarization with $E = +1$ kV cm^{-1} (bold curve on $i(E)$): (**a**) At $T = 122$ K near the transition, texture changes occur in a single movement and a few LES peaks appear during rearrangements of domain sets; (**b**) at $T = 121$ K, the texture no longer changes continuously; collective phenomena mix with abrupt movements of walls and domain tips; (**c**) at $T = 112$ K, the widths of needle domains are quite regular and interactions between their domain tips are observed. The maximum emission of light signals occurs for an electric field greater than that of the maximum of $i(E)$. The light signals, in addition to the notable peaks due to the creation and disappearance of domains, are attributed to displacements of needle domain tips and to micro-lateral displacements of walls (Bornarel and Kvitek, 1993), which would trigger, for example, local ionizations and luminescence pulses (Sidorkin and Fedosov, 1981)

ferroelectrics (Poprawski and Kolarz, 1989), acoustic emissions in ferro-elastics (Lupascu et al., 2002, 2003; Salje et al. 2019) and electron emissions (Biedrzycki, 1991) especially those concerning avalanche phenomena (Harrison and Salje 2010, 2011) (Annexe A8). A good understanding of microscopic phenomena can help interpret macroscopic results enriched with these "pulse" distributions.

- Before revisiting in Sect. 3.7 the ferro-ferros within the various theoretical models proposed for switching, the existence of three temperature regions of very different importance from one crystal to the other is retained:
- A region very close to the transition similar to what is observed in many ferroelectric crystals: the free energy changes as indicated in the figures of Chap. 1 with, as the temperature decreases, a growth of the energy gap that must be crossed to go from one minimum to another; the "single-domain" model seems almost sufficient to qualitatively understand the experimental results.
- In the plateau region, despite the processes of multiple interactions, collective phenomena, the reversal dynamics of the polarization-deformation is generally summarized by friction processes except for low-intensity excitations.
- The strong growth at low temperature of E_c and of t_m illustrates the so-called "domain freezing" phenomenon with a non-linear response to the excitation (studied in detail using dielectric constant and loss constant studies in Chap. 4).

3.7 Switching (Polarization and Deformation): The Ferro-Ferros Within the Ferroic Crystals

3.7.1 The Experimental Results

Given the technological interests, the polarization switching of ferroelectrics has been widely studied to obtain crystals with good dielectric properties at room temperature, capable of withstanding a large number of switching. The chosen crystals are predominantly non-piezoelectric in the para phase and obviously become so in the ferro phase. When the optical properties varied little between the domains, researchers encountered some difficulties in observing them. Some observations were made by applying pulses of the same polarity with chemical attack and observation between each pulse, that is to say under experimental conditions very different from those of switching between a single-domain state of a determined polarity and the single-domain state of the opposite polarity. In some cases, especially in TGS crystals and perovskites, the authors observed situations with many growing nuclei following the ferroelectric axis (Binggeli and Fatuzzo, 1965), so numerous that their dimension perpendicular to this axis did not have time to change (Merz, 1954). In this context, many studies have focused on electrostatic aspects and the depolarizing field (Landauer, 1957). The creation of these nuclei, for these materials, sometimes occurred randomly (Stadler and Zachmanidis, 1964a). Other authors, on the contrary, observed the creation of a few fine domains growing rapidly along the ferroelectric axis, then the lateral extension of these nuclei in the plane perpendicular to this axis (Miller and Savage, 1958, 1959; Jaskiewicz and Przeslawski, 1979) with isotropic or anisotropic wall displacement speeds (Miller and Savage, 1959; Stadler and Zachmanidis, 1963, 1964b). The phenomenon of "collapse", the meeting of domain walls that transforms two isolated domains into a larger domain, was thus studied (Chynoweth and Abel 1959).

But the most frequent electrical studies of polarization switching were carried out without simultaneous observation of domain textures. The experimental results obtained for the usual parameters, switching current i_m and inverse of the switching time t_m^{-1} as a function of the amplitude of the applied electric field E, are often of the form $e^{-\alpha/E}$, when E is of low value, whether it concerns nucleations or wall displacements. When the applied electric fields are larger, classical friction laws are published with speeds proportional to the amplitude of the excitation field E or variations in E^n, n varying from 1 to 1.4 (Stadler and Zakmanidis, 1963). The majority of experimental results, therefore, come from electrical measurements, switching currents or hysteresis cycles, taking more or less into account the effects of "surface layers" interpreted in different ways by different authors (Drougard and Landauer, 1959; Miller and Savage, 1960, 1961). Some optical observations allow us to know the evolution of domain textures in certain massive crystals, but simultaneous electrical and optical studies are mainly developed recently with thin layers. The wide variety of measurements carried out is of interest in itself since almost always similar law forms are found, the one in $e^{-\alpha/E}$ for i_m and t_m^{-1} (Pulveri and Kuebler 1958; Miller and Savage, 1958) rather for low electric field values, and variations proportional to E or to E^m (with m between 1 and 2) for high values of E (Wieder 1958a, b). What models are proposed to explain these results?

3.7.2 Switching Models

Modelling the reversal starting from the phenomenological description of Landau only involves to pass the crystal in Fig. 1.16 and following, from a potential well with the order parameter value $-q_S$ to the other value $+q_S$ (for example). The implicit assumptions of this model should not be forgotten: the crystal is considered to be of infinite dimensions and thermal agitation, fluctuations are neglected.

In the work of Ishibashi and Takagi (1971), the Landau-Khalatnikov equation (1954) is used:

$$\gamma \, dP / dT = - \; aP - bP^3 + E \tag{3.9}$$

For a second-order transition (Ishibashi and Iwata, 2007; Ricinschi and Okuyama, 2007; Loh et al., 2013), based on the relation (1.9) and with γ viscosity coefficient, a theoretical value of coercive field E_c is obtained:

$$E_c = 2\left(A^3 / 27b\right)^{1/2} \; \left(T_C - T\right)^{3/2} \tag{3.10}$$

One can calculate the associated switching current and its maximum value over time with a macroscopic polarization of $P_S/3^{1/2}$. This maximum value of the current depends on a term function of the temperature, of the form $(T_C - T)^{3/2}$ and a linear term with the electric field E; the variation in field was predictable since a friction term was introduced in the Landau-Khalatnikov equation. The other results, especially concerning the time at which the maximum occurs, are far from experimental results, which is consistent given the assumptions of the model (crystal of infinite dimensions, etc.). In addition, the values obtained for E_c are about two orders of magnitude higher than those observed experimentally. We find ourselves in a classic situation of solid state physics: microscopic phenomena allow the initiation of a dynamic with excitations significantly weaker than that predicted by a global macroscopic model. However, some cases can be cited where the reversal takes place without the creation of domains with values of E_c close to the theoretical value predicted on the basis of the Landau-Khalatnikov model (Highland et al., 2010). It appears paradoxical that it is an ultra-thin layer of a few nanometres that seems in agreement with a model of an infinitely sized crystal; but in fact, the increase in the energy of the depolarizing field and the disappearance of the walls leads at the nanometric level to this maximum value of the coercive field.

The experimental variations observed in $e^{-\alpha/E}$ had little chance of being reproduced by the macroscopic model. The modelling attempts at the local level using three-dimensional nuclei also failed: are initially used the work of Kolmogorov (1937) and Avrami (1939, 1940, 1941) on the crystal growth, reformulated for the domains (Ishibashi and Takagi, 1971). An extensive bibliography exists as syntheses on the models (Fatuzzo and Merz, 1959, 1962, 1966; Tagantsev et al., 2010), particularly concerning the very studied thin layers currently (Scott, 2016), outside the scope of this book.

Consider, for example, nucleation as a thermal activation phenomenon with a probability of appearance of a nucleus of energy ΔF in the form $e^{-\Delta F/kT}$. Figure 3.49 proposes three situations of nuclei with shapes that take into account the anisotropy of the crystal: A is created in volume, B stuck to one of the faces of the sample perpendicular to the ferroelectric axis, and C on a sample ridge.

We can write for the creation of each of these nuclei the necessary energy increase:

$$\Delta F = -2EP_S bld + \alpha\, lb\gamma_x + \beta\, ld\gamma_y + \delta\, db\gamma_z + ½NP_S^2 dlb \qquad (3.11)$$

α, β, δ take the values 2, 1, or 0 depending on the situations of the nuclei, and γ_x, γ_y, γ_z are the energies per unit surface of nuclei in the planes perpendicular respectively to the axes Ox, Oy, and Oz. If we neglect the effect of the depolarizing field, by minimizing ΔF with respect to the various dimensions, we find simple relations such as:

$$b/l = \beta/\delta\left(\gamma_y/\gamma_z\right) \qquad (3.12)$$

which express that the smallest surface of the nucleus will be the one whose energy per unit surface is the largest. The nuclei will thus be elongated along the ferroelectric axis and will grow rapidly along this axis to suppress the effect of the surface

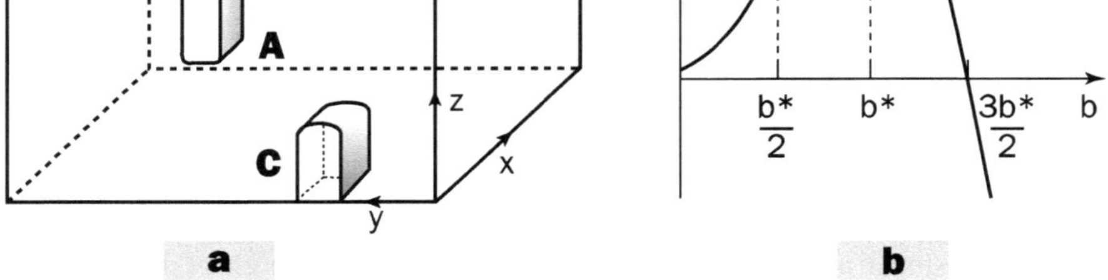

Fig. 3.49 (**a**) Examples of phase 2 nuclei created in a phase 1 sample. The nuclei are arbitrarily chosen to be parallelepiped in shape. They can be created inside the volume like A, on a face (001) like *B*, or on an edge of the sample like *C*. (**b**) Variation of free energy during the creation of a nucleus as a function of a dimension, here *b*: if *b* is less than a value *b**, the nucleus disappears; it grows if *b* is >*b**

perpendicular to it, a surface which includes electric charges, and in our cases, quasi-dislocation knees, etc. We also obtain:

$$\Delta F = \alpha\beta\delta\, \gamma_x \gamma_y \gamma_z / \left(EP_S\right)^2 \tag{3.13}$$

a relation that does not allow to obtain models in agreement with the results (of i_m and inverse of t_m) in $e^{-\alpha/E}$. Comparable models assuming a creation of nuclei in the volume with isotropic surface energies (Merz, 1954) lead to variations of ΔF in $E^{-5/2}$ also not reconcilable with the experimental results.

To obtain speed and nucleation laws in agreement with a variation of type $e^{-\alpha/E}$, the simple solution is to consider, by hypothesis, that only two dimensions play a role in the phenomenon. We thus encounter models that assume the creation of very numerous nuclei of identical sections that propagate along the ferroelectric axis (Merz, 1954; Fatuzzo 1960a, 1960b). But, the model most compatible with the many observed results, with different orientations and domain shapes, is based on the existence of two-dimensional nuclei, with a constant, very small compared to the others; this is what the authors impose a priori in one way or another, often by fixing for this smallest dimension, that of an elementary cell or a few cells (Miller and Weinreich, 1960). This result is not specific to ferroelectric nuclei; these forms of nucleation and speed laws are known in mechanics and elasticity with growth probabilities in $e^{\alpha/\sigma}$, σ applied stress (Gilmann, 1965; Nabarro 1967a, b) and in magnetism with the $e^{-\alpha/H}$, H applied magnetic field. These laws encountered for the nucleation processes, for the domain wall speeds, for the characteristic values of the switching current, thus suggest that two-dimensional nuclei may be at the base of different phenomena: a nucleus could initially form on the faces of the samples and then, during the reversal process, nuclei on the walls of existing domains to allow their lateral movement (Wieder 1958a, b). Figure 3.50 summarizes proposals with two-dimensional triangular-shaped nuclei (Miller and Weinreich, 1960) or with other shapes that take into account the anisotropy along the ferroelectric axis (Nettleton, 1967; Stadler, 1966; Molotskii 2003 TACM van der Put, 2011), introduce charged curved walls (Darinskii and Sidorkin, 1987), assume nucleations on the electrodes or in the middle of the walls (Burtsev and Chervonobrodov, 1983), etc. Interesting developments on this point can be found in the work of Tagantsev et al. (2010).

Continuing this approach and assuming for applied fields of increasing value the creations of nuclei with one or more elementary cell thicknesses which occur simultaneously or successively, some authors have proposed to explain the results in $E^{4/3}$ by the sum of a series of terms in $e^{-\alpha/E}$ whose α values vary with the height of the nucleus (Stadler and Zachmanidis, 1963).

To model the switching process in its entirety, the proposed models rely on simple assumptions that can be solid for non-piezoelectric ferroelectric crystals in the para phase with W_∞-type walls:

- Random creation of nuclei on faces perpendicular to the ferro axis.
- Very high growth rate of nuclei along this axis (this assumption remains solid in all cases) (Molotskii, 2003).

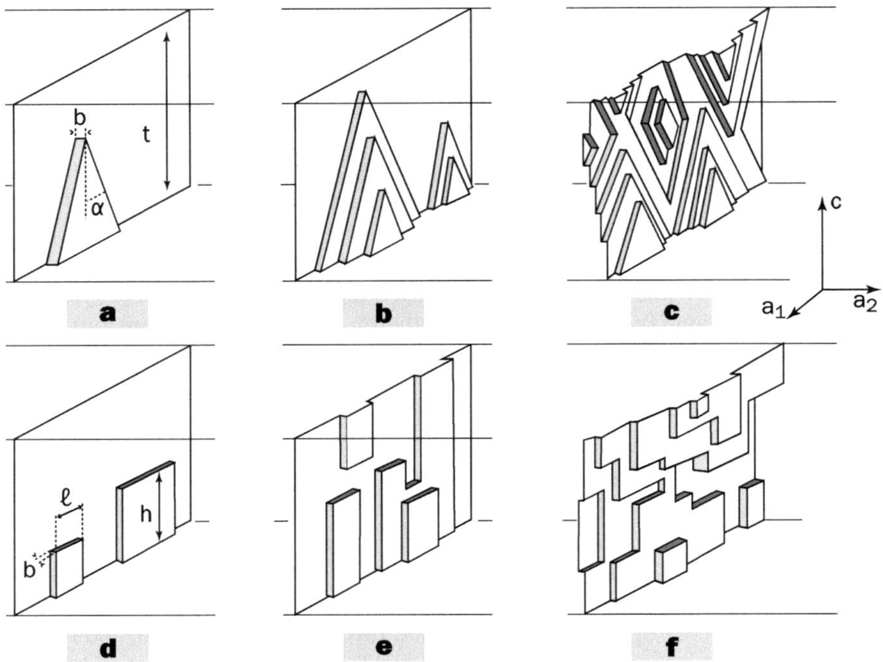

Fig. 3.50 The model frequently used for nucleation laws and domain wall speed of type $e^{-\delta/E}$ as a function of the applied electric field E is based on the existence of two-dimensional nuclei; authors usually choose nuclei of one or several lattice thicknesses. (**a**) Some authors have proposed triangular-shaped nuclei (Miller and Weinreich, 1960; Hayashi, 1972). (**b**) If we assume a superposition of nuclei with thicknesses of one lattice, two lattices, three... etc., the sum of the laws in $e^{-\delta/E}$ with values of δ proportional (one, two, three, etc.) tends towards variations in E^x (Stadler and Zachmanidis, 1963). (**c**) A multiplication of nuclei by layers allows us to imagine, walls as described in Fig. 3.37b (Schmidt et al., 2000) with "roughness" phenomena, especially under the action of alternating excitation fields (Domenichini et al., 2023). (**d**) A rectangular shape of the nuclei could be slightly more favourable than the triangular shape (Nettleton, 1967), but the dimension l is critical. (**e**) If the energy per unit surface of the "bl" parts is very large compared to that of the "bh" surfaces, the nuclei will very quickly cross the crystal in the direction of the ferroelectric axis. (**f**) If the energy per unit surface of the "bl" parts is relatively comparable to that of the "bh" parts, we can imagine a situation comparable to that of (**c**) and, as far as our ferro-ferro crystals are concerned, for temperatures higher than those of the gel, domains (quite wide walls) in coherence with the model of Fig. 3.37b (Schmidt et al., 2000)

- Isotropic growth of domains perpendicular to the ferro axis.
- Absence of local variation in polarization and deformation within the domains during reversal.
- Movement of the walls governed by the same law as that of nucleation for weak electric fields ($e^{-\alpha/E}$); this assumption is questionable since the nucleus may have a size of the order of a micrometre while the growth nuclei on a wall would have a thickness of a few elementary cells.

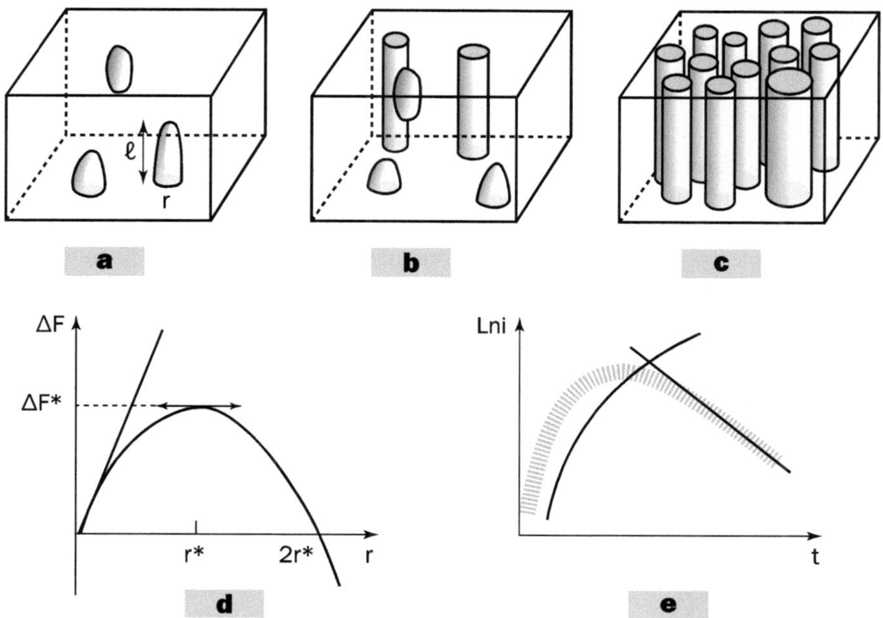

Fig. 3.51 In the Chynoweth model, the nuclei created in volume and on the faces perpendicular to the ferro (**a**) axis, grow along this axis until they cross the entire sample (**b**) and (**c**); then the cylinder domains (**c**) widen laterally until their walls "collapse" while other domains are created. (**d**) Typical variation of free energy during the creation of a nucleus as a function of the radius *r* of the cylinder. (**e**) The variation of the switching current in this model is a combination between the growth curve of the nuclei and that of "collapse" of the walls (in continuous lines). Experimental results are indicated in grey

- Wall displacement speeds proportional to E for sufficiently high E values and without the existence of a critical field for movement, thus with wall movement from the creation of the nuclei.
- Consideration of the energy of the depolarizing field or not: the energy of the depolarizing field does not seem notable for nucleation itself. However, it acts to elongate the nuclei in the direction of the ferroelectric axis (see Fig. 1.40).

With such assumptions, the switching can be summarized in three phenomena, a first of nucleation, a second of lateral extension of domains located quite randomly, and a third of "collapse" between the walls of the domains until a completely monodomain sample is obtained; this is the basis of many models such as, for example, the one proposed by Chynoweth (1958) illustrated in Fig. 3.51. The growth of the nuclei and their widening after having crossed the entire sample along the ferroelectric axis as well as the variation of the energy of a nucleus as a function of its dimension r are represented. With the classical theoretical approach with nuclei (Jacobs, 1985) and isotropic lateral extension (Wieder, 1960) is obtained the total surface area of the domain section S. The variation of the switching current i(t) is proportional to dS/dt (Chynoweth, 1958): i(t) from which is deduce the variation of the switching current proportional to d*S*/d*t*:

$$i(t) \approx k_1 t - 1 + \exp(-k_1 t) \tag{3.14}$$

with k_1 of the form $g\ e^{-\alpha/E}$, g being a frequency coefficient.

After a certain time t_1, the average value of S is too large to assume free growth of each domain and the domains meet; the calculation of variation of the section S of the sample by a purely geometric treatment of this "collapse" phenomenon Avrami (1939, 1940, 1941) allows to write again the variation of $i(t)$:

$$i(t) \approx \exp(-k_1 t) \tag{3.15}$$

represented by the straight line with negative slope in Fig. 3.51e.

Some experimental results obtained in perovskites have been correctly described by this model. Many other experimental results have been compared with this law in exponential decay with time; the agreement is satisfactory when only the part of the switching curves with wall movements without interactions is considered, that is to say in situations where these purely geometric models can be applied (Nogami and Imasaki, 1973).

Other comparable models are found for crystals with permitted W_∞-type walls (Rumyantsev, 2020) or those that exhibit low anisotropies. It is always the slowest phenomenon that determines the variation of the reversal over time and its characteristic magnitudes.

Figure 3.52 gives an idea in arbitrary units (for the switching current i and for time t) of the situations that can be encountered depending on the values of the probability of creation of small nuclei, the speed of lateral displacement of the walls and, the process of "collapse". The authors, to account for their experimental results, use numerical simulations, by playing on the parameters, usually satisfactorily during the growth of the current up to the maximum value i_m and, much less convincingly for the end of the switching. In this context, a parameter k is usually defined which compares the ability of lateral displacement of domain walls to that of creating new nuclei. For example, for Fatuzzo (1962), k is written in the form

$$k = 2R^* / R \tag{3.16}$$

in the case of a displacement of the walls by creating nuclei in two dimensions on the wall (R* is the probability of creating micronuclei of fixed thickness on the wall of an existing domain and R the probability of creating new domain nuclei per unit of time) or:

$$k = \mu E / R r_C \tag{3.17}$$

in the case of a block translation of the wall (with a mobility μ under the action of an electric field E; the quantity r_C defines the size of the nucleus at the start). In the

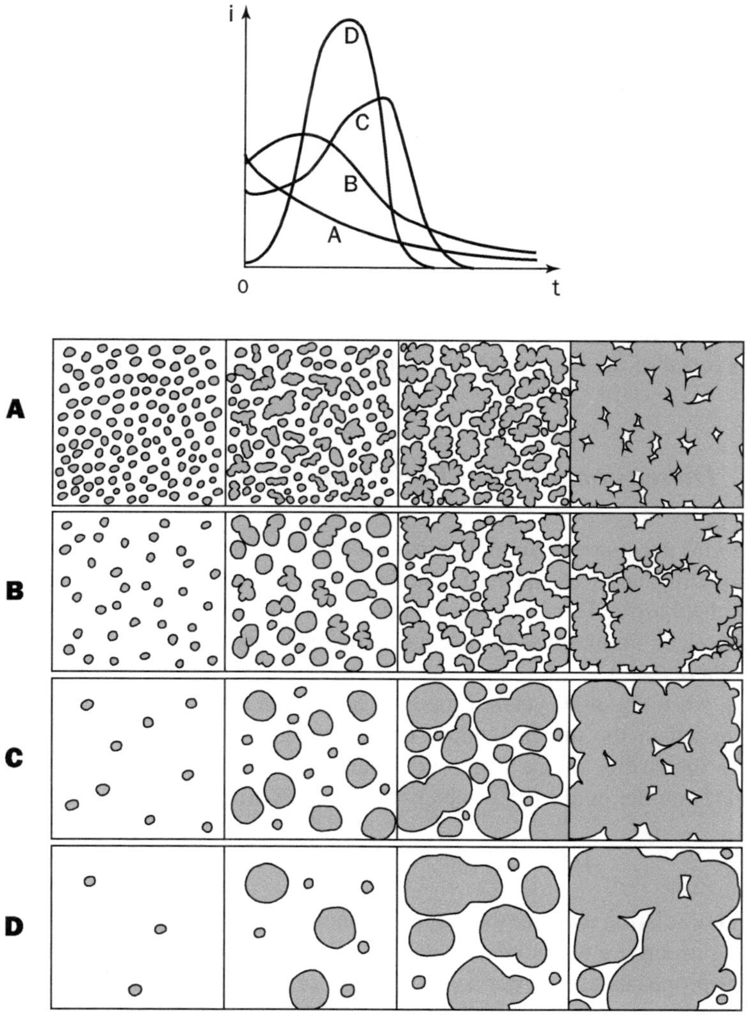

Fig. 3.52 The $i(t)$ curves of various shapes (top figure) in arbitrary units for i and t, correspond to the situations schematized by the sequences of textures (bottom figure): (**a**) The nucleation probability is so great that the speed of lateral wall displacement can be considered negligible (example in Shishkina et al. (2022)): at $t = 0$ the number of created nuclei is such that i is maximum; the switching is carried out by creating other nuclei which "plug the holes"; the $i(t)$ law is a decreasing exponential. (**b**) We are quite close to situation (**a**), but the first created nuclei have time to grow a little before the "collapse" with their nearest neighbours. (**c**) The speed of creation of the nuclei is significantly lower than that of growth by lateral wall displacement, and this phenomenon takes on a certain importance. (**d**) We are in the situation opposite to (**a**): very few nuclei are created, and the switching is almost exclusively commanded by the lateral displacement of the walls in the plane perpendicular to the ferroelectric axis and the "collapse" of the rare domains of the crystal

representations of Fig. 3.52, situation A corresponds to a value of *k* close to zero (*R* very large) and situation *D* to a value of *k* much >1 (R low).

It is easy to compare experimental results and theoretical models for ferroelectric crystals with W_∞ walls in situations of type D for which the value of the current *i* at $t = 0$ is zero while in other cases, the non-zero value is difficult to appreciate in the capacitive peak observed at the origin of time.

The results of calculations from these models qualitatively account for the shapes of *i*(*t*) laws, but the fit is still not very good at the same time, and on the increasing part, and on the decreasing part of these curves, including for certain crystals with W_∞-type walls. The most obvious criticisms concern the assumptions: absence of consideration of the finite dimensions of the samples and constancy during the entire reversal of the nucleation probability and of the speed of the walls.

3.7.3 Dimensionality and Finite Dimensions of the Sample

To take into account the finite dimensions of the samples, some authors have proposed modifications to the basic model, often titled KA or KAI (Kolmogorov, Avrami, Ishibashi). The proposals involve adding additional parameters that facilitate numerical simulations for a good agreement between calculated and experimental curves.

Thus, since about 20 years, proposals have been made on the basis of the KAI model in which the dimensionality coefficient n is considered as non-integer to integrate the reality of the different situations observed at various stages of the reversal (Dimmler et al., 1987) which allows the writing of *i*(*t*) in the form:

$$i(t) = 2P_S\left(An/t_0\right)\left(t/t_0\right)^{n-1}\ \exp\left(-t/t_0\right)^n \tag{3.18}$$

this kind of relation used for thin layers phenomenologically provides services with n varying according to thickness between 3.1 (very small thicknesses) and 1.6.

Other proposals take better account of the experimental reality by detailing the changes in dimensionality during the reversal as illustrated in Fig. 3.53a with $n = 3$ for a nucleus that grows in volume, $n = 2$ for growth in a plane section, and $n = 1$ for a lateral displacement of a plane wall (Shur et al., 1998). Thus, interesting matches between calculated and experimental curves are obtained (Fig. 3.53b, c, d). The idea that these successive phases of reversal produce various contributions in i(t) allows to explain certain forms of variation i(t) less usual (Shur et al., 2013).

The models have therefore gradually taken into account the limited dimensions of the crystal, either by the play of dimensionality as we have just illustrated, or by imposing size constraints on domains (Ishibashi and Iwata, 2012), but these considerations remain predominantly geometric (Duiker and Beale, 1990). It is usually considered that there is no interaction between the domains, nor between the domains and the sample edges of the crystal. Thus, the phenomenon of "collapse" remains governed by Avrami's theorem, which deals with the problem geometrically, without physical interactions of the walls. The same geometric assumptions in

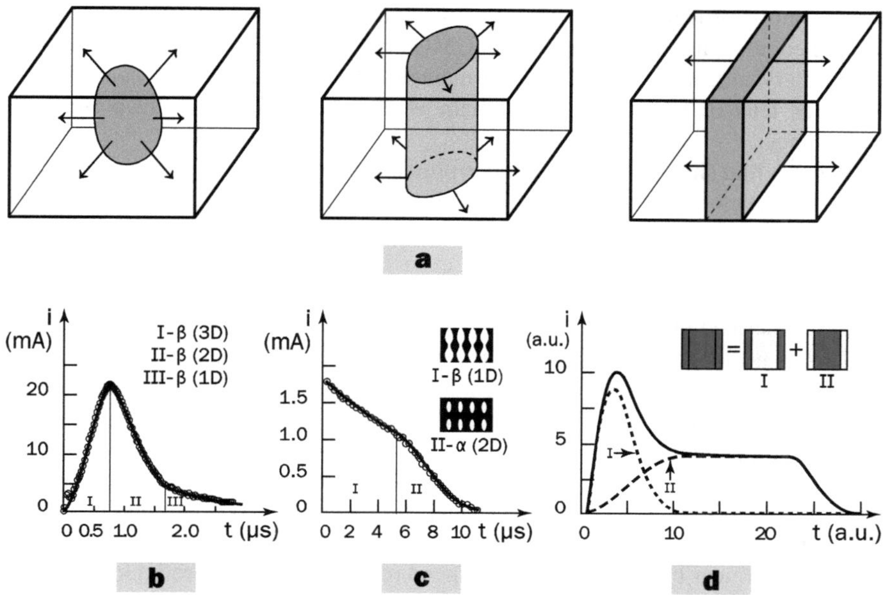

Fig. 3.53 The concept of dimensionality is introduced to model switching of polarization (deformation): (**a**) The variation of the domain is carried out with a dimensionality $n = 3$ for the scheme on the left, $n = 2$ for the one in the centre, and $n = 1$ for the one on the right. (**b** and **c**) These variations of $i(t)$ clearly show the different values of dimensionality (3 different values in (**b**) and 2 in (**c**) (Shur et al., 1998). (**d**) The overall result $i(t)$ is considered as the sum of two stages of evolution: the first occurs in the proximity of the electrodes with $n = 2$, and the second in the volume with $n = 1$ (Shur et al., 2013)

more complicated and numerous models for thin layers return to the situations schematized in Fig. 3.52: the fact of supposing small size grains (periodic for example) leads to situation A which neglects the lateral displacements of walls, and that of choosing large grains for situation D with wall displacement speeds that take precedence over the nucleation phenomenon.

3.7.4 Models for Ferro-Ferro Crystals?

Are previous works applicable to ferro-ferro crystals?

Some assumptions stated above for W_∞ type walls are still suitable in situations where the shape anisotropy of the domains is not too significant: for example, when domain walls can easily exit the wall planes allowed in the case of W_S walls. Thus, wedge-shaped domains (in section perpendicular to the ferro axis) whose apex angles can easily vary during the reversal (in the context of Fig. 3.4b) are not very far from situations C and D of Fig. 3.52. The energy of the quasi-dislocations of the wedge type distributed regularly in the wall does not obviously prevent wall disorientations due to thermal gradients, the dynamics of reversal, or the state of the sample edges. This can be observed in GMO crystals at room temperature. Usually, in these crystals, it is only after being maintained in the same experimental conditions for tens of hours that the domain walls take position in the permissible wall planes.

We also find situations close to those previously modelled for ferroelectrics with W_∞ walls when domain nucleation is very easy, as observed in the range of a few degrees below the transition of crystals from the KDP family (see Chap. 4 on dielectric properties). In this particular case, even if the anisotropy is evident, we can make a link with situation A of Fig. 3.52: the creation of domains is very fast. The models described previously can still give relatively satisfactory results.

This is no longer the case when the anisotropy of the domain walls becomes strong; the wall surfaces are then essentially in the permissible wall planes, and the probability of creating nuclei is not the only parameter. Two extreme situations have been developed in this chapter:

First, the one where only one wall or a few walls exist in lateral movement during the switching (Sects. 3.2 and 3.3). The shape of the hysteresis cycles and that of the switching currents can be very simple: hysteresis cycles are square or geometric shapes with linear branches if the wall speed reaches the limit speed due to friction phenomena as soon as it starts moving. In such cases, the local deformation (local polarisation) changes little during the reversal (see Fig. 3.8 with GMO). The cycle becomes more complex in the opposite case (as demonstrated by Fig. 3.9 with a Tanane crystal). Usually, the speed of lateral displacement of the walls is the essential magnitude, with laws of the form $k(E - E_S)$ thus with the existence of a critical field for setting the wall in motion (unlike the results and models developed for ferroelectrics with W_∞ walls), a critical field that can evolve with the average polarization-deformation.

A very different situation is given by polarization reversals with needle domains, far from the transition temperature:

- Domain nuclei, at a given temperature, are created with almost constant dimensions on the sample edge, and after a few reversal cycles, almost at the same places from one cycle to another.
- A domain tip does not start moving as soon as the nucleus is created, and its speed for sufficient E values is of the form $k(E - E_S)$ as described in Figs. 3.22, 3.25, and 3.26,
- Domain walls and especially the ends of needle domains interact with each other with collective phenomena.
- Domain tips interact with the sample edges through the play of image forces.

Taking into account the phenomena previously illustrated in this chapter, we have the keys to predict the shapes of the $i(t)$ curves. Figure 3.54 illustrates the essential differences observed in the $i(t)$ curves of a ferro-ferro crystal (the KDP at 113 K at various E values in Fig. 3.54a) compared to a "proper" ferroelectric: the rise of $i(t)$ is faster but especially when the number of ends in simultaneous movements is important (here for E values >300 V cm^{-1}) the fall is very fast (Nogami and Imasaki,

1973) because of the collective phenomena between domain tips and the image forces of attraction of the domain tips by the edges of the crystal. We note that except with intense electric fields, this reversal is not as simple as the sequence "nucleation-displacement-collapse": the creation of domain nuclei in 1 of Fig. 3.54b is followed by a new creation of nuclei in 5 before their attraction and disappearance in 7. The models that take into account the difference during the switching between the two periods of opposite polarities (as shown in Figs. 3.40, 3.42, 3.43) are not very numerous (Sikora and Parlinski, 1999). Figure 3.55 further proposes a simplistic modelling using the information from Figs. 3.51e and 3.54b. For high electric fields and temperatures close to transitions, the $i(t)$ variations of ferro-ferro resemble those of pure ferroelectrics while at lower values of electric field and temperature, variations with practically two maxima can exist. Thus, if some statistical results (i_m, t_m) can be easily modelled, it is sometimes difficult in ferro-ferro to accurately explain the shapes of the reversal curves if the domain texture is not simultaneously observed. Studies in this sense allow to quantify the phenomena of

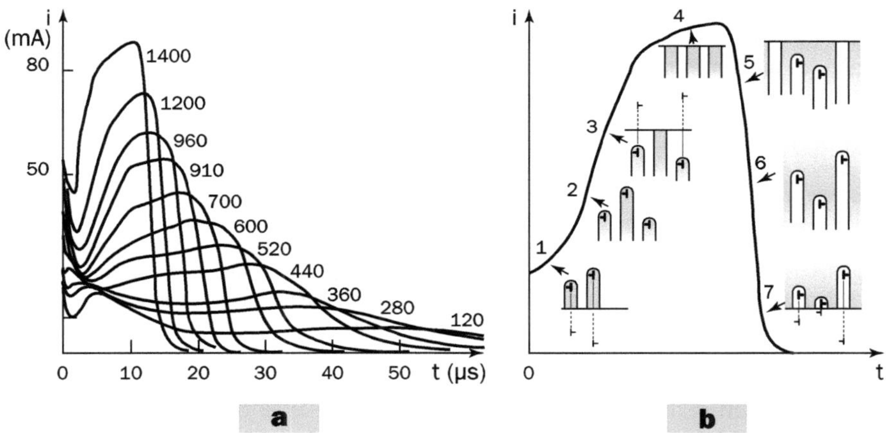

Fig. 3.54 (**a**) Variation law $i(t)$ for a KDP sample ($T = 113$ K, thickness = 2.52 mm, $S = 12.5$ mm^2; the ratio between the dimensions of the sample along axes a_1 and a_2 and the thickness along c corresponds to an easy reversal case). The electrical voltage applied to the sample is indicated in Volts on each curve. The maximum electric field of 555 V cm^{-1} corresponds to variation laws of i_m and t_m^{-1} of the form $k(E - E_S)$, whereas for smaller E values, the laws are of the form $e^{-\alpha/E}$ as it is possible to verify on Fig. 3.47a. (**b**) At various points on the $i(t)$ curve, the diagrams explain the variations in its slope:

1. Nucleations on the sample edges; the image forces of the Burger vectors summarized from the domain tips delay their movements towards the middle of the crystal
2 The moving domain tips interact, and collective phenomena occur
3 Besides their own interactions, the domain tips are attracted by their image forces of the edge which accelerates their movement
4 When all the domain tips have disappeared, small lateral displacements occur in a more collective and rapid way as we are close to the phase transition
5, 6 The domain tips created are initially held back by interactions with their images in relation to the edge, then "ejected"
7 The domain tips are attracted to the edge before disappearing

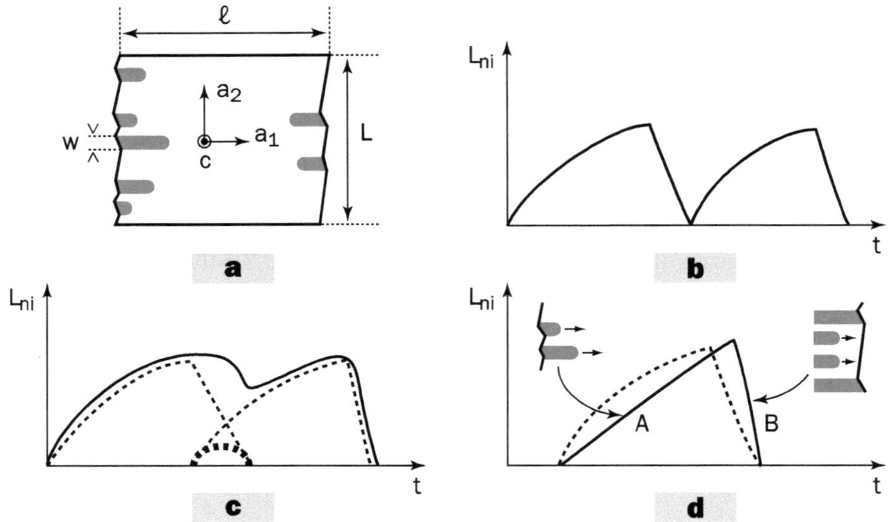

Fig. 3.55 (**a**) Representation of a needle texture with only one wall direction allowed. (**b**) In a purely geometric model, the variation of *i*(*t*) can be broken down into two times (repeating the curve of Fig. 3.51e): passage from the mono-domain to the maximum domain density, then passage from the maximum domain density to the mono-domain of opposite polarization. (**c**) During a reversal of the polarization-deformation near a null macroscopic polarization of the sample, it is difficult to see the "exchanges" of polarity of the needles taking place at every point of the sample at the same time *t*, hence the superpositions of curves: some regions of the sample are, on average, at a time *t*, of positive polarization and others of negative polarization. (**d**) As illustrated by Fig. 3.54b, the effect of image forces on the domain tips is not the same at the beginning of the switching (they slow down the departure of the tips) and at the end (they accelerate them)

avalanches (Annexe A8) at the domain level, especially at the tips and, more difficultly considering the very short time intervals, at the level of the walls.

It is clear that the link between macroscopic and microscopic properties needs to be improved, including through theoretical simulation work. These, currently due to thin layers, not treated here, are developing in the particular context of sets of a few or tens of cells, especially in ferroelastics (Lu et al. 2019). However, these works on materials with other elastic parameters are interesting to go further in the phase fronts of Chap. 6 for example, concerning flexoelectric effects (Majdoub et al., 2008).

Memo Chapter 3

The changes in global polarization *P*, deformation u of the sample, correspond to lateral displacements of permissible walls and longitudinal changes of needle domains that they can constitute. Are differentiated the partial reversals due to experimental processes as in ferro-elastics and those from a mono-domain to a mono-domain of opposite polarizations (switching). Only the latter contain all types of interactions between domains and sample edges.

Hysteresis cycles with linear parts often characterize a reversal by lateral displacement of permissible walls. These walls can interact by repelling each other ((−+) and (+−)), or attracting a wall by a constituted domain with relay phenomena. Their dynamics depend on the values of the local parameters (*P* and *u*) within the domains they delimit, parameters that change in some crystals during the reversal contrary to the hypotheses often assumed. If the wall density is low, relay phenomena can be described as translation of a polarity front perpendicular to the walls. On the contrary, for the highest wall density, the lateral displacements are small compared to the domain width but violent, with the slope of the hysteresis cycle almost vertical at the coercive field.

For other experimental conditions (nature of the crystal, temperature, state of the sample edges), needle domains are created with at their tips, assemblies of quasi-dislocations of wedge type for the retained ferro-ferro crystals. The dynamics of these tips is an illustration of the dislocation theory with variation of the tip shape, interactions of tips with each other and with the sample edges. The shape of the quasi-static hysteresis cycles allows to dissociate the microscopic phenomena (nucleation, interactions, displacements).

The phenomena at the level of elementary cells, of cell sets in domain walls, of domains and sample edges, and the accumulation of these phenomena explain the macroscopic results. Thus, the macroscopic laws of reversal in e^{-E/E^*} are explained by the dynamics of two-dimensional nuclei on the walls, with small electrically charged knees of quasi-dislocations. The macroscopic laws observed in $k(E - E^{''})$ cumulate friction laws at the tip level, different tip critical fields, interactions between tips themselves and between tips and sample edges, avalanche phenomena (accelerations and brakes) at the microscopic level, to result in macroscopic friction laws. Overall, depending on the temperature and the amplitude of the conjugated fields, either macroscopic responses in e^{-E/E^*} appear, or others in $k(E - E^{''})$. In addition, the switching time t_m as a function of the coercive field E_c corresponds to a linear law for low t_m and E_c, or non-linear (t_m as E_c^3) for larger values at low temperatures towards the "freezing" of domains.

Simultaneous studies at the microscopic and macroscopic levels also explain the end of switching faster than predicted by classical models: the interactions between obligatory singularities in the walls and the sample edges create avalanche phenomena. These phenomena are linked to the emissions of electrical (Barkhausen) and to sound and optical pulses observed in the parts corresponding hysteresis cycles.

The link between macroscopic properties (hysteresis cycles and switching current) and microscopic phenomena is established, but questions remain:

- Why is the coercive field E_c almost constant over a significant temperature range (the "plateau")?
- How to explain its sudden increase at low temperatures (freezing of domains)?
- How to dissociate in the macroscopic results what is due to the domains and the walls?

- How to explain the minimum width of a needle domain?
- What phenomena can exist within a wall?
- What are the impacts of the deformations u_{xx} and u_{zz} despite their low value compared to u_{xy}?

…

Some answers to these questions are given in the next chapter dedicated to susceptibilities, with applied excitations of small amplitudes, at various frequencies, to test the dynamics of domains and walls without significantly modifying the overall domain textures.

Chapter 4
Susceptibilities and Domains

4.1 Experimental Procedures and Susceptibilities

4.1.1 Sensitivity and Non-linearity

Chapter 3, dedicated to the switching phenomenon of the macroscopic polarization and deformation of a ferro-ferro crystal, has helped to understand the phenomena at the level of domains, local deformations, and interactions between walls. To understand what is happening at the wall level better, the application of small excitations using periodic electric fields (or mechanical stresses) is appropriate. Dielectric susceptibilities (or elastic constants and quality factors), polarization responses (and deformation) with related losses, are usually defined "at zero field". This assumes the use of conjugate fields of sufficiently low amplitude to evaluate the slope of the $P(E)$ (or $u(\sigma)$) variations. Experimentally, these amplitudes are the smallest that can be used with significant sensitivity. This is not a problem if changing the amplitude of the excitation does not significantly change the result. This is the case in para phases, rarely in ferroelectric and ferroelastic phases. For example, Fig. 4.1a shows the variation of the dielectric constant measured in KDP at various temperatures, in the ferro phase, as a function of the amplitude of the electric field of measurements, which varies over four orders of magnitude (Bornarel, 1972). Even if the logarithmic scale "flattens" the variations, it can be seen that the measurement results are little changed when the amplitude of the measurement field is between 10^{-3} and 10^{-1} V/cm, but an effect beyond these values is evident and very variable depending on the temperature. As the response of a ferro-ferro is often non-linear, the experimenter must ensure that the measurement process does not make the exploitation of the result erroneous. In this sense, Fig. 4.1b proposes a measure of non-linearity which encourages caution in certain temperature regions. Such variations in non-linearity have been studied by some authors (Choi and Lockwood, 1990; You and Kim, 1997; Mueller et al., 2004; Mokry, 2008).

J. Bornarel, *Ferroic Crystal Domains, Walls, and Phase Fronts*, Springer Series in Solid-State Sciences 209, https://doi.org/10.1007/978-3-031-98099-2_4

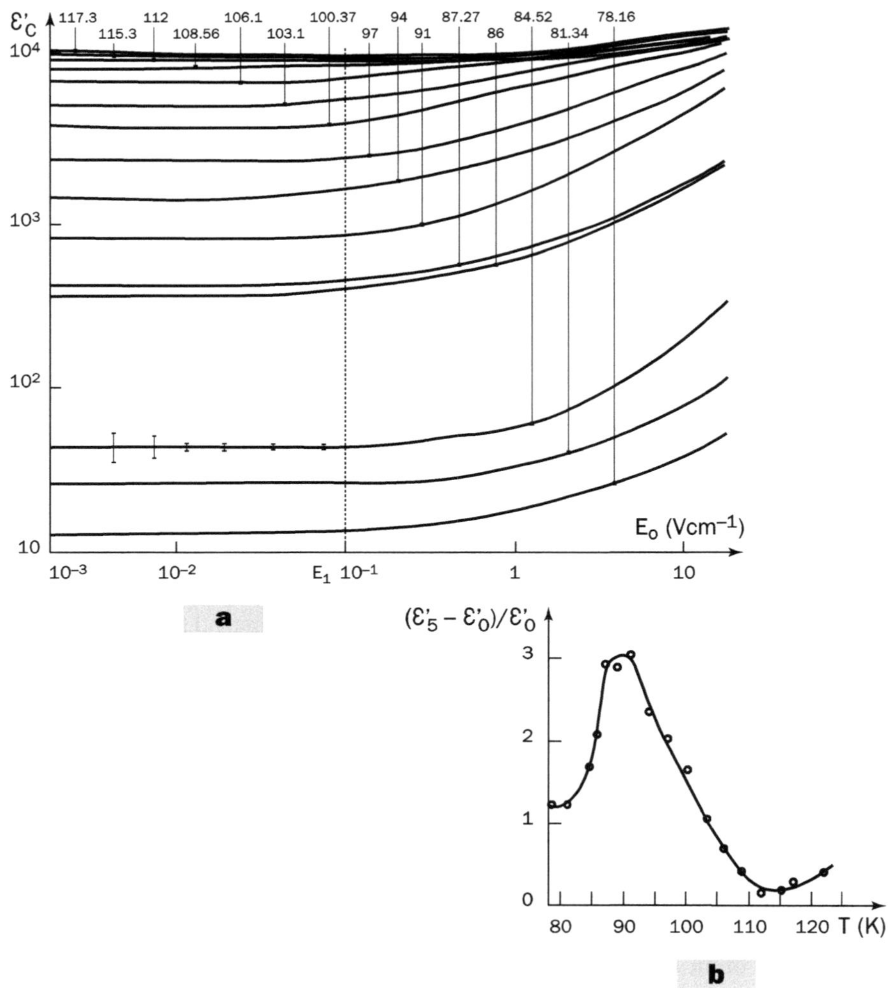

Fig. 4.1 (**a**) Variation of the dielectric constant ε'_c as a function of the amplitude of the measurement electric field E (at 1 kHz) at various temperatures indicated on the graph in Kelvin degrees (for the KDP crystal). In this logarithmic scale representation, its variation is notable for a measurement electric field amplitude >10^{-1} V cm^{-1}. (**b**) The variation of $(\varepsilon'_5 - \varepsilon'_0)/\varepsilon'_0$ with temperature is an indication of non-linearity of the dielectric response (ε'_5 is the dielectric constant measured with an electric field amplitude of 5 V cm^{-1} and ε'_0 is the value considered "at zero field")

In addition, the operating principles of susceptibility measuring devices deserve attention. This is particularly true for dielectric measurements and the comparison of results between measurement bridges and other impedance meters, on a sample whose loss constant $\boldsymbol{\varepsilon}''$ varies in significant ratios. For example, the impact of calculations in parallel resistance or in series resistance in the case of Shering type bridges to qualify the dissipation factor must be analysed; it is also necessary to know the principle of automatic devices before concluding to an effect deemed specific to the studied crystal. The danger can come from an impedance meter which automatically performs numerous measurements, calculating average values, and offering multiple possibilities of programmed calculations: To achieve this result, the device with its "artificial intelligence" is quite capable of automatically increasing, to ensure good sensitivity, the amplitude of the excitation on the studied crystal without announcing the value actually used!

4.1.2 *Sample Quality*

The quality of the sample and that of its orientation play a major role: Fig. 4.2 presents a poorly oriented sample in relation to the tetragonal planes whose dielectric constant measurement reveals the existence of a significant latent polarization. Such samples are to be excluded as those deformed by their mode of fixation (Fousek et al., 1976).

The quality of the polishing and metallization, as noted previously, can modify by a factor of 1000 the value of the susceptibility deduced from capacity measurements. For example, in KDP at the transition, under the best conditions, the measured value for ε' in the direction of the ferroelectric axis, relative dielectric constant, can be greater than 10^6. But this measurement can evolve over time, and the calculated value loses a factor of 10 after a few cycles of temperature, electric field, etc.

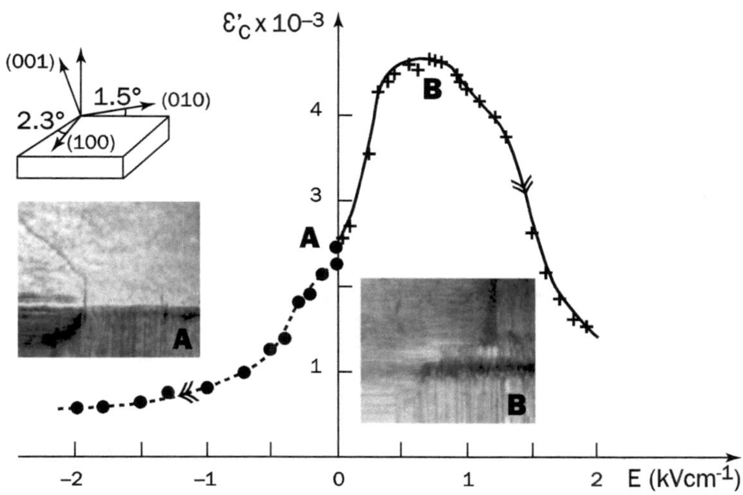

Fig. 4.2 This KDP sample at 87 K is disoriented with respect to the tetragonal planes, which creates a latent polarization with, at zero applied field, mono-domain zones in (**a**). The application of an electric field allows to leaf through the texture (and to obtain a maximum dielectric constant value in **b**), but the texture in domains will retain several domain sets and poorly reproducible dielectric properties

The important thing is not to momentarily reach records, with remarkable conditions that could not last during the duration of the experiments, but to obtain reproducible experimental conditions. In this sense, gold electrodes evaporated under chrome vapour often reconcile the quality of electrical conductivity of gold and the property of adhesion of chrome on crystals. The result is lower in calculated dielectric constant value, but of good quality concerning reproducibility.

The effect of "surface layers" has been studied with many crystals such as TGS, $BaTiO_3$ (Wieder, 1957; Drougard and Landauer, 1959; Nettleton, 1967; Mokry et al., 2001) and also regarding ferro-ferro crystals (Bjorstam and Oettel, 1967; Höchli, 1975; Mansingh and Prasad, 1976). These works will not be developed here. Two aspects are to be dissociated: the understanding of the "surface layer" and the incidence of surface layers on the macroscopic dielectric results which concern us here. The precautions described below ensure the robustness of the experimental results in which the existence of domains is involved. We must not forget that the thickness of a "surface layer" is estimated at one-tenth of a micron. Similarly, only low-temperature crystals are studied here to abstract from problems of conductivity and space charges as observed in the GMO (Tikhomirov 1996, 1997).

4.1.3 *Thermal History and Quenching*

The temperature history is a major parameter for the susceptibilities of ferro-ferro crystals, with their domains and their twins. It modifies the domain textures as illustrated in Fig. 4.3: the dielectric constant measured at decreasing temperature is always greater than that measured during a rise in temperature, unless the

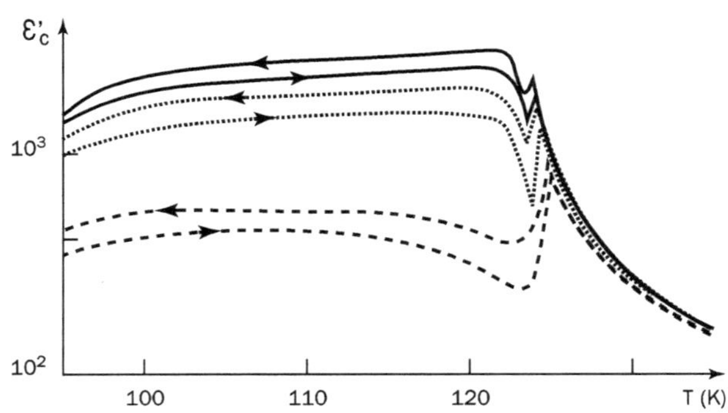

Fig. 4.3 Variation of the dielectric constant measured along the ferroelectric *c* axis as a function of temperature during several para-ferro and ferro-para temperature cycles: the values obtained during cooling are always higher than the values measured during heating, the latter corresponding to domain textures closer to equilibrium textures (KDP with $t = 0.9$ mm, E_0 measurement field amplitude at 1 kHz of 2.5 V cm^{-1}, with a continuous field E applied at 1.1 kV cm^{-1} (continuous lines); 2.2 kV cm^{-1} (dotted lines); 6.6 kV cm^{-1} (dashed lines))

permanent application of a strong electric field continuously makes the crystal mono-domain. This is true even for low rates of temperature change. The domain texture at decreasing temperature is indeed progressively quenched and, by successive rearrangements, more or less frequent, it simplifies. The application of a continuous external excitation (E or σ) can facilitate rearrangements towards a simpler texture, but that of a suitable alternating field is more efficient. We can obtain a texture close to the equilibrium texture without application of external electric field if the cooling rate remains below a value that depends on each crystal (10^{-2} K min^{-1} for KDP and its family). But, despite this precaution, a texture comprising several sets of domains ("species") may exist and complicate the phenomena and the domain rearrangements.

Figure 4.4 summarizes the phenomena observed in a crystal cooled to a certain temperature, of good quality but comprising several sets of domains. The first polarization curve leads to notable rearrangements. A few cycles of complete quasi-static hysteresis are then necessary to obtain a simplified texture and reproducible results.

Even when domain rearrangements allow the achievement of a simple texture, with only one wall direction, the result illustrated in Fig. 4.3 remains: at increasing temperature, the measured value of the susceptibility is lower, and the domain texture is more stable than that obtained at decreasing temperature.

Thus, the domain texture with only one permissible wall direction, obtained during a simple cooling cycle, rarely corresponds to the equilibrium texture; this can be sought at a regulated temperature T by measuring the dielectric constant which decreases over time until a certain stabilization, in a few tens of minutes or a few hours depending on the temperature (the "after-effect"). Annealing, carried out by

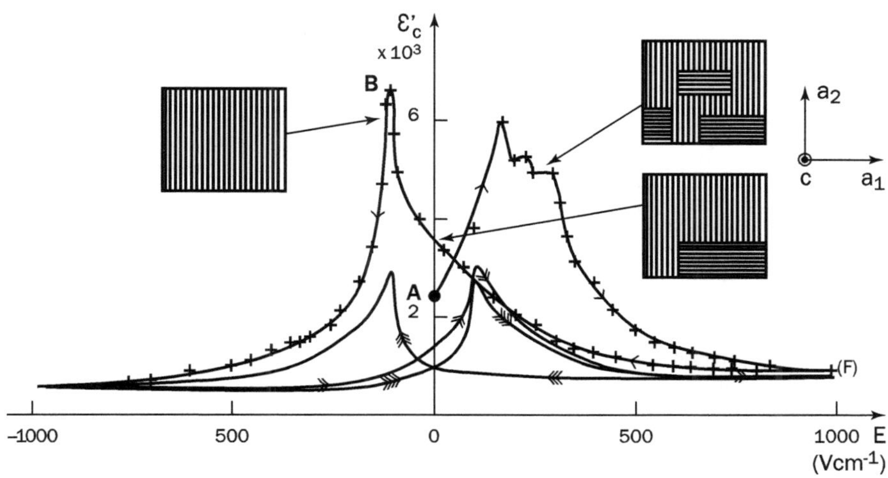

Fig. 4.4 Variation of dielectric constant as a function of the applied electric field during quasi-static hysteresis cycles: the first application of the field between A, F, and B simplifies the texture. The domain texture only becomes regular and reproducible after two or three electric field cycles (KDP at 90 K)

slow temperature evolutions, also contributes to improving the stability of the texture and the results of dielectric measurements.

4.1.4 Treatment to Obtain a Domain Texture at Equilibrium

To obtain at a given temperature a stable domain texture and reproducible dielectric results, it is effective to perform an adiabatic depolarization similar to the adiabatic demagnetization in ferromagnetism. The process is presented in Fig. 4.5: an alternating electric field of sufficient amplitude to trigger small rearrangements of the texture is applied to the sample (low-frequency electric field from a few Hz to a few tens of Hz) with an amplitude decreased over time following a law that depends on the temperature (Fig. 4.5b). Thus, the mosaicity of the crystal, initially of average quality (Fig. 4.5a), is improved as shown by the γ diffraction peak becoming finer and more symmetrical (Fig. 4.5c).

A combination of thermal and electrical treatments allows to obtain textures that include only one direction of domains and values of dielectric constant and loss constant stable at a certain temperature. Figure 4.6 presents a principal diagram of

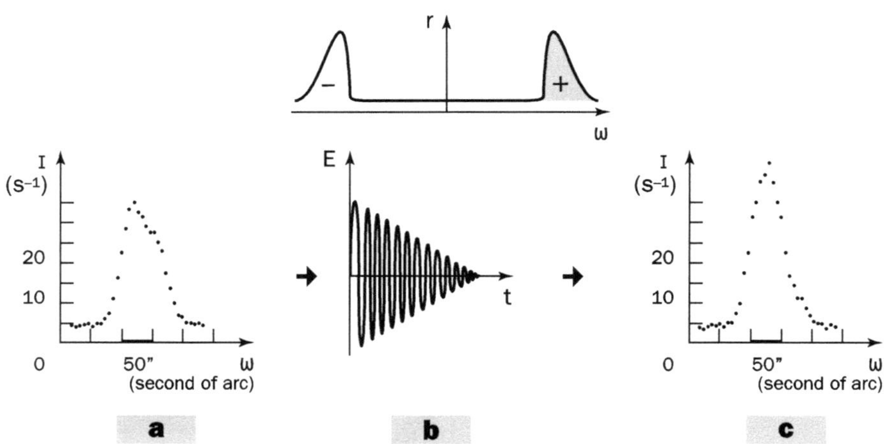

Fig. 4.5 The γ diffraction curve corresponds to a domain texture already simplified with a single wall direction (Annexe A3); the two peaks correspond, for the one on the left, to the domains of polarity, and for the one on the right, to the domains of polarity+. (**a**) The peak corresponding to the + domains, obtained by simple thermal treatment, is not symmetrical; a certain mosaicity is observable. (**b**) Electrical treatment by application of a 50 Hz frequency field and decreasing amplitude until it cancels out in 3 h (starting amplitude of 100 V cm^{-1}). (**c**) The peak of the + polarity domains becomes symmetrical after the adiabatic depolarization treatment; its full width at half maximum (FWHM), initially 90 arc seconds, is reduced to 60 arc seconds

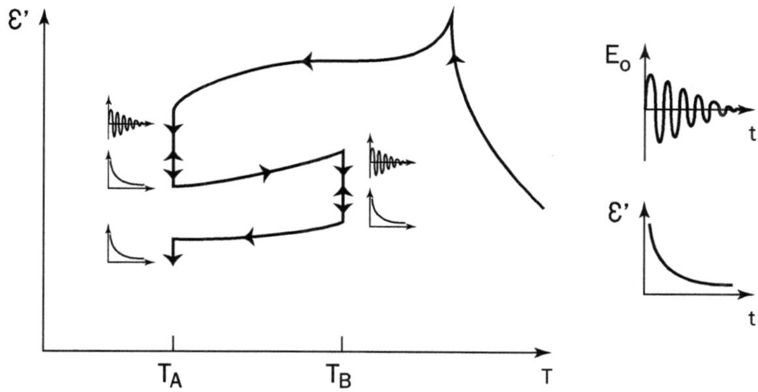

Fig. 4.6 Representation of a succession of thermal and electrical treatments that allow to obtain, for example, at temperature T_A, "stabilized" domain textures, i.e., whose dielectric values remain unchanged for a time >15 min. The pictograms on the curve and on the right (enlarged diagrams) indicate an adiabatic depolarization electrical treatment (top) and a thermal regulation until stabilization of the dielectric constant with the evolution of the "after-effect" (bottom)

such treatments. This process is used especially for the study of macroscopic dielectric properties as a function of the total lateral area of domain walls, presented in Figs. 4.26 and 4.31.

Obtaining a texture at equilibrium is not easy during measurements of elastic susceptibility at low and very low frequencies. Indeed, the methods used, by the application of a single force ("parallel plate stress PPS") or by the "three-point" ("three-point-bending TPD") device with dynamic analyser (such as DMA Perkin-Elmer), combine the application of a variable force and that of a static force slightly higher than the amplitude of the sinusoidal excitation so that the exciting element does not lose contact with the sample. This widely used method (Schranz and Havlik, 1994; Schranz, 2011; Puchberger et al., 2018) certainly allows interesting measurements of Young's modulus.

Comparisons with dielectric measurements are numerous, but the difference in experimental conditions is obvious: the crystal always keeps the same sign of polarity (deformation). It is therefore not in a situation of equilibrium of the domain texture "at zero field", but rather in that of a first polarization curve as illustrated in Figs. 3.3 and 4.2. Thus, when the experiment is carried out over a long enough time, the crystal approaches a mono-domain situation. To be close to the equilibrium texture, the setup described in Fig. 1.3a must be used as illustrated by Figs. 1.39 and 2.3; it allows for the alternation of positive and negative shear stresses. The similarity of the results obtained by the application of an electric field (following *c*) and by a shear stress perpendicular to this axis can be verified.

4.2 Domain Textures

4.2.1 Responses of a Mono-Domain or Poly-domain Sample

To study the phenomena related to domains by measurements of dielectric constant and loss constant, a small amplitude alternating field is applied. The materials chosen in this work are piezoelectric in the para phase and therefore subject to resonances induced by this periodic electric field (or this mechanical stress). Figures 4.7 and 4.8 illustrate the situation: the dielectric measurements correspond to the parallelepiped *abcda'b'c'd'* of Fig. 4.7, and resonances occur each time the geometric conditions are met, observable between crossed polarizer and analyser for light propagating along the crystallographic *c* axis. The characteristic wavelengths of resonances vary with elastic properties, and therefore with temperature (Torche, 1986). The same type of observation can be made with mono-domain samples in the ferro phase. Figure 4.8 gives the variations of this electric capacity as a function of the excitation frequency (KDP sample). The lower curve corresponds to a crystal in the para phase with piezoelectric resonances detected at frequencies between 10^5 Hz and 10^6 Hz. With other sample geometries, these frequencies could be slightly lower (a few 10^4 Hz). The upper curve corresponds to measurements made in the ferro phase, with a domain texture "at equilibrium". New resonances and relaxations are observed: some recall those observed in the para phase, others appear at lower frequencies, related to the existence of domains. This result explains the frequency of the measurement field most often used, of about one to a few kHz at the same time, to observe the dynamics of the domains (by stroboscopy) and avoid the piezoelectric resonances related to the shape of the sample. The correlations between domain texture and dielectric results will first be described with low-frequency excitations. Then, the frequency variations will be presented and interpreted at the microscopic level.

4.2.2 Sample Shape and Domain Texture

What is the impact of domain texture on the measured susceptibility? Domain textures encountered in crystals subjected only to temperature cycles were presented in Chap. 2, with the role of thermal history and that of sample preparation. During dielectric measurements, the making of electrodes and the electrical measurement circuit lead to a situation comparable to that of short-circuited samples; these, in ferroelectric crystals, should not present domains at equilibrium. This is not the case for ferro-ferro crystals as illustrated by the experiments described in Fig. 2.3: the textures are also ferroelastic twins that allow to obtain a quasi-null macroscopic deformation of the sample; a notable depolarizing field effect is observed, especially when the crystal becomes mono-domain (Fig. 1.39). So, is there an effect of

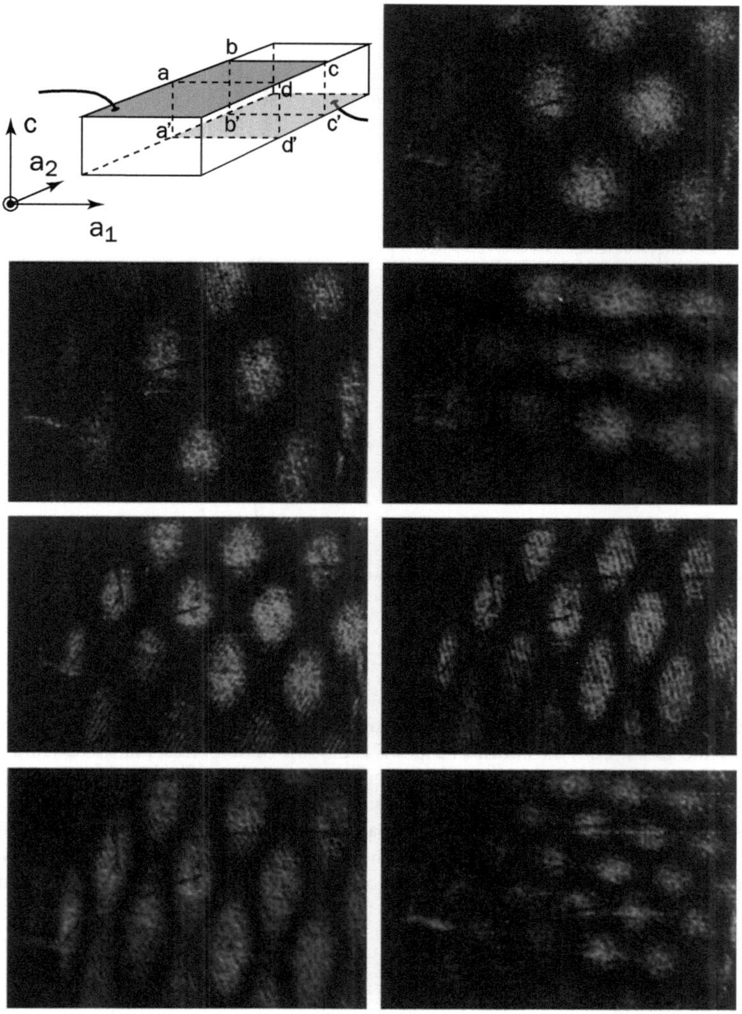

Fig. 4.7 Crossed electrodes deposited on the (001) faces of a "perfectly" oriented sample in the tetragonal planes define a region *abcda′b′c′d′* in which a small alternating electric field is applied. The photographs are obtained in the para phase between crossed nicols for light propagating along the *c* axis (in the KDP). Successive resonances are observed at various temperatures: the temperature decreases in the sequence of photographs (from left to right and from top to bottom)

sample shape on the domain texture and the measured susceptibility? Figure 4.9a summarizes the observations on a few hundred crystals of different origins and cuts: the samples that are too small may not contain domains (Fig. 2.4a) as demonstrated with colloidal particles of dimensions $<10^{-1}$ μm (Jaccard et al. 1953) (the situation of thin layers is different). Domains can exist in a sample of dimensions greater than

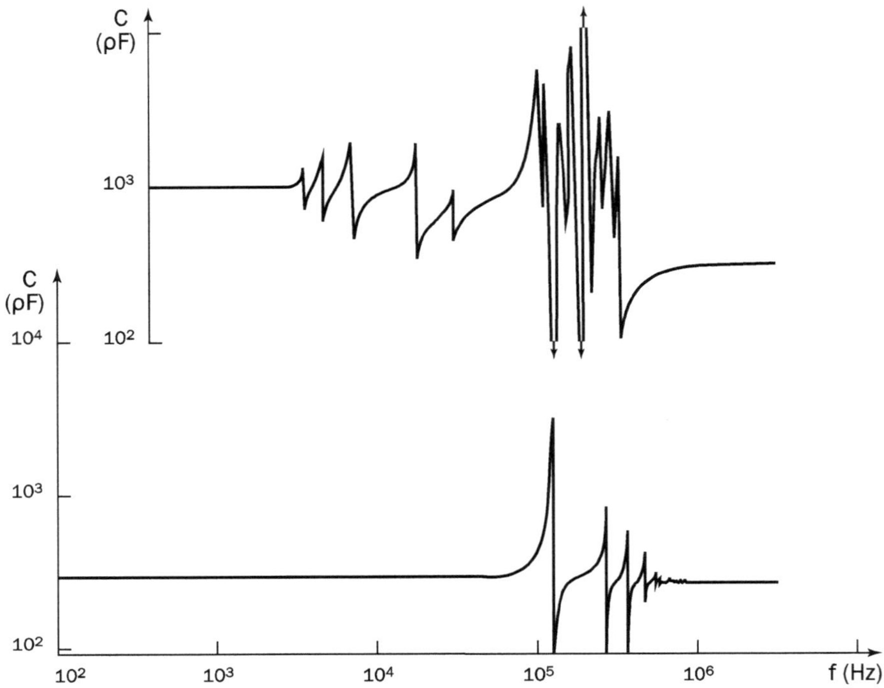

Fig. 4.8 Electric capacity measured as a function of frequency for a KDP sample (with electrodes following the faces (001) as shown in Fig. 4.7): the top curve corresponds to a temperature of 115 K (ferro phase) and the bottom one to a temperature of 129 K (para phase) which allows measuring the resonances due to the shape of the sample without domain. The response in the ferro phase is more complex since to this response related to the shape of the sample are added those due to domain sets, to a given set, to domains, to their tips, to moving walls, and to correlations between these causes. It is more complex than the classic responses of resonators with simple domain textures (Erhart et al., 2008)

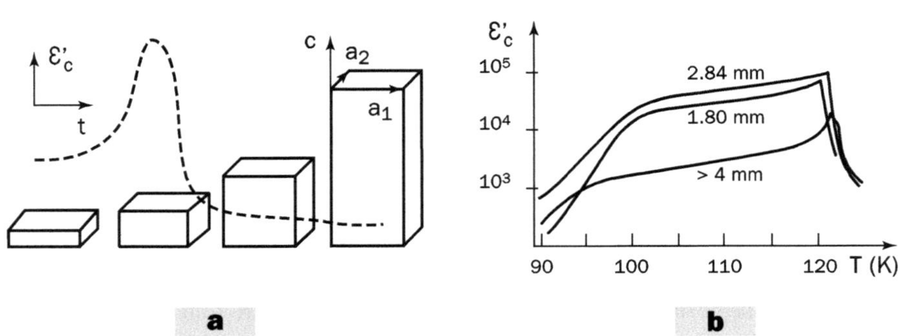

Fig. 4.9 Representation of the variations of the dielectric constant measured with samples of the same surface perpendicular to the ferro axis (a^2) and of different thicknesses t. In coherence with the textures of Fig. 2.4, a maximum of measured dielectric constant is observed for a thickness of about half of a, intermediate situation between the phenomena of "pinning" with small t and those with mono-domain regions in volume if t is significantly greater than the other dimensions

a few micrometres (Gilchrist, 1998; Motegi et al., 1985). If the thickness in the direction of the ferroelectric axis is small compared to the other dimensions, the sample is well described by the platelet of Fig. 2.4b. From an electrostatic point of view, the effect of the depolarizing field, if it exists, can contribute to the presence of an extremely layered texture at equilibrium. From a mechanical point of view, the quasi-dislocations necessary for the movement of a wall, even in the hypothesis of apparently resting walls, undergo a phenomenon of "pinning" on the (001) faces which decreases their mobility. Furthermore, the interaction between the ends of needle domains is observed at distances of the order of the thickness of the sample (cut-off radius). The application of an alternating excitation can break the "pinning" of the ends of quasi-dislocations and facilitate the lateral movement of the walls as the longitudinal movement of the ends. The situation is totally different if the sample is very elongated along the ferroelectric axis (Fig. 2.4d). One can expect a negligible effect of the depolarizing field in all situations (Fig. 1.40), and the existence of domain textures is not justified by electrostatic reasons. Experimentally, regions of poly-domain volume and others of mono-domain are observed (Fig. 2.4d), which create inside the sample, at their limit surfaces perpendicular to the ferroelectric axis, charged sets whose energy contributes to limit their importance. These areas have, from a mechanical point of view, knees and grain boundaries of notable energy, which limits their probability of presence. The poly-domain regions are mainly located near the electrodes, and the central parts of the samples are often mono-domain. However, sometimes we can observe some lamellar domains whose ends traverse the entire sample along c; the mobility of the ends is then very low; their balance is mechanically difficult: these lines of quasi-dislocations interact with their image forces in relation to the four faces (100) and (010) of the sample and with all the other lines of quasi-dislocations of the sample (the cut-off radii are larger than the size of the sample along a_1 and a_2). This is shown by the application of a small excitation: their field of movement is high. Furthermore, crystals in the form of rods elongated along c encounter torsion problems; thus, there are torsions at the transition whose amplitude increases as the temperature decreases, and in addition, significant thermal hysteresis if the textures are complex (Krainyuk and Otko, 1990; Binder et al., 2000; Gridnev and Kravchenko, 2000; Gridnev and Kosilov, 2002).

This quick overview of the effects of shape and dimension of the samples on the measured susceptibility encourages us to take a certain distance from the physical measurements published without knowledge of the domain textures: dielectric measurements are often carried out with samples in the form of platelet; those of diffraction for example γ diffraction, rather on samples of shape close to the cube; others for torsion measurements or X-ray measurements, neutrons, with samples in the form of rods. We must question, in each situation, the specificity of the texture in domains.

4.2.3 Domain Species and Susceptibility

The results proposed in this book correspond, except for certain bibliographic citations, to relatively simple textures. Thus, a sample whose shape is between the optimal platelet for dielectric measurements ($t = a/2$) and the cube interesting for γ diffraction measurements. Simultaneously carried out are stroboscopic optical observations of textures, dielectric measurements, and γ diffractions.

The domain texture represented in Fig. 4.10a includes two sets ("species"). The γ peak corresponding to the + polarity domains of region II under an alternating excitation of + polarity and opposite polarity – is specially studied using synchronized detectors (see Annex A3) (Bornarel and Bastie, 1980). The frequency of the alternating field of measurement is 1 kHz and its amplitude varies from 0 to 125 V/cm. The shape and position of the diffraction peak vary significantly with the amplitude of the applied electric field. Its width increases with the amplitude of the field. To accurately know the variation of the deformation inside the domains given by the

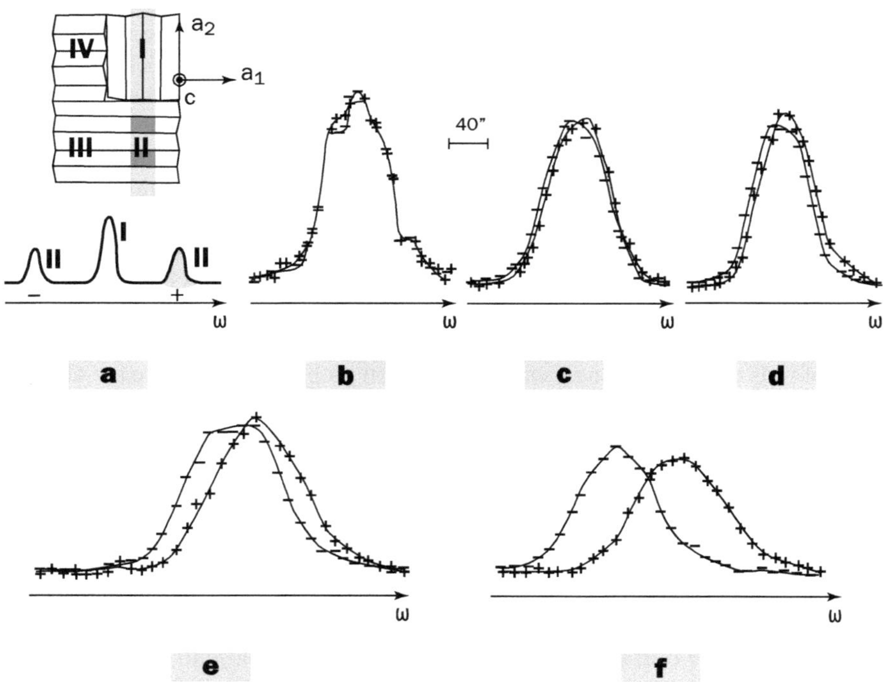

Fig. 4.10 (**a**) Schematic representation of a KDP crystal with two sets of domains. For the beam of γ rays indicated in grey, we obtain by diffraction on the lattice planes, a scanning curve that corresponds to the domain family I (central peak), to the domains – (on the left) and + (on the right) of the family II. (**b–f**) The peak corresponds to the + domains (on the right in **a**) of the II region; the crystal is subjected to an alternating field whose amplitude is varied at a frequency of 55 Hz; as explained in Annex A3, it is possible to obtain a plot for the – polarity and one for the + polarity of the applied electric field; the amplitudes of the alternating field are: (**b**) 0 V cm^{-1}; (**c**) 62 V cm^{-1}; (**d**) 75 V cm^{-1}; (**e**) 100 V cm^{-1}; (**f**) 125 V cm^{-1}

position in ω, it is necessary to take into account the possible rotations of the sample. Very precise measurements were carried out simultaneously at the transition, on the same lattice planes, in γ diffraction at small angles and in neutron diffraction at large angles (Bastie et al., 1987). The rotation phenomenon that can accompany the shear is complicated in crystals with several species of domains by the symmetry operations specific to the crystal concerned as illustrated by Figs. 4.11 and 4.12: because of the inversion operation lost at the transition in the KDP, in Fig. 4.11a, the

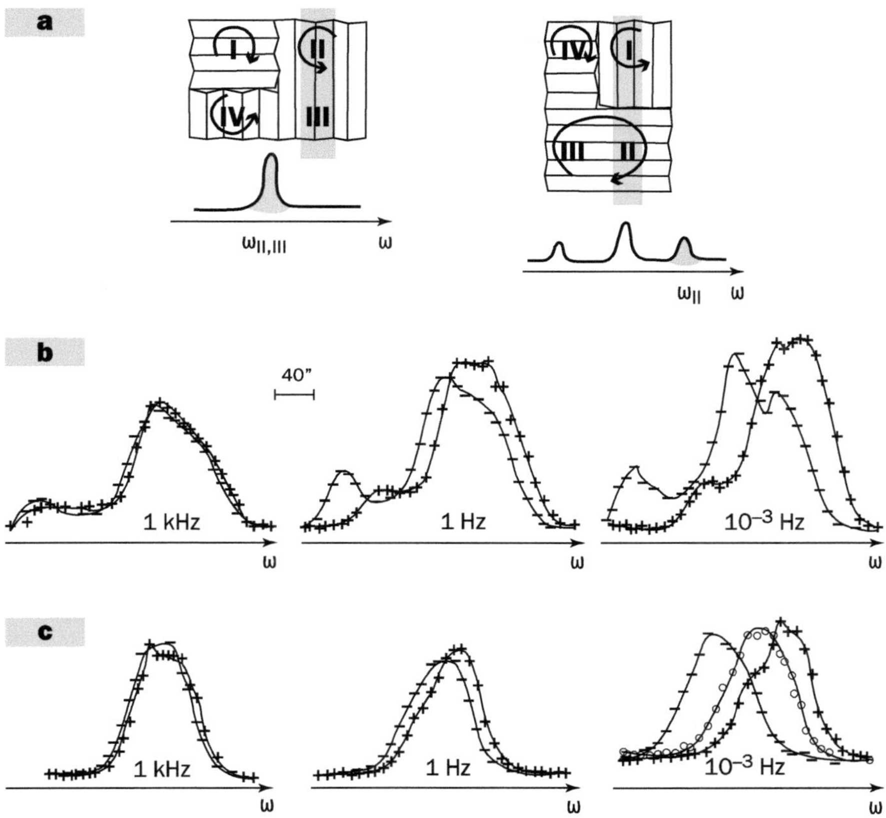

Fig. 4.11 (**a**) The diffraction curves correspond to the beam of γ rays indicated in grey; arrows remind on the sample that an applied electric field of one polarity will cause shear-rotation of opposite directions between sets of domains whose permitted domain walls are almost perpendicular (the inversion in the symmetry elements). (**b**) Variation of the shape of the $\omega_{II,III}$ peak of the left diagram in (**a**); the lattice planes concerned correspond both to domains of – polarity and to domains of + polarity; the curves indicated – correspond to the measurements made during the – alternation and those indicated + to the measurements made during the + alternation of the alternating field whose frequency is indicated: the induced rotation effect is highlighted. (**c**) Variation of the shape of the ω_{II} peak of the representation (**a**) on the right. This peak is plotted for each of the polarities of the applied electric field (with 75 V cm^{-1} amplitude for the indicated frequencies) and without applied field. It is observed that it is more difficult to increase the deformation than to decrease it as illustrated by the curves at 10^{-3} Hz in which the same peak is plotted in the absence of applied field (curve with o) (Bornarel and Bastie, 1980)

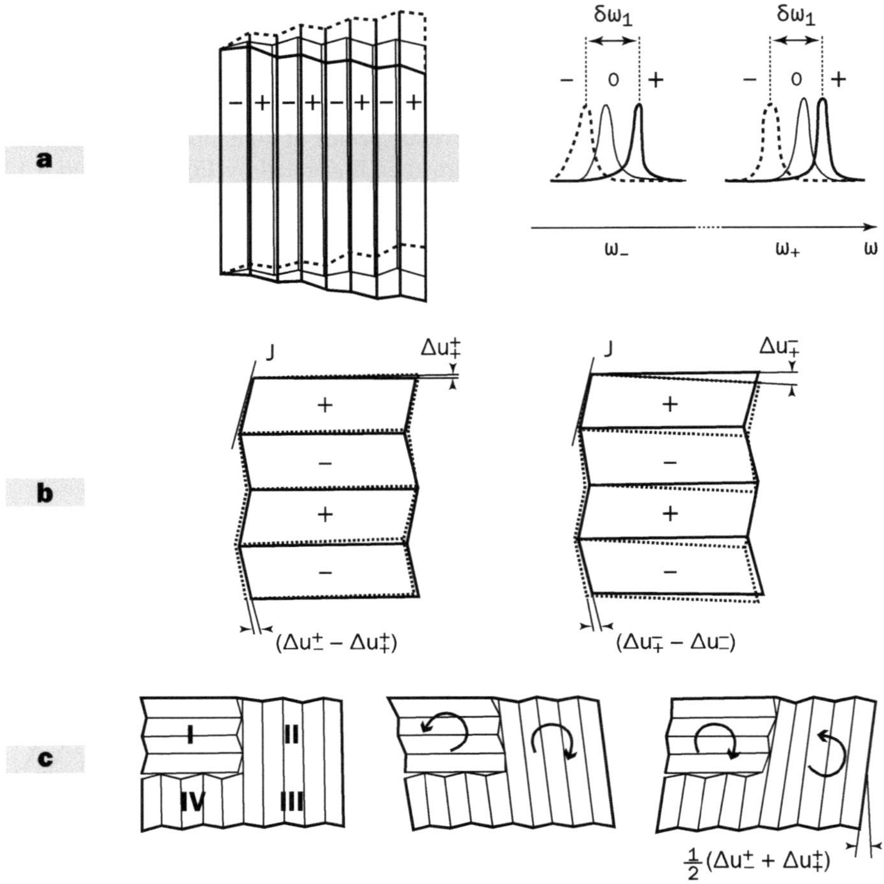

Fig. 4.12 (**a**) Representation of a simple domain texture (only one permitted wall orientation) and diffraction curves that correspond (on the right) to the beam of γ rays indicated in grey; for each of the peaks of the – polarity domains (ω_-) and the + polarity domains (ω_+), the plots are shown at zero field (o) and for each polarity of the applied alternating field. We call u_+ the average deformation of the + polarity domains. The variations are in the upper index, the direction of the applied field in (**b**) and (**c**). (**b**) If the preserved lattice plane is the J plane, the domain wall planes will rotate with the angles indicated for each polarity (+ on the left and – on the right). (**c**) The sample now includes two sets of domains with perpendicular wall plans (left diagram). The application of an alternating field causes deformation-rotations as indicated for the alternation of polarity + (in the centre) and of polarity – (on the right) with the overall rotation indicated

application of a field of a polarity called + induces the rotation indicated by the arrow in region 1, and an inverse rotation in regions II, III, and IV. The simultaneous study of the $\omega_{II,III}$ peak and the ω_{II} peak, which only concerns the + polarity domains, allows the following conclusions:

- It is usual to obtain for a sample with several species of domains antagonistic rotations of the permissible wall planes, rotations which are added to the shears in the measurements.
- For the KDP, the variation of the deformation angle u_+ of + polarity domains, in an electric field of the same polarity, about 10–15 arc seconds in our example, is in modulus about twice less important than its variation in a field of opposite polarity (about 25 arc seconds). Figure 4.12 summarizes the effects of a small amplitude alternating field (and low frequency), on the one hand, on a domain texture with only one permitted wall orientation (Fig. 4.12a, b), and on the other hand, on a texture with two species of domains (Fig. 4.12c). We can continue these representations with a larger number of species. The result is complicated by the internal deformations of the twins. Everything is simpler if these deformations are considered constant, then only the volume modifications of the twins of opposite deformations intervene (Ozaki and Nakamura, 1982).

The studies of many textures of ferro-ferro crystals demonstrate that the incoherent twin boundaries, border regions between species of domains, can provide a significant dielectric contribution in the case of simple species like those schematized in Fig. 4.12c. We understand this result since these regions are in situations far from spontaneous polarization (and deformation), close to a para situation, therefore in metastability. On the other hand, if the number of species becomes important, the incoherent twin boundaries and the domain textures are constrained; the result in the macroscopic measurement of the susceptibility is then a decrease (we are in a constrained context, as in a quenched metal). We can thus encounter paradoxical situations: a very good crystal cut in the tetragonal planes with very good precision, during a very slow cooling cycle, with a beautiful thermal homogeneity of the sample, may include nucleations of domains on the lateral edges of the sample at very many places at the same time, and the resulting texture may, in some cases, be very complex. If the rearrangements are not effective within a few degrees below the transition, the texture, quite complicated, is likely to be constrained when the temperature decreases.

From a mechanical point of view, a texture comprising many species can, during a measurement of macroscopic deformation, give results of deformation of the sample almost null, and it is difficult in some cases to exploit the results, especially static ones, if the texture is unknown.

4.3 Susceptibility of the Mono-Domain Crystal

To study the impact of domain texture on the susceptibility of a ferro crystal, it is necessary to know what its value would be if the sample was mono-domain under the same experimental conditions. The parameters are temperature T, hydrostatic pressure p, and the quantities conjugated to the order parameter, electric field or/and mechanical stress. These impose their amplitude and frequency. The experiments reported in the literature often focus on variations in temperature and frequency without information on the amplitude, which implicitly assumes quasi-linear responses, and should prohibit the use of anharmonic, non-linear models afterwards. Attention to this point should not be relaxed.

In this section, results concerning a sample in the para phase are presented, then those of a mono-domain in the vicinity of the transition.

4.3.1 *Susceptibility and Conductivity in Para Phase*

Let's first recall that there is a correlation between the melting-solidification temperature and the para-ferro transition temperature for the crystals of the KDP family. When the temperature decreases below the melting point, different phase changes occur in the hydrogenated and deuterated compounds. DKDP thus shows a transition tetragonal-monoclinic (Blinc et al., 1975) which illustrates the complexity of the equilibrium in these materials, between the tetrahedral groups held by hydrogen bonds and other ions. In addition, the history of the crystals plays a notable role. This parameter, too often neglected, has now become a fashionable theme of study for about a decade under the general term of "aging".

The conductivity changes according to the temperature, for example, by following an activation law with an energy of 0.58 eV for DKDP (Schmidt and Uehling, 1962), which corresponds to the jump of deuterons from one bond to another. This phenomenon of proton (deuteron) hopping from one bond to another, common at high temperature, becomes difficult, even negligible below a temperature that can be determined for each of the compounds (O'Keeffe and Perrino, 1967). This phenomenon can be forgotten in the context of this book. Other experiments reveal changes in quality factor at temperatures close to room temperature for KDP and around −50 °C for DKDP, interpreted as creations of new phase inclusions in the tetragonal phase, but the role of impurities or defects is not clear (Gridnev and Kravchenko, 2000).

4.3.2 *Curie's Law and Collective Phenomena*

In the temperature region located above the transition, over several tens of degrees or even a hundred, the measured susceptibility χ is usually correctly described by Curie's laws $\chi = C/(T - T_0)$ with values of T_0 that vary a lot depending on the deuteration of the crystals as for the KDP-DKDP pair (about a hundred degrees Kelvin). As illustrated in Fig. 1.37, the temperature T_0 varies according to the frequency of the measurement field. Unfortunately, the accuracy of the experimental results of T_C and T_0 varies from one author to another. The reason is often the difference in temperature between the measuring element and the sample itself, but also a poor evaluation of the thermal homogeneity of the sample. The relative temperature deviations in a given experiment give more usable values. The Curie constant itself varies with the deuteration of the crystal and, for the same crystal, with the frequency.

Refer to Annex A5 to understand the para-ferro transitions of the crystals of the KDP family. For example, concerning the DKDP, the observed Curie's law varies according to the frequency of the measurement field very clearly: concerning the value of T_0, from 218.36 K (1 kHz) to 211.36 K (35 GHz). The Curie constant also evolves with the frequency. The authors explain these results by a relaxation of the dipoles involved with an average characteristic time inversely proportional to the temperature deviation from the transition (from 5×10^{-11} s at 100 K above the transition, to 5×10^{-7} s at 10^{-2} K from the transition) (Hill and Ichiki, 1963a, b).

Let's recall the origin of Curie's law: If the application of an electric field E allows to polarize elementary dipoles of a system without any interaction between them, the electric polarization, the response of the system to this excitation, can be written as:

$$P = \chi_0 E$$

in which χ_0 is the susceptibility of the base element. In a real material, the base elements interact with each other. If we accept the molecular field approximation, we can assume that the local field that summarizes the action of all these interactions is proportional to the overall polarization of the system and write thus (Annex A1):

$$P = \chi_0 \left(E + \alpha P\right)$$

α accounts for the local field. Therefore:

$$P = \left(\chi_0 / \left(1 - \alpha\chi_0\right)\right)E = \chi E,$$

χ susceptibility of the overall system. If we assume for χ_0 a form comparable to that produced by rotating dipoles under the action of an electric field E, or C/T, then the overall susceptibility takes the form

$$\chi = C / \left(T - T_0\right) \text{ with } T_0 = \alpha C.$$

Given the simplicity of the assumptions made, it is surprising to encounter this Curie's law so frequently, a sign of a system in which there are collective interactions between the elementary dipoles. Understanding the phenomena behind these interactions often allows us to model the phase transition that will occur at T_0 if the transition is of the second order.

4.3.3 Susceptibility of the Mono-Domain Crystal Around the Transition

For a crystal in the ferro phase, we can simplistically pose the dielectric susceptibility calculated using macroscopic measurements as the sum of a contribution due to what we would encounter if the crystal were single-domain and another linked to the existence of domains. If we retain the relative dielectric constants, therefore dimensionless, we can by hypothesis consider that the measured value ε_c' is the sum of two contributions:

ε_m' contribution to the dielectric constant due to the mono-domain crystal
ε_d' contribution to the dielectric constant due to the existence of domains

$$\varepsilon_c' = \varepsilon_m' + \varepsilon_d' \tag{4.1}$$

An experiment, summarized in Fig. 4.13, supports this approach: the measurement of the dielectric constant with simultaneous observation of the domains is carried out for various continuous electric fields during temperature cycles. In all experimental situations, the domains appear quickly during cooling, and sometimes even before the maximum of the dielectric constant is reached. This contribution of the domains depends on the ratio between the thickness of the sample and the surface of the electrodes, as illustrated in Fig. 4.14. With the methods illustrated in Figs. 4.5 and 4.9, it is possible to obtain domain textures close to equilibrium. After stabilizing at 20 or 30° below T_C for several hours, the measurements of the dielectric constant then carried out at increasing temperature with a temperature change rate $<10^{-2}$ K min^{-1} give for any temperature, results lower than those obtained at decreasing temperature and, an important phenomenon occurs between points E and F on Fig. 4.13: the sample is usually mono-domain in this temperature range close to the transition, even for relatively weak continuous fields, except in particular cases such as that of Fig. 4.14c. Thus, using the variations in Fig. 4.15a, we can calculate $\varepsilon_m'(T)$ by ensuring the absence of domains in the sample near the transition, then $\varepsilon_d'(T)$ (Fig. 4.15b).

We use the previous process to plot Fig. 4.16, the maximum variation of $\varepsilon_m'(T)$ following c as a function of the external continuous field applied (the amplitudes of the electric field measurements are sufficiently small not to modify the result of the measurements). Simple calculations concerning the maximum of the single-domain dielectric constant and the temperature of this maximum can be carried out using the Landau development, which corresponds to this mono-domain crystal situation. We can, for example, write $G_2 - G_0$ in a very simplified form with, in addition to the term in P^2 (or q^2), only one other term of the development in even powers of the order parameter, assumed to be very much greater than all the others:

$$G_2 - G_0 \approx (a/2)P^2 + (b_{2n}/2n)P^{2n} - EP \tag{4.2}$$

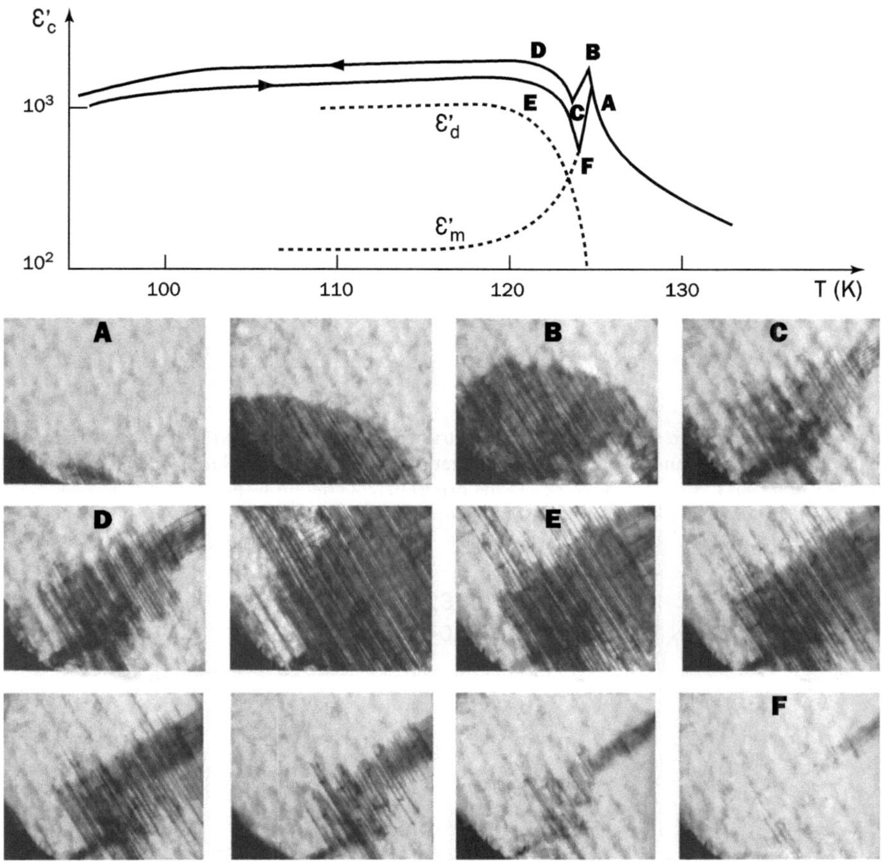

Fig. 4.13 Variation of the dielectric constant following c during para-ferro and ferro-para cycles (KDP sample with $t = 0.9$ mm under a continuous electric field of 2.2 kV cm^{-1} and a measurement field of 2.5 V cm^{-1}, 1 kHz, and a temperature T variation rate of 3×10^{-2} K min^{-1}). During cooling, the domain texture appears in (**a**), grows until (**b**), decreases until (**c**), and then slowly increases. During warming, the texture remains the same, from low temperatures up to (**e**). The domains then gradually disappear up to (**f**). At higher temperatures, the peak of dielectric constant, and its maximum ε_m', occurs with a mono-domain sample. The study of this peak allows to estimate what would be the contribution of the mono-domain as a function of temperature in the hypothesis where the presence of domains in the sample would not change the value of the local polarization and of the dielectric constant of a small volume of the crystal

After minimizing $G_2 - G_0$, the inverse of the susceptibility $\delta E/\delta P$ is calculated, and then the conditions for obtaining the maximum of this susceptibility (isotherm under the conditions of measurements chosen). It will be considered that the maximum of the susceptibility corresponds to that of the measured dielectric constant, given the significant values of the latter. If χ_M is the value of the maximum

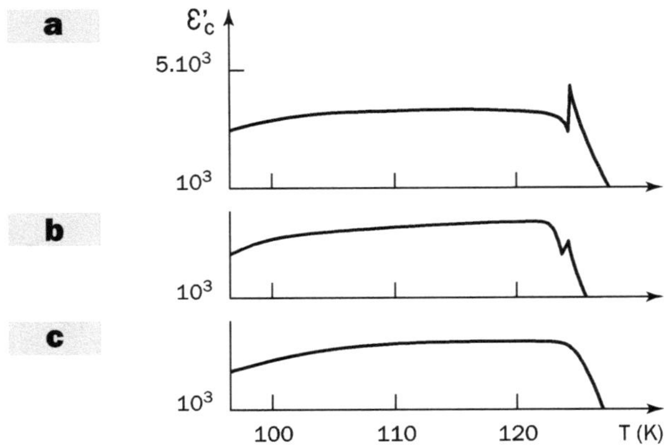

Fig. 4.14 Temperature variation of the dielectric constant measured with electrodes following (001) for three KDP samples that differ by the ratio X between their thickness t and the surface of their section (001): $X_a < X_b < X_c \approx 2X_a$. In the experimental conditions presented ($E_{//} = 1$ kV cm^{-1}, $E_0 = 2.5$ V cm^{-1}and $f = 1$ kHz), the contribution due to the domains ε_d' increases with X (Valla, 1967)

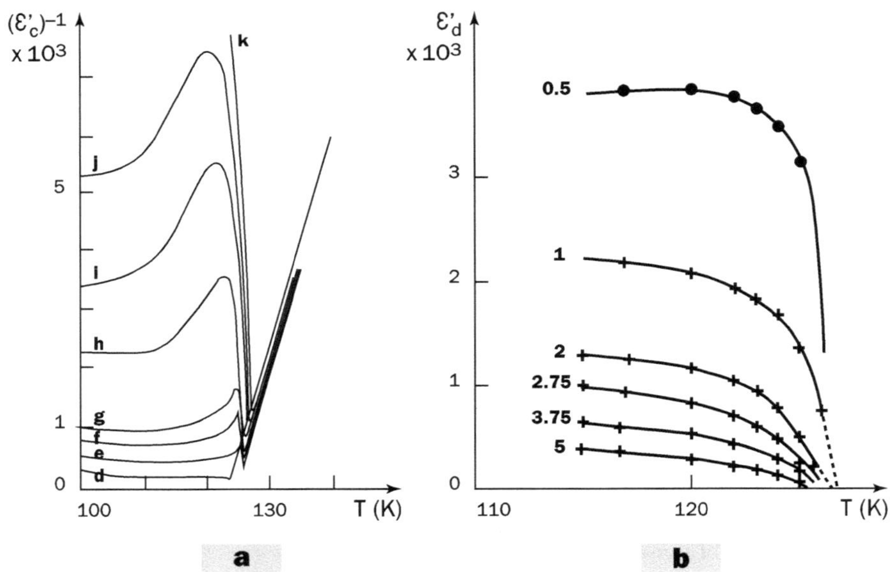

Fig. 4.15 (**a**) Inverse of the dielectric constant $(\varepsilon_c')^{-1}$ deduced from macroscopic measurements as a function of the temperature T of a KDP sample, during temperature rises under applied continuous electric field $E_{//}$ (values of $E_{//}$ in kV cm^{-1}: $d = 0$, $e = 1$; $f = 2$, $g = 2.75$, $h = 5$, $i = 6.25$, $j = 7.5$, $k = 8.75$). The linear parts correspond to a mono-domain sample. (**b**) If we consider the dielectric constant measured during a temperature rise as the sum of a contribution from a mono-domain crystal ε_m' and another linked to the existence of domains ε_d', we can deduce from the curves of (**a**) the values of ε_d'. The values of the applied continuous electric field are indicated in kV cm^{-1} on the curves

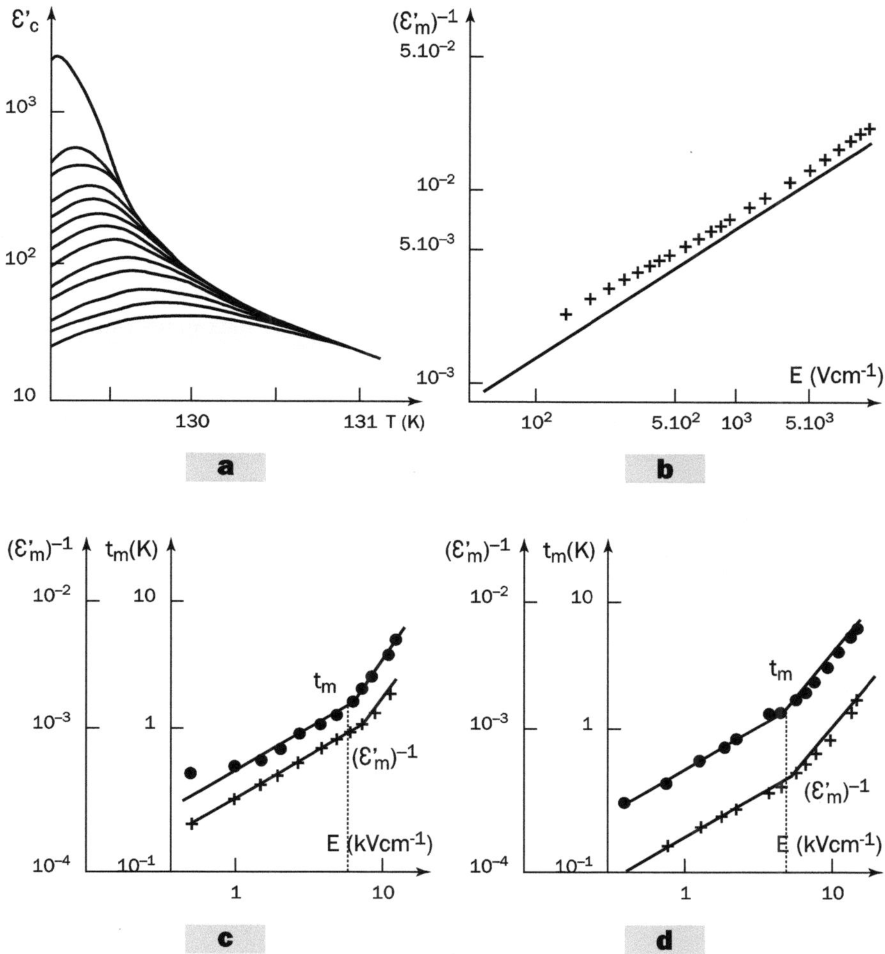

Fig. 4.16 The maximum of the dielectric constant ε_m' at the transition of a para-ferro crystal and the temperature at which this maximum occurs t_m can vary depending on the field conjugated to the order parameter, here the electric field. These variations, if the effect of the domains is negligible, provide information on the order of the transition as shown by the calculation in Chap. 1 using a classic Landau development. (**a**) Variation of the dielectric constant as a function of the temperature with the continuous field applied in the TSCC crystal whose transition is of the second order (Bornarel and Schmidt, 1981). (**b**) Variation of $(\varepsilon_m')^{-1}$ with the continuous field applied E in logarithmic scales corresponding to the results of the TSCC from (**a**). A straight line with a slope of 2/3 with a continuous line shows that the terms of the development in P^2 and in P^4 describe this transition well. (**c**) Our measurements and (**d**) those of Baumgartner (1950, 1951) for KDP carried out under different experimental conditions lead to quasi-similar results concerning $(\varepsilon_m)^{-1}$ and $t_m = T - T_0$ as a function of the continuous field applied E (T_0 is the temperature at which the dielectric constant is maximum when $E = 0$). We observe the particular situation of KDP (the continuous lines in these logarithmic scale representations indicate the slopes 2/3 and 4/5): it can be described for low values of the continuous field using a Landau development with terms in P^2 and P^4 (variation in $E^{2/3}$), and for larger E values with terms in P^2 and P^6 (variation in $E^{4/5}$)

susceptibility, with $t_M = T_M - T_0$ (T_M temperature at the maximum of susceptibility), the following relations are obtained:

$$1/\chi_M = a\left(2n/(2n-1)\right)t_M \tag{4.3}$$

and

$$t_M = E^{(2n-2)/(2n-1)}\ f\ \left(a,b_{2n},n\right) \tag{4.4}$$

f being a function of a, b_{2n}, and n. In this simple model, the temperature of the maximum of the dielectric constant t_M, and the inverse of the maximum value of this latter vary with the applied field E following a law in $E^{2/3}$ if $n = 2$, in $E^{4/5}$ if $n = 4$, in $E^{6/7}$ if $n = 6$, etc.

It is observed in Fig. 4.16a, b for a well-known ferroelectric for its second order transition, the TSCC crystal (Bornarel and Schmidt, 1981), the obtaining of slopes in 2/3 in the representations plotted in logarithmic scales of the maximum inverse dielectric constant as a function of E.

Regarding KDP, the application of an electric field E less than 5 kV/cm induces a transition comparable to a second-order transition, but for values of E greater than this value, the term in P^6 becomes predominant compared to the term in P^4 which is only possible if the coefficient of this latter is very small. This result confirms the proximity of the tri-critical point of this para-ferro transition (Troussaut et al., 1981) and explains the situations often described concerning KDP in which the authors argued for a first-order transition somewhat masked by long-distance constraints that suggested a second-order transition (Reese and May, 1967). We will return to the transition in Chaps. 5 and 6 by choosing a crystal whose transition is clearly of the first order, the DKDP, to study the interfaces, that is to say, the phase fronts and the domains, centres of interest of this book.

4.3.4 Relevance of the Relationship $\varepsilon_c' = \varepsilon_m' + \varepsilon_d'$

The strategy of dissociating the contributions to the susceptibility of the mono-domain crystal, on the one hand, and the domains, on the other hand, seems easy to implement. But the idea of removing from the measured susceptibility the contribution of what it would be if the sample were mono-domain remains a convenience approach to be handled with caution. Indeed, the presence of domains, of walls, can sometimes lead to local variations in physical characteristics and, in other situations, modify them throughout the sample. The following sections illustrate this by focusing each on a particular temperature range: the few degrees below the transition, the so-called plateau region, the domain freezing region and finally the often neglected, very low temperature region. This highlights the richness of the phenomena that can occur in a ferro-ferro crystal subjected to a small excitation.

4.4 Contribution of the Domain Texture to the Susceptibility in the Vicinity of the Transition

4.4.1 Domains Under the Transition: "A Solid Modulation of a Distorted Structure"

In the few degrees below the transition temperature, rapid variations with temperature are observed, of the coercive field and the polarization-deformation-birefringence for light propagating along the ferroelectric axis. Simultaneously, the domain texture changes noticeably. The differences can be significant between crystals with a clear first-order transition like DKDP, and others with second-order transitions or close to the tri-critical situation, for example, second-order RbDP and first-order KDP. In the first case, we observe at various temperatures sudden rearrangements that change the number of domains and therefore the total surface area of the walls. In the second case, in addition to these changes, domain textures appear in optical observation, with a low contrast just below the transition, contrast which improves as the temperature decreases, therefore when the "polarization-deformation-birefringence" increase. The phenomenon is easily understood given the causes of domain visibility in polarized light as in white light, previously explained (Fig. 2.8). Laser beam diffraction studies confirm direct observation: in this temperature region, the diffraction pattern is wide, disordered, and stretched, and it is only at lower temperatures that it becomes truly periodic (Hill and Ichiki, 1964; Velichkina et al., 1969; Ha and Kim, 1985). Similarly, we note the low contrast of X-ray topographies (Authier and Petroff, 1964; Aleshko-Ozhevkii 1982) and the estimated widths of domain walls using X-ray diffraction (Afonikova et al., 1987; Andrews and Cowley 1986) which increase when $(T_C - T)$ decreases (Fig. 2.11b). The use of X-ray diffraction results on the assumption of a continuum (Bruce, 1981) excludes the hypothesis of rough transition in the walls (Lajzerowicz, 1980) and leads to wall widths of about ten cells at $T_C - 2K$ below the transition for KDP, and to even larger widths closer to the transition.

Gamma diffraction confirms these observations: a good quality crystal shows notable extinctions concerning the characteristic reflections in the para phase. These extinctions disappear at the transition (Bastie and Becker, 1984). The same is true in neutron diffraction (Balagurov et al., 1979). The gamma diffraction scan curve of RbDP shown in Fig. 1.13 (Bastie et al., 1978a, b) illustrates, in the case of this second-order transition, the orientation deviations of the lattice planes almost parallel to the wall plane allowed near the transition. The situation is less extreme for KDP; however, the width of the gamma diffraction peaks can be attributed to a mosaicity of these planes of some 20–30 arc seconds. The X-ray topographies of DKDP can give even more notable results: from 2 to 5 arc minutes (Aleshko-Ozhevkii, 1982).

This set of results demonstrates that the domain texture just below the transition, appears as a "solid modulation of a distorted structure" to use the expression of Andrews and Cooley (Andrews and Cowley 1986), in a context that is all the more " soft" as the crystal temperature is closer to T_C. In this temperature range, the treatment of dislocations related to walls and their effects by continuous medium models are justified (Combs and Yip, 1983; Fisher and Szpilka 1987; Gröger et al., 2008).

4.4.2 A Maximum of Losses Is a Landmark Often Present

As illustrated by Figs. 4.3 and 4.13, the values of the dielectric constant as a function of temperature differ between cooling and reheating. The contribution of domains to the dielectric constant ε_d' increases when $(T_C - T)$ grows below the transition (Fig. 4.15b). From one crystal to another in the KDP family, the variations of $\varepsilon_c'(T)$ remain qualitatively similar (Fig. 4.17), with different "plateau" widths before the so-called gel drop of domains at lower temperatures. The loss constant $\varepsilon_c''(T)$ shows more diversified temperature variations.

We will sometimes observe a small narrow peak at the transition but more systematically a few degrees below the transition, a rapid variation with a more or less marked maximum. The peak at the transition may not be detected under optimal conditions of crystal quality, sample and experimental procedures, especially homogeneity and temperature stability. It is not correlated to the existence of domains and

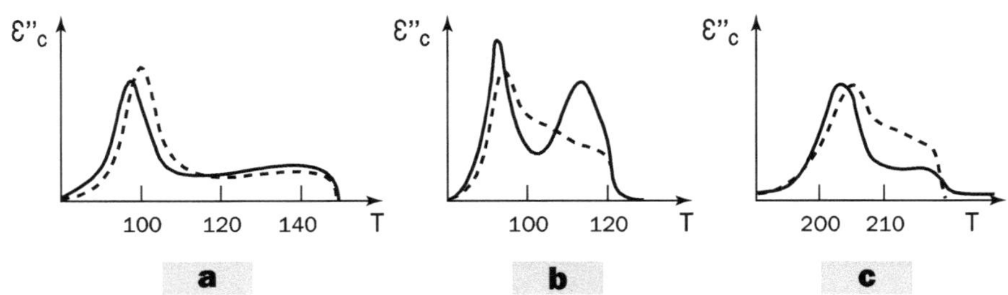

Fig. 4.17 Typical shapes of $\varepsilon'_c(T)$ curves at the top, and $\varepsilon_c''(T)$ at the bottom, for RbDP (**a**), KDP (**b**), DKDP (**c**) crystals. The units along the ordinate axes are arbitrary. The solid line curves correspond to a frequency of 1 Hz and those in dotted lines to 1 kHz. It is noted that the variations of $\varepsilon'_c(T)$ maintain similar shapes between the various crystals while those of $\varepsilon_c''(T)$ notably change in shape

can be observed at the critical point (T_C) of some second-order or tri-critical transitions; it is attributed to the fluctuations that may exist during these transitions.

Except for this narrow peak, the shape of $\varepsilon_c''(T)$ below the transition is related to the domain existence and varies from one crystal to another (Fig. 4.17): the application of a sufficient continuous electric field leads to the almost disappearance of $\varepsilon_c''(T)$. The value of ε_c'' is impacted by that of ε_d' which varies particularly with the shape of the sample as illustrated in Fig. 4.14; it depends on the crystal and on the order of its transition. In KDP, the phenomenon of maximum losses is clearly observed (Fig. 4.18a) with a thermal hysteresis between the para-ferro and ferro-para cycles (Mueller et al., 2003). When the frequency of the measurement field increases, the value of the maximum decreases and the temperature at which it occurs increases (Gridnev and Darinskii, 1978; Huang et al., 1992, 1997; Gridnev, 2002). When the amplitude of the measurement field increases, so does the maximum loss (when, in the published results, the value of this field is comparable to that of the coercive field near T_C, we are obviously far from the definition of dielectric constants "at zero field"!).

Finally, if a moderate continuous field is applied simultaneously to dielectric measurements, which simplifies the texture, the value of the maximum loss and the temperature of this maximum decrease (Kumar and Awasthi, 2013). The

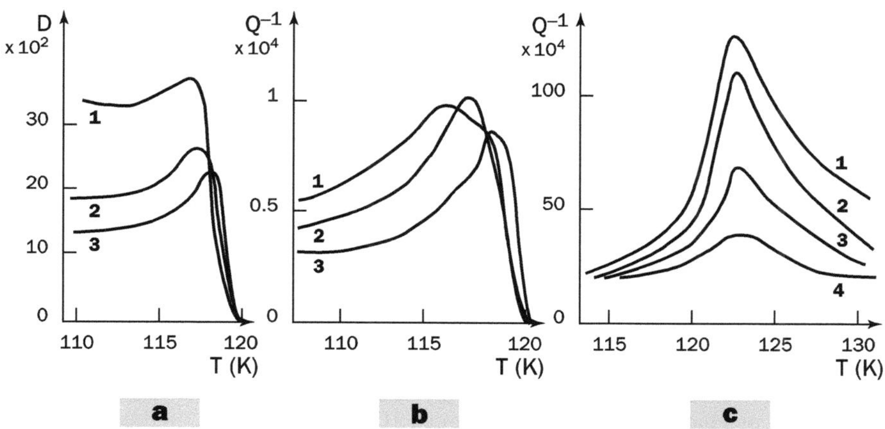

Fig. 4.18 (**a**) Variation of the dielectric dissipation factor D with the temperature at various frequencies (1: 1.2 kHz, 2: 2 kHz, 3: 3 kHz) for a KDP crystal (of dimensions 10 × 10 × 1 mm^3) during cooling at a speed of 1 K min^{-1} (Huang et al., 1992). (**b**) Variation of the quality factor (internal friction) Q^{-1} as a function of temperature for a KDP crystal (of dimensions 10 × 10 × 1 mm^3), at various frequencies (1: 1.7 kHz, 2: 2.2 kHz, 3: 4.1 kHz) during a warming cycle (Huang et al., 1992). (**c**) Variation of the quality factor (internal friction) Q^{-1} measured with the torsion pendulum method, as a function of temperature for a KDP crystal (of dimensions 25 × 2 × 2 mm^3), at various frequencies (1: 2.6 Hz, 2: 3.2 Hz, 3: 5.5 Hz, 4: 16.6 Hz) during a warming at 0.5 K min^{-1} (Gridnev and Darinskii, 1978). We notice the change in the shape of the graph and of the temperature of the maximum compared to (**b**) due to the very elongated shape of the sample along the *a* axis

quantitative comparison of published results demonstrates a significant number of parameters:

- Quality of the crystal and crystal growth, region in which the sample was obtained, density of impurities, or defects by chemical or irradiation which significantly change the values and positions of the maximum loss (Gridnev et al., 1974)
- Preparation and shape of the sample
- Values of physical quantities that change significantly with ($T_C - T$) and thermal history

Mechanical quality factor measurements complement dielectric studies at low and very low frequencies. Either the pendulum method between 10^{-2} and 10^2 Hz (Gridnev et al., 1974, 1975) or the three-component resonator method between 10^2 and 10^5 Hz (Gridnev et al., 1974, 1975; Gridnev and Darinskii, 1978; Huang et al., 1992) is used. It is obviously necessary to take into account the very different shapes of samples in the interpretation of the results: platelets perpendicular to c (or at maximum, cube-shaped samples) for dielectric measurements and resonator methods, very elongated rods following a for pendulum methods. Figure 4.18(b and c) summarizes the typical results concerning KDP family crystals for the quality factor Q^{-1}: decrease in the maximum loss when the excitation frequency increases, usually following a ($1/f$) law (Gridnev et al., 1975; Gridnev and Darinskii, 1978; Wang et al., 1997). Figure 4.19 summarizes the variations of the maximum dissipation factor D_m in all situations.

This phenomenon of maximum loss as a function of temperature under the transition is observed in many other ferroelastic and ferroelectric crystals such as RDA (Kamysheva and Drozdhin, 1987; Gridnev et al., 1975), TMO (Gridnev and Trukhachev, 2013), in pure ferroelastics such as $KH_3(SeO_3)_2$ (Shuvalov et al., 1984), CuAlZnNi (Wang et al., 1996), ferroelastic and superionic crystals such as $LiKSO_4$ (Cach et al., 1985; Gridnev and Khodorov, 1996), superconductors such as $YBa_2Cu_3O_7$ (Arzhavitin et al., 1989; Staines et al., 1988), non-piezoelectric ferroelectrics in the para phase, such as pure or doped TGS (Huang et al., 1997), and doped or irradiated materials. It exists in situations with displacements of electric charges or dipole moments in interfaces, domain walls or twins, grain boundaries, phase fronts. It also exists when mechanical imperfections move in the interfaces. It is found in mechanical experiments between 10^{-3} and 10^2 Hz and in dielectric measurements between a few Hz and 10^5 Hz (Gridnev, 2002, 2007). Generally, the value of the maximum loss increases linearly with the rate of temperature change (Fig. 4.19c). On the other hand, the temperature of this maximum T_m does not vary in such a simple way: when the rate of temperature change increases, this temperature T_m is measured increasing in KDP (Huang et al., 1992), RDA (Gridnev et al., 1975), CsH_2AsO_4 (Gridnev and Kravchenko, 2000), and $BaTiO_3$ (Gridnev, 2002) but sometimes decreasing in the same KDP (Gridnev and Darinskii, 1978) or TMO (Gridnev and Trukhachev, 2013). This shows how detailed knowledge of crystals, samples, and experimental conditions is crucial for any quantitative exploitation of results over time, in this temperature range where a small variation of a physical

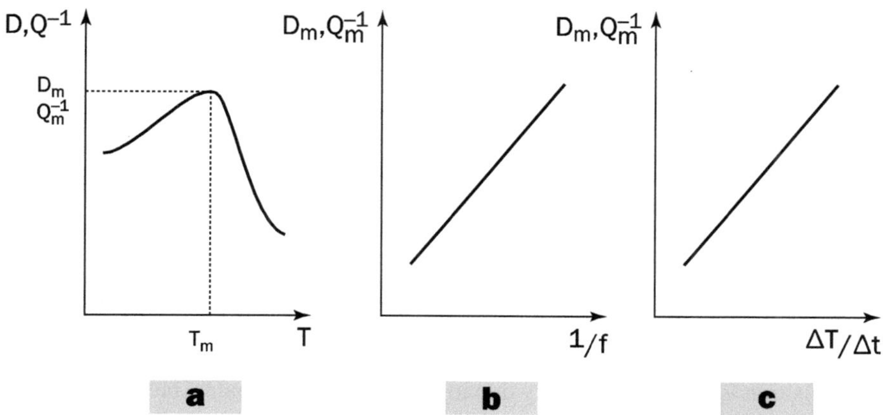

Fig. 4.19 (**a**) Typical variation of losses as a function of temperature in an interval of a few degrees below the transition in the family of KDP type crystals, either dielectric (dissipation factor *D*) or elastic (elastic quality factor Q^{-1}): maxima D_m and $Q_m{}^{-1}$ can be detected at a temperature T_m. The value of these loss maxima usually varies as the inverse of the excitation frequency *f* (**b**) and linearly with the rate of temperature change, whether it is increasing or decreasing (**c**)

quantity induces a profound modification of the domain texture, thus of the response to an excitation. It is clear, for example, that the use for quality factor measurements, of very elongated samples whose physical parameters vary from one point to another in the torsion pendulum method, is enough to explain the different T_m variation laws depending on the rate of temperature change. Then all results called "aging" require a rigorous definition of the parameters and of the experimental conditions to envisage understanding the phenomena involved.

4.4.3 The Hypothesis of a Continuum Below the Transition

The existence of the maximum loss just below the transition in various situations argues for fairly general models. For this reason, the dielectric relaxation phenomenon observed at very low frequencies (from 10^{-3} Hz to a few Hz) was initially attributed to the central heat diffusion peak for KDP, the TGS (Chaves et al., 1993) and in various situations when the frequency and amplitude of the measurement field vary, for RbDP, DKDA (Fally et al., 1995, 1997), KSCN (Schranz and Havlik, 1994), $SrTiO_3$ (Kityk et al. 1996), even if some suggested for the KDP family that it could not be the only explanation (Araùjo et al. 1998). In fact, frequency studies (Kim and Kim, 1997, 1999), depending on the amplitude of the excitation, and the shape of the samples (Fally et al., 2001), have demonstrated the direct link between these phenomena and the movement of domain walls.

The results obtained between the temperature of maximum loss and that of the transition are characteristic of a continuum. This is no longer the case for temperatures lower than that of maximum loss. The observation shows that for these lower temperatures, there are better dissociable domains: they interact laterally and their ends also interact; the concept of continuum is no longer suitable. It is this transition between an almost continuous medium and a set of discrete interacting domains that seems to correspond to the peak of losses. Moreover, if a clear change in the number of domains is observed at decreasing temperature, warming can occur without noticeable change in the number of domains up to the transition. Certainly, the phenomenon of maximum loss occurs with a thermal hysteresis and slightly lower losses than those observed at decreasing temperature, but it remains of the same order of magnitude. The main cause of the loss anomaly in the few degrees below the transition is therefore not only due to the number of domains, but to the overall instability of this dense set of domain walls. Given the significant variations of all physical quantities with temperature, it is difficult to consider the texture in domains as a situation of equilibrium. This is even more true when the temperature is changed rapidly with thermal inhomogeneities that amplify the losses.

To explain the variations in Fig. 4.19c, some models focus on the nucleation of domains (Gridnev and Darinskii, 1978), assuming local inhomogeneities in the sample, especially temperature, and the existence of small volumes of nucleation of one phase in the other, which have, indeed, been observed, but during a particular first-order para-ferro transition (KDA crystal (Kirikov et al., 1976)). This hypothesis can be broadened by considering that any wall displacement, even when the number of domains does not vary, requires the creation of nuclei of opposite polarity, for example, two-dimensional nuclei on the walls. Thus, the laws of variation of D_m and Q_m^{-1} can include a term proportional to the derivative of the number of nuclei with respect to time (Gridnev, 2002) and therefore, to the rate of temperature change. However, the most general model must take into account sets of interacting walls, restoring forces attributed to defects, viscosity forces, Peierls forces (usually neglected), and forces due to the action of the conjugate measurement field (Huang et al., 1992). We thus obtain classical equations of motion with all the ingredients to describe resonance and relaxation. The essential parameters are expressed as a function of the interval ($T_C - T$) and the expressions obtained account for the losses and quality factors obtained experimentally. As the number of walls remains almost constant during a cycle at increasing temperature, it can be considered that the phenomenon is mainly due to the non-linear responses of an easily deformable and polarizable quasi-periodic medium whose properties vary greatly during a small temperature change. The fact that proposed models are unrealistic by considering the walls as infinitely thin, and the number of walls identical at increasing or decreasing temperature, demonstrates that as all the previously mentioned ingredients (restoring, friction, interaction, external action) are present to ensure losses and non-linearity, the walls of the models should be considered as average positions of regions with high polarizability and low shear modulus located between other regions whose polarization and deformation remain larger without however reaching the values that would be measured in a mono-domain. This phenomenon is not the result of a thermal activation process but that of a "modulated ensemble with a distorted structure" (Andrews and Cowley 1986) under the action of an excitation field of pulsation ω and amplitude low enough not to make the domains disappear. These models propose loss factors and quality factors of the form

$$D = g(T_C - T)\omega\tau_0 / (1 + \omega^2\tau_0^2),$$
$$Q^{-1} = f(T_C - T)\omega\tau_0 / (1 + \omega^2\tau_0^2) \quad (4.5)$$

in which g and f are increasing functions of $(T_C - T)$ with a characteristic time

$$\tau_0 = A_0 \exp(-B / (T_C - T)) \quad (4.6)$$

for example, about 10^{-6} s for $T_C - T = 1$ K in KDP.

Other studies have assumed the introduction of defects, created chemically or by irradiation, which change the forces of Peierls, the restoring forces and those of friction. Generally, the more defects there are, the more the losses would increase in value under the transition.

4.4.4 *The Effects of the Excitation Frequency in the KDP Family*

Figure 4.20 illustrates the dielectric parameters obtained in crystals of the KDP family at a determined temperature under the transition, between very low frequencies (10^{-3} Hz) and piezoelectric frequencies (which start around 10^4 Hz) (Mueller et al., 2002b). Two dispersion phenomena are observed, often simultaneously and sometimes, one supplanting the other:

- A logarithmic variation of losses with frequency (Mueller et al., 2002a), well illustrated by the example of RbDP (Fig. 4.20 a, b). This variation, very important towards the maximum of losses, is observed in various fields of solid physics with interfaces: in magnetism (Préjean and Souletie, 1980), in piezoelectric ceramics (Damjanovic, 1997), and in other ferroelectrics. To explain the generality of this type of variation, a process of local interactions between interfaces and randomly distributed defects is usually assumed (Nattermann et al., 1990; Ioffe and Vinokur, 1987). It can be shown that it exists both in a contribution that follows Rayleigh's law, therefore with weak pinning, and in other contributions due to interface roughness (Damjanovic and Demartin, 1996).
- Debye type relaxations (Paul et al., 1976; Kim and Kim, 1997, 1999; Mueller et al. 2002a, b, c) therefore induced by the existence of a phenomenon with a relaxation time τ: one of the physical quantities concerned, for example the

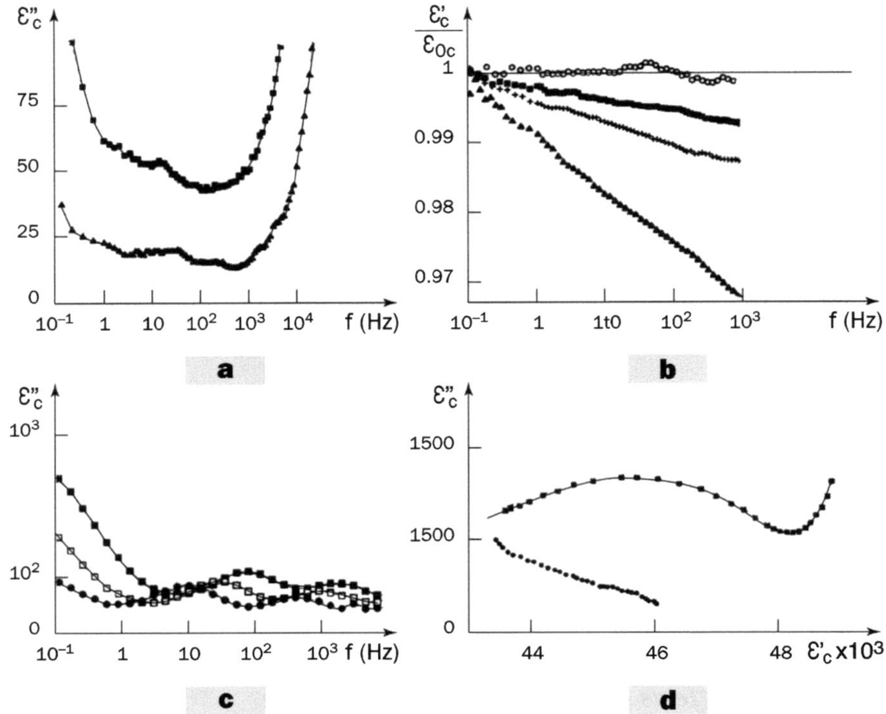

Fig. 4.20 (**a**) In the RbDP crystal, the variation $\varepsilon_c''(f)$ at 144.7 K (for a transition at 146 K) illustrates the existence of two phenomena that are quite generally found: a relaxation phenomenon at very low frequencies (a few Hertz and below) on the one hand, and piezoelectric dispersions in the kHz and beyond on the other hand. The two curves also demonstrate the evolution of measurements over time: the one with black squares corresponds to measurements made after half an hour of stabilization of temperature and that with triangles after 59 h. These changes do not qualitatively alter the observed dispersion laws (Mueller et al., 2002b). (**b**) The measured value of the dielectric constant ε_c' of RbDP as a function of frequency remains constant in the para phase (curve at 148.7 K (circles) for a transition at 146 K). It can be described under the transition by a law type $\varepsilon_c' = A\,(1 - B \ln \omega\tau_0)$, ω is the excitation pulsation, τ_0 is the relaxation time, and A and B are constants. It is observed that τ_0 varies with temperature (130.7 K (squares), 144.7 K (+), 145.7 K (triangles)) and reaches a maximum a few degrees below T_C (Mueller et al., 2002b). (**c**) Depending on the experimental conditions (nature of the crystal, shape and size of the samples, rate of temperature change, temperature) the variation $\varepsilon_c''(f)$, in the example of KDP, illustrates the existence of Debye type relaxations, often only one, sometimes two or three in the frequency range presented (Kim and Kim, 1997; Kim et al., 2009; Mueller et al., 2002c). (**d**) Even when Debye relaxations are evident, they can hide a logarithmic dispersion as shown in (**b**): a good example is given by this Cole-Cole, which highlights a Debye relaxation and a weak logarithmic variation in KDP at T_C-7K (Mueller et al., 2002a)

deformation u, after the change of one of the experiment parameters, will evolve over time as

$$u = (u_0 - u_\infty)(1 - e^{-t/\tau}) \tag{4.7}$$

in which u_0 and u_∞ are the values of u, respectively, at time $t = 0$ and at infinity. A relaxation curve of the real susceptibility will then be observed, for example, of the dielectric constant ε, and a maximum of the loss factor of the form

$$(\varepsilon_0 - \varepsilon_\infty)\omega\tau / (\varepsilon_0 - \varepsilon_\infty)\omega^2\tau^2 \tag{4.8}$$

with the characteristic Cole-Cole diagrams (see Fig. 4.20d).

In the crystals of the KDP family, either the logarithmic dispersion (RbDP in Fig. 4.20a, b), or Debye type relaxations (KDP in Fig. 4.20c) are predominant, relaxations that do not completely mask the logarithmic dispersion (Fig. 4.20d). If the excitation frequency increases beyond 10^4 Hz, the piezoelectric resonances appear as illustrated in Fig. 4.8. They depend on the domains, the species of domains, and the dimensions of the sample. Beyond that, a relaxation occurs in the frequency range between MHz and a few GHz.

Some piezoelectric resonances can be described as resulting from the vibrations of the macroscopic sample whose average properties vary with temperature. Thus, the classic resonance laws of a plate in the para and ferro phases have been verified with the lowest frequency f_r varying according to the relation (Nakamura and Kuramoto, 1988):

$$f_r = (2l)^{-1}\left(c_{66}{}^E / \rho\right)^{1/2} \tag{4.9}$$

l is the smallest dimension of the sample following a_1 or a_2, $c_{66}{}^E$ is the elastic constant at a constant electric field, and ρ is the volumetric mass (Mason, 1946).

The contribution to the dielectric constant can then be described by a relationship of the form:

$$\varepsilon(\omega,T) = (\varepsilon_0(T) - \varepsilon_\infty(T))(\omega_0^2 / (\omega_0^2 - \omega^2 + i\gamma\omega) \tag{4.10}$$

in which $\varepsilon_0(T)$ and $\varepsilon_\infty(T)$ are respectively the limiting values of the dielectric constant at zero frequency and at infinite frequency, γ is the damping constant, and ω_0 is the resonance frequency.

If the lowest frequency piezoelectric resonance depends on the size of the sample, other resonances depend on the size of the domain species and, ultimately for the highest frequency, on the minimum dimension of the domains, therefore their width. Figure 4.21 allows us to imagine the various resonances that overlap.

An important macroscopic result is illustrated by the relationship between the dielectric constant at constant stress and the elastic constant at a constant electric field in Fig. 4.22a (Nakamura and Kuramoto, 1988). The experimental points cover

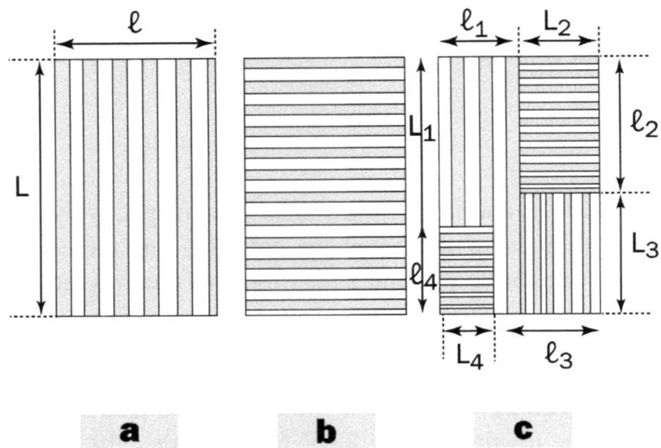

Fig. 4.21 (**a–c**) The dielectric constant calculated from the dielectric measurements varies with the shape of the sample and with the domain texture (numbers and shapes of domain sets); the same is true for the resonance frequencies. The problem is not only geometric since many domain sets can constrain such a region in its overall deformation

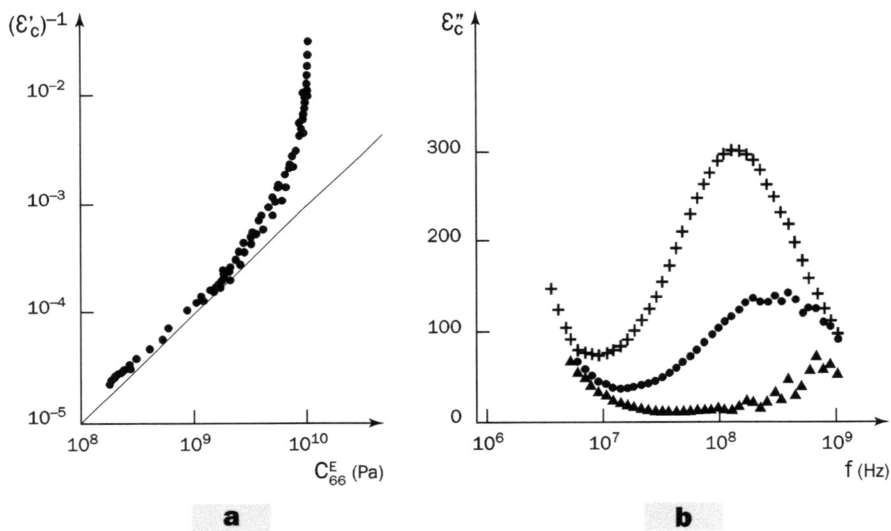

Fig. 4.22 (**a**) For the same sample, all the results of the measurements of $(\varepsilon_c')^{-1}$ and $c_{66}{}^E$ in the various experimental situations are positioned on the graph: for the lowest values (close to the transition temperature), their magnitudes are clearly proportional (the solid line is of slope 1)); on the contrary, at low temperatures, the rigidity tends towards a limit value while the dielectric constant drops (the domains and the walls are no longer mobile) (Nakamura and Kuramoto, 1988). (**b**) Frequency variation of ε_c'' near the transition for KDP at temperatures of 120.4 K(+), 122.98 K(•), 123.03 K(triangles) (Nakamura and Kuramoto, 1988)

all experimental situations for a sample, in para phase and in ferro phase. The range of a few degrees below the transition corresponds to the smallest values of $(\varepsilon_c{}'^X)^{-1}$ and of $c_{66}{}^E$, the region of the graph in which these quantities appear proportional (slope equal to 1 on Fig. 4.22a). Thus, this region can be modelled as a continuous set with the average values of the susceptibilities. It is observed that the inverse mechanical susceptibility $c_{66}{}^E$ is reduced by the existence of walls: these make the set more "soft". It is in this interval of a few degrees below T_C that the amplitude of variation of the dielectric constants and the loss constants during piezoelectric resonances are the most important, and this is in all the crystals of the KDP family. When the frequency of the excitation increases further, a relaxation phenomenon appears beyond 10^6 Hz. It varies significantly in the few degrees below T_C as illustrated in Fig. 4.22b. When $(T_C - T)$ decreases, the characteristic frequency increases and the maximum value of the losses decreases until it cancels out at the transition. The usual Cole-Cole plot in KDP shows a fairly clear situation in the few degrees below the transition with good quality semi-circles; however, when $T_C - T$ increases, the arcs of the circle usually move away from the semi-circle, illustrating a more complex situation with a distribution of relaxation times.

Taking into account both the piezoelectric resonances and the relaxation at high frequencies, Kuramoto proposed for the dielectric constant an equation of the form:

$$\varepsilon(\omega,T) = ((\varepsilon_0(T) - \varepsilon_{cl}(T))$$
$$\left(\omega_0^{\,2}\left(\omega_0^{\,2} - \omega^2 + i\gamma\omega\right)^{-1} + \varepsilon_{cl}(T) - \varepsilon_\infty(T)\right)\int g(\tau,T)(1 + i\omega\tau)^{-1}\, d\tau / \tau \tag{4.11}$$

in which $\varepsilon_0(T)$, $\varepsilon_\infty(T)$ and $\varepsilon_{cl}(T)$, are, respectively, the limiting values of the dielectric constant at zero frequency, at infinite frequency ("clamped"), γ is the damping constant, ω_0 is the resonance pulsation, and τ a relaxation time which is assumed to have a certain distribution in this expression (Kuramoto, 1987). The relaxation time τ changes rapidly under the transition: when $(T_C - T)$ is small in the ferro phase, the medium is soft and, at the same time, the response to an excitation is fast. Whether this is at very low frequencies ($<10^{-2}$ Hz) or on the contrary at the highest frequencies before domain relaxation (10^9–10^{10} Hz), the characteristic times of response to excitations decrease considerably when $(T_C - T)$ varies between a few degrees and zero with mechanical and electrical rigidities that drop.

The proposed models consider the variations of deformations (and polarizations) directly linked to the order parameter (Fig. 2.11). They rarely take into account the distance to the sample edge (Fig. 2.15) (Novak and Salje, 1998); yet depending on this distance, the potential in which atoms and ions will have to move with the wall, varies (Lee et al. 2003a, b, c). But to limit oneself to shear deformation, while the deformations u_{zz} and u_{xx} vary under the transition (Figs. 1.36 and 2.3.9), may be a simplistic assumption in this context where the continuum seems to be accepted especially if many domains coexist.

This region under the transition is an extremely rich subject of study to link, at the nanoscopic level the probabilities of existence of Slater and Takagi ions (depending on $|T - T_C|$, at the microscopic level the Monte Carlo simulations (Schmidt et al., 2000) (Fig. 3.37) which show how high the density of Slater and Takagi defects can be far from the centre of the wall when $(T_C - T)$ decreases (Fig. 3.37) and the macroscopic models of potentials. Thus, modulated regions can be constructed with ion sprays (Schmidt, 2020) (Annex A5), and statistical simulations can be proposed in this continuum context.

4.5 The "Plateau" of Susceptibility as a Function of Temperature

A particularity of the dielectric responses of crystals from the KDP family, except those with a marked first-order transition, is the existence of what we have become accustomed to calling "the plateau" between the "continuum" zone and the domain freezing zone. What is this phenomenon?

It concerns the "zero field" susceptibility, therefore with excitation fields that do not modify the domain texture, but cause small variations in the existing local polarization and small wall displacements as observed in stroboscopy. Figure 4.23 outlines these phenomena concerning the ends which "vibrate" differently whether the temperature is close to that of freezing (the "bottom" of the plateau) or close to the region of the continuum (Daratchiev et al., 2007). In all situations, estimates of contribution to the dielectric constant by these end modifications give results ten to a hundred times too small compared to the measured values of $\Delta\varepsilon_c'$ on the "plateau". Therefore, another explanation must be found.

This plateau phenomenon does not concern all crystals from the KDP family. Figure 4.24 uses work developed in Chaps. 5 and 6 on DKDP and its clear first-order transition: when the polarization inside the domains is well defined from the transition, the plateau phenomenon does not exist, and the dielectric constant measured in the ferro phase, certainly varies with the domain texture but, remaining proportional to the dielectric constant of the mono-domain at all temperatures.

How to explain this phenomenon of significant dielectric contribution from domain walls? Does it exist in whatever the situations of the domains outlined in Fig. 4.25 and well studied in Chap. 3? To answer these questions, the methods described in Sect. 4.1 and in Fig. 4.6 for obtaining a "stable" texture at the domain level are used. Thus, in each situation, during a thermal cycle or during a hysteresis cycle, are obtained for such a configuration of domains considered as "stable" values of ε_c' and ε_c'' related to the total wall surface of the sample. Figure 4.26 illustrates the results obtained on the plateau (Bornarel and Torche, 1987):

$$\Delta\varepsilon_c' \approx \Delta S^n, \Delta\varepsilon_c'' \approx \Delta S^m \tag{4.12}$$

swelling known in the world of twins with dislocations of larger Burgers vectors that move more than the others. For their part, Figures (**e** and **f**) are typical of plateau temperatures closer to freezing with a compact tip, less deformed during movement, and smaller lateral wall displacements. These tips can create stresses and interactions at distances of 200 μm in the sample (Bornarel and Lajzerowicz, 1968), therefore of the order of the thickness of the sample. It will be remembered if one simply models the walls as simple surfaces, that besides movements in the direction of the ferro axis, they move, on the "plateau", longitudinally and laterally for a domain in wedges or needles. Figures (**g**–**j**) present the profiles obtained in continuous medium by optimizing anisotropy, curvature, and Peierls energies. (**g** and **h**) correspond by hypothesis, to geometries of stresses of a parallel plate and (**i**), (**j**) of three-point bending (Kityk et al., 2000). The profiles correspond to the situation at $t = 0$, start of application of a continuous stress that increases the length y of the twin of width w, and the other to the profile at $t = 100$ s. In (**g** and **i**), the anisotropy energy is the most important; in (**h** and **j**), it is the curvature energy that predominates

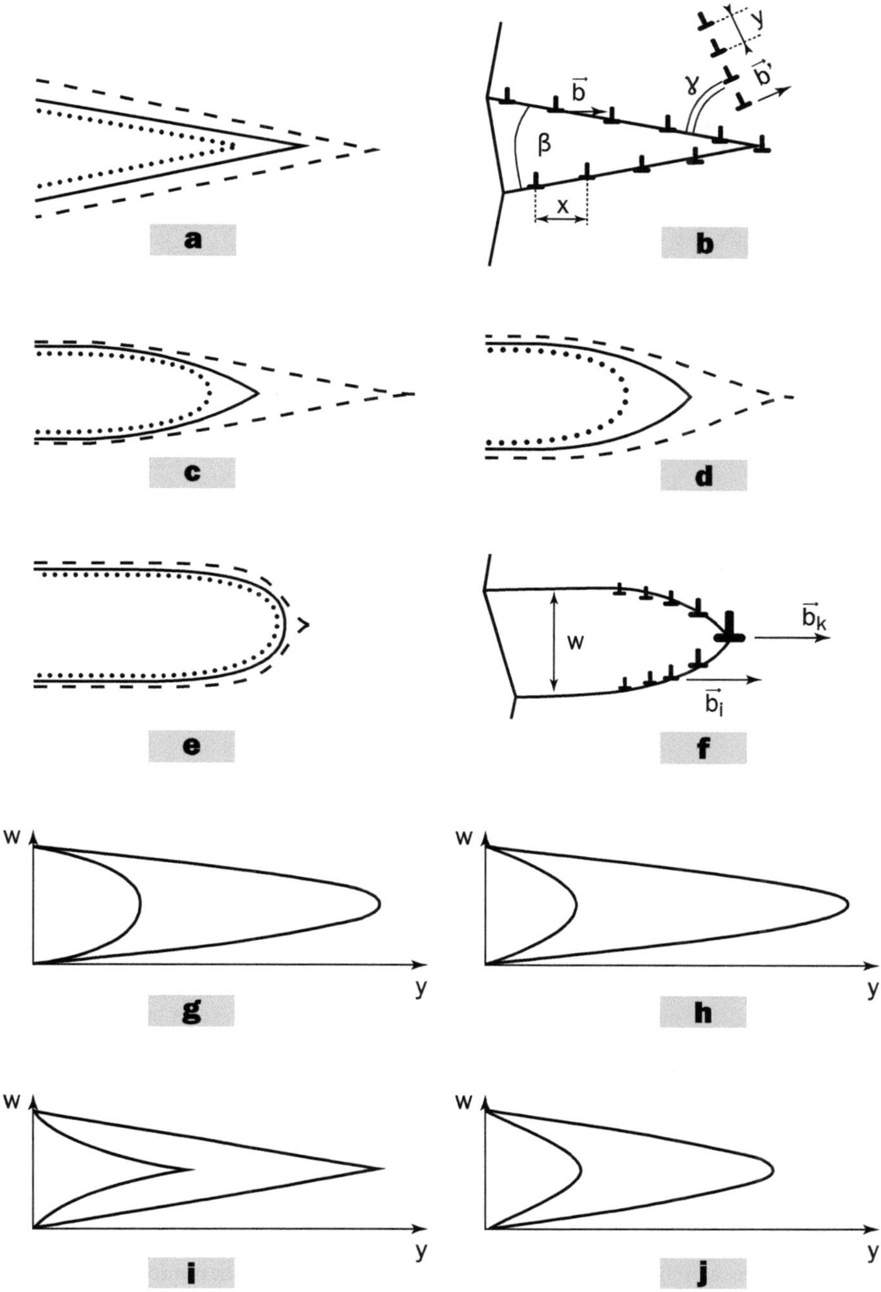

Fig. 4.23 If the domain is wedge-shaped, wall translations are illustrated in Figures (**a** and **b**) and are consistent with identical translations of quasi-dislocations of the wedge type. For a temperature among the highest on the "$\varepsilon'_c(T)$ plateau", Figures (**c** and **d**) illustrate the stretching and then the

(continued)

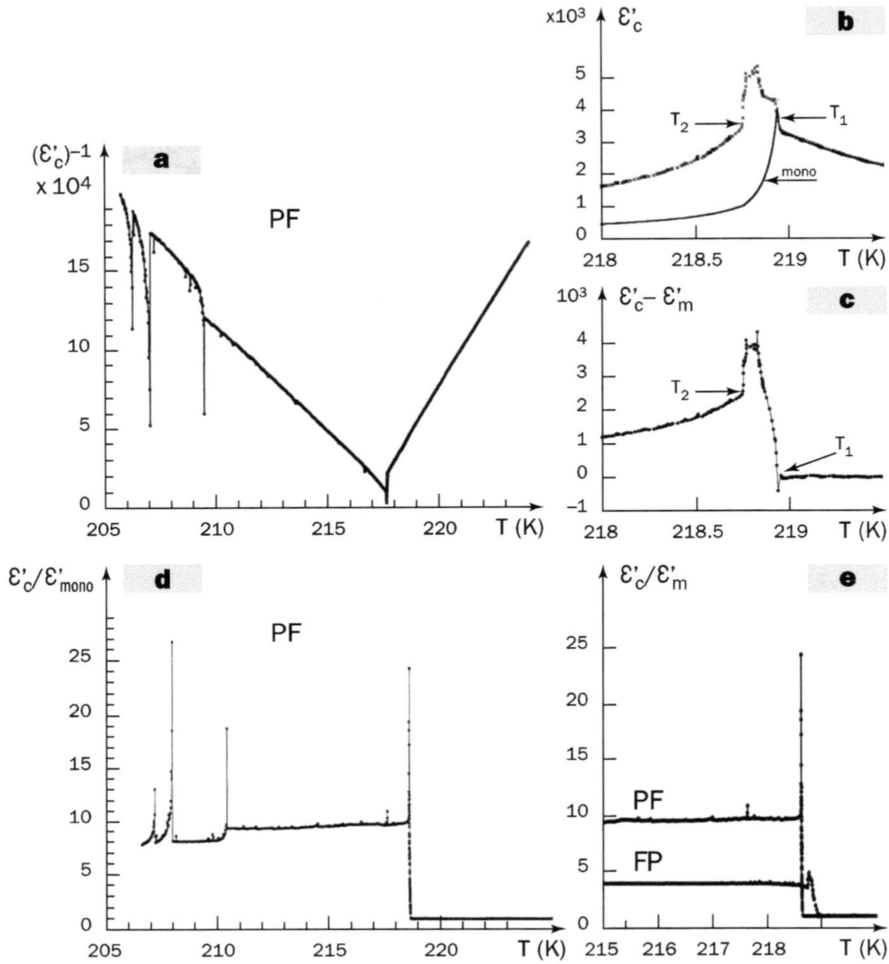

Fig. 4.24 The phase coexistence of DKDP is described in Chap. 5 in samples without applied continuous field. The measurement of the dielectric constant along *c*, except in the coexistence zone, gives simple laws during a Para-Ferro (PF) cycle (**a**), (**b**). The value measured after deduction of the mono-domain contribution can be plotted (**c**). In this crystal with a first-order transition, the contribution of domains under the transition is proportional to the mono-domain contribution with a proportionality factor modified at each rearrangement of the domain texture (**d**). This factor during a cycle FP is always lower than that of the previous PF cycle (**e**) (Bornarel and Cach, 1999)

Therefore, for a given temperature, the variations of ε_c' and ε_c'' follow simple and increasing laws with the total surface of the walls. These laws are not proportional to this surface which demonstrates the non-linearity of the phenomena. We understand qualitatively the result: if the sample only had one needle domain of length l, the lateral displacements of walls would necessarily be easier if both tips exercised their "pinning" action at a greater distance. The values of the exponents n and m

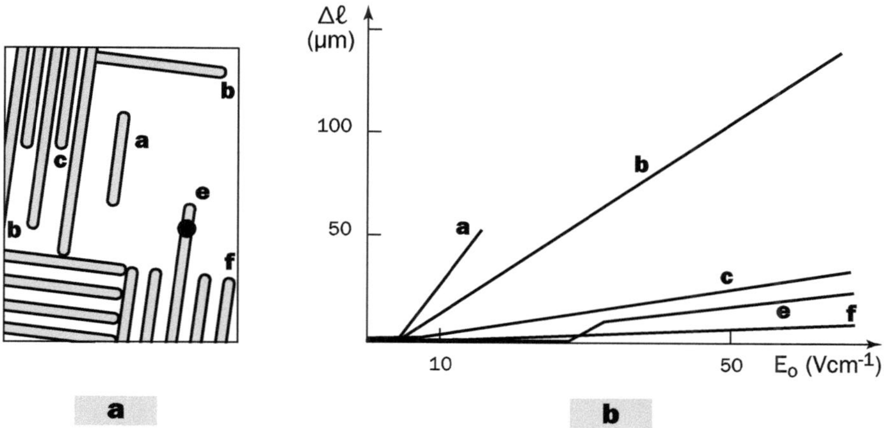

Fig. 4.25 (**a**) Different situations of needle domains tips observed in section (001). (**b**) Typical variations in needle length under the action of a small low-frequency alternating electric field: significant for free tips (**a** and **b**), weak when they are at the front (**f**) or slowed down by a defect (**e**) (at a typical temperature of 90 K in KDP); the ratio of speeds between an isolated domain tip and a front of tips is about 10 (Bornarel, 1972))

vary depending on the temperature on the plateau between 1.3 and 1.7 for *n* and 1.3 and 1.45 for m (Torche, 1986). This result is consistent with numerous studies of non-linear dielectrics (Matsuda and Abe, 1974; Mueller et al., 2004) in which a development of the dielectric constants and power losses of the alternating measurement field allows detection up to the term of exponent 5 on the plateau (You and Kim, 1997), and up to 9 in the region of domain freezing.

The illustrations in Fig. 4.26b allow for some comments:

- For small values of *S*, we are in a context of sparse textures, and all interactions between domain tips and edges of the sample explain the dispersion of results; nevertheless, the relations (4.12) are solid for a nice variety of textures.
- When *S* is maximum, we observe for the same value of the total surface, the largest values and a notable jump depending on the context: the sample is in the situation of maximum density of domains, towards the situations X of Fig. 3.40. Any variation in this case results in small abrupt and collective displacements (Bornarel, 1975, 1987) such as those of avalanches in Fig. 3.44 (Salje et al. 2011) (Annex A8). The lateral interactions between walls are strong enough that a small variation in temperature or excitation maintains the dielectric and mechanical instability. This situation of metastable equilibrium explains the high values of dielectric constant resulting from capacitive measurements, different in a ferro-ferro compared to a ferroelectric (Darinskii et al. 1989; Kopal et al., 2000) without hypothesis on the model of flat wall or with slightly inclined and therefore charged parts (Fedosov and Sidorkin, 1976). These domain walls remain a field of study to be exploited by various strategies, for example, with simultaneous action of magnetic field for magnetoelectric effects (Fetisov and Srinivasan, 2024).

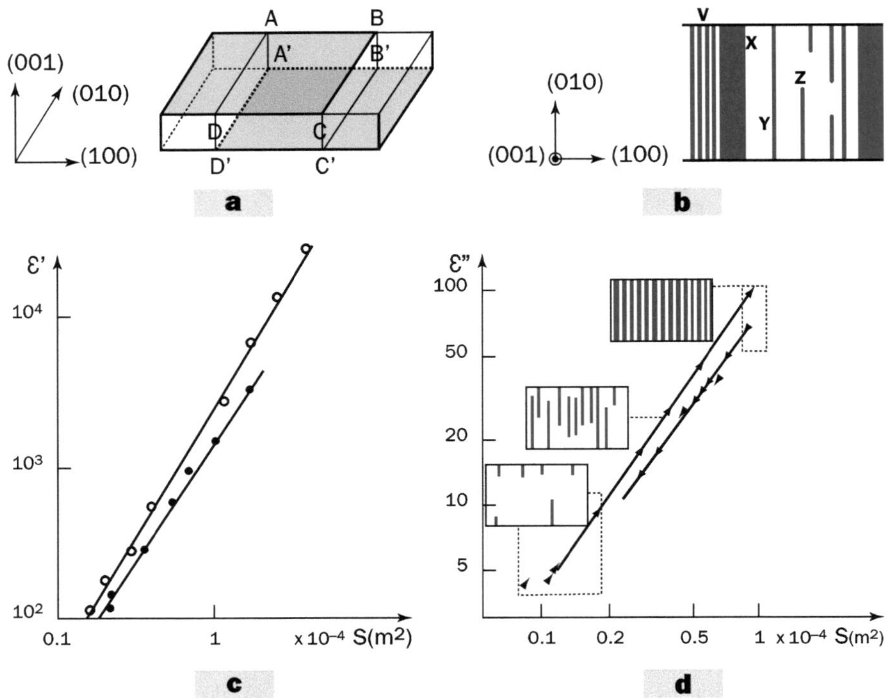

Fig. 4.26 (**a**) Schematic representation of a KDP sample with crossed semi-transparent electrodes that simultaneously allow dielectric measurements and observation in the volume ABCDA'B'C'D'. (**b**) Diagram of the needle domains observed in a section (001). (**c** and **d**) Variations of ε_c' and ε_c'' as a function of the total surface area of the walls S in the sample (at a temperature of 91.65 K)

In conclusion, the so-called plateau phenomenon, in crystals with transitions close to the tri-critical, corresponds to situations of metastability of dense domain textures. These textures cannot be significantly modified under the action of too weak excitations but the small modifications of walls induce significant responses due to their high polarizability and the importance of interacting wall surfaces; the phenomena related to domain ends exist but are not of the same order; they are all the weaker as the temperature decreases and approaches that of the freezing of the domains presented in the following section. The plateau is an ideal region for testing the many models of "aging", "after effect" (Sidorkin, 1998; Lee et al. 2003a, b, c; Schranz et al., 2012; Hecht et al., 2017), etc. of migration of charged defects to be compared with the work of Schmidt in annex A5 with the ions of Takagi and, in a different order of idea, the use of phenomena at short and long distances between the walls to create exceptional macroscopic properties (Deluca and Taniguchi, 2022; Domenichini et al., 2023). The field of applications with textures blocked in a number of domains by the qualities of sample edges, coupled with adapted histories to create macroscopic elements of great dielectric or/and superconducting properties, remains open (Van Aert et al., 2013; Salje, 2013; Deluca and Taniguchi, 2022).

4.6 Domain Freezing

4.6.1 Observation of Textures and Domains

The phenomenon of domain freezing was detected as early as the first publication on KDP (Busch and Scherrer, 1935). At first, it was confused with a structural transition, but doubt was quickly dispelled. There are, on the crystals of the KDP family, several hundred publications on the subject, and our goal is not to draw up a monograph but to best understand the phenomenon. Domain freezing has also been observed in ferroelectric crystals such as TGS (Huang et al., 1997), $BaTio_3$ (Wang et al., 2007) and ferroelastic crystals such as ALHS (Zimmermann and Schranz, 1996, 1997) CuAlZnNi (Huang et al., 1997). Overall, the authors often refer to the Vogel-Fulcher law (Nakamura et al., 1989), but the parameters of this freezing phenomenon are not always completely explained (Garcia-Colin et al., 1989). For the KDP family, let's take stock of the phenomena and their interpretations to better understand the physical reality.

The value of the dielectric constant is high in the "plateau" as shown in Sect. 4.5, then decreases with temperature with, at the same time, the passage of the loss constant through a maximum often important. Stroboscopic observations made simultaneously with low-frequency dielectric measurements show that the maximum of the loss constant corresponds, in a needle texture, to the progressive immobilization of domain tips (Bornarel, 1972). The phenomenon is not abrupt, because as illustrated in Fig. 4.25b, the interactions between domain tips play their role: the amplitudes of displacement of the tips of an isolated domain and those of the tips of a front are very different, and the first immobilize at lower temperatures than the second. Thus, the maximum of the loss constant is the indicator of "average freezing" of the tips in a needle texture.

4.6.2 The Role of Crystal Quality and Sample Shape

Regarding the quality of the crystal, experimental work has shown that the pyramidal regions of the crystals contain fewer accidental dislocations and local defects than the prismatic regions: the freezing of the domains occurs at lower temperatures for samples cut in the first region than in the second. Similarly, the creation of point defects can facilitate freezing, which then occurs at higher temperatures. These defects can be created by irradiation or by a chemical process, but the result is qualitatively the same if the crystal lattice is not deeply modified. The creation of defects that increase the proportion of hydrogen and alter the dynamics of Slater or Takagi ions, on the other hand, can lower the freezing temperature.

Therefore, we choose a crystal of the best quality, with cuts and polishes as described in Chap. 2, to obtain a platelet sample with faces in the tetragonal planes. For samples with the same section perpendicular to the c-axis, the thickness along this axis plays a major role: Figure 4.27 shows the variations of the loss constant with thickness t (Bornarel and Torche, 1992). The differences in values are largely due to the previously explained variations of the dielectric constant (Fig. 4.9). These variations with temperature might seem disordered, but the temperature of the maximum loss follows a very simple variation law (Fig. 4.27b) due to that of the dissipation factor, variation of the form $T_f = A - B \, \mathrm{Ln}(t/t^*)$: the thinner the sample, the higher the freezing of the domains occurs (Bornarel and Torche, 1992).

4.6.3 *Domain Freezing as a Function of Excitation (Amplitude and Frequency)*

The measuring electric field, as when tracing a hysteresis cycle, imposes forced variations on the samples. However, unlike the hysteresis cycles analysed in Chap. 3, the domain texture remains approximately the same if all precautions are taken to do so, and therefore the amplitude of the excitation remains limited. The simplest is obviously to optically observe the domains simultaneously with the measurements to ensure this. It is under these conditions that the curve networks of Figs. 4.28 and 4.29 were drawn, the first by regulating the temperature to a determined temperature and waiting for the stabilization of the values of the dielectric constant and losses, the second at increasing temperature with a heating speed of about 10^{-2} K min^{-1}. The qualitative results are the same: the greater the amplitude of the excitation field,

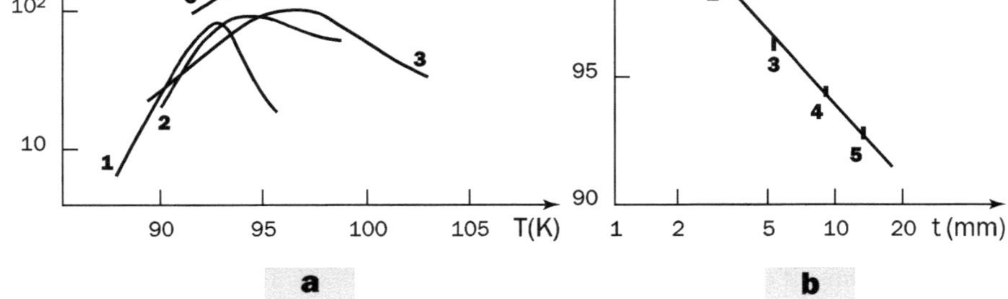

Fig. 4.27 (**a**) Variation of the loss constant as a function of the temperature for a KDP sample subjected to a measurement field of 1 V cm^{-1} and 1 kHz. The parameter is the thickness along the ferroelectric axis t of the samples (Bornarel and Torche, 1992). (**b**) Variation of the freezing temperature T_f (temperature of the maximum of the loss constant) in semi-logarithmic scales as a function of the thickness of the sample; the figures correspond to the curves of (**a**)

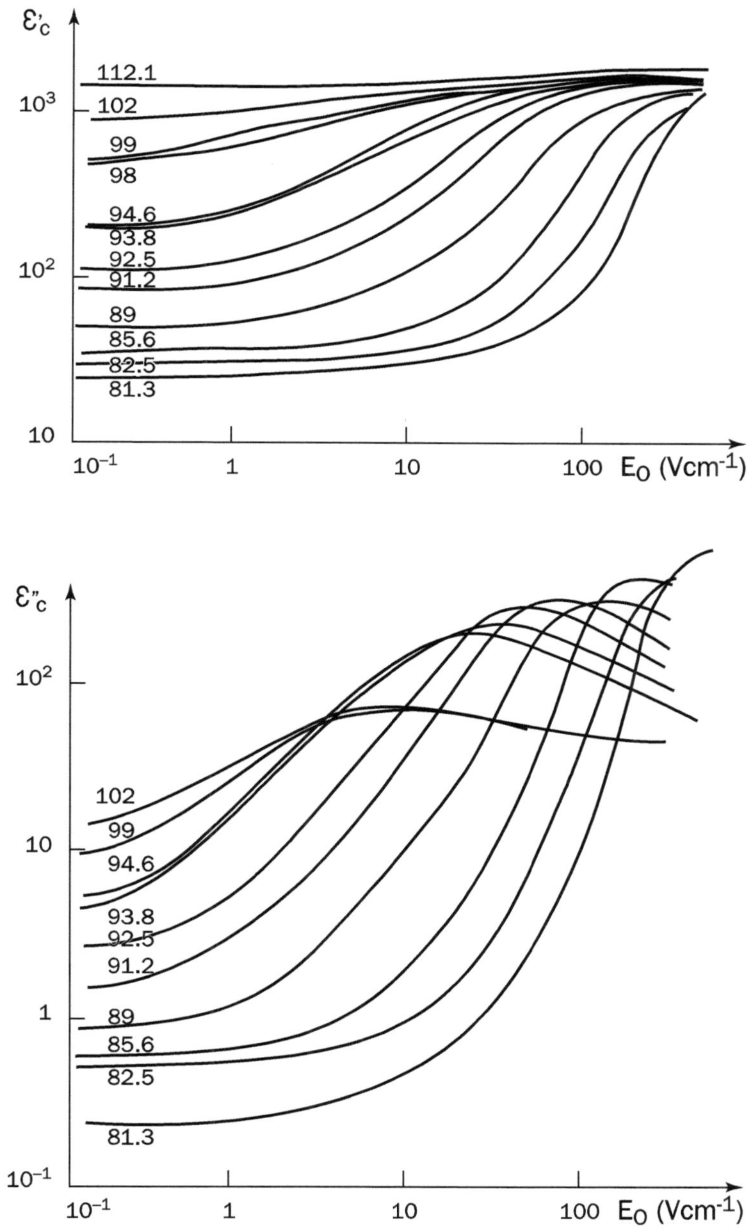

Fig. 4.28 Variation of the dielectric constant $\varepsilon_c'(T)$ (top) and of the loss constant ε_c'' (T) (bottom) as a function of the amplitude of the measurement field E_0 ($f = 1$ kHz) at the temperatures indicated for a KDP sample (Bornarel, 1972)

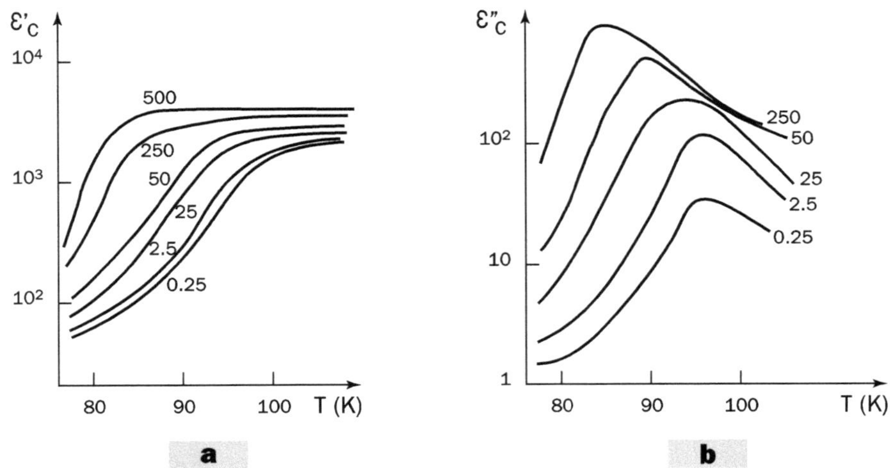

Fig. 4.29 (**a**) Variation of the dielectric constant $\varepsilon_c'(T)$ and (**b**) of the loss constant $\varepsilon_c''(T)$ with the amplitude of the alternating measurement field E_0 (in V cm^{-1} at 1 kHz) as parameter: the larger E_0 is, the lower the freezing temperature (maximum of $\varepsilon_c''(T)$) is. The highest values of E_0 are comparable to those of the coercive field: the domain textures are therefore significantly modified in this context

the lower the temperature of the maximum of the loss constant, therefore the more the freezing of the domains is pushed towards lower temperatures (Bornarel, 1972; Sidorkin et al., 1984). Figure 4.30 summarizes the results obtained with different samples, three for the measurements at increasing T and five for the measurements at constant T, results corrected for variations as a function of thickness given in Fig. 4.27. It can be seen that these corrections allow a grouping of the data which confirms the quasi-exponential variation of the freezing temperature as a function of the amplitude of the excitation, with nevertheless a certain curvature of the cloud of points.

An important result concerns the surface of the domains. The same experiments as those described in Sect. 4.5 and summarized in Fig. 4.26 for a chosen temperature on the "plateau" were carried out for temperatures significantly lower than T_f: as illustrated in Fig. 4.31, the values of the dielectric constant now vary linearly with the total lateral area of the domain walls, and, the loss constant is directly proportional to this total surface S. The situation is very different from that observed on the plateau where non-linearity was the rule. It can be concluded that the domain walls, with immobile ends, vibrate laterally under the effect of the alternating field, less than on the plateau, but especially without interaction between them, intrinsically (Bornarel and Torche, 1987).

Before obtaining equilibrium values, the modification of the dielectric constants following a change in excitation occurs more or less quickly, but always with notable differences between temperatures on the plateau and temperatures lower than T_f. Figure 4.32 summarizes the "after-effect" phenomena due to the domains for

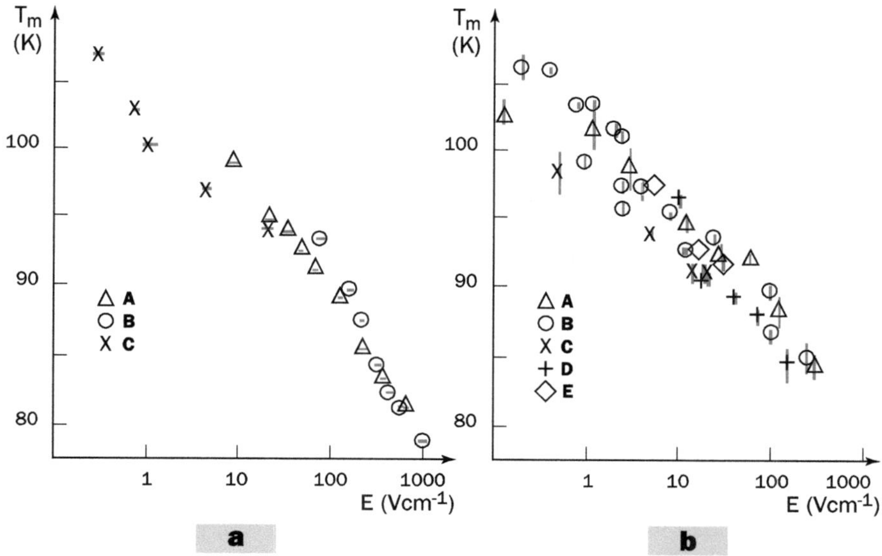

Fig. 4.30 Variation of the freezing temperature T_f (maximum of $\varepsilon_c''(T)$) as a function of E_0 for various samples. The results of Fig. 4.27 have been used to perform a correction that takes into account the thicknesses of the samples

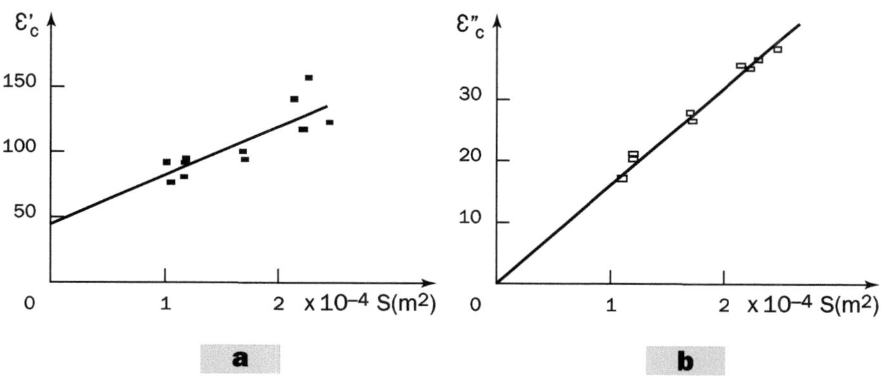

Fig. 4.31 Measurements in KDP at a temperature of 89.87 K as a function of the total surface area of the walls S as defined in Fig. 4.26: (**a**) $\varepsilon_c'(S)$, (**b**) $\varepsilon_c''(S)$. These linear variations demonstrate that the interactions between domains and walls that existed on the "plateau" have become negligible

temperatures higher than the freezing of the domains and for lower temperatures (Torche, 1986). It should be noted that above the freezing temperature, the domains interact with each other, both during rapid variations due to a step of E_0 and during slower evolutions to the quasi-equilibria of the measurements. Below the freezing temperature, only intrinsic phenomena to the domain walls can explain the results both in terms of amplitude variations (capacitive and losses) and relaxation times.

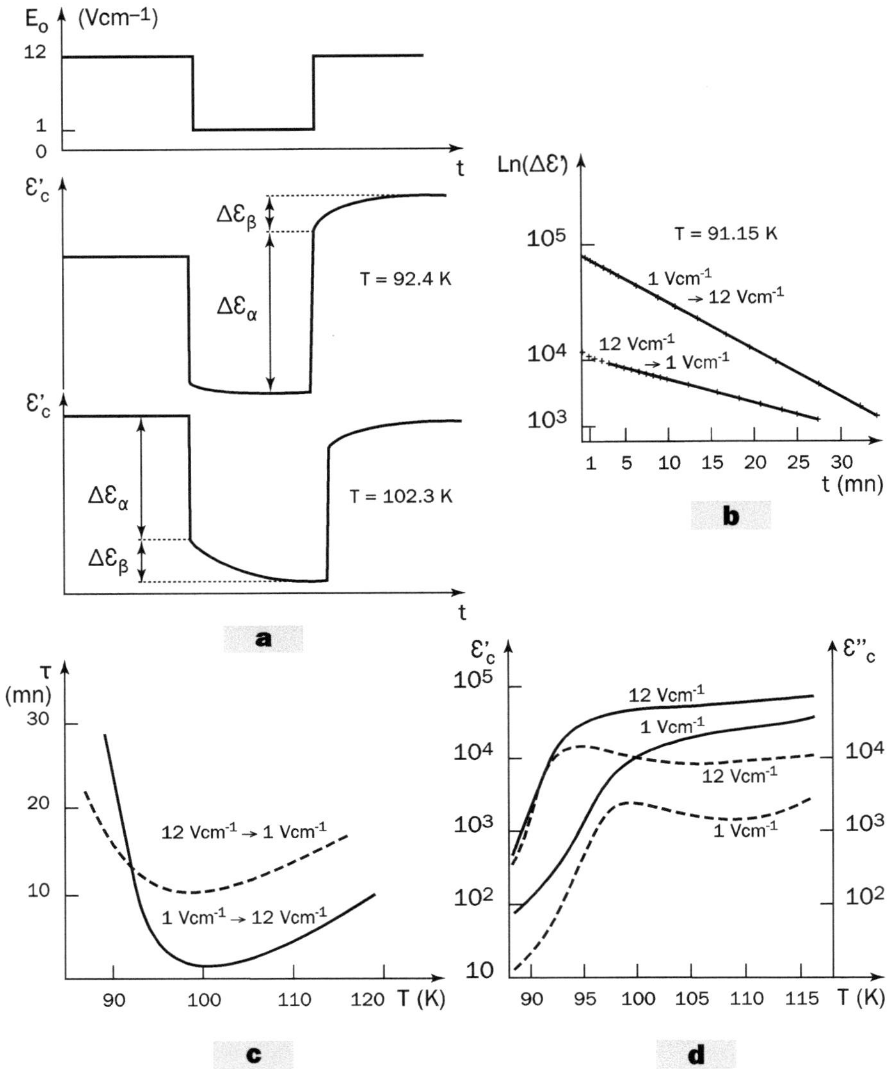

Fig. 4.32 If the amplitude E_0 of the measuring electric field $E_0\, e^{i\omega t}$ is changed from E_1 to E_2, the value of $\varepsilon_c{}'$ (and that of $\varepsilon_c{}''$) changes in less than a second from a value $\Delta\varepsilon_\alpha{}'$, then an "after effect" normal (of the same sign) of amplitude $\Delta\varepsilon_\beta{}'$ occurs as illustrated in Figure **a**) (sample thickness 1.80 mm). Two temperatures are chosen, one higher and the other lower than the freezing temperature T_{fi}. The results of these "after effects" around the freezing of domains are summarized below ($E_1 = 1$ V cm^{-1}, $E_2 = 12$ V cm^{-1}):

– An increase in E_0 results in a variation $\Delta\varepsilon' = \Delta\varepsilon_\beta{}'\,(1 - e^{-t/\tau})$ with a single relaxation time τ_a, while a decrease in E_0 initially shows several characteristic times and then a single relaxation time τ_d (Figure **b**)

(continued)

– If T is higher than T_{fi}, τ_d is larger than τ_a (Figure **c**) with a sample thickness of 2.84 mm: the system of collective interactions between domains, tips and walls requires more time to reach a quasi-equilibrium when the excitation amplitude decreases than when it is forced by the E_0 growth; this result is consistent with relaxation time measurements with continuous electric field jumps that alter the texture: imposing a texture by increasing the field or constraint is faster than letting it regain its equilibrium at zero field when the domain tips have been moved (Kamysheva and Drozdhin, 1987 and earlier publications)

– If the temperature is lower than T_{fi}, τ_d is smaller than τ_a (Figure **c**): the domain tips are blocked, and the walls are isolated from each other. These quickly block if the field decreases and require more time to get moving when it increases

– If the temperature is between the two temperatures T_{fi} corresponding to the starting and arrival field values, the system is more complex, with the largest observed ratio of the report $\Delta\varepsilon_\beta'/\Delta\varepsilon_\alpha'$ and relaxation times τ_d and τ_a of comparable values

– As Figure (**a**) illustrates, $\Delta\varepsilon_\beta'$ is not the same at field E_0 increasing or decreasing at a given temperature, and the ratios are reversed above and below the T_{fi}

Figure (**d**) recalls the variations of $\varepsilon_c'(T)$ (continuous lines) and $\varepsilon_c''(T)$ (dashed lines) in quasi-equilibrium contexts (after stabilization of measurements)

These experiments were carried out with different rates of temperature change and different sample thicknesses. The relaxation times are longer as the values of dielectric constant are higher. The results presented remain qualitatively correct for all domain textures obtained and consistent with the results, depending on the thickness of Fig. 4.27

4.6.4 Models to Explain the Freezing of Domains

Numerical Exploitation of Experimental Results (Vogel-Fulcher)

For about 20 years, the freezing of domains at low temperatures has often been the subject of data analysis using the Vogel-Fulcher relation:

$$X = X_0\, e^{A/(T - T_{\mathrm{VF}})} \tag{4.13}$$

X is the quantity studied with three adjustable parameters X_0, A, and T_{VF}. In addition to the ease of adjusting three parameters, the relationship takes into account the failure of Arrhenius' law to account for the experimental results obtained. The approach began in the 1920s (Vogel, 1921; Fulcher, 1925), in chemistry for the reaction kinetics of certain liquids. It was then used in melting transitions (Angell, 1988), glasses, polymers (McCrum, 1967), spin glasses and dipoles (Bhattacharya et al., 1982), and closer to the materials that concern us, in ferro-antiferro mixtures (Courtens, 1984). The authors justified it by cooperative rearrangements of randomly distributed dipoles, drawing a parallel between dielectric relaxation and viscosity (Adams and Gibbs, 1965; Samara, 2001; Ben Ishai et al., 2004; Pirc and Blinc, 2009). But other authors used it for the growth and percolation of polar nanoregions in relaxors with this time very anisotropic dipole distributions (Néel, 1949; Pirc and Blinc, 2007). There is therefore a large bibliography to justify its use in situations where very local movement phenomena and more global phenomena exist.

In uses for domain freezing, parameter A is usually introduced in the form of a distribution around a mean value, either symmetric or asymmetric. It is assumed that there are sets of dipoles that would turn in different places for different τ. Thus, for the dielectric dispersion of domain freezing, a Gaussian law of activation energy distribution is chosen (Kuramoto, 1987):

$$f(E) = (2\pi\sigma)^{-1/2} \exp\left(-(E - E_0)^2 / 2\sigma^2\right) \tag{4.14}$$

With E_0 and σ constants independent of temperature.

The dielectric constant is then described by formulas with resonances and relaxations, and for the latter, the parameters allowing the closest possible representation of the experimental results are sought numerically. These exploitations have been carried out for the crystals in this book: KDP (Kuramoto, 1987; Nakamura and Kuramoto, 1988; Huang et al., 1997; Kim et al., 2009), DKDP (Huang et al., 1997), KDA (Kubinec et al., 1995), DKDA (Fally et al., 1995), CsDA (Hayashi et al., 1988). The simulations are based on experimental values in sufficient frequency and temperature ranges, excluding resonances. The intermediate parameters used rather follow the relation (4.14) in the temperature range of the plateau, while the simulation is built to account for domain freezing. These simulations are based on macroscopic measurements that usually do not take into account domain configurations and wall shapes.

Phase Transition Models in the Domain Wall

Domain freezing is presented by some authors as a phase transition internal to the wall. For example, a study of short-range pseudo-spin interactions taking into account tunnelling effects and interactions with the network through heavy ions (Fedosov and Sidorkin, 1977a) leads to two wall configurations. These energies are in the classic representation of a saddle whose temperature change changes the shape. The inversion in the energy ratio would occur at the freezing temperature. But these types of models can hardly resist experimental results, especially the variation of the freezing temperature with the thickness of the samples. They all rely on a certain contradiction of considering as continuous the parameters of a wall of thickness of one or a few elementary cells (Fedosov and Sidorkin, 1977b).

Interesting simulations are proposed at the nanoscopic level (Schmidt et al., 2000; Sinitski and Schmidt, 1996). With the tetrahedron configurations already illustrated in this book (Figs. 1.32 and 2.10), the energies of Slater defects E_0 and Takagi defects E_2 are taken into account. In addition, the energy E_d of interactions between parallel hydrogen bonds located at different ordinates along c is also considered (Ishibashi et al., 1998). This energy is essential to obtain the coherent displacement of the walls. The Monte Carlo simulation developed on more than 16,000 hydrogen bonds gives the results illustrated in Fig. 3.37: the density of Slater defects and Takagi defects is significant at notable distances from the centre of the wall for temperatures close to T_C and on the "plateau". But, as the temperature decreases, the probability of existence of Slater defects and Takagi defects becomes increasingly weak; it remains significant only in a "width" of the wall that itself tends towards zero. The simulation (Fig. 3.37) demonstrates the importance of Ishibashi's energy $E_{d,}$ which greatly facilitates the attainment, at low temperatures, of a flat wall, close to the wall permitted by symmetry. The authors call transition the passage from a wall situation with Slater and Takagi defects to a wall situation without these defects. These simulation results are in agreement with the experimental results: the application of an electric field (or a mechanical constraint) slightly widens the walls, but above all increases the probability of the presence of Takagi and Slater defects, thus the mobility of the wall. Thus, the freezing temperature of the domains decreases when the external field applied increases. Similarly, freezing only occurs when the wall becomes increasingly flat, that is to say, when it is becoming increasingly difficult to find surfaces with sets of Takagi (and Slater) defects: the thicker the samples are, the greater this probability will be, and freezing will therefore occur at lower temperatures. It is interesting to note that these simulations are consistent with previous models of quasi-dislocations with or without knees and with the following model of critical field for setting in motion.

Critical Field for Setting Dislocations in Motion

To explain the freezing of domains, Barkla and Finlayson (1953) were the first to suggest the role of dislocations. About 15 years later, their long-distance effect (200 μm) and their existence in any wall that deviates from the theoretically permissible wall were demonstrated (Bornarel and Lajzerowicz, 1968; Bornarel, 1971, 1972). A fortiori, any wall displacement, even the width of a half-elementary cell, at absolute zero, generates lines of quasi-dislocations of the wedge type as presented in Chap. 2. The whole can obviously be more complex, with knees, involving a more or less important set of elementary cells, and when approaching the transition, with a volume distribution of ions that can be described as well as elements of dipoles (Fig. 3.37) as knees of quasi-dislocations (Fig. 3.6). A simulation allows to predict that their density and the wall width decrease as the temperature drops (Schmidt et al., 2000). The presence of lines that cross the entire sample along the axis (001) is well demonstrated by the interactions between domains, especially between domain tips: the theory of infinite length wedge type dislocations is indeed sufficient to properly model the observed displacements and interactions. The questions to be deepened are of various orders:

- Is the setting in motion of these quasi-dislocations the primary cause of domain freezing?
- How to reconcile all the experimental results obtained in freezing with this modelling?

To answer the first question, the simplest model was considered: as the experimental results have shown, at sufficiently low temperatures, lines of wedge quasi-dislocations concentrated at the tips of the needle domains move in blocks (Bornarel, 1972). In this context, the Peierls model (Peierls, 1940) used in its most general form (Nabarro, 1967a, b) allows to calculate the critical stress σ_S for setting such a dislocation line of Burgers vector **b** and width A in motion:

$$\sigma_S = 2\mu\left(1-\nu\right)^{-1} e^{\left(-\,4\pi A/b\right)} \tag{4.15}$$

with μ shear modulus and ν Poisson's ratio. This relation is only valid if $e^{(4\pi A/b)}$ is much greater than 1, which is the case in the considered crystals. The amplitude of the electric field that would correspond to the critical value for setting in motion would be, in this hypothesis,

$$E_S = 2\mu\left(d_{36} / \chi_{33}\right)\left(1-\nu\right)^{-1} e^{\left(-4\pi A/b\right)} \tag{4.16}$$

Very small variations in the (A/b) ratio are enough to cause significant changes in E_S: in KDP, a value of the ratio of 1.6 would correspond to a critical field for setting in motion of 3.7×10^{-2} V/cm while a value of 0.8 would correspond to a critical field of 865 V/cm. These values are consistent with the choice to neglect the Peierls forces to explain the phenomenon of maximum losses observed in the few degrees below the transition (small b and quite large A). On the other hand, it is obvious that the Burgers vector b increases as the temperature decreases and that the wall width decreases with the increasing rigidity. Thus, E_S increases drastically when (A/b) decreases with temperature.

If the considered dislocation is the smallest imaginable, with a lateral wall displacement of one cell (Fig. 2.24) or a half-cell (Fig. 2.25), the electric field for setting in motion E_S will remain low and the blocking process in this hypothesis of global displacement of the dislocation line is unlikely, even at temperatures lower than those of domain freezing: it is not the Peierls phenomenon that for a wall located in a plane (100) or (010) can prevent displacements lateral by sliding of quasi-dislocation corners in the wall plane.

However, regarding the needle domain tips, what values of A and b should be retained to summarize them? They depend on the shape of the end (Fig. 4.23). The quasi-dislocation of Burgers vector $b_m = 2w_m\, tg\, u_{xy}$ that will block the entire domain tip will be the one whose ratio A_m/b_m will be the largest (with $2a\, tg\, u_{xy} < b_m < B = 2w\, tg\, u_{xy}$). A_m is usually written (Nabarro 1967a, b) $(K_m\, a)/2(1-\nu)$, and as ν and a vary little with temperature, the ratio w_m/K_m is decisive for defining the freeze (Bornarel, 1972). Figure 4.33 thus presents the variations of this ratio calculated with the data collected at the maximum of the loss constant, for various crystals of the KDP family, in various experimental situations, as a function of the deformation u_{xy}, reference mechanical parameter. The temperatures at which the freezing phenomenon occurs differ by several tens of degrees, and yet, the correlation is evident, which argues for this freeze model based on the mechanical properties of the crystal. Also note that in Fig. 4.30, the slight curvature observed in the logarithmic scale variation is consistent with the convexity of the law $u_{xy}(T)$.

This simple model is in agreement with all the published results, especially with the variation of the freezing temperature as a function of the thickness of the samples (Bornarel and Torche, 1992): dislocation lines are not infinite as the model

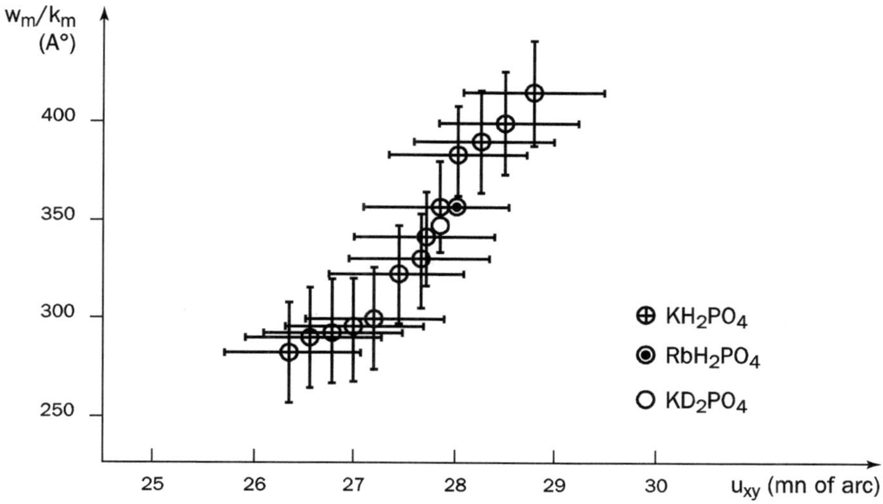

Fig. 4.33 Variation of the ratio w_m/k_m (Bornarel, 1972) as a function of the spontaneous deformation u_{xy} for different crystals of the KDP family; w_m is the lateral wall displacement that would correspond to the average Burgers vector b_m as defined in the text based on the relation (4.15)

considers, and the pinning on the surfaces of the sample facilitates the freeze the smaller the sample thickness is. This is what is experimentally observed in Fig. 4.27. The fact that the freezing temperature rises as the frequency increases goes in the same direction of a mechanical explanation (Guyon and Lajzerowicz 1966; Huang et al., 1997; Kim and Kim, 1999).

4.6.5 Brief Synthesis Regarding Domain Freezing

Domain freezing is observed in ferroelectric and ferroelastic crystals with sufficient anisotropies for there to be preferential orientations of walls. The phenomenon also exists in non-piezoelectric ferroelectric crystals in para phase, such as TGS (Huang et al., 1996). But in these crystals, whose domain texture is generally made up of S_∞ walls, it only occurs when the crystal contains impurities that create anisotropies and therefore domains in the form of corners or needles (Huang et al., 1997).

The freezing phenomenon is systematically present in ferro-ferro crystals for which the mechanical and dielectric properties are difficult to separate. Those of the KDP family, with low conductivities due to their temperature ranges, are prime candidates for reducing certain electrical effects. In these crystals, any wall modification generates Takagi ions and Slater ions with other short-distance interactions. These modifications (see Fig. 3.50) require the creation of ions, of quasi-dislocation lines or dislocations with or without knees. At sufficiently low temperatures, the flatter the walls, the more the dislocation lines are defined and concentrated at the tips of needles with notable Peierls forces of dissociable dislocations. The Burgers vectors of such assemblies of quasi-dislocations vary according to local deformations. Many models proposed to explain freezing pay little attention to the wall itself but rather to distributions of defects, often in volume, which by "pinning" would hinder its movement and explain the freeze. There are always accidental volume defects, and they have been estimated in some of our experiments (10^3 defects per cm^3 is a critical density). But if the defect density obviously modifies the freezing temperature, the phenomenon does not disappear in the most perfect crystals. We must therefore take into account the wall itself and the systematic imperfections created during its movements. The Peierls forces, the distributions of Burgers vectors, the knees, and the charges of the associated ions are sufficient elements to explain freezing, the variation of its reference temperature as a function of the applied field (amplitude and frequency) and that as a function of the sample thickness.

Indeed, the Peierls forces of dissociable dislocations of domain ends, those of flat walls, depend on their size and are linked at the nano level to atomic movements such as hydrogen jumps on bonds. Furthermore, at the microscopic level, the responses of the domains depend on the sets they constitute in the sample. We therefore have various levels of definition of the "local" and the "global", with accumulation at the macroscopic level of these responses. This explains the tendency to use the relation of Vogel-Fulcher and its numerical optimizations with several parameters. Our goal is to explain the phenomena at play. The nano-micro link still needs to be deepened.

4.7 Susceptibilities at Very Low Temperatures

Measurements of susceptibilities of ferroelectric materials at temperatures, typically below 30 K, show anomalies compared to the predictions of models for pure crystals and powders. These results are sometimes correlated with those observed in glasses, and to understand them, variations in specific heat and thermal conductivity as a function of temperature are valuable (Annex A9). These anomalies are often linked to a change in parameter such as the amplitude of a continuous field applied, or to a thermal shock. These "after effects" exist either in "normal" form (when there is a sudden change in the field, the dielectric constant changes rapidly in the direction induced by the field, then varies slowly over time in the same direction), or "abnormal" (a rapid variation of the dielectric constant in one direction, followed by a slow variation in the opposite direction). Better specifying these phenomena and linking them to domains, walls, atomic, and/or electronic displacements allow for a better understanding of the elementary processes in the walls. What "anomalies" exist concerning our ferro-ferro crystals?

4.7.1 Anomalies of Dielectric Properties

Dielectric measurements of KDP, below 30K, were carried out on pure crystals and on powders as early as the 1970s (Holste et al., 1976; Motegi et al. 1983a, b, 1985; Kuramoto et al., 1984). Figure 4.34 illustrates the anomalies observed in addition to a regular variation of absorption and dispersion along the ferroelectric axis *c* (Fig. 4.34a) and along the axis *a* (Fig. 4.34b). In both cases, the maxima of loss constants cannot be attributed to conductivity, negligible at these temperatures. But there is an essential difference according to the two axes: the application of a continuous electric field along the *c* axis during cooling, with a value large enough to make the crystal mono-domain does not change the results along the *a* axis, and on the other hand, leads to the disappearance of the anomaly along the *c* axis, demonstrating a direct link with the existence of domain walls. The effects are more significant in powders, especially for the peak around 12 K with grains oriented along *a* and along

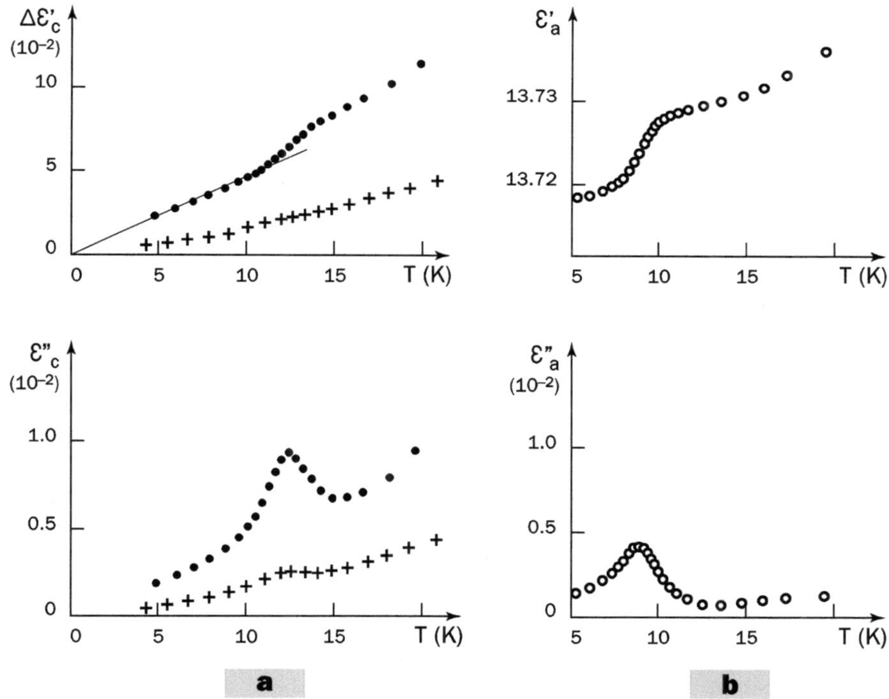

Fig. 4.34 (**a**) Temperature variation, for a KDP crystal of $\Delta\varepsilon_c'$ (dielectric constant along the *c* axis decreased by its value at 0 K) (top) and of ε_c'' (bottom), without continuous field applied (•), and when the crystal is subjected to a strong continuous field applied during cooling from the para phase, thus without domains (+). (**b**) Temperature variation of the dielectric constant ε_a'(top) and the loss constant ε_a''(bottom) measured along the *a*-axis for a pure KDP crystal with or without a continuous electric field applied along the *c*-axis

c, thus with a greater dielectric measurement incidence for the dielectric constants of average values compared to very large values; this peak occurs at close temperatures for crystals and for powders. The relaxation frequencies f_r are determined by taking as a reference the maximum of loss constant. They follow Arrhenius laws

$$f_r = f_0\, e^{(-U/kT)} \tag{4.17}$$

with for the measurement along c, $T_{max} = 12.3$ K, $f_0 = 45$ GHz, $U = 19$ meV and for that along a, $T_{max} = 8.5$ K, $f_0 = 3.5$ MHz, $U = 6$ meV (Motegi et al., 1984, 1985). The plotting of Cole-Cole diagrams is carried out using the classical relation (Cole and Cole, 1941)

$$\varepsilon(\omega) - \varepsilon_b = (\varepsilon - \varepsilon_\infty) / \left(1 + (i\omega\tau)^\beta\right) \tag{4.18}$$

in which ε_b is the basic dielectric constant. The value of β is about 0.90 for the anomaly along a, therefore a value close to that of the Debye model ($\beta = 1$) (Annex A9). This value for measurements along c linked to the existence of domains is about 0.72, therefore less well described by this model.

The work carried out on powders with grain sizes between 50 μm (Motegi et al. 1983a, b, 1985) and 3 μm (Gilchrist, 1998) show the existence of domain walls, whereas they had been considered absent for grain dimensions of a tenth of a micron (Jaccard et al., 1953). The authors have demonstrated that the anomaly around 12.5 K cannot be due to the Maxwell-Wagner effect (Daniel, 1967) and cannot be linked to grain surface effects in the case of powders; this anomaly is linked to the existence of domain walls.

4.7.2 The "Normal" After-Effect

As previously illustrated, the dielectric permittivity changes abruptly after a sudden change in the applied electric field ΔE, then afterwards, more slowly following power laws approximated by a relation in Lnt. Studies at various temperatures, 4.2 K, 20.4 K, 77.8 K, and 300 K (Zimmer et al., 1987), have allowed us to analyse these variations of "normal" after-effect.

These studies have been compared to those of dielectric glasses, especially around 1 K (Salvino et al., 1994), even though it has been demonstrated that the situation of KDP with a clear transition is different from those of ferroelectrics with diffuse transition. To study the small variations observed of ε' and ε'' with time t, we took the habit of defining:

$$s' = d\varepsilon' / d(\text{Lnt}) \text{ and } s'' = d\varepsilon'' / d(\text{Lnt}) \tag{4.19}$$

Figure 4.35a provides an example of variation at 25 K of s' and s'' in relation to ΔE, or rather $|\Delta E|$ (only the amplitude of ΔE plays a role, its polarity does not affect the

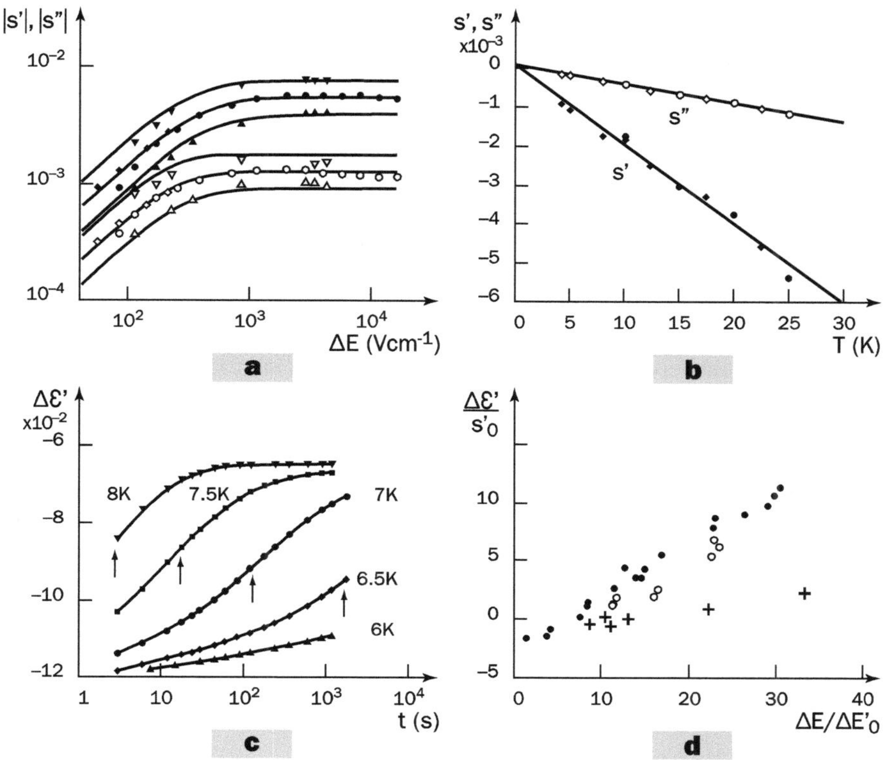

Fig. 4.35 (**a**) Variation of s' and s'' (defined by the relation (4.19) in relation to the amplitude of the step of the continuous electric field ΔE. The chosen temperature of 25 K is deliberately distant, both from the freezing of domains and from dispersions at very low temperatures. We observe the existence of two situations for each curve (parameter the frequency): linear variation below a critical value of ΔE and quasi-stability above. The first corresponds to the "normal" after-effect and the second to the "abnormal" after-effect (Gilchrist, 1999). The solid lines are the results of calculations with the empirical formula $s' = -s_0'(1 + (\Delta E_0'/\Delta E)^2)^{-1/2}$ and the similar formula for s_0''. (**b**) The linear variation of the logarithmic decrements of ε' and ε'' as a function of the temperature T of a KDP crystal (28 V cm^{-1} and 1.2 kHz) continues up to about 60K (Gilchrist, 1999). (**c**) Variation of $\Delta\varepsilon'$ (43 V cm^{-1} at 1.2 kHz) over the time t for a pressed KDP pellet, cooled from a temperature above 123 K to the given temperatures after a sudden change ΔE of 5.7 kV cm^{-1} at time $t = 0$. The arrows indicate the time constants obtained by the Arrhenius laws (Kuramoto et al., 1984) and correspond to the inflection points of the curves. (**d**) The relative variations of the abnormal after-effect are linear, if we normalize $\Delta\varepsilon'$ by s_0', value defined by the curves (**a**), and ΔE by the threshold value $\Delta E_0'$. The result is general, if we take into account the quenching cycles and the indicated normalizations (Gilchrist, 1999). The experimental points come from numerous samples classified into three groups, according to the ratio between the value of ε_c'' maximum (at the peak around 12 K) and the minimum value around 15 K (100–110% (+); 110–120% (o); 125% (•))

results). The variations of s' and s'' are comparable when ΔE increases: proportional to ΔE for values lower than critical values $\Delta E_0'$ and $\Delta E_0''$, and almost constant for higher values. The critical values $\Delta E_0'$ and $\Delta E_0''$ decrease with T and increase with the frequency; the saturation value s_0' varies like T and like a weak power (−0.15) of f. We can dissociate the two situations of Fig. 4.35a, the first corresponds to the "normal" after-effect and the second to the "abnormal after-effect".

The normal after-effect exists in KDP from 1.4 to 300 K. It resembles phenomena observed in glasses and is often attributed to the existence of micro-domains. We assume the existence of notable dipole moments, with minor orthorhombic distortions. These micro-domains would be easily reversible, and the after-effect would be caused by a rearrangement within the walls with activation energy distributions. The effect would depend on the number of domains that increase as the temperature rises. In KDP, this would seem to be in agreement with the domain freezing described in Sect. 4.6. We know that the application of ΔE can modify the shape of the domain tip and thus locally modulate walls. The situation in the grains is a bit different due to different constraints from those of the pure crystal. We obviously find a wide distribution of situations attributed to interactions between nearby walls and long-range constraints. The after-effect would be the overall relaxation resulting from these wall incidents. Experimental demonstrations of these models remain to be supported by measurements of wall surfaces and the number of tips, simultaneously with dielectric measurements.

4.7.3 The "Abnormal After-Effect"

The abnormal after-effect has been demonstrated in dipolar glasses (Saint-Paul and Gilchrist, 1986), anti-ferroelectrics (Ema et al., 1988), spin glasses (Lundgreen et al., 1983), crystals with high permittivities (Gilchrist, 1989), ferroelectrics doped with defects such as $BaTiO_3$ (Robels and Arlt, 1993), and PZT ceramics (Hall and Ben-Omran, 1998). Studies on KDP (Gilchrist, 1998) show that the abnormal after-effect occurs at temperatures of about 6–8 K (Fig. 4.35c) under conditions of activation energy comparable to that of the anomaly of dispersion of the dielectric constant along the c axis at 12 K (Fig. 4.34a), which argues for a close connection between the two phenomena. The link with the existence of domain walls is established. As the phenomenon exists for values of ΔE higher than the threshold values, arrangements of wall or micro-domains are possible, but the nature of the dipoles involved is different from that of the dipoles of the normal after-effect, with smaller dipole moments, therefore easily reversible. This reversal of dipoles can interact with the wall environment without deeply modifying it. The domain texture is considered stable with walls locally in metastable situations, therefore including certain types of defects. These defects would not be able to migrate but rather be in potential configurations with multiple wells, in the simplest case, a double well. We easily think of the two possibilities of a hydrogen bond at low temperatures and their impact on neighbouring tetrahedra. This double well is strongly coupled to network deformations and, in particular, to the shear deformation that accompanies the ferro state. A defect would thus correspond to a double well of about 16 meV in depth. The application of a sufficient ΔE would cause the defect to move from one well to the other, and we would observe over time the relaxation of this local phenomenon, internal to the wall. The results obtained in the dielectric measurements at 12 K, on the one hand, and during the abnormal after-effect at 7 K, on the other hand, show similarities in terms of characteristic times (about 60 s) and activation energy of about 19 meV, very reproducible.

There are arguments to suggest a coupling between a proton displacement on the hydrogen bond and a network mode; Gilchrist has proposed HPO_4^- or (and) $H_3PO_4^+$ ions. Unlike the statistically numerous helix paths in the temperature range known as the "plateau" and even at the beginning of domain freezing, these ions would not be coupled to each other and would be randomly distributed within a wall, which would not prevent long-distance effects on the walls when they are reversed by the applied field. It is understandable that as the temperature rises, these direct interactions between dipoles would disappear with the creation of cascading proton movements. Experimental results comparable to those obtained for KDP are obtained in KDA (KH_2AsO_4) and RbDP (RbH_2PO_4) for comparable characteristic times (60s) and similar temperatures: 10.5 K. For KDA, the activation energy is 31 meV, but the ratio of temperatures of the abnormal after-effect and the Kuramoto effect (*E*//*c*) is the same in KDA and in KDP-RbDP (10.5/17 = 7.2/12) which reinforces the hypothesis of the same type of defects, with comparable results in different atomic environments.

Memo Chapter 4

Stable and reproducible susceptibility measurements require electrical and thermal treatments adapted to the shape of the samples and the chosen measurement temperature. We can then consider the macroscopic result as the sum of a mono-domain contribution and another due to the domains. Four temperature intervals are delimited by maxima of the loss constant:

- Except in clear situations of first-order transition, just below the transition, the crystal can be described as a "continuous, modulated and distorted medium" with a proportional relationship between inverse mechanical and dielectric susceptibilities. A description for the chosen crystals using statistics of Slater and Takagi ion chains (Schmidt) works better than other models. In this temperature range, the low value of coercive fields can lead to confusion between dielectric susceptibility measurements and hysteresis cycle measurements.
- In the so-called "plateau" interval, if the domains and domain walls are densely packed without being able to easily change their number, significant susceptibilities demonstrate metastable equilibria, with the possible existence of avalanche phenomena and strong interactions between separable walls and domains. The susceptibilities and losses vary as S^n (S total area of the permissible walls) with n between 1.3 and 1.7.

- If the domain walls are increasingly narrow, separable singularities are created, and dislocations are subject to the temperature freezing phenomenon. This phenomenon, often treated by numerical simulations using the Vogel-Fulcher formula, corresponds to Peierls forces that block the domain tips. The susceptibility at lower temperature becomes proportional to the total area S of the plane permissible walls, isolated from each other ($n = 1$): below these freezing temperatures, the walls no longer interact with each other.
- At even lower temperatures, between a few degrees K and 20 K, low measurement fields only cause changes within the walls on the ionic imperfections that exist here and there.

The temperature intervals described above are found in many contexts under different names (Harrison et al., 2004; Novik et al., 2023) with the "continuum" and the gel as landmarks (Goian et al., 2023) and maxima of losses that delimit them (Salje et al. 2005). In each type of situation of these non-linear systems, we find at the macroscopic, microscopic, and nanoscopic levels hysteresis effects and "after-effect" characteristic of the phenomena concerned.

Chapter 5
Phase Fronts and Domains

5.1 Introduction

The previous chapters have allowed to link macroscopic results to microscopic phenomena at the level of domains and walls. These links can be transposed into other crystals than those presented and in other situations. However, our choice of these transitions with a change in tetragonal-orthorhombic symmetry, to obtain simple polarizations and deformations, and visible competitions between elastic and electrical properties should not be forgotten. In addition, the proximities with the tri-critical situation in RbDP (second-order transition) and in KDP (first-order transition) have also been exploited.

To study and understand phase fronts, the choice is necessarily different: that of a clear first-order transition with the appearance of a notable value of the polarization–deformation as soon as phase coexistence occurs. To do this, the DKDP compound is chosen because of a comfortable temperature range to carry out measurements and observations. The concern to broaden the discussion remains the same.

The phase front, the interface between a para phase and a ferro phase, is the main subject of this chapter without forgetting its possible interactions with the domains. There are some errors in the literature about the order of a transition when a photograph showing a clear interface between two phases is presented as proof of a first-order transition. The reason is often the existence of significant thermal gradients in the samples studied; the misinterpretation is common if we are dealing with crystals close to the tri-critical transition (Nakamura and Ihara, 1969, 1970). The aim is to limit ourselves here to the study of real phase fronts. Previous work carried out at the transition on ferroelectric or ferro-elastic crystals have shown that the Fousek-Janovec-Sapriel model on "permissible interfaces" can work for phase fronts (fronts in which the deformation remains the same in both phases). Regarding the chosen crystals, Fig. 5.1 illustrates this with TMO whose (110) planes are the permissible

J. Bornarel, *Ferroic Crystal Domains, Walls, and Phase Fronts*, Springer Series in Solid-State Sciences 209, https://doi.org/10.1007/978-3-031-98099-2_5

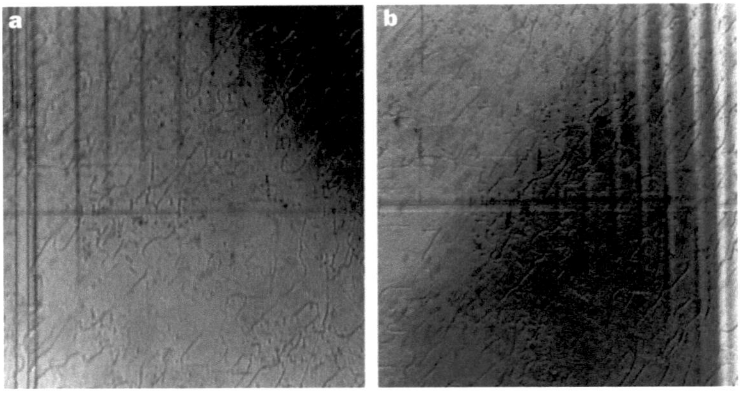

Fig. 5.1 Cross-section observation (001) of a TMO crystal, during the tetragonal-orthorhombic phase transition (the horizontal and vertical directions of the photograph correspond to the tetragonal axes a_1 and a_2). As predicted by the model of interfaces allowed by identical deformation in both phases (Fousek and Janovec, 1969), the phase front between the para phase and the ferro phase appears between the black region and the illuminated one, between crossed polarizers, with an angle of 45° relative to the tetragonal planes. The photograph (**a**) corresponds to a PF transition and the photograph (**b**) to a FP transition respectively

interfaces between the para and ferro phases. In this case of an improper ferroelectric crystal, with a tri-critical transition and various types of walls including anti-phase walls, domains extend through the interface between the para and ferro phases. A monograph on such a theme can be found in "Domains in ferroic crystals and thin films" (Tangantsev et al., 2010).

For the study of phase fronts, the approach previously chosen is continued by the analysis of solid experimental results preceding models and discussions. The phase front and the domains in a DKDP crystal are first studied with samples subjected to various temperature protocols, without the application of a conjugate field. Observations along three perpendicular axes are carried out simultaneously with laser brush diffractions to detect the domains and with dielectric measurements. Section 5.2 presents the results obtained with crystals under optimal thermal homogeneity conditions. Specifically, the temperature at various points in the crystal differs by less a few thousandths of a degree. In Sect. 5.3, the use of a constant thermal gradient allows the analysis of the shapes and possible orientations of the phase front. The exercise is simplified by choosing typical thermal gradient orientations relative to the crystallographic axes. In Sect. 5.4, models are proposed and compared with experimental results. The correlations between phase front shapes, domain texture, and macroscopic properties are summarized in Sect. 5.5. Section 5.6 allows us to imagine what can happen depending on the context: the impact of thermal inhomogeneities, the influence of the sample edges and mechanical defects, the evolution over time towards the shape of the most energetically favourable phase front, and the causes of hysteresis are illustrated. The only external parameters considered in this chapter are temperature and time; the effect of conjugate fields will be taken into account in Chap. 6.

5.2 Phase Fronts of a Crystal at Homogeneous Temperature

5.2.1 *Natural Orientation of the Phase Front in DKDP*

Before the neutron diffraction work of Zeyen and Meister (1976a, b), scientists interested in the properties of KDP and its family were essentially ferro-electricians. For them, the orientation of a phase front was not in doubt and was indeed little studied: the phase fronts had to be necessarily surfaces parallel to the ferroelectric axis *c* to prevent them from being electrically charged (Fig. 5.2, top and bottom).

The proposal of phase fronts perpendicular to this *c* axis was initially mocked. Then, neutron (Zeyen et al., 1976), X-ray (Aleshko-Ozhevskii, 1982, 1992), optical and γ diffraction studies (Bastie et al., 1980; Bachheimer et al., 1981) have demonstrated the relevance of the assertion even if the front shapes varied significantly depending on the thermal experimental conditions. The decisive experiments were obtained by optical observations in three directions, simultaneous with other optical and dielectric measurements (Bornarel and Cach, 1999) for samples of great thermal homogeneity. The coexistence at almost homogeneous temperature occurs in coherence with Fig. 1.15: the transition, during a para-ferro (PF) cycle as during a ferro-para (FP) cycle, begins practically at the same temperature, that of Curie (Zeyen, 1975) and during the entire coexistence, the temperature of the crystal remains the same (see Fig. 1.15b). As the temperature measured during an experiment is not that of the crystal, but usually that of an element close to the crystal, the evolution of the parameters measured as a function of this temperature is in fact a variation over time during coexistence. In all cases of coexistence with very good thermal homogeneity of the sample, we observe as illustrated in Fig. 5.3:

- At the beginning of coexistence, the creation of nuclei of the new phase at the eight corners of the sample
- Then, the creation of two quasi-planar phase fronts approximately perpendicular to c that move towards each other in the sample
- The disappearance of the initial phase with phase fronts that approach each other in the c direction.

5.2.2 *Phase Front, Domains, and Dielectric Properties*

The dielectric property curves of Fig. 5.4 are typical of the results obtained in the near-absence of a thermal gradient. We must not forget that the phase front is perpendicular to the *c* axis and therefore a charged interface with its characteristics (Sluka et al., 2016). Let's highlight the similarities and differences between the PF and FP cycles: In both cases, the loss factor, therefore ε'' measured, takes significant values when a dense texture in domains occupies the part of the coexistence zone close to the ferro phase. These textures exist if the domains can cross the entire sample along the ferroelectric *c* axis. Their density decreases if they encounter a

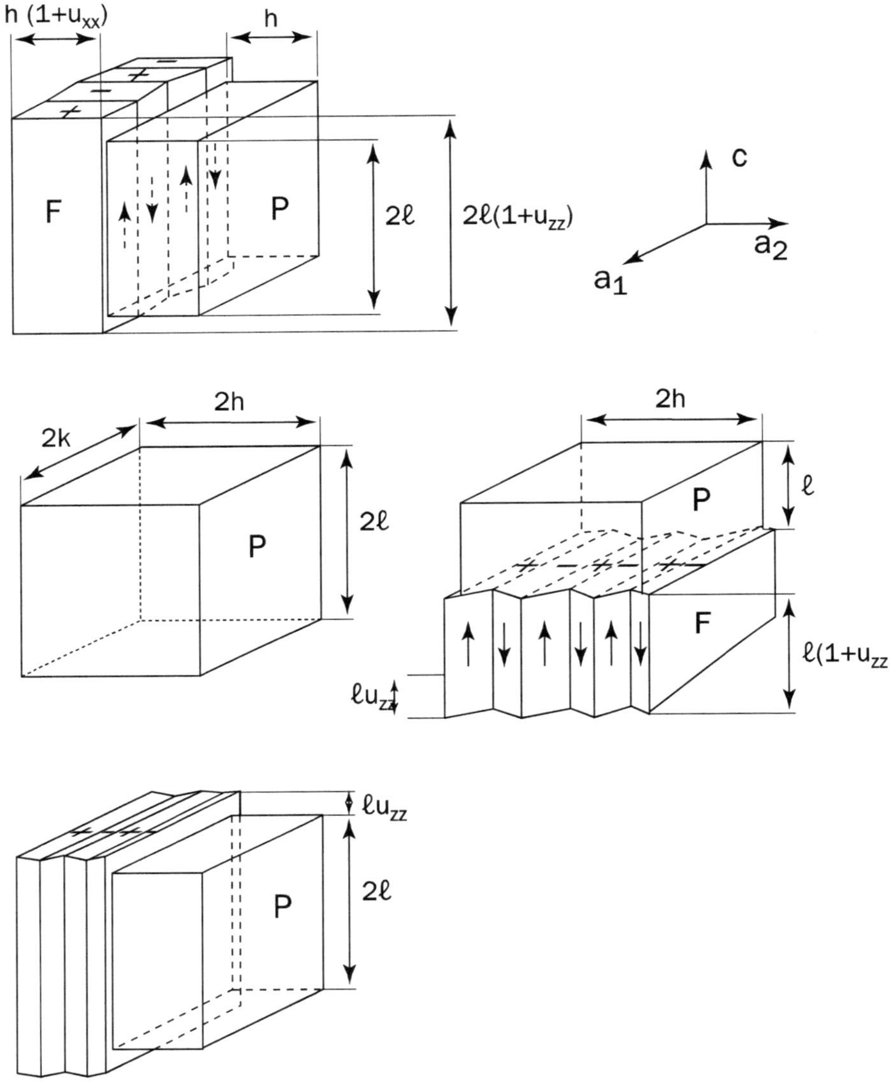

Fig. 5.2 Schematic representations of a DKDP crystal sample, of tetragonal shape, volume $2k.2h.2l$, in the para phase (P) as presented in the centre left. The deformations between the para phase and the ferro phase (F) are very schematically represented: they are amplified about 10000 times! u_{xx} is about ten times smaller than u_{zz}.

If we assume the phase front in a plane (001) as in the figure on the right in the centre (with a para volume $2k.2h.l$ and a larger ferro volume), this front will be electrically charged and deformed due to the deformations $u_{xx} = u_{yy}$ and u_{xy}. However, the deformation u_{zz} will not directly intervene.

If now, we assume a phase front in a plane (010) as in the figures on the left, top, and bottom, it will not be electrically charged but will mainly undergo the variation of the most significant deformation u_{zz}. We note that there are two possibilities concerning the domains: in the figure at the bottom left, the wall plane is parallel to the phase front, whereas in the figure at the top left, the deformations u_{xy} of the domain ends add to the variations due to u_{zz} and u_{xx}. This orientation of the domain walls perpendicular to the front plane, on this very theoretical diagram, seems less likely than the previous one

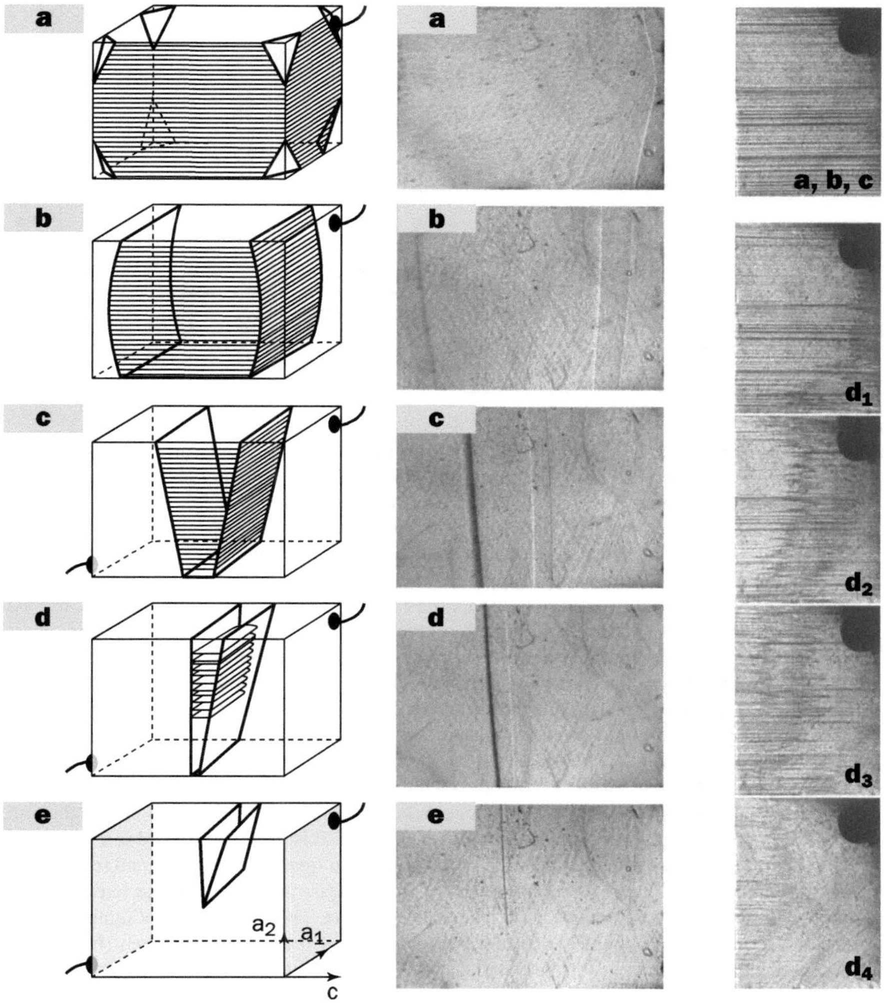

Fig. 5.3 A DKDP sample (dimensions: $a_1 = 5.5$ mm, $a_2 = 4.4$ mm, $c = 7.4$ mm) is suspended by electrode wires in a quasi-isothermal gas exchange cell: temperature regulated with an accuracy of 2×10^{-3} K, modified at a maximum speed of 10^{-4} K min^{-1}. The domains of the diagrams on the left correspond to the photographs of texture in domains in section (001), presented on the right; the phase front is visible in section (100) (photographs in the centre)

In such a thermal context, the shape of the phase front is the same during a PF cycle and during a FP cycle. The situations *a*, *b*, *c*, *d*, *e* are recalled in the dielectric property curves as a function of temperature in Figs. 5.4a, b. As illustrated by the present figure of a transition (FP), the transition begins at the same time in the eight corners of the sample (**a**); then, two quasi-planar phase fronts are formed (**b**); move towards each other (**c**); join on one of the edges of the sample (**d**); and withdraw (**e**) in such a way that a part of the new phase crosses the entire sample along the ferroelectric *c* axis. The domains disappear when the distance between the phase fronts becomes <400 μm as illustrated by the photographs d_1, d_2, d_3, d_4 (Bornarel and Cach, 1999)

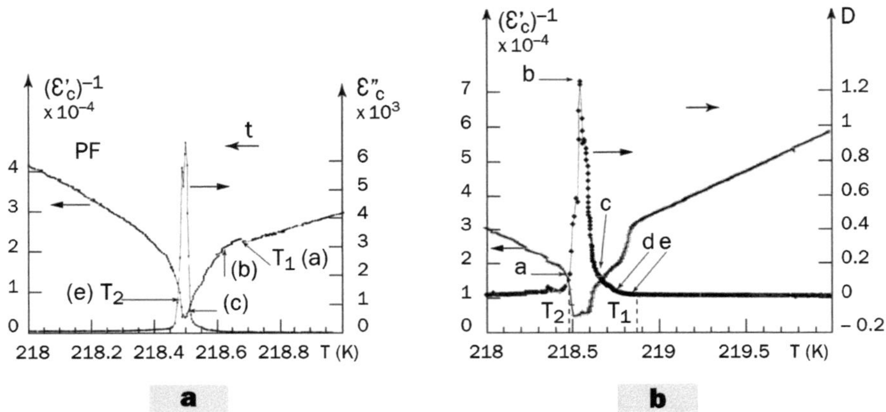

Fig. 5.4 Variation of the inverse dielectric constant and the loss constant during a phase coexistence with very good thermal homogeneity of the DKDP crystal (Bornarel and Cach, 1999). The indications *a*, *b*, *c*, *d*, and *e* correspond to the situations presented in Fig. 5.3. The large values of the loss constant ε″ correspond to textures in dense domains. The reference temperature is that of a platinum probe near the sample. (**a**) Dielectric results during a PF cycle, with the situations for the phase front marked: (**a**) appearance of phase front nuclei in the eight corners of the sample, (**b**) creation of two fronts slightly distorted but approximately perpendicular to the ferroelectric axis (few domains are observed in the ferro phase), (**c**) the two fronts join and a dagger (of para phase) begins to withdraw allowing the establishment of a dense domain texture, as it withdraws; in (**c**) the domain texture invades the entire sample, totally ferro in (**e**). An arrow with *t* recalls the evolution of time during the experiment; the temperatures T_1 and T_2 are measured as closely as possible, outside the sample. The losses are significant when the domain texture is dense in most of the sample. (**b**) Dielectric results with observations of phase fronts and domains presented in Fig. 5.3 during a FP cycle(the arrow with t recalls the evolution of time) with the same reference indications: (**a**) start of coexistence, nuclei in eight corners of the sample, (**b**) creation of two distorted phase fronts approximately perpendicular to the ferroelectric axis; between (**b**) and (**c**), the volume of the domains decreases with the approach of the two phase fronts; between (**c**) and (**d**), the dagger front withdraws and the mono-domain ferro region appears in the dagger; the ferro dagger is totally mono-domain and withdraws in (**e**): coexistence is over. The loss factor *D* is significant only when the volume of the domains occupies a notable part of the sample (between (**b**) and (**c**)). Between (**c**) and (**e**), there is a contribution to the dielectric constant beyond that, classic, of Curie's law (Bornarel and Cach, 1999; Bornarel et al., 2000)

phase front along the *c* axis, and the observed walls almost disappear if two phase fronts are close to each other (critical distances of 200–400 μm). It is difficult in these observations to determine whether we are dealing with domains or simple walls, given the weak optical contrasts. We find during coexistence the differences demonstrated in Chap. 4 regarding the PF and FP cycles: great instability with rearrangement of the domains during a PF cycle (Bursian et al., 1977) and a stable domain texture during an FP cycle. In Fig. 5.3 (FP cycle), we observe almost a disappearance of the domains in a ferro region limited by close fronts, with a kind of sweeping of the existing texture. Conversely, we observe a few rare mobile domains at the highest temperatures of a PF coexistence. The values of ε'' are directly correlated to the volume of the regions in dense domains; the values of ε' also depend on it but also have other origins as we can see in the regions of the coexistence zone close to the para phase.

The end of a coexistence with a negligible thermal gradient always ends with the "collapse" of the two fronts that come closer. Consider the simple FP case with a domain texture stabilized by a stay at low temperatures and a decrease in its volume without internal rearrangements.

The thermal homogeneity of Figs. 5.5 and 5.6 is of lesser quality than that of Fig. 5.3 but sufficient to observe the nucleations at the eight corners of the sample. In Fig. 5.5, the phase that disappears, here the ferro phase, is totally included in the phase that is established, here the para phase, a beautiful illustration of a first-order transition. Figure 5.6 proposes another situation of fairly good thermal homogeneity with a difference in thermal input due to the different conductivities of the surrounding environment. In such a case, the disappearance of the ferro phase, due to the different front speeds on the two faces, takes place in the form of a dagger that slowly withdraws. These experimental results demonstrate that the optimal position of the phase front is the (001) plane. Even slight local thermal variations can somewhat modify this orientation.

The previous observations shed light on numerous works carried out in good conditions of thermal homogeneity, for example, during calorimetric measurements with or without associated dielectric measurements. Their authors often highlight the values of the dielectric constant higher than those expected by classical calculations of the Landau development type, with supercooling jumps often attributed often to domains that would remain present above the transition (Reese, 1969a, b). But, the maximum of the dielectric constant is not necessarily the marker between a naturally mono-domain para sample and a poly-domain ferro sample. The macroscopic results in the temperature range around T_C in which the Curie laws are not followed (including for the KDP, of the order of 10^{-4} to 10^{-2} K) can be truly understood if we know the phenomena of phase coexistence and the interactions between phase fronts and domains.

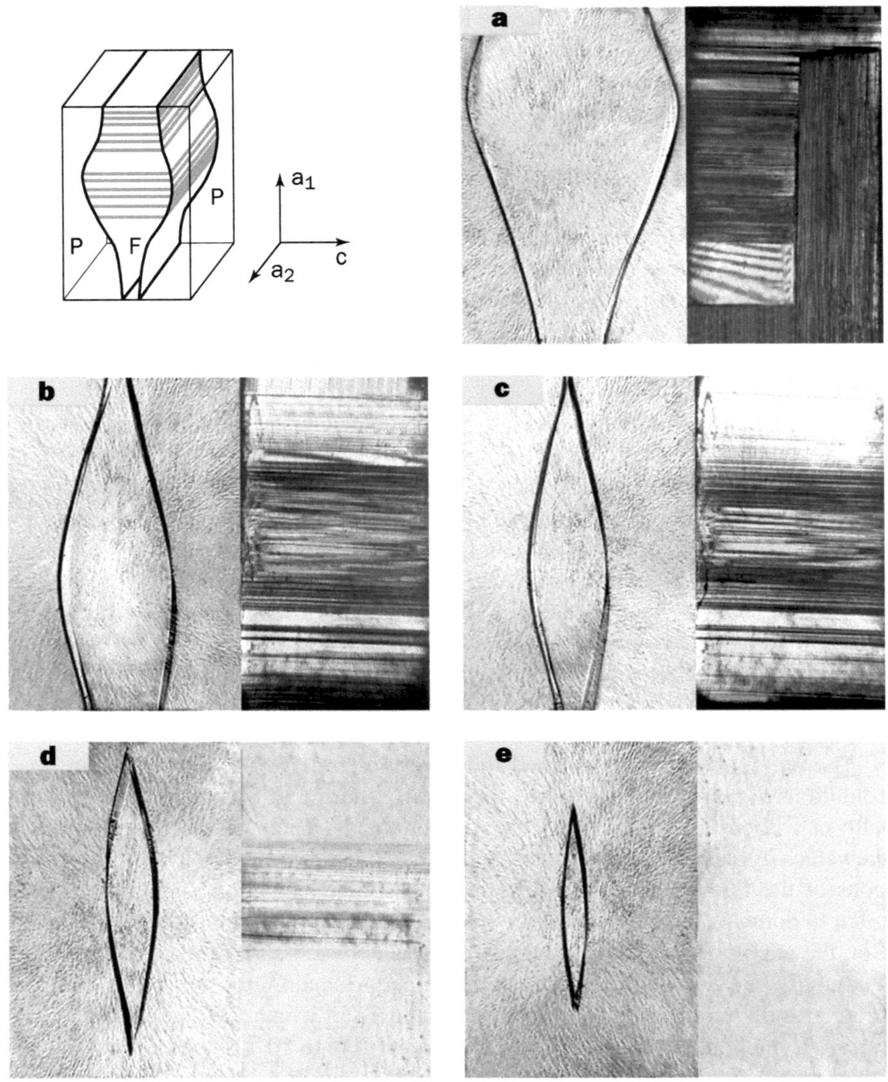

Fig. 5.5 Phase fronts and domains during a FP transition in DKDP with a thermal inhomogeneity in the sample G_i <5 m Kmm^{-1}. The two phase fronts move towards each other and create an inclusion of the ferro phase in the para phase. When the distance between the phase fronts becomes <400 μm, the domain texture disappears (sample dimensions: a_1 = 6.75 mm, a_2 = 3.85 mm, c = 4.98 mm; coexistence occurs with a temperature difference measured next to the sample of 0.253 K) (R.Cach). This figure, like Fig. 5.6, illustrates the difference between domain walls and the phase front: the walls between domains of deformations proportional to the order parameter (u_{xy} and $-u_{xy}$) are coherent interfaces in crystallographic planes, and changing their orientation requires a lot of energy with the creation of quasi-dislocations: they remain very close to the (100) and (010) planes. The phase front is an incoherent interface; its energy is minimal in a (001) orientation, but it is observed to deviate by a few degrees or a maximum ten degrees depending on the isotherms and local normal stresses u_{zz}

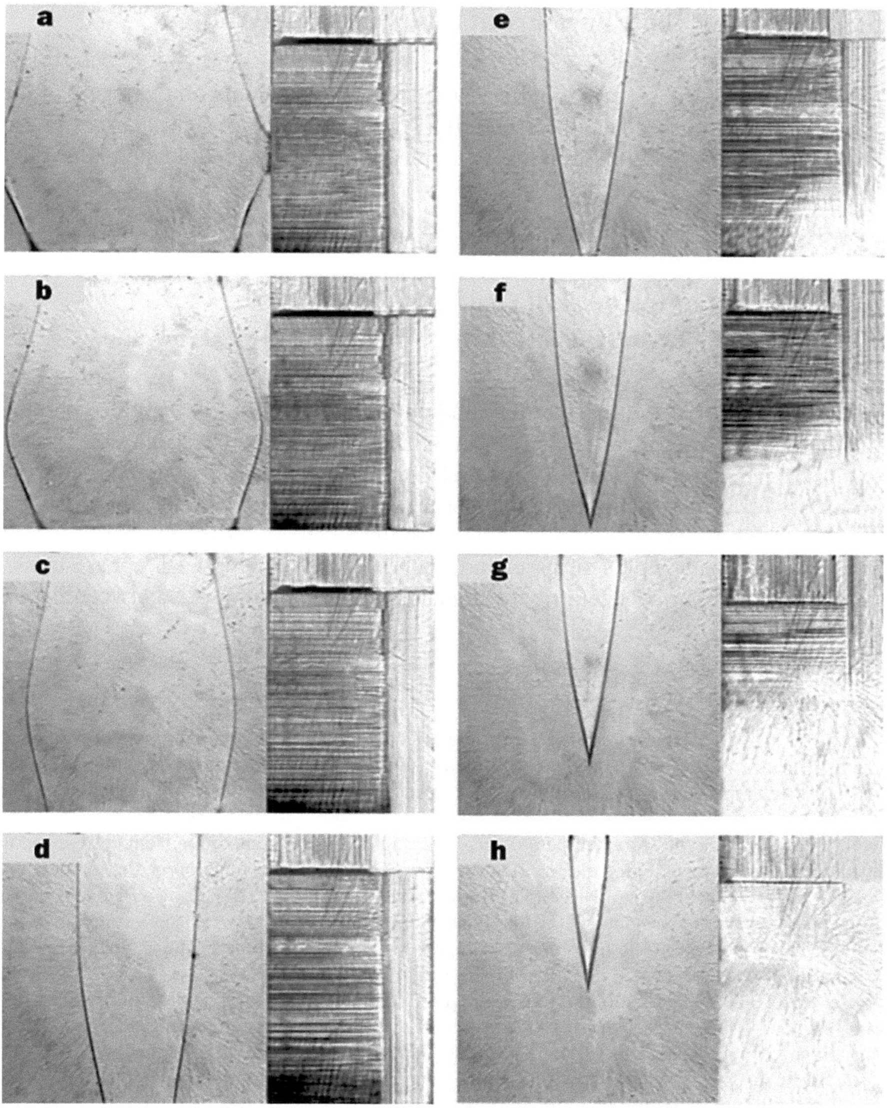

Fig. 5.6 FP transition in DKDP: the thermal conductivity of the crystal and that of the window on which the sample is placed (lower part of the photographs which is the hottest face in this FP cycle) are 15 times higher than that of the exchange gas helium, hence a faster phase front speed on the "window" face, and the creation of a ferro phase dagger that withdraws upwards. The creation of fronts in the corners imposes a very deformed phase front (**a**, **b**); from (**c**), the difference between the front speeds at the top and bottom is evident: front speed three times higher at the bottom (typically 4–5 μms^{-1}). The domain texture disappears in the dagger when the distance between the phase fronts is about 500 μm (sample dimensions: a_1 = 6.30 mm, a_2 = 4.05 mm, c = 6.96 mm) (R.Cach)

5.3 Shape of the Phase Front and Thermal Gradient

Section 5.2 showed the effect of very small thermal gradients on the shape of the phase front. In this section, the effect of a significant and constant thermal gradient is studied, still with samples without applied electric field and mechanical stress.

5.3.1 *Effect of a Thermal Gradient Parallel to the Ferroelectric* **c** *Axis*

In a sample with almost no thermal gradient, the phase front is therefore perpendicular to the ferroelectric *c* axis. We can reinforce this result by placing the sample in an external thermal gradient $\mathbf{G_e}$ parallel to this axis. Then, as illustrated in Fig. 5.7 with $G_e = 0.2$ K mm^{-1} and negligible perpendicular components, the phase front is almost planar, perpendicular to c, with a position defined by the temperature law, here linear in the sample (Bornarel and Cach, 1994). We find, during phase transitions, the differences concerning the domains described in Chap. 2 between the PF

Fig. 5.7 Phase coexistence in a DKDP crystal subjected to a constant thermal gradient parallel to the ferroelectric axis. This gradient can be calculated by the temperature measurement made at the base of the sample during the movement of the phase front. (**a**) Schematic representation of the sample and the external thermal gradient. (**b**) The phase front, of almost plane shape, translates with the isotherm at the Curie temperature T_C. The slight deformations of the front correspond to local stresses. (**c**) Variation of the position d of the phase front as a function of the temperature measured on the lower face of the sample (Bornarel and Cach, 1994). (**d**) During a PF cycle, variations of the dielectric constant and the loss constant measured along the *c* axis as a function of the volumetric proportion of the ferro phase V_F relative to the total volume of the sample V_0. At the beginning of coexistence, the domains are almost absent and the loss constant is low as observed in Fig. 5.7e. The dielectric constant grows regularly with the ratio V_F/V_0; the loss constant also grows with V_F/V_0: jumps accompany the creation of a domain or a rearrangement of texture. These variations change during the various PF cycles depending on the textures. (**e**) Variation of the dielectric constant ε'_c and the loss constant ε''_c during a PF transition (sample: $a_1 = 2.41$ mm, $a_2 = 4.71$ mm, $c = 9.02$ mm; at 4 kHz with 10^{-1} V cm^{-1}; $G_i = 8\times10^{-3}$ K cm^{-1}). The ferro phase is mono-domain in the first part of the coexistence and the losses remain low; jumps are noted in the losses during the growth of the volume of the domains and peaks during the rearrangements of texture; the thin vertical lines delimit the coexistence interval (Bornarel and Cach, 1993, 1994). (**f**) Variations of the dielectric constant ε'_c and the loss constant ε''_c during the FP cycle for the same sample ($G_i = 7\times10^{-2}$ K cm^{-1}). The texture in regular domains stabilized at low temperatures is swept by the phase front; the crystal becomes mono-domain towards the end of the coexistence as illustrated by the change in slope of $\varepsilon''_c(T)$

cycles with unstable (and sparsely populated during coexistence) domain textures and the FP cycles with a dense and stabilized domain texture. However, the existence of the phase front changes the context with an additional difficulty if the domains cannot traverse the entire sample following c. The dielectric variations are, like the textures, always more regular during FP cycles than PF cycles, and the loss constant remains the "detector" of the domains.

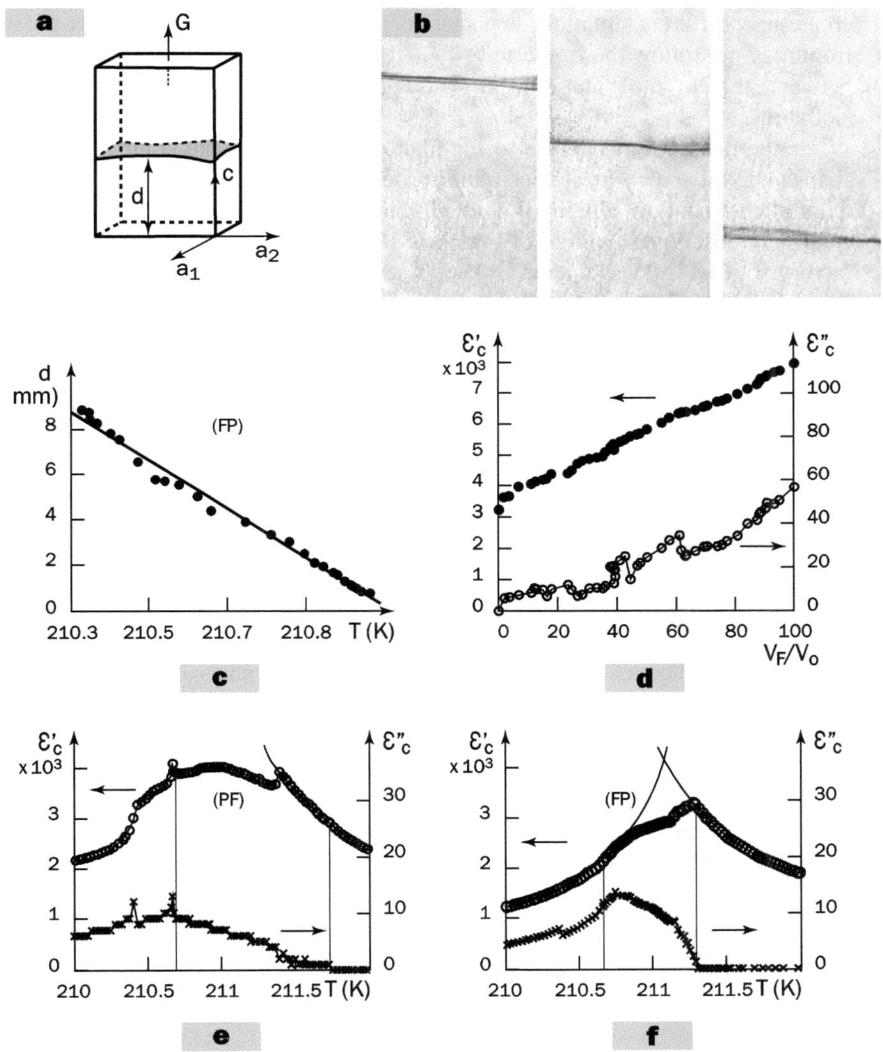

5.3.2 *Phase Front of a DKDP Crystal Subjected to Any Thermal Gradient*

To understand the impact of temperature distributions in the crystal, a DKDP sample is placed at different angles in a homogeneous external thermal gradient (G_e = 0.2 K/mm) as illustrated in Fig. 5.8. The results can be summarized as follows:

- A sufficiently weak external thermal gradient in the exchange gas is enough to create very different situations in a sample, particularly in its corners. The phase front may not follow the isotherm $T = T_C$. It usually takes intermediate positions between that of the "phase front without thermal gradient" and that of an isotherm.
- When the phase front is close to the plane (001), it can take slightly rounded shapes, modulated by local constraints, around this plane.
- On the other hand, in a thermal gradient that imposes a direction of coexistence zone very different from that of the plane (001), the front takes the shapes of a factory roof with flat elements whose angles relative to the plane (001) do not exceed a critical value, about 25° in the case of DKDP. It is understood how the results of macroscopic measurements can differ in ferro-ferro depending on the

Fig. 5.8 A DKDP sample is placed in helium gas with a constant temperature gradient G_e = 0.2 K mm^{-1}. The position of the sample is marked by the angle α between the *c* axis of the crystal and $\mathbf{G_e}$. The temperature distribution in the sample is not simple, especially in the corners which are privileged places of thermal exchanges between the crystal and the outside. But if the sample is massive enough to be approximated to an ellipsoid, it can be shown that the internal thermal gradient in the crystal $\mathbf{G_i}$ is almost constant in the middle of the sample. $\mathbf{G_i}$ and $\mathbf{G_e}$, in general, are not parallel, except in the particular cases α = 0° and α = 90° (Kvitek, 1995; Bornarel et al., 1996). In these two cases, the calculations simplify and we can write:

$G_i/G_e = ((\kappa_C \cos\alpha)^2 + (\kappa_A \sin\alpha)^2)$, and $\beta = \tan^{-1}((\kappa_A/\kappa_C) \tan\alpha)$,

κ_A and κ_C are functions of the dimensions of the sample and the ratio *r* between the thermal conductivity of the crystal and that of the external medium (In the presented case, we can retain the values $r = 20$ and $\kappa_A/\kappa_C = 0.5$). β is the angle between the plane (001) and the phase front if it is planar (•) or its envelope if it has a more complicated shape like a zigzag (■). In this latter case, β′(□) and β″(Δ) are the angles of plane parts of the phase front with (001) as indicated in the diagram with α = 81°

When the angle α is less than about twenty degrees of arc, the phase front observed in section a_1 is almost flat in the middle of the sample. It can take particular zigzag shapes if it is close to the corners of the sample. As the angle α increases, the part of the almost flat phase front, in the middle of the sample, narrows (the diagram for α = 55° illustrates the phenomenon) and the zigzag-shaped front also appears in the middle of the sample for α = 60°. The variations of β, β′, β″ are plotted as a function of α. The "factory roof" front is located between two envelope surfaces whose inclination with respect to (001) increases with α; the angle β′ (or β″) of a plane part of the phase front with (001) always remains <25 degrees of arc. The values of the angles β′ and β″, different when α is between 60° and 80°, become practically equal for α greater. At 90°, the factory roof is regular with two slightly curved envelopes close to the (010) planes. The dotted curve corresponds to the isotherm $T = T_C$ while the solid line is only a guide for reading the variation β(α). The phase front is not confused with the isotherm except for α = 0° and α = 90°: its shape is intermediate between the isotherm surface $T = T_C$ and the plane (001) thermal homogeneity of the samples. Figure 5.9 illustrates this with the dielectric properties measured simultaneously with the observations (Bornarel et al., 1996, 2000). We thus note the temperature translation of the inverse dielectric constant in accordance with the simple model presented with $\mathbf{G_i}$ linear (Kvitek, 1995), translation different from those due to pressure, constraints, etc. (Smolenskii et al., 1984). Moreover, the Curie law is not strictly found in the para-electric phase because of the existence of this thermal gradient as explained by a model integrating the laws of temperature distribution in the sample (Kvitek, 1995). We especially notice the change in the shapes of curves and the values of $(\varepsilon_c')^{-1}$ and

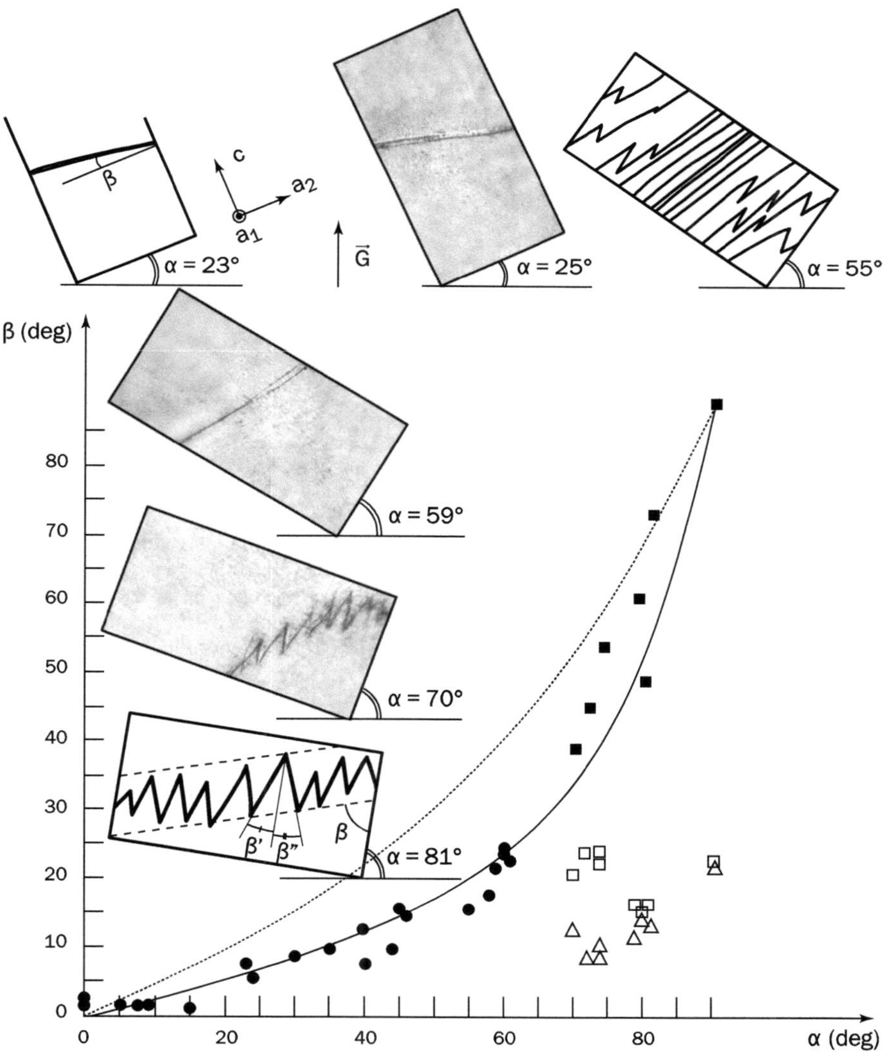

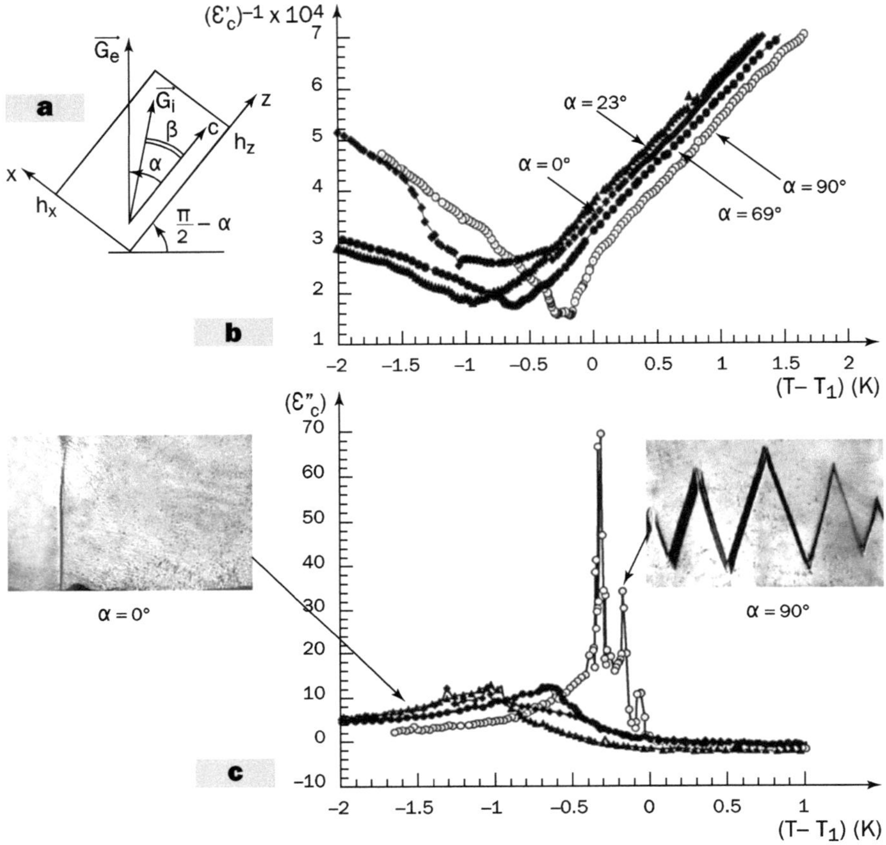

Fig. 5.9 Dielectric measurements carried out simultaneously with the observations of Fig. 5.8 (Bornarel et al., 1996): (**a**) Schematic representation of a sample placed in a constant thermal gradient $\mathbf{G_e}$ which generates inside the crystal a thermal gradient $\mathbf{G_i}$ homogeneous in the middle of the sample; the angles of the gradients with respect to the c axis are indicated. (**b**) Variations of $(\varepsilon'_c)^{-1}$ and (**c**) of ε''_c as a function of $T - T_1$ during a PF cycle for various values of the angle $\alpha(T_1)$ indicated on the curves (T_1 is the appearance temperature of coexistence on the lower a_1 face). Two photographs in section (010) illustrate on the graph $\varepsilon''_c(T - T_1)$ the shape of the phase front in the extreme situations $\alpha = 0°$ and $\alpha = 90°$

ε_c'' in the coexistence interval depending on the angle α, and for example, for low values of α, the relative regularity of ε_c'' with minor rearrangements of the domains, or even their near-absence (Sect. 5.2). On the contrary, the variation peaks for α = 90° announce significant contributions from the domains during coexistence. Their studies in this simple situation, with $\mathbf{G_i}$ constant and parallel to $\mathbf{G_e}$ in the middle of the sample, are the subject of the following paragraph. Even in this case, Fig. 5.10 alerts us to the importance of the shapes of the sample and the risk of hasty conclusions by considering a "mean temperature" of the sample during a macroscopic measurement.

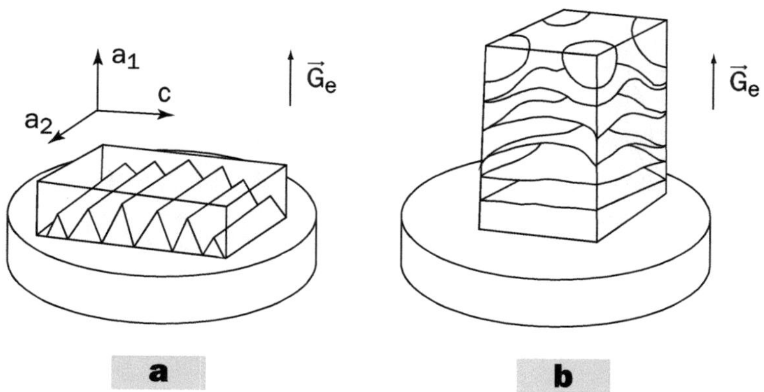

Fig. 5.10 (**a**) Sample in the form of a platelet placed on a glass window in a gas exchange cell with a homogeneous vertical external thermal gradient. The curvature of the factory roof envelopes results from the relaxation of constraints on the lateral sample edges a_2 and the thermal transfers of the two faces a_1, different because of the external environments (helium gas and glass). Given the exchanges in the corners, the isotherms will not be flat there. (**b**) Schematic representation of the isothermal surfaces in a good size sample; we understand the variety of results obtained by various techniques using different sample shapes

5.3.3 Phase Coexistence with Thermal Gradient Perpendicular to the Ferro c *Axis*

The photographs of the phase front traces reproduced in Figs. 5.11 and 5.12 show a sample cut along its tetragonal planes, the evolution of the "factory roof" front shape with a thermal gradient perpendicular to the ferroelectric axis, cooling or heating speeds $<5 \times 10^{-4}$ K/min, and a nice linearity of the thermal gradients $G_{\rm i}$ in the middle of the samples. The observation of the domains and the phase front is carried out simultaneously with the dielectric measurements. These studies in PF cycles (Fig. 5.11) as in FP cycles (Fig. 5.12) confirm the previously presented results: the coexistence zone (the factory roof region) can remain almost monodomain as demonstrated by the optical observations, the absence of diffraction of a laser beam, and the very low loss constant between situations a and c (in Fig. 5.11) and between situations e and f (in Fig. 5.12) for which the ferroelectric region exists in the sample only in the coexistence zone. On the contrary, as soon as the domains traverse the entire sample following c, ε_c'' takes large values with jumps at each rearrangement of domain texture.

For its part, the dielectric constant does not follow the classic laws of a monodomain and evolves quite regularly when the factory roof translates in the sample. In this context of well-defined thermal gradient, the relationships between G_e and G_i, the dimensions of the zigzags and the thermal hysteresis at the transition as a function of G_i have been systematically studied. Figures 5.13 and 5.14 illustrate the variations of the factory roofs at equilibrium in the middle of the sample with faces in the tetragonal planes: the flat parts of the phase fronts make with the plane

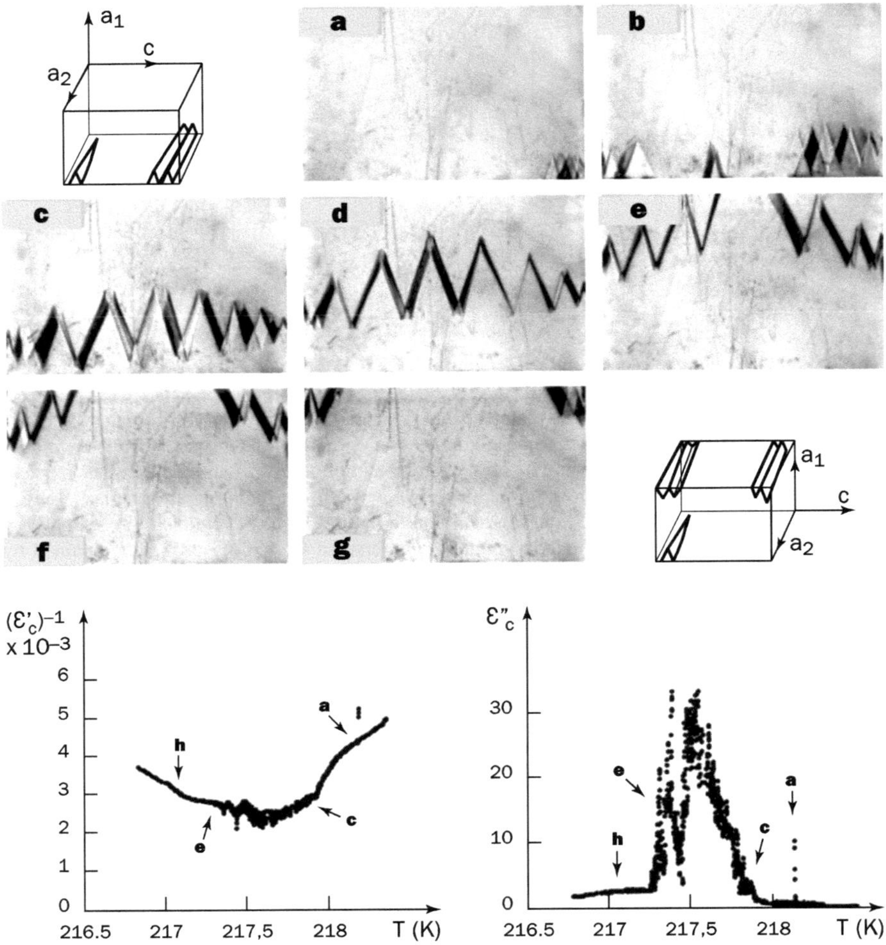

Fig. 5.11 A DKDP sample cut along its tetragonal axes (a_1 = 5.5 mm, a_2 = 4.4 mm, c = 7.4 mm), placed in a homogeneous thermal gradient oriented along a_1, is cooled during a PF cycle (cooling rate of 5 ×10^{-4} K min^{-1}). The phase coexistence is illustrated by the appearance of the phase front on one face a_1 (situations (**a**) and (**b**)) and its disappearance on the other (situations (**f**) and (**g**)). The phase front translates practically without deformation in the central part of the sample. The photographs are marked on the variation curves of $(\varepsilon'_c)^{-1}(T)$ and $\varepsilon''_c(T)$. The sample includes very few domains in the coexistence zone (the zigzag region) (Kvitek and Bornarel, 2002)

(001) a maximum angle of 25°± 5° (β = β″ defined previously) as demonstrated by the graph in Fig. 5.16b (Kvitek and Bornarel 2002). The same studies were carried out under similar experimental conditions with samples cut according to the orthorhombic planes (Fig. 5.15). The angle between the flat traces of the phase

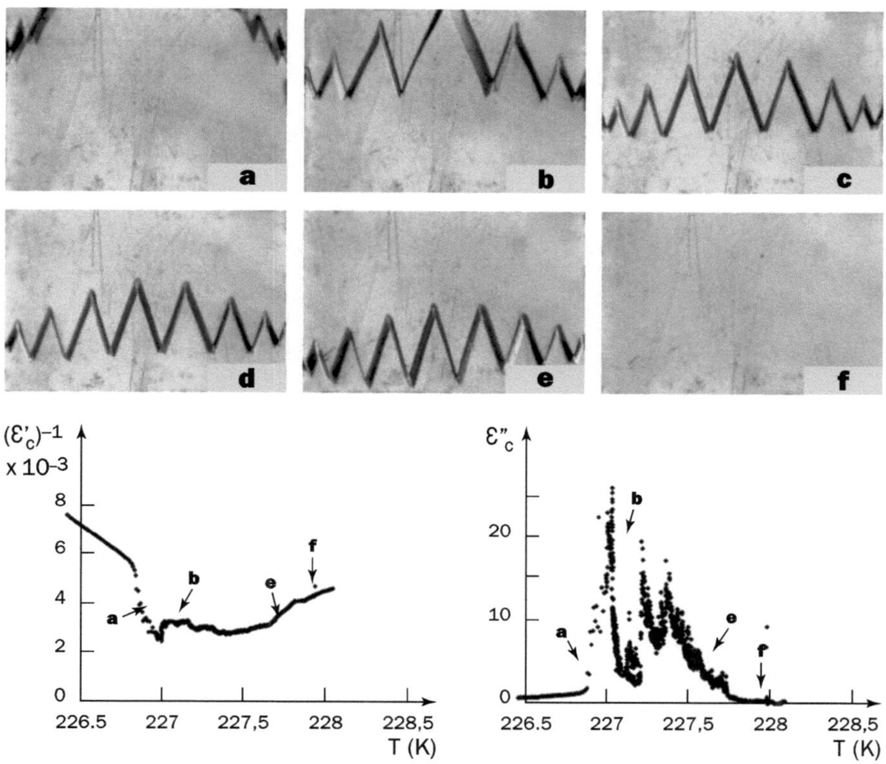

Fig. 5.12 The same DKDP crystal as in Fig. 5.11, placed in a constant thermal gradient oriented along a_1, is heated during a FP cycle (heating rate of 5 ×10^{-4} K min^{-1}). The increase in temperature leads to the appearance of the phase front on one face a_1 and its disappearance on the other. The phase front translates without deformation in the central part of the sample. The illustrated situations are marked on the curves $(\varepsilon'_c)^{-1}(T)$ and $\varepsilon''_c(T)$ (Kvitek, 1995)

front and the plane (001) is less than that of the sample cut in tetragonal planes: 18°± 4° as the graph of Fig. 5.16c demonstrates. Any theoretical model concerning the phase front must account for this reality: the phase front is in an optimal situation in a plane (001) and does not deviate from this plane beyond a limit angle. This limit angle varies according to the orientation and the context of the sample used, but in the DKDP, never exceeds 25 degrees of arc. The thermal gradient imposes a coexistence zone perpendicular to that of the plane of incoherent interface that the phase front would take without a thermal gradient; this coexistence zone with a factory roof front occupies a volume limited by two curved envelopes of maximum size in the centre and minimum at the sample edges (Fig. 5.17). The envelope close to the ferro phase is always less curved than that close to the para phase.

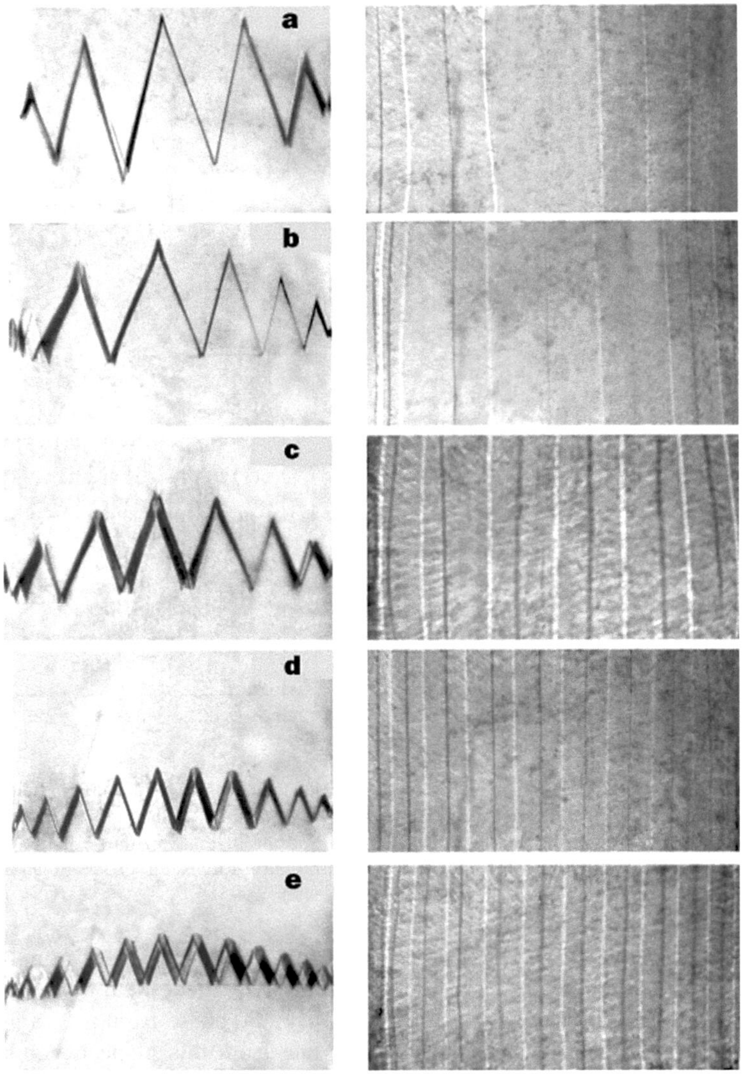

Fig. 5.13 The phase front in the middle of the DKDP sample with faces in the tetragonal planes is photographed in section a_2 (on the left) and in section a_1 (on the right) during a PF cycle, as illustrated by Fig. 5.11. The external thermal gradient G_e oriented along the axis a_1 takes the following values: (**a**) 0.2 K mm^{-1}, (**b**) 0.5 K mm^{-1}, (**c**) 1 K mm^{-1}, (**d**) 2 K mm^{-1}, and (**e**) 3 K mm^{-1}. The law between $\mathbf{G_i}$ and $\mathbf{G_e}$ is clearly defined in this particular configuration (Kvitek, 1995; Kvitek and Bornarel, 2002). "Factory roof" interfaces appear in various contexts of solid physics when two intensive parameters are in competition (Kosevich and Boiko, 1993)

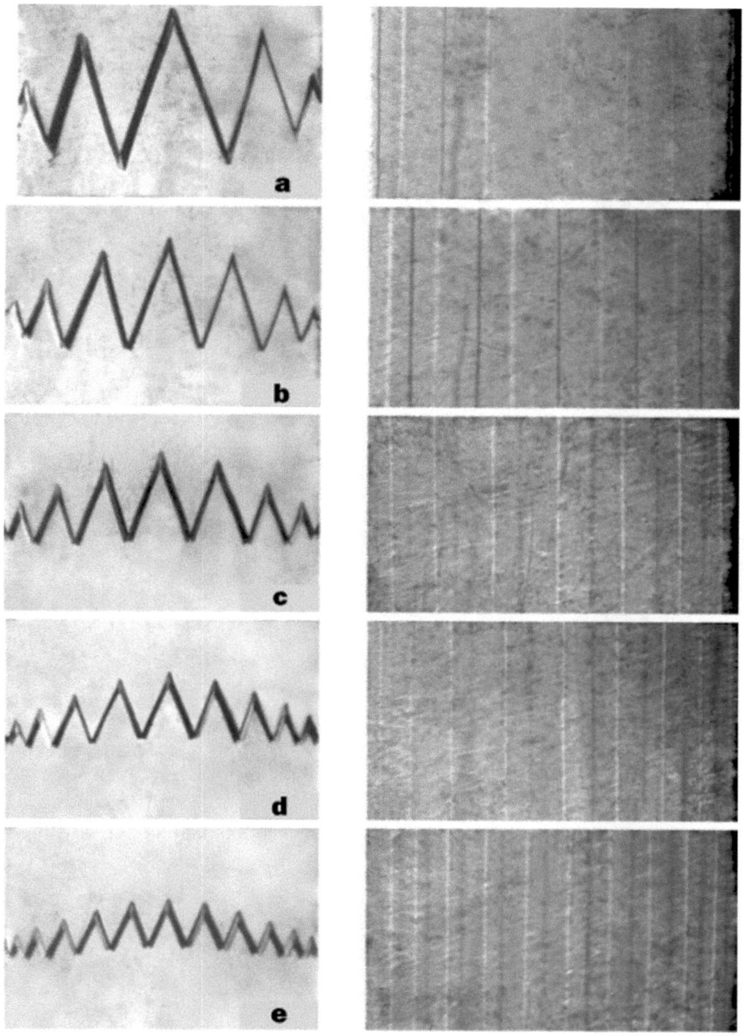

Fig. 5.14 The phase front in the middle of the same DKDP sample as in Fig. 5.13 is photographed in section a_2 (on the left) and in section a_1 (on the right) during a FP cycle, as illustrated in Fig. 5.12. The values of the external thermal gradient G_e oriented along the axis a_1 are as follows: (**a**) 0.2 K mm^{-1}, (**b**) 0.5 K mm^{-1}, (**c**) 1.2 K mm^{-1}, (**d**) 2 K mm^{-1}, and (**e**) 3 K mm^{-1}. The law between $\mathbf{G_i}$ and $\mathbf{G_e}$ is clearly defined in this particular configuration (Kvitek, 1995; Kvitek and Bornarel, 2002)

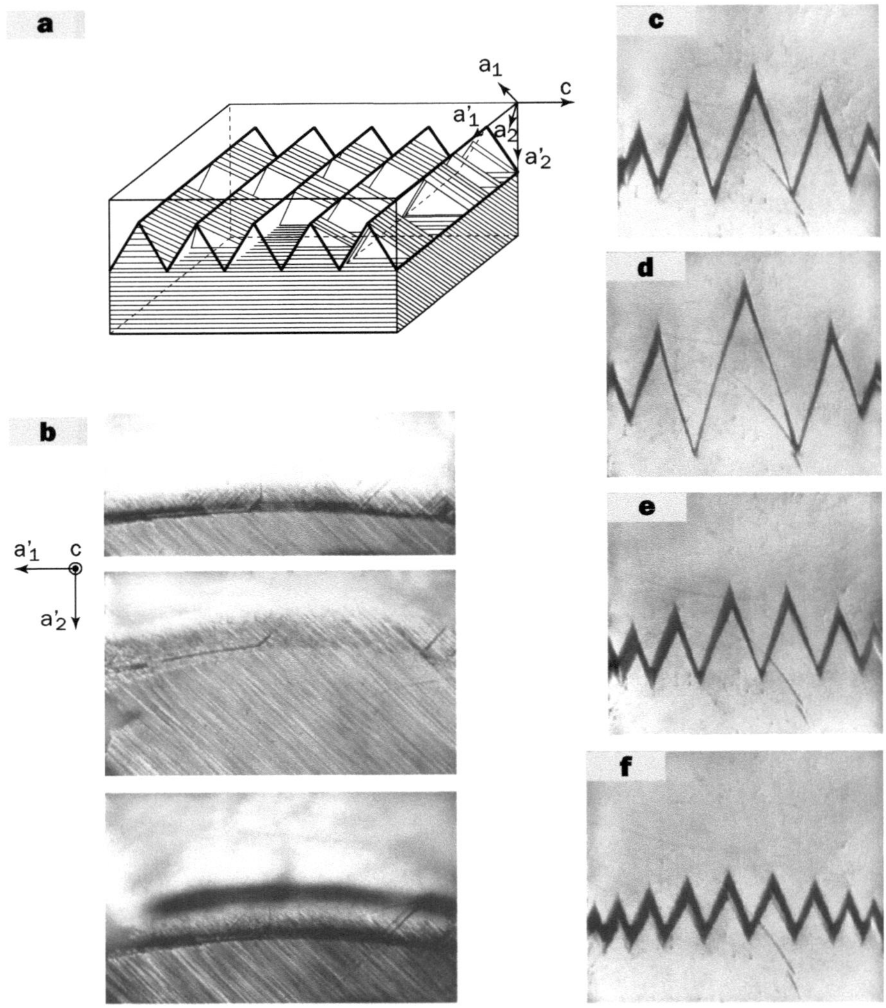

Fig. 5.15 (**a**) Schematic representation of a DKDP sample cut according to the orthorhombic planes and subjected to a thermal gradient along the vertical axis parallel to a'_2 ($a'_1 = a_1 + a_2 =$ 9.9 mm, $a'_2 = 8.9$ mm, $c = 6$ mm). (**b**) Observations along the c axis with polarizations and orientations of light that highlight the limits of the coexistence region, the domains, and the variations of constraints near the front. The dark bar in the photograph below corresponds to the singularities of the factory roof edges close to the para region. (**c**–**f**) Photographs of the phase fronts in section a'_1 in the middle of the sample for various values of the external thermal gradient oriented along a_2': (**c**) 0.12 K mm^{-1}; (**d**) 0.33 K mm^{-1}; (**e**) 0.8 K mm^{-1}; (**f**) 2.4 K mm^{-1} (Kvitek, 1995)

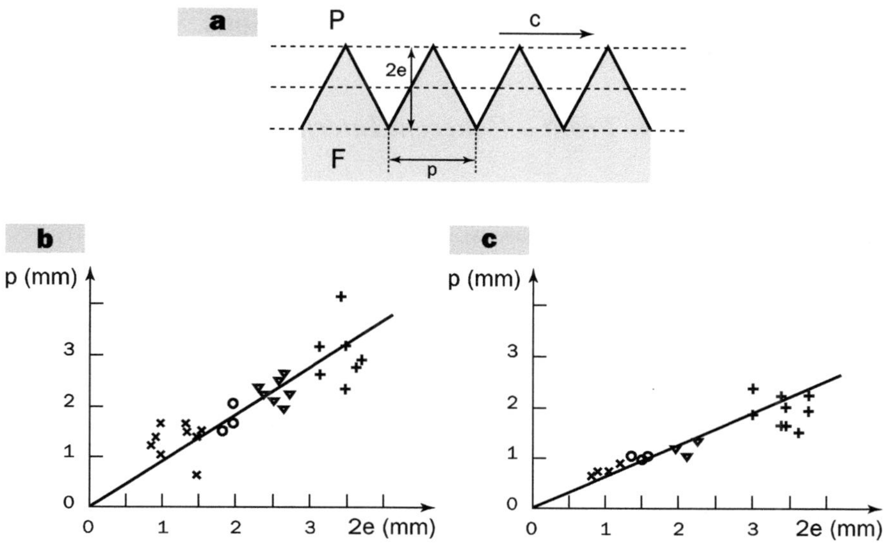

Fig. 5.16 (**a**) Schematic representation of the trace of a phase front in a "factory roof" in the middle of the sample as illustrated in Figs. 5.11, 5.12, 5.13, 5.14 and 5.15; 2e is the height of the coexistence zone and p is the period of the zigzag. (**b** and **c**) Variation of p as a function of 2e for a sample whose faces are (**b**) in tetragonal planes as in Figs. 5.11, 5.12, 5.13 and 5.14 and (**c**) in orthorhombic planes as in Fig. 5.15 (Kvitek and Bornarel, 2002)

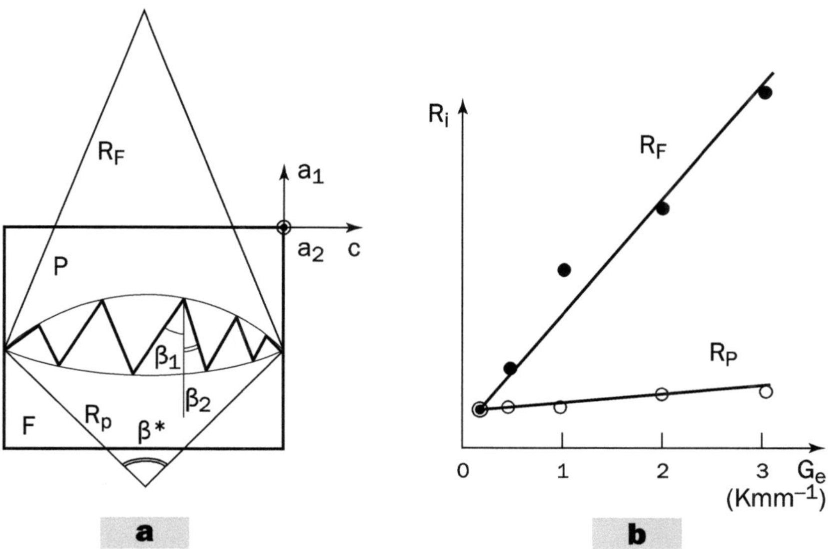

Fig. 5.17 (**a**) Schematic representation of a phase front in a factory roof in section a_2 in a sample subjected to a thermal gradient perpendicular to c. It is limited by two envelope surfaces with radii of curvature R_F and R_P. R_F is always greater than R_P due to the contact of the lower face (100) with a glass window of good thermal conductivity. (**b**) R_F varies as indicated depending on the importance of the thermal gradient and becomes infinitely large if a significant electric field is applied along c as shown in Chap. 6: the envelope of the coexistence zone becomes a plane parallel to c, such as (100)

5.4 How to Explain the Shapes and Orientations of the Phase Front?

5.4.1 Is the Phase Front a Coherent Interface?

This question comes down to looking for the planes in which the deformation of the ferro phase and that of the para phase would be equal. As with the walls of ferroelectric domains (Fousek and Janovec, 1969) and ferroelastic domains (Sapriel, 1975), this was achieved for phase fronts (Khachaturyan, 1967; Roytburd, 1974; Vallade and Dolino, 1982; Dec, 1993). Such a coherent interface would not create mechanical sources of stress and, therefore, no long-range mechanical deformation energy in the sample. It is quickly understood that this solution cannot exist in the DKDP when going from a para phase to a mono-domain ferro phase (with $u_{xy} = 75u_{xx}$ approximately), but solutions can be sought with a poly-domain ferro region, macro-deformation defined by a parameter m, ratio between the volume V_+ of the domains+ (for example), and the total volume V of this phase (Roytburd, 1974, 1983). In the case of the DKDP crystal (Kvitek and Bornarel, 2002), coherent interfaces would then be possible if the following conditions were met:

$$u_{xx}.u_{zz} < 0 \text{ and } u_{xx}^{\ 2} - u_{xy}^{\ 2}\left(m - \tfrac{1}{2}\right)^2 = 0 \tag{5.1}$$

As all experimental results give u_{zz} positive and of good value, u_{xx} must be negative for a coherent interface to be possible. The values of u_{xx} published are usually positive but very small; therefore, the possibility of a coherent interface should not be excluded a priori. However, studies carried out on samples with faces in tetragonal planes (Fig. 5.14) and on others in orthorhombic planes (Fig. 5.15) exclude this hypothesis of coherent interfaces: the experimental results of Fig. 5.16 with angles between front and plane (001) higher in the "tetragonal" sample than those of the "orthorhombic" sample contradict the theoretical models which lead to a ratio in the other direction. The phase front is therefore not a coherent interface.

5.4.2 Why Is the Phase Front in a (001) Plane?

The phase front in the DKDP is therefore an incoherent interface whose optimal orientation must be defined. Work has been developed on this theme taking into account the twinned aspect (Metrat, 1980), often in complex situations in terms of domain possibilities like in perovskites; the continuity conditions when passing from one phase to another have been a central point (Roytburd and Kurdjumov, 1979; Roytburd 1983). We are looking here for a fairly simple situation in agreement with experimental realities.

Figure 5.3 shows, for a sample of homogeneous temperature, a phase front generally perpendicular to the ferroelectric axis which demonstrates the weak role of electrostatic energy on the shape and orientation of the front. We therefore neglect, in a first step, this component of energy to focus on the other forms of energy, chemical energy, mechanical deformation energy, and surface energy of the phase front. The changes in shape of the front, quite flexible around a plane (001), suggest that this last energy is weak compared to the other two.

Therefore, in the absence of an applied external conjugate field, the main roles seem to be governed by the mechanical energy in volume created by the phase coexistence and by the chemical energy as illustrated in Fig. 5.8. The inclusion model described by Khatchaturyan (1967, 1993) seems suitable for experimental conditions and observations of front shapes: with it, we consider the local energy of the interface between the two phases, weak compared to the mechanical energy in volume, the elastic constants comparable in the two phases, and the spontaneous deformation in ferro phase without notable change in the sample with temperature. If we also retain the hypothesis of an infinite medium, the mechanical energy can be written in the form (Khatchaturyan, 1983):

$$E = ½\int\int\int d^3k/(2\pi)^3 \ B(n)|\vartheta(k)|^2 \tag{5.2}$$

where $\mathbf{n} = \mathbf{k}/k$, normal to the plane of inclusion, and

$$B(n) = \lambda_{ijkl} \ u_{ij} \ u_{kl} - n_i \sigma_{ij} \Omega_{jl} \sigma_{\mathrm{lm}} n_m \tag{5.3}$$

where $\sigma_{ij} = \lambda_{ijkl}\, u_{kl}$ and $\Omega_{il}^{-1} = \lambda_{ijkl}\, n_j\, n_k$, with λ_{ijkl} elastic constants, u_{ij} mechanical deformations and σ_{ij} stresses. The value of $\vartheta(\mathbf{k})$, Fourier transform of the function $\vartheta(\mathbf{r})$ expressing the shape of the inclusion is 1 in the phase II and 0 in the phase I. The energy will be minimal if $B(\mathbf{n})$ is very small when $|\vartheta(\mathbf{k})|^2$ is large. In the case of an infinite phase I matrix and a homogeneous crystal in temperature, the inclusion will be an infinitely thin platelet with its normal perpendicular to **k** as illustrated in Fig. 5.20a. The particular case where $B(\mathbf{n})$ would be zero would correspond to a coherent interface.

The term $B(\mathbf{n})$ which characterizes the elastic properties is major in the case of an infinitely thin platelet defined by the angle between its normal and the c axis. Limiting ourselves to the case where **n** is in the xOz plane, the variation of $B(\beta)$ is presented for a sample cut in the tetragonal planes in Fig. 5.18c, d and for a sample cut in the orthorhombic planes in Fig. 5.19c, d (Kvitek and Bornarel, 2002). In both cases, in theory, a minimum of energy could occur for a non-zero value of the angle β if u_{xx} was negative, with a value of B being zero in the case of faces cut in the orthorhombic planes (coherent interface) (Fig. 5.19c); but, in the present experimental case, with u_{xx} positive, the only solution corresponds to $\beta = 0$ and B non-zero, therefore to an incoherent interface. We thus find the experimental results of a phase front located in the plane (001) due to, at the transition, the preponderance of u_{zz} as suggested as early as in 1980 (Bastie et al., 1980).

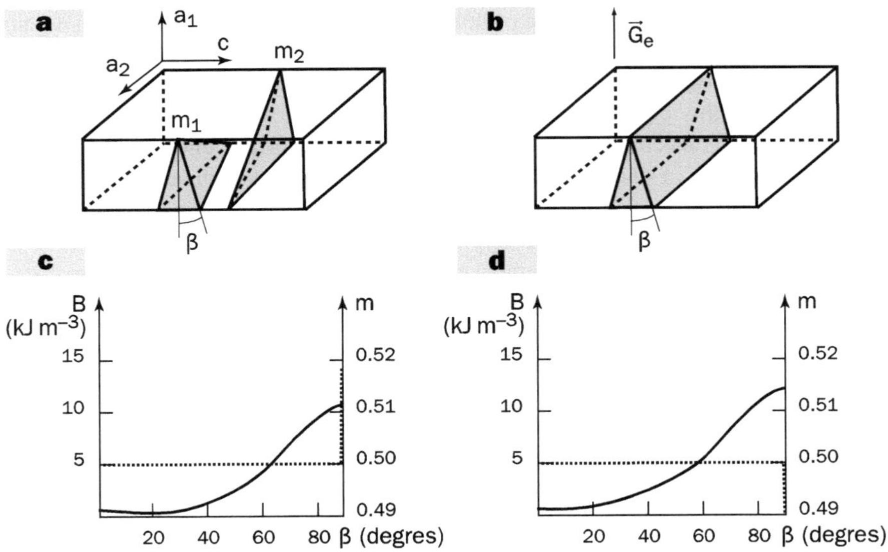

Fig. 5.18 Theoretical studies of plane phase fronts in a sample of DKDP with ferro poly-domain phase (Kvitek and Bornarel 2002). The faces of the sample are tetragonal planes in Fig. 5.18 and orthorhombic planes in Fig. 5.19. m is the proportion of volume V_+ of polarity + compared to the total ferro volume $V = (V_+ + V_-)$. The domains presented are assumed to have walls in tetragonal planes (100)

Given the relationship (5.1) of the text, in the absence of a thermal gradient, a coherent interface would only be possible if u_{xx} was negative. The solutions are as follows: (**a**) for the "tetragonal" sample, $m_1 = 0.49$, $m_2 = 0.51$ with the angle $\beta = \text{arc } tg(u_{xx}/u_{zz})^{1/2}$ of 21.4 degrees of arc, and for the "orthorhombic" sample, $m_1 = 0.51$ and $m_2 = 0.49$ with the angle $\gamma = \text{arc tg}[2|u_{xx}|/u_{zz}]^{1/2}$ of 29 degrees of arc. These theoretical results are not consistent with the experimental results of Fig. 5.16 in which the angles vary inversely (about 25° and 18° respectively). This excludes the hypothesis of a coherent interface with u_{xx} negative

Thus, this simple model of infinitely thin inclusion accounts for the orientation of the phase front in the plane (001) for an infinite crystal, homogeneous in temperature during a slow and regular thermal cycle. It is not suitable for regions close to the sample edges and especially the corners, but it remains plausible in the middle of a sample. It can be deepened, taking into account the texture in domains (introduction of m), the finite size of the sample, and the appearance of PFPFPF multilayer structures obtained in other thermal contexts. Such developments (Roytburd, 1974; Aleshko-Ozhevskij, 1992; Khachaturyan, 1993) do not alter the essential result.

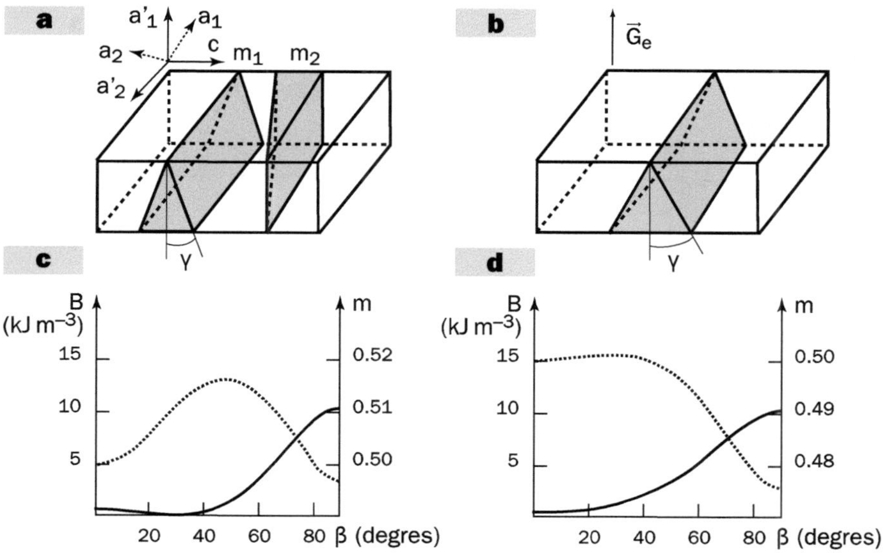

Fig. 5.19 Kvitek proposed to define the orientation of a plane phase front, incoherent interface, a description of the mechanical energy in volume (Kvitek and Bornarel, 2002). The term B, energy density per unit length along c, characterizes the elastic properties (see relations (5.2) and (5.3) of the text). In the absence of a thermal gradient, (**c**) shows the variation of B as a function of the angle β (solid line) in the hypothesis u_{xx} negative and that of the values of m retained in the calculation (dotted lines). For u_{xx} positive, we find with the same representations in (**d**) the minimum energy for $\beta = 0$ which corresponds to the experimentally observed results of phase fronts perpendicular to the ferroelectric axis c, with an incoherent interface ($B \neq 0$). For these calculations, the formulas (5.2) and (5.3) were used with the following numerical values (as usual, the notations are such that $x = 1$, $y = 2$, $z = 3$):

$u_{11} = u_{22} = 1 \times 10^{-4}$ (Aleshko-Ozhevskij, 1992); $u_{33} = 6.5 \times 10^{-4}$ (Zeyen and Meister, 1976b); $\lambda_{1111} = \lambda_{2222} = 7.2 \times 10^{10}$ N m^{-2}, $\lambda_{1122} = -1 \times 10^{10}$ N m^{-2}, $\lambda_{1313} = \lambda_{2323} = 1.2 \times 10^{10}$ N m^{-2}; $\lambda_{3333} = 5.4 \times 10^{10}$ N m^{-2}, $\lambda_{1133} = \lambda_{2233} = 1.2 \times 10^{10}$ N m^{-2} (Landolt-Börnstein, 1981)

5.4.3 Explanation of the "Factory Roof" Phase Front?

To study the front orientation, Kvitek (Kvitek, 1997) studied the energy of a factory roof inclusion as presented in Fig. 5.20b. This model is adapted to the phase fronts observed in the middle of a sample in the experimental context described previously (Figs. 5.13, 5.14, and 5.15). The energy density per unit length of front following z is written as:

$$\rho = \rho_{\text{elas-strain}} + \rho_{\text{chem}} + \rho_{\text{surface}} + \rho_{\text{coin}} \tag{5.4}$$

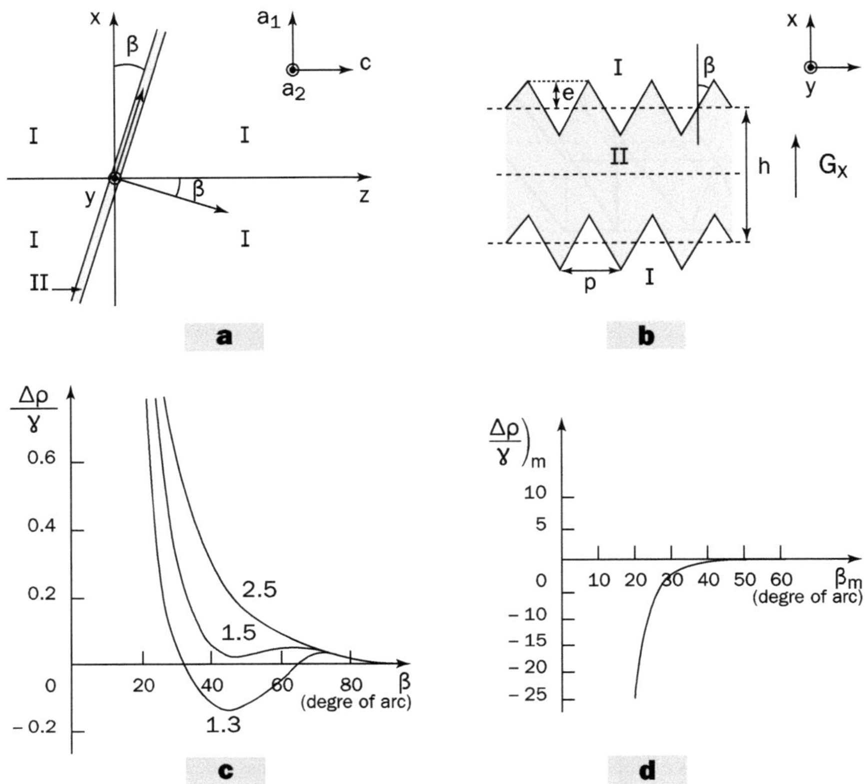

Fig. 5.20 (**a**) In an infinite and homogeneous crystal in temperature, phase II is assumed to appear in phase I in the form of an infinitely thin planar inclusion. The quantities characterizing its orientation are defined. (**b**) An infinite inclusion in the form of a factory roof of phase II appears in an infinite matrix of phase I. The crystal is subjected to a constant thermal gradient G_x while the mechanical energy is optimized by a front perpendicular to z. The infinite inclusion is represented in section in the plane perpendicular to the y-axis, with the main deformation at the transition following z. The angles of a flat part of the phase front with the plane xOy, e the half-height, and p the period of the zigzag, constant over the entire front, are called β′ (or β″). The phase fronts in the form of factory roofs are periodic following z and identical following y. The distance h between the two fronts is assumed to be large enough to neglect the interaction between them. (**c**) Variation as a function of the angle β, of the ratio between $\Delta\rho = \rho(e, \beta) - \rho(0{,}90°)$ variation of the overall energy density, and γ surface energy density of the phase front. The parameter δ indicated on the curves is the ratio between, on the one hand, the product of the surface energy and the chemical energy and, on the other hand, the square of the elastic deformation energy; for δ = 2.5, only one flat front is possible with β = 90°; for δ < 1.4, the zigzag front is stable; intermediate metastable situations of factory roof front can exist for δ between 1.4 and 1.7 as illustrated by the curve with δ = 1.5 (Kvitek, 1997). (**d**) The existence of the zigzag phase front corresponds to the minima in β_{min} of the curves in (**c**) other than the minimum at β = 90°. The graph of $(\Delta\rho/\gamma)_{min}$ as a function of β_{min} demonstrates how the energy density becomes low for angles β_{min} <25° in agreement with experimental observations (Kvitek, 1997)

where $\rho_{elas\text{-}strain}$ is the energy density of elastic deformation in the crystal due to the existence of the front, ρ_{chem} is the chemical energy density, $\rho_{surface}$ is the surface energy density of the front, and ρ_{coin} is the energy density due to the zigzag edges (i.e. corresponding to the singularities of factory roof edges). The evaluation of the various components of energy density per unit length of front following z (Kvitek, 1997) gives the following results:

$$\rho_{\text{front-elas-strain}} = \left(-A(\beta)e\right)/6 + B\left(\beta = 90°\right)h/4 \tag{5.5}$$

equal to half the energy density of elastic deformation assuming that the two fronts of the inclusion are sufficiently distant from each other to neglect their interaction, with $A(\beta) = B(\beta = 90°) - B(\beta)$ reduced in the assumption of a supposed isotropic medium to $2\mu(1-\nu)^{-1}(1-\sin^4\beta)u_{zz}^2$ (ν is the Poisson's ratio, μ is the shear modulus equal to $E/2(1+\nu)$ with $E = 5\times10^{10}$ J/m^3, Young's modulus, in the assumption of an isotropic medium) (Kvitek, 1997); the two terms of the relation (5.5) are truly independent and illustrate a reality of this energy contribution which varies according to the volume of the inclusion, on the one hand, and the orientation parameter, on the other hand:

$$\rho_{\text{chem}} = G_i \left|\Delta S\right| e^2 / 6 \tag{5.6}$$

where $|\Delta S|$ is entropy change at the transition;

$$\rho_{\text{surface}} = \gamma\left(\beta\right) / \sin\beta \tag{5.7}$$

where $\gamma(\beta)$ is the energy per unit plane surface of the front;

$$\rho_{\text{corner}} = \text{E}_{\text{corner}} \;/\left(2e\,tg\beta\right) \tag{5.8}$$

where E_{corner} is the energy of a corner.

By neglecting this last component of the energy and, with simple assumptions (isotropic medium, spontaneous deformation limited to u_{zz}, $\gamma(\beta)$ considered as constant, electrostatic energy and corner energy neglected), Kvitek showed (Kvitek, 1997) that the shape of the phase front depends on the ratio δ between, on the one hand, the product of the surface energy and the chemical energy, and on the other hand, the square of the elastic deformation energy. If this ratio δ is sufficiently large, for example when G_i is very important, the only possible phase front shape is a plane perpendicular to G_i, therefore in our case, a plane parallel to the z axis. If on the contrary, the elastic deformation energy prevails, the factory roof front shape is stable. In intermediate situations, the factory roof phase front can exist in a metastable way. Figure 5.20c illustrates this result. Figure 5.20d for its part shows how much the value of the "normalized potential" (see legend) as a function of the angle β' decreases drastically towards small angles, which is consistent with experimental observations (orientations β′, β″ of the plane parts of the front always <25 degrees of arc). The simplicity of the model and the consistency with the experimental results demonstrate how much the shape of the front depends on the competition between the elastic deformation energy throughout the volume of the sample and the chemical energy, obviously in the absence of an imposed external conjugate field (the effect of the application of a continuous electric field will be studied in Chap. 6).

In this model, even if the most obvious energies are the elastic deformation energy (the value of $\Delta\rho_{\text{elas-strain}}$ in our experimental situations varies between 1 and 10 J/m^2) and the chemical energy ($\Delta\rho_{\text{chem}}$ between 0.1 and 2 J/m^2), the other energy sources that limit the number of zigzags are nevertheless very present. The photographs in polarized light from Fig. 5.21 illustrate the situation: obvious singularities

Fig. 5.21 Photographs taken along the a_2 axis of a factory roof phase front of DKDP (R.Cach): the ferroelectric axis is horizontal and the thermal gradient vertical. The optical setup allows to visualize the variations of constraints. The para phase appears perfectly black, the ferro poly-domain phase is dark. The colour variations of the ferro parts of the coexistence zone illustrate the deformation variations within them. Moreover, the projections of the very bright corners characterize these lines of singularities. The links between the domain density and the gradients of local properties remain to be deepened (Tsatskis et al., 1994; Alpay and Roytburd, 1998)

at the zigzag corners and inclined and flat phase front surfaces of notable size but without significant local constraints. It is difficult to evaluate these energy densities $\Delta\rho_{surface}$ and $\Delta\rho_{corner}$ in relation to each other in a very solid way. The easy way is to neglect one, then, to neglect the other. This gives ranges of values: between 0.1 and 2 J/m^2 for $\Delta\rho_{surface}$ and for $\Delta\rho_{corner}$ values that correspond to E_{corner} of 10^{-3} J/m. The localized aspect of the factory roof edges easily explains the interaction of these with the defects as illustrated at the end of the chapter.

Experiments like the one illustrated in Fig. 5.22 allow the estimation of the previous energy densities by comparing the situations (in samples with faces in tetragonal planes) of zigzags of height 2e, located one in the middle of a sample (Fig. 5.13 with β' angles of about 22 degrees of arc) and the other with the edges in the faces of the sample (Kvitek and Bornarel, 1995), thus with a relaxation of the corner energies which become negligible. In this latter case, the angles β', β'' take values of 10–12 degrees of arc. This gives the estimates of E_{corner} and γ previously cited. Some assumptions such as that of a constant γ as a function of β are harsh, but at the same time, the major importance of long-distance elastic energy compared to the local energies of the front is demonstrated. Thus, the optimization of the inclusion energy can explain, at least qualitatively, the observed phenomena of phase fronts in factory roofs and notably why the angle β' (or β'') of a flat part of the phase front with the plane (001) never exceeds 25 degrees of arc.

5.4.4 *Description Using Dislocations and Disinclinations*

The phase front, incoherent interface, can be modelled by dislocations. With the same assumptions as before of the primacy of the deformation u_{zz}, the phase front becomes a plane of slippage of quasi-dislocations of Burgers vectors $b = u_{zz}k$ as illustrated in Fig. 5.23d. To search for the conditions of stability of such a plane of quasi-dislocations, the forces exerted on a quasi-dislocation by all the others are considered. The method has been used for a long time for ferroelectric domains (Chervonobrodov and Roytburd, 1988) taking up an idea from Lifchitz (1948). Both rises following x and slips following z are taken into account. The result, general for any assembly of edge dislocations, gives a possible stability for an angle β of 21.5 degrees of arc (Kvitek, 1995), thus compatible with experimental observations. This set of quasi-dislocations accounts for the long-distance deformation energy.

If we retain this concept of quasi-dislocations for the flat parts of the phase front, we can account for a zigzag phase front by also using the models of dipoles of quasi-dis-inclinations as rotation defects. Two successive edges then correspond to disinclinations of opposite Frank angles (Li, 1972; de Witt, 1973a, b; Kleman and Schlenker, 1972) with a dipole per zigzag period which allows to find the quasi-dislocation of Burgers vector $\mathbf{B} = 4\mathbf{b}\mathbf{u}_{zz}$.The mechanical energy of the whole can be written (Li, 1972):

$$\left(-\mu B^2 / 4\pi\left(1-v\right)\right).\ln\left(2e / R_0\right) + 4\gamma e / \cos\beta' \tag{5.9}$$

with R_0 core energy of the quasi-dislocation, and γ surface energy of a flat part of the front. In this representation, the Burgers vector $\mathbf{B}$ accounts for the long-distance

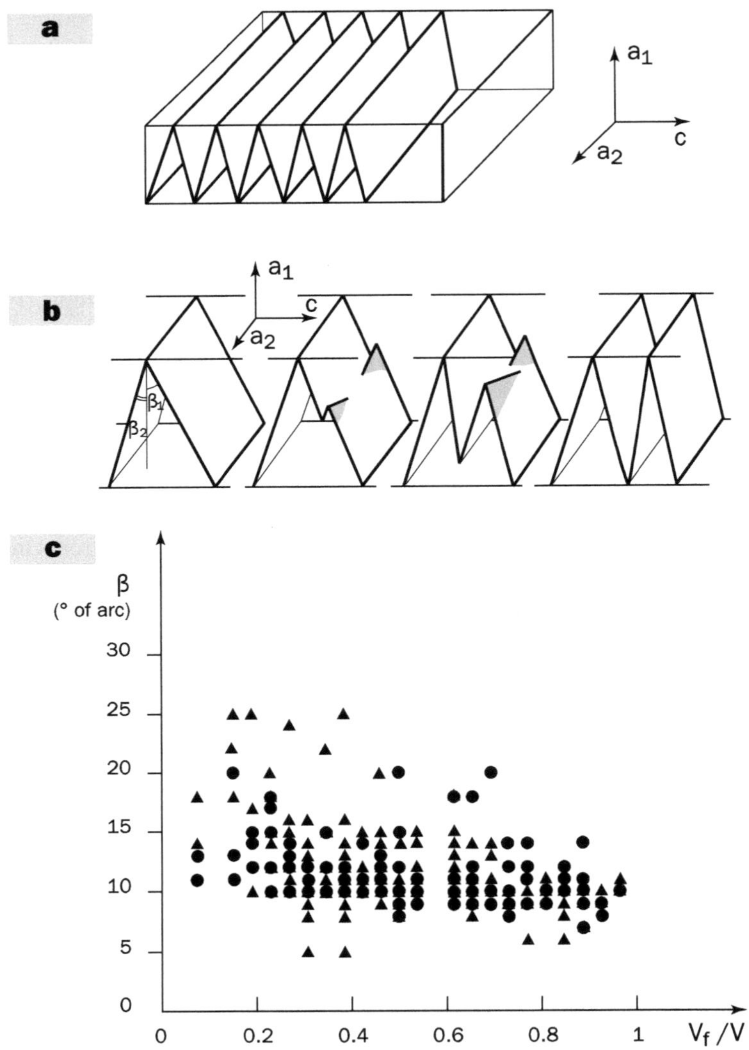

Fig. 5.22 In a sample with tetragonal plane faces, with thermal gradient along one of the a_i axes, the flat parts of the factory roofs in the middle of the sample make with the plane (001) angles β_i between 21 and 25 degrees of arc (Figures 5.11, 5.12, 5.13 and 5.14). To evaluate the role of the singularities of the factory roof corners, the thermal gradient is slowly decreased until the corners are in contact with the upper and lower faces of the sample. After some rearrangements, the angles β_i take values much lower than those observed with the phase front in the middle of the sample: about 10–11 degrees of arc (instead of 25°). In this situation, the stresses created by the lines of singularities have been relaxed and it is considered as a first approximation that the energies of the corresponding corners have disappeared. This process allows them to be evaluated (Kvitek and Bornarel, 1995). (**a**) Situation just before the entire sample is occupied by the zigzag front. (**b**) Example of creating an additional edge when the value of the angle β_i is too high compared to that which minimizes the overall energy of the front. (**c**) Distribution of the β_i angle values, for a sample with factory roof corners of the phase front in the faces a_1 depending on the ratio between the ferro volume and the total volume, (this angle would be 20 and 25 degrees of arc if the zigzag front was in the middle of the sample)

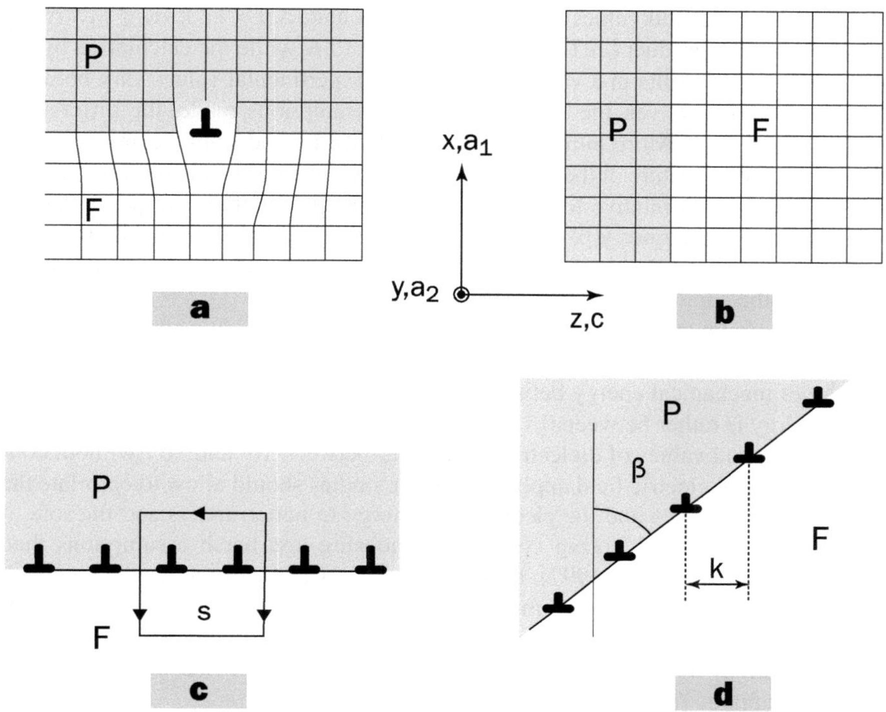

Fig. 5.23 (**a**) Schematic representation of a Frank wedge dislocation in the hypothesis of preeminence of the deformation u_{zz} and of a phase front in the plane (100). The diagram exaggerates the context with 10 para phase cells for 9 in ferro phase, hence a deformation u_{zz} 10^4 times greater than it actually is. (**b**) If the phase front was in a plane (001) and if the deformations other than u_{zz} were neglected, this front would be a coherent interface. (**c**) If the phase front is in a plane (100), it can be described by a set of quasi-dislocations of the type of the dislocation represented in a); the closed contour of dimension s along z allows to reveal the Burgers vector of the corresponding quasi-dislocation oriented along z and of module $b = u_{zz}\, s$. *d*) When the phase front makes an angle β with the plane (001), the stress created by the front can be described by a distribution of equidistant wedge quasi-dislocations with Burgers vector $b = u_{zz}\, k$, k being the distance indicated on the diagram

deformations of all the quasi-dislocations of a front period while the local energy of the front (thickness 2e following x) is expressed by the dipole of quasi-dislocations. Thus, with the simplest previous assumptions of the inclusion model (mechanical energy and chemical energy), the energy per unit length following z can be written (Kvitek, 1995):

$$\rho = \left(-\mu e \left(tg\beta'\right) u_{zz}^{2} / \pi \left(1-\nu\right)\right) \ln\left(1/2tg\beta'\right) + \gamma / \sin\beta' + \left(\left|\Delta S\right| G_i e\right)/6 \quad (5.10)$$

and, by minimizing ρ with respect to e, we obtain

$$e = \left(\left(3\mu \left(tg\beta'\right) u_{zz}^{2}\right) / \left(\pi\left(1-\nu\right)\left|\Delta S\right| G_i\right)\right) \ln\left(1/2tg\beta'\right) \quad (5.11)$$

With the experimental values $\beta' = 22$ arc degrees and $|\Delta S| = 17$ kJ/Km^{-3} (Strukov et al., 1972), the product ($2e\ G_i$) takes the value 0.17 K while the calculation by the inclusion model results in a value of 1.1 K. The experimental values vary between 0.1 K and 0.6 K. Given the very simplifying assumptions made, the differences remain satisfactory. More in-depth work should take into account the anisotropy of the crystal, the curvature of the phase front envelope, and the zigzag periods depending on the position relative to the edges of the sample to clarify in particular the variations in the energies $\gamma(\beta)$ and E_{corner} as a function of the angles β' and β'' of the part of the phase front with the plane (001). These restrictions cannot undermine the solidity of the approach summarized below:

The experimental results suggested neglecting the electrostatic energy compared to the mechanical energy. The models and these results allow us to estimate the density of mechanical energy between 1 and 10 kJ/m^3 while the density of electrostatic energy is rather between 0.1 and 1 kJ m^{-3} for $P_S = 4.25 \times 10^{-2}$ C/m^2 (Strukov et al., 1972) and values of dielectric constant ε_c' between 10^3 and 10^4 (without continuous external electric field applied). Further studies should allow to correlate the front surface energies and the electrostatic energy to better understand the role of the singularities of the zigzag corners by choosing less harsh assumptions than those used (Kvitek, 1995, 1997). However, the consistency of the results obtained, despite the simplicity of the assumptions, confirms the predominant role of mechanical energy on the shape and orientation of the phase front in competition with the chemical energy in this phase coexistence of a DKDP-type crystal without external excitation applied (electric field or mechanical stress).

5.5 Contribution of Phase Coexistence in Susceptibility

One of the objectives of this book is to establish the links between macroscopic properties and microscopic and nanoscopic phenomena. Regarding the dielectric properties, does phase coexistence make a significant contribution and what are its characteristics?

For example in the Chap. 4, il has been possible to dissociate in the macroscopic capacitive measurements of the ferro phase the contribution of the single domain crystal from what is due to the existence of domains. The different textures thus give contributions to the dielectric constant and to the loss constant according to their characteristics (number of domains, lateral surface of the walls, number of domain sets). Figure 4.24 thus provides information about DKDP, different from that obtained for the KDP, regarding the impact of domain textures. But, does the existence of a coexistence zone make a specific contribution?

5.5.1 Domains in the Coexistence Region

The systematic observation of domain textures along the *c* axis, demonstrates the difference in domain texture during coexistence, between a para-ferro PF cycle and a ferro-para FP cycle:

In the case of a thermal gradient parallel to c (Fig. 5.7), as when the thermal gradient is negligible, during a PF cycle, the ferro region of the sample in the coexistence zone is often almost single domain with a few very mobile domains or walls. For a sample subjected this time to a thermal gradient perpendicular to the *c* axis as illustrated by Fig. 5.24, again during a PF cycle, we observe, also in the ferro factory roofs, a few numerous and extremely mobile domains.

The situation is totally different in FP cycle as illustrated in Fig. 5.25 with $\mathbf{G_e}$ perpendicular to c. A domain texture inherited from the ferro phase exists in the coexistence zone, remarkably stable, in a large part of its ferro region. However, the tops of ferro factory roofs are clearly single domain when the distances between phase fronts are <200 or 300 μm. This phenomenon is verified in Figs. 5.3, 5.5, 5.6, and 5.21.

Figure 5.26 summarizes all the situations encountered: the domains of the ferro factory roof in a PF cycle (Fig. 5.26a) are in permanent instability, while those of the FP cycle (Fig. 5.26b) are immobile during coexistence when moving the factory roof in the sample. Studies by various methods of the part of the factory roofs that appear mono-domain are to be developed. The regions of great instabilities of the PF cycle could contain element precursors of the domain structure such as the shear bands observed in atomic force microscopy in lead phosphate (Li et al., 2016). In DKDP, the orientations of these textures mobiles can be perpendicular to the domains of the ferro phase as was the case between the shear bands and the twins.

5.5.2 Dielectric Susceptibility During Phase Coexistence

The dielectric measurements of Figs. 5.4, 5.7, 5.11, and 5.12 illustrate the diversity of variations in dielectric constant and loss constant measured following c depending on thermal situations. Figure 5.27 synthetizes the essential difference between a PF cycle and an FP cycle: It is common with high-quality crystals and samples, under the same thermal conditions (same gradient and same cooling speed), to observe during PF cycles significant differences from one cycle to another; one will include several sets of domains that rearrange continuously and the other will undergo a momentary blocking of the sets of domains with sudden rearrangements.

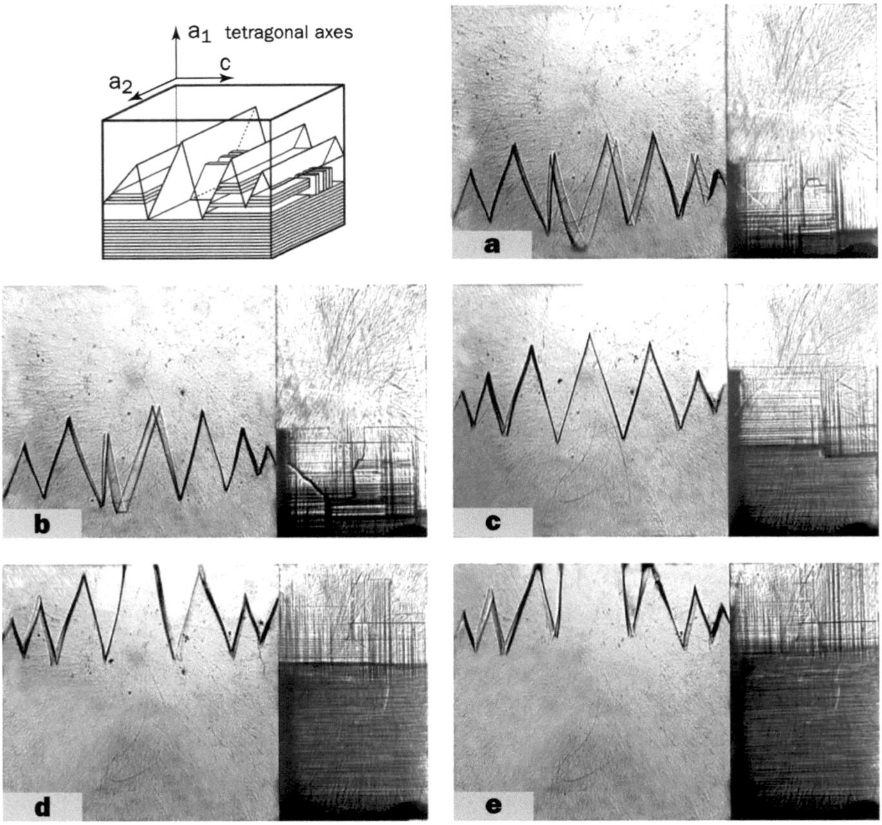

Fig. 5.24 The DKDP sample is schematically represented. The photographs on the left correspond to sections (010), and those on the right to sections (001). During this PF cycle (with $\mathbf{G}_e$ along the axis a_1), we observe the factory roof phase fronts and the domain texture. In this sample without applied electric field, the domain texture of the ferro phase is dense with domain walls in planes (100). In the coexistence zone, the domains are few, unstable and predominantly oriented in planes (010)

As illustrated by Fig. 5.27 (top), the systematic phenomena are the arrival and departure in the sample of phase fronts with unstable situations and jumps in dielectric constant; moreover, the appearance of domains in sufficient density and their movement lead to a growth of the loss constant. Conversely, during an FP cycle, the phenomena occur without instabilities with fairly regular variations of ε_c' and ε_c''. This context is developed below.

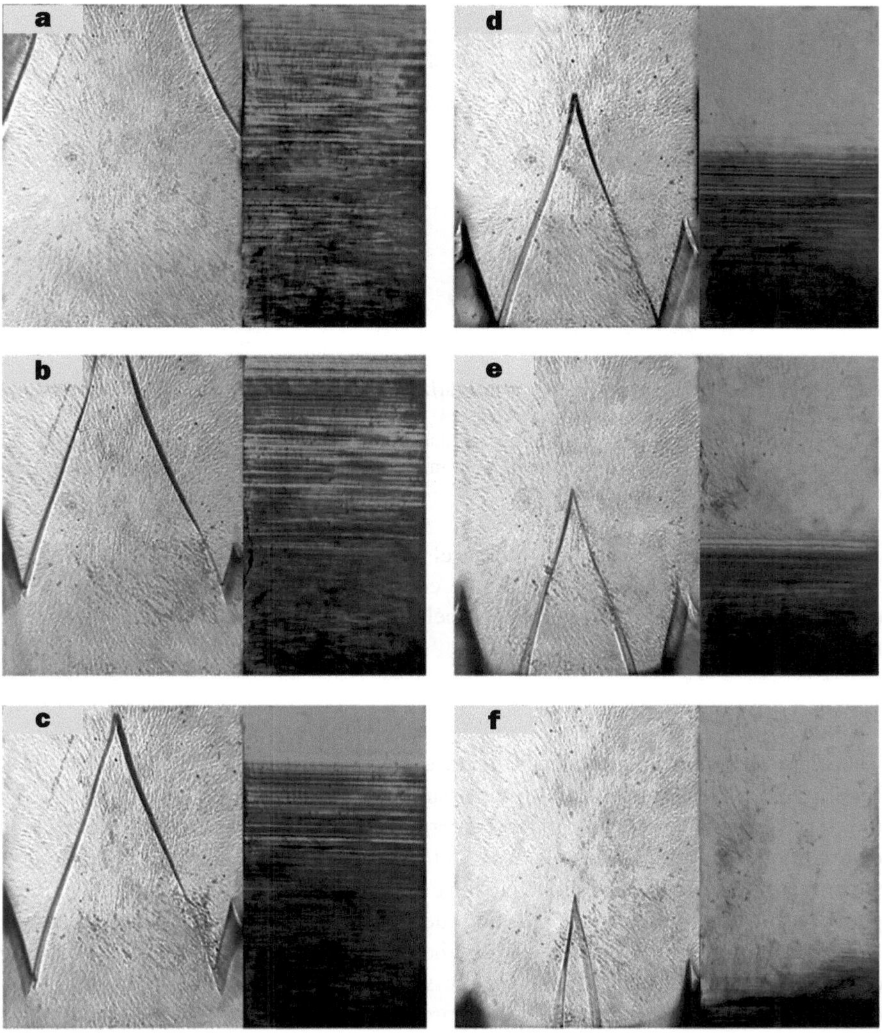

Fig. 5.25 Evolution of the shape of the phase front (photographs on the left in section (010)) and domain textures (photographs on the right in section (001)) during an FP cycle of DKDP (R.Cach). A thermal gradient G_e of 0.74 K mm^{-1} is applied along the axis a_1. The domain texture of the ferro phase only includes one wall orientation (100). The domains of the ferro parts of the coexistence zone have the same orientation. The upper part of the ferro daggers is mono-domain

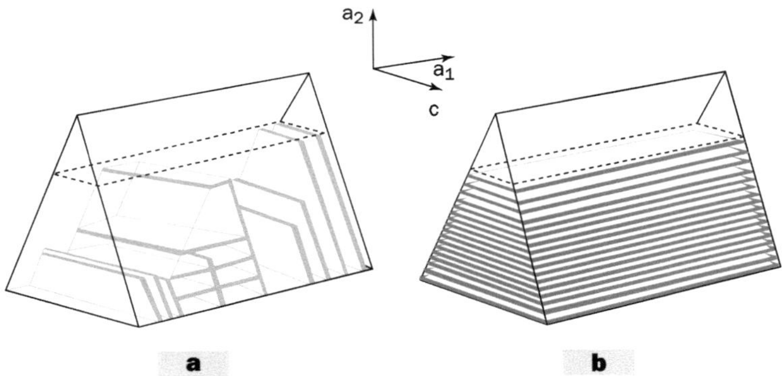

Fig. 5.26 The schematic representations summarize the situations observed regarding the domains in the factory roofs of the phase coexistence: (**a**) During a PF cycle, very mobile, unstable, few domains, and/or walls. (**b**) During an FP cycle, stable domains in regular texture In both cases, the upper part of the ferro factory roofs is mono-domain when the phase fronts are at distances less than a few hundred micrometres

As previously demonstrated (Bornarel and Cach, 1999, 2003; Bornarel et al., 2000), in the ferro phase of DKDP, the contribution of domains to the dielectric constant following c, in the few degrees below the transition, can be described using the following expression:

$$\varepsilon_d' = (\Lambda - 1)\varepsilon_m' \tag{5.12}$$

The parameter Λ accounts for the texture in domains as illustrated in Sect. 4.4. Figure 5.28 uses these results with the assumption, demonstrated in Fig. 1.15b, that the coexistence zone remains at the temperature T_C (the indicated temperature is the external temperature measured closest to the lower a_1 face of the sample). The dotted curve in Fig. 5.28b corresponds to the calculation made with the Curie law of the para phase from the formula (5.12) and the classic relation (1.17) illustrated by the law in continuous line. It is noted that the experimental results are higher during coexistence than those predicted by the simple sum of mono-domain and poly-domain contributions, a result observed previously (Reese, 1969a, b; Sidnenko and Gladkii, 1973a, b). A notable contribution ε_{pc}', appears during the phase coexistence (Fig. 5.29):

$$\varepsilon_{pc}' = \varepsilon_c' - (\varepsilon_m' + \varepsilon_d') \tag{5.13}$$

To better understand the phenomenon, a detailed study of volumes and textures in domains in the ferro phase in the sample, of ferro and para volumes in the coexistence zone, and also of the phase front surface was conducted in the "simple" situations of factory roofs (stable zigzags in the middle of the sample).

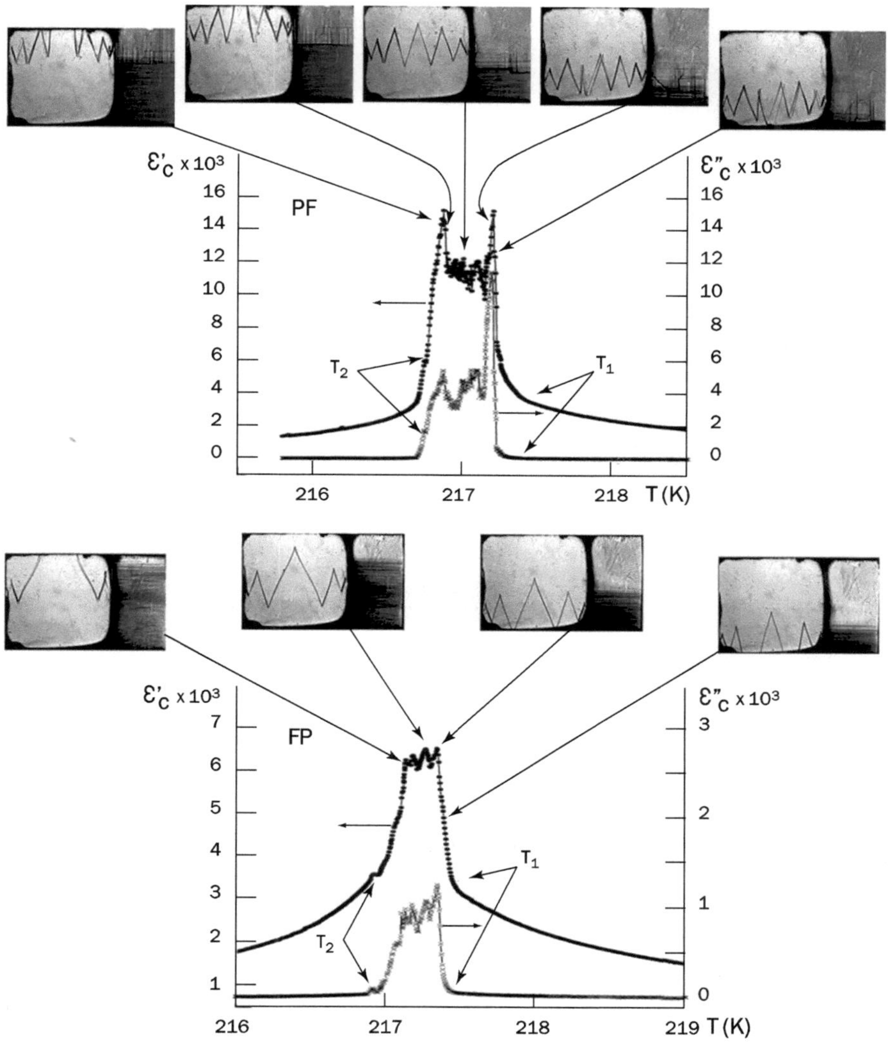

Fig. 5.27 Correlation during phase coexistence between shape of the phase front (photographs on the left in section (010)), domain texture (photographs on the right in section (001)), and dielectric measurements along c in a DKDP sample with a thermal gradient along a_1. The top results correspond to a PF cycle (G_i = 0.109 K mm^{-1}) and those at the bottom to an FP cycle (G_i = 0.076 K mm^{-1}). The domains during the phase coexistence of the PF cycle to consider from right to left are few and very mobile, while the texture is fine and very stable during the phase coexistence of the FP cycle to consider from left to right. In both cases, the ferro region in the coexistence area is mono-domain if the two fronts of a factory roof are <200–300 μm apart (Bornarel and Cach, 2003)

The scales are not the same in the PF and FP figures: the values of the dielectric constant and that of the losses are much higher during the PF cycle than the FP cycle. This result concerning the dielectric contribution of the domains is known (Chap. 4 with the KDP, RbDP) but it is even more marked during the coexistence with a loss constant >10^{-2} at the beginning of PF coexistence

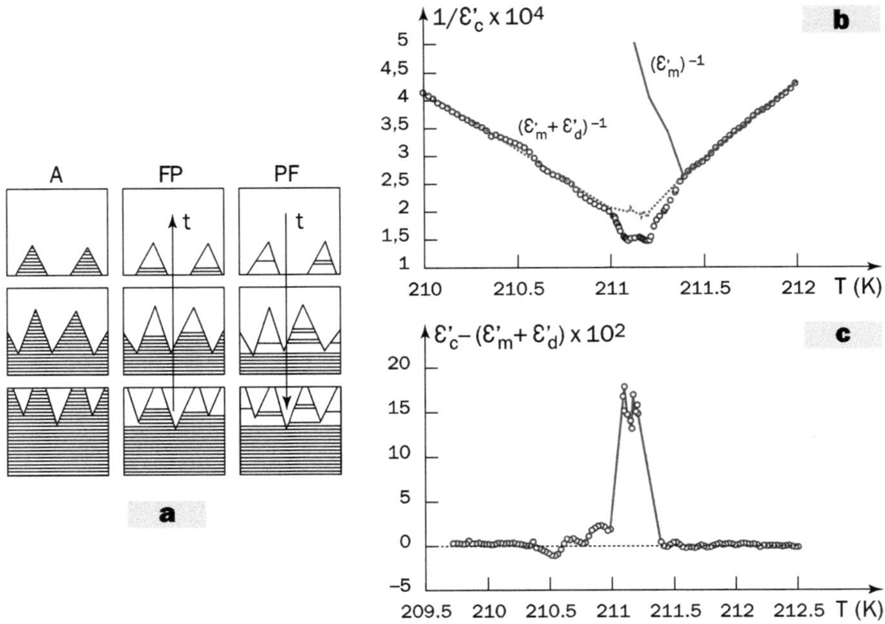

Fig. 5.28 (**a**) Schematic representation of the phase front and ferro domains in section a_2 in various situations: In theoretical assumption A on the left, the ferro region of the coexistence zone is assumed to be entirely occupied by a dense domain texture. In the FP and PF columns, the experimental observations are reproduced. Arrows indicate the direction of time evolution t, opposite in the two cases. (**b**) Experimental results of $(\varepsilon_c')^{-1}$ as a function of temperature during an FP cycle: (empty circles). The quantity calculated for a mono-domain crystal $(\varepsilon'_m)^{-1}$ is plotted as a function of temperature (continuous curve). Also plotted in dotted lines is the variation of $(\varepsilon'_m + \varepsilon'_d)^{-1}$. It is noted that during coexistence, the experimental result cannot be described as the simple sum of a mono-domain contribution and one due to the domains: a phase coexistence contribution appears. (**c**) The difference between the measured dielectric constant and the sum of the calculated contributions of the mono-domain and the domains is notable during phase coexistence

The results of Fig. 5.30, from different samples in very different thermal contexts, demonstrate two systematic characteristics of ε_{pc}':

- ε_{pc}' is more important during a PF cycle than during an FP cycle.
- The ε_{pc}' part of the overall dielectric constant is all the higher as the volume occupied by the phase coexistence zone is larger in the concerned sample.

The calculation method of ε_{pc}' from the relation (5.13) with the calculated value of ε_d' subtracted from ε_c', even when no dense domain texture is observed, minimizes the value of ε_{pc}'. This contribution exists in all situations. It is all the more important as the ferro region impacted by the phase coexistence is large, for example with a weak thermal gradient perpendicular to c.

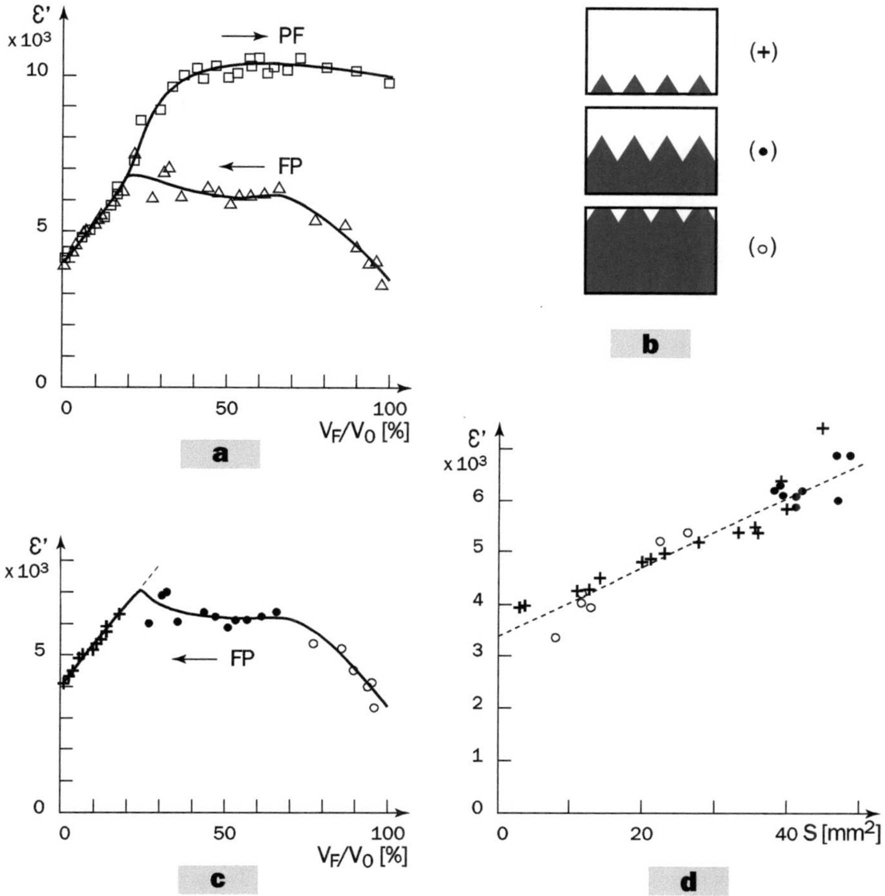

Fig. 5.29 (**a**) During a phase coexistence with a factory roof front (thermal gradient following a_2), dielectric constant measured along c as a function of the ratio between the volume of the ferro phase (V_F), and the total volume V_0 of the sample, during a PF cycle and during an FP cycle. The value during the FP cycle is always lower than that during the PF cycle. (**b**) Different shapes of the front, during its appearance in the sample, during its translation with the same zig–zag section, and during its disappearance from the sample. (**c**) Variation of the dielectric constant in the FP case as a function of V_F/V_0. The signs defined in (**b**) allow to relate the results to the microscopic situation. (**d**) Considering the simple translation of the coexistence zone without modification of the domain texture in this region, this figure demonstrates a quasi-proportionality between the value of ε′ and the surface of the phase front, itself proportional, to the volume of the coexistence zone

This is obviously the case with very good thermal homogeneity, without ferro region in which the domains can cross the whole sample following c as in Figure 5.3b. The contribution in the dielectric properties is relatively minimal when a thermal gradient parallel to c minimizes the coexistence zone with a very thin front thickness.

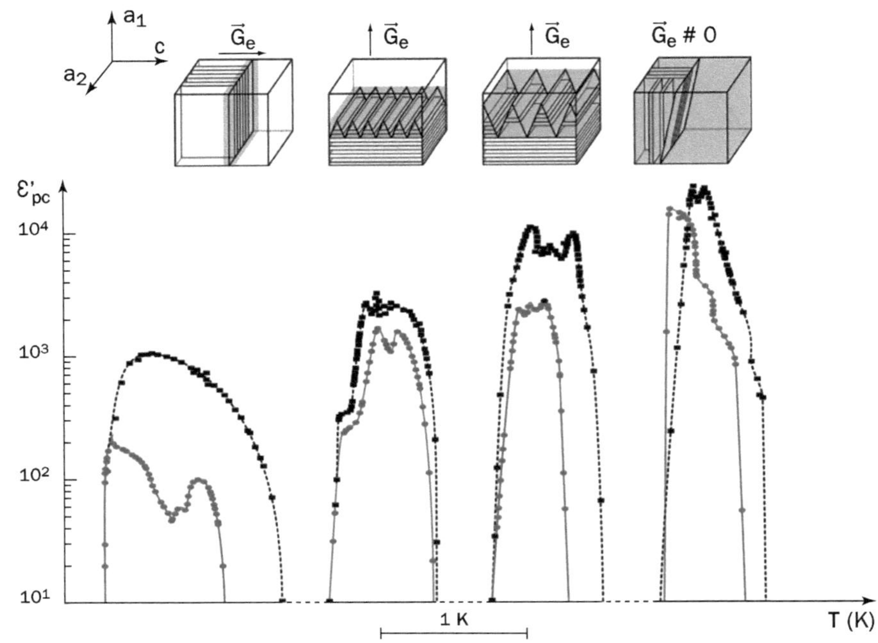

Fig. 5.30 Temperature variation of the ε'_{pc} contribution due to phase coexistence, calculated using relation (5.13), in the various experimental cases schematized. The indicated temperature is measured by a platinum probe located outside the lower face of the sample. It is a relative indication between the various examples since during the experiments, the gradients induce quite different variations from one case to another, especially in para and ferro. It is observed that, for a given situation, the same value of measured temperature corresponds, for the PF cycle and for the FP cycle, to the limit temperature between the situation of the sample with coexistence of the two phases and the entirely ferroelectric sample. The values of ε'_{pc} calculated during PF cycles (full circles) are always higher than those calculated during FP cycles (empty circles). These values depend on the volume occupied by the phase coexistence region in the sample, or more precisely, on the size of the ferro region of the coexistence zone in which either a few unstable domains or a mono-domain is observed

The set of these results shows that the volume of coexistence of almost mono-domain phase makes a significant contribution to the measured susceptibility. Various explanations remain to be substantiated:

- Near the transition, within the framework of the Schmidt model, multiple paths of Takagi ions and Slater defect lines are possible. Then a domain-free region can be the site of a large dielectric instability with a significant impact on the dielectric susceptibility measured during phase coexistence, even in a situation of very good thermal homogeneity.
- It can also be considered that this ferro region, blocked between two para regions at a relatively short distance (100–300 μm), undergoes stresses σ_{xy}, σ_{xx}, and σ_{zz}, whose effect can only move away from the situation of minimum thermody-

namic energy, with therefore an instability which would account for the growth of susceptibility, but experiments are lacking for these proposals to go beyond the level of conjecture.

To conclude, the contribution to the dielectric properties of phase coexistence is evident. As observed for example in martensitic transitions, susceptibilities and losses are greater during a cooling cycle (creation of domains-creation of the front and search for balance of the whole) than during a heating cycle (stabilized domain texture swept by coexistence). This is the case in all the results produced in this work (with ratios from 3.5 to 12 depending on the situations). In addition, losses increase depending on the rates of change of the temperature with the existence of critical temperatures as in ferro-elastic alloys: below these values, low and continuous losses are observed, and above, more significant losses with jerks, peaks that can be described as avalanche processes (Annex 7) (Gallardo et al., 2010).

5.6 Usual Phase Front Shapes and Orientations

In the previous sections, simple experimental conditions were chosen to facilitate the understanding of phenomena and their modelling: high-quality crystals, sample cuts in the parts of the crystal with low dislocation density, samples with faces in crystallographic planes, simple thermal conditions in terms of homogeneity or constant temperature gradients and appropriately oriented. These conditions are not always met. The majority of publications contain little information on these issues. What then becomes of the phase front, the domain textures during phase coexistence? By keeping the example of DKDP let's illustrate what can happen depending on the thermal context or the mechanical realities due to the sample edges.

5.6.1 Effect of Temperature Distribution in the Sample

Thermal inhomogeneities modify the region of the sample in which phase coexistence exists. Figure 5.31 shows the effect of a rate of temperature change too high to hope for a simple thermal context: the sample in the form of a platelet with the large faces in planes (010) is subjected to a thermal gradient of main component following a_2 and weaker components in directions perpendicular to this axis. We note the appearance of factory roof phase fronts in the corners of the face (010) where coexistence begins, followed by a growth of the daggers of the new phase that move towards the centre of the sample to win it entirely. The same phenomenon, reversed from the point of view of the phase, occurs towards the end of the coexistence.

Another example frequently encountered during measurements made at transitions whatever the physical quantity measured is that of phase fronts in the shape

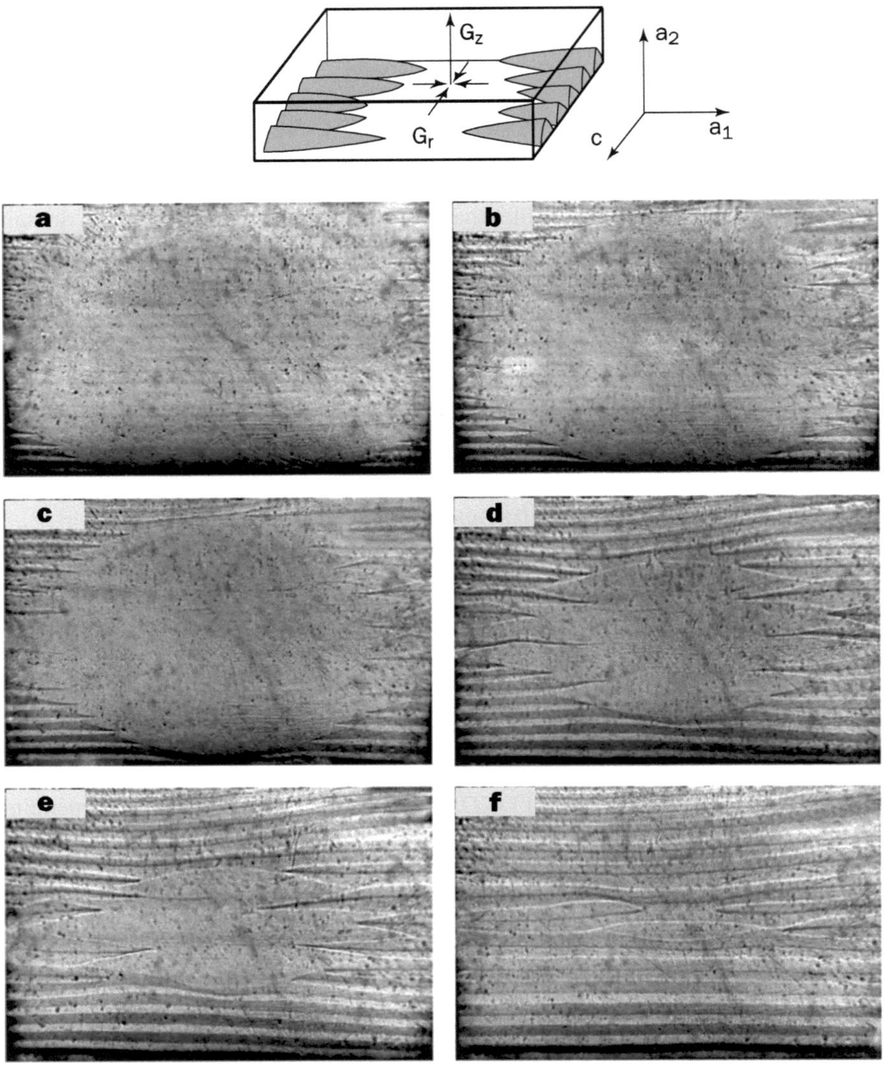

Fig. 5.31 The DKDP platelet depicted at the top of the figure is subjected to a significant thermal gradient along the a_2 axis with lesser radial components around this axis. The photographs in section (010) illustrate the situation of the phase fronts at the beginning of coexistence of a PF transition therefore with decreasing temperatures from (**a**) to (**f**). The factory roofs created initially in the corners develop towards the center of the sample until they fill the entire section

of daggers as in Fig. 5.32. If the thermal gradient includes a component in the direction of certain faces of the sample and another smaller one in a direction approximately perpendicular, daggers will appear which, unlike regular factory roofs, will have edges inclined with respect to the crystallographic axes. These daggers of various shapes will mix during coexistence, especially towards the end

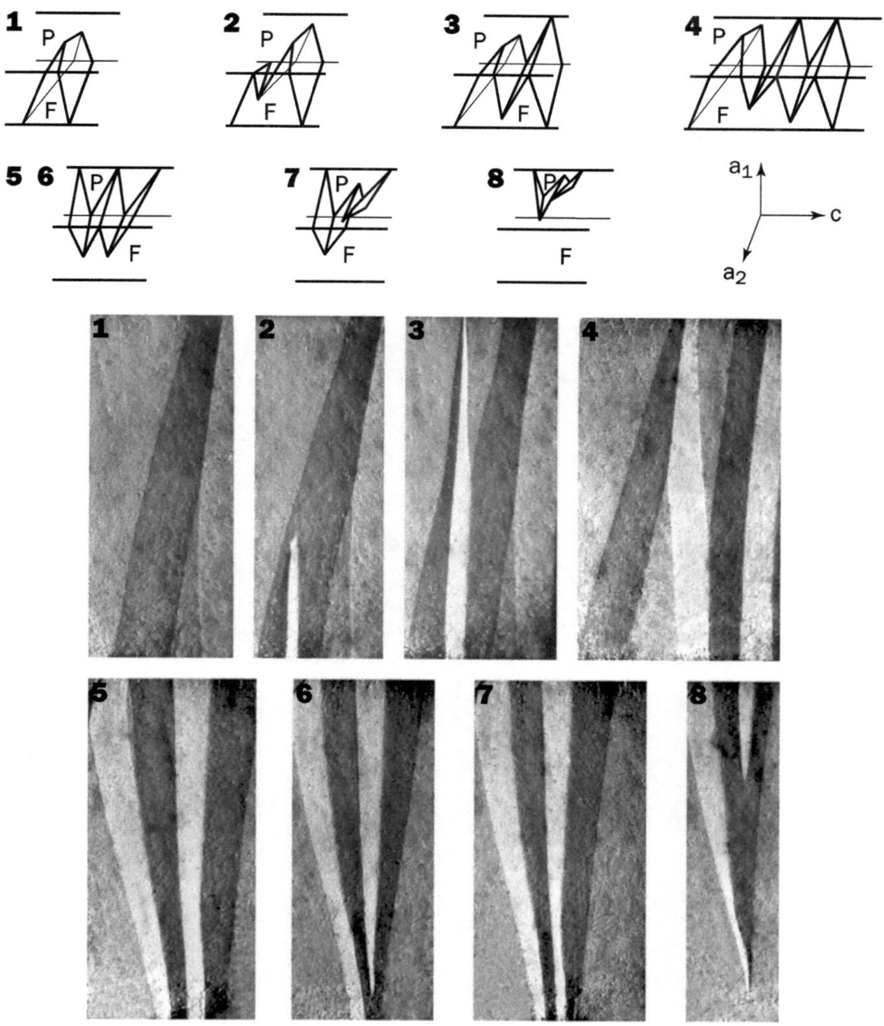

Fig. 5.32 Coexistence zone in the form of daggers: Schematic representations and photographs (in section a_1) of phase fronts during a PF transition of DKDP. The component of the external thermal gradient along a_1 is ten times greater than that along a_2. The temperature decreases regularly from photographs 1 to 8

of the process (Bornarel and Cach, 1993). Depending on the mode of investigation (various topographies, observation in one direction only), the experimenter will conclude to the existence of needles, daggers, or even lamellae para-ferro-para-ferro. The result confirms above all the existence of thermal inhomogeneities in the sample.

Even when the gradient allows to define a factory roof in a coexistence zone, the simple translation of the front towards the middle of the sample modifies the conditions as illustrated by Figs. 5.10 and 5.33: the most stable front shape in one position is no longer in another position or at another time. The most energetically favorable values of angles β′ and β″ change during the process and the number of edges can also change. Such evolutions over time can occur successively usually to reduce the number of edges and, therefore, increase the average values of β′ and β″.

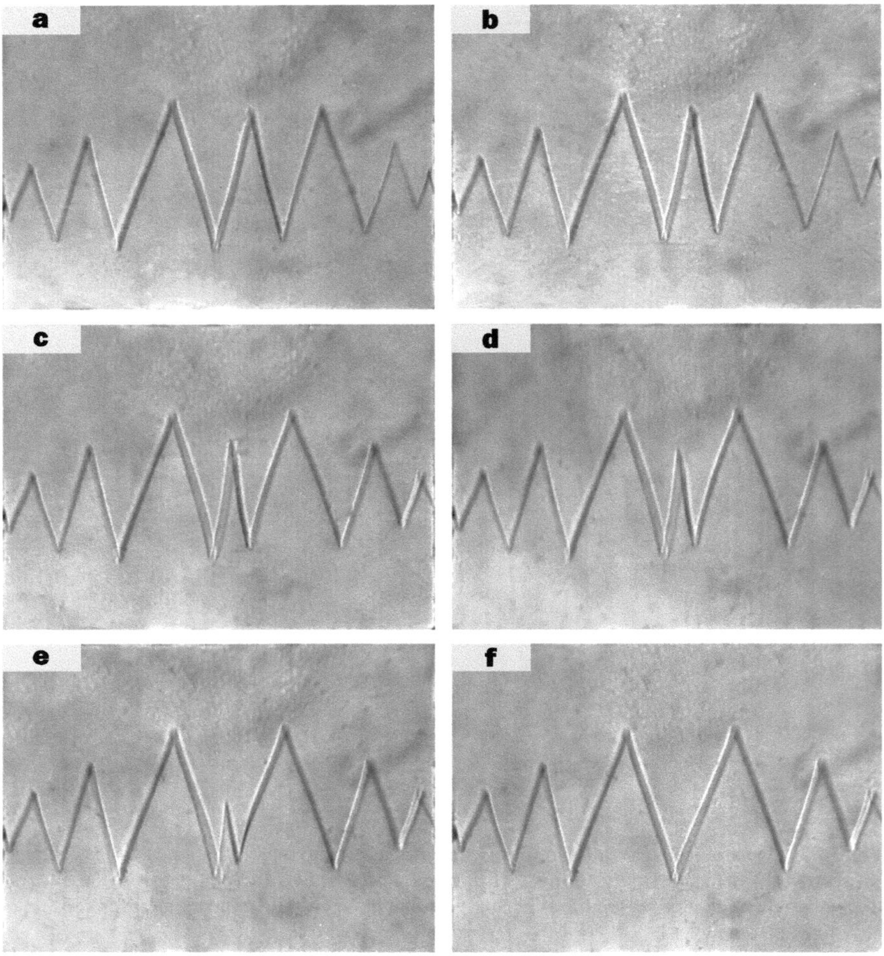

Fig. 5.33 The factory roof phase front in the middle of the sample is the result of a PF cycle such as those described in Fig. 5.11 with a linear and stable internal thermal gradient G_i. The temperature is regulated to a few thousandths of K which blocks the front in the middle of the sample. In the hours that follow, the front shape optimizes through a process opposite to that of Fig. 5.22: the front initially comprised 14 factory roof edges; a "roof" edge withdraws in the form of a dagger and the front balances after a certain time (from a few tens of minutes to a few hours) with 12 edges and β′ and β″ angle values larger than initially

5.6.2 Phase Fronts and Mechanical Limits

The results obtained with the minimum possible defects allow to propose models and understand fundamental physical phenomena. However, knowing what such a defect induces as an effect is not useless and, in all cases, the limits of the sample must be taken into account.

Figure 5.34 illustrates a situation with a thermal inhomogeneity which induces fronts in the form of daggers deformed by growth defects in the crystal. The daggers deform quite easily to fit the shapes of the defect in a flexible way, which confirms the lesser importance of energy per unit of surface area of the front (except for singular lines such as edges) compared to volume mechanical energy and chemical energy. However, these interactions between phase front and defects can hold back the front in its development and increase the temperature range of phase coexistence.

The influence of the sample edges is very different if the front is parallel or perpendicular to them.

The result presented in Fig. 5.35 allows us not to be confronted at the same time with the problem of thermal conduction as in the case of phase front envelopes (see Fig. 5.31): fractures located perfectly in planes (001) release the crystal's constraints and the phase transition in this good thermal homogeneity situation occurs starting from the fracture. The flat phase fronts thus created then move the remaining flat, on either side of the fracture, in a coexistence scheme consistent with that illustrated in Figure 5.3. This example confirms that a phase front in a plane (001) corresponds to an optimal situation.

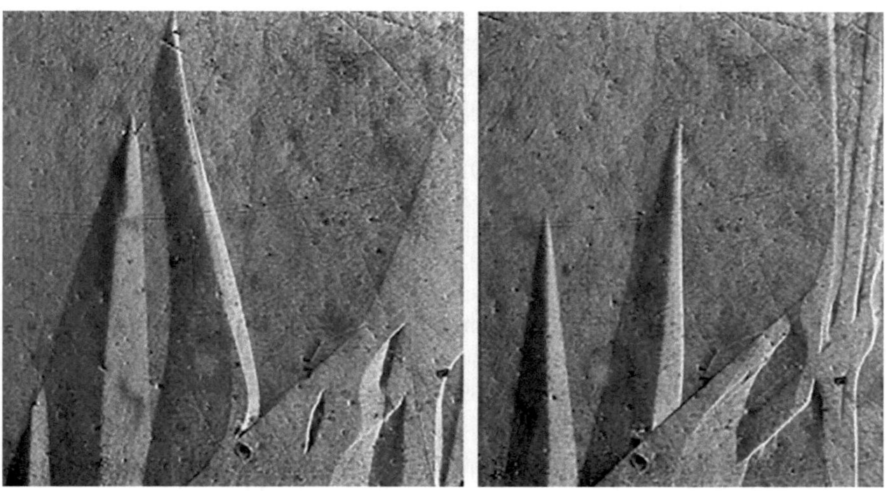

Fig. 5.34 Phase front shapes during a PF transition of a DKDP crystal in a context close to that of Fig. 5.32. A thermal gradient of 0.2 K mm^{-1} exists in the direction a_2 (the photographs are in a section (100)) with weaker components along a_1. The phase front encounters growth defects and deforms by adopting their shape without singularity

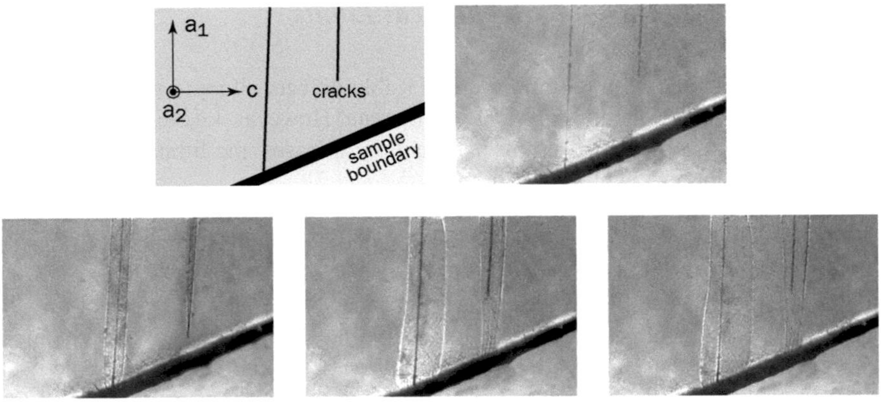

Fig. 5.35 Breaks have been created in a DKDP crystal, perfectly in planes (001). In a sample with good thermal homogeneity, the new phase appears primarily on the edges of the break, where the stresses are released and the modification of u_{xx} and u_{zz} is easier. Two almost planar phase fronts then move laterally while remaining in these planes (001) (the width of the photographs corresponds to 2 mm at the crystal level) (Kvitek, 1995)

The situation is quite different when the phase front is shaped like a factory roof. The total deformation energy is greatly impacted and the sample boundaries play their role. Figures 5.36 and 5.37 highlight the long-range constraints created by a phase front in the case of samples cut, respectively, in tetragonal planes and in orthorhombic planes. It is understood how certain results, obtained with samples cut according to planes other than the tetragonal planes, can be difficult to interpret and not very reproducible. This illustration chosen deliberately with almost cubic sample shapes mainly helps to shed light on numerous works, some obtained with samples in the form of platelets and others elongated along the c-axis or another axis. Many of these experimental results could be exploited again by taking into account the impact of their shape. Moreover, the long-range deformations created by the sample edges must be exploited in the future to create situations favourable to applications of ferro-ferros (Manaka et al., 2022).

Figures 5.36 and 5.37 also shed light on the work on flexoelectricity, modification of polarization with a gradient of stress, or temperature and the experiments carried out specifically to better understand and exploit the phenomenon (Arias et al., 2022; Zhang et al., 2023; Huang et al., 2023): a first-order transition of a ferro-ferro with the phase front and the sample edges offers a rich variety of studies with thermal gradients and electric field and applied stresses.

Memo Chapter 5

In a theoretical model of first-order transition, an infinite-sized crystal with a homogeneous temperature and without external excitation should switch from one phase to another (para and ferro) at temperature T_C. It is conceivable that at a decreasing temperature, a coexistence could start at T_C and persist until T_0 (and similarly at

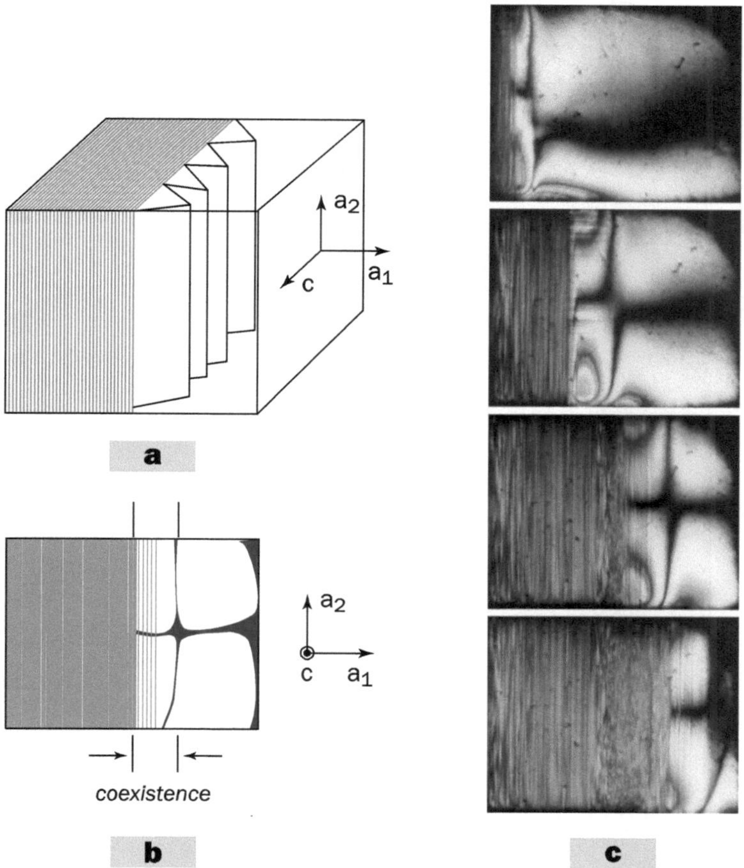

Fig. 5.36 (**a**) The faces of the DKDP sample are tetragonal planes. Coexistence during a PF cycle is carried out with a thermal gradient $\mathbf{G_i}$ directed in the direction a_1. (**b**) Diagram corresponding to the photographs in section (001) between crossed polarizer and analyzer, which highlight the domains, the phase front (the very visible projection of the phase front edges corresponds to an area of strong modifications of u_{xx} and u_{zz}) and the stresses created in the coexistence zone and in the para phase. (**c**) In addition to the variation of stresses during the translation of the coexistence zone in the sample, one can observe in white light in this zone the walls of some domains (very transparent); some are located parallel to a permitted plane (010) in the second photograph; they are replaced in the third photograph from the top by domains, walls in the permitted plane (100). The stresses observed in the sample recall the old observations of flexo-optic effect made in DKDP (Vasilevskaya et al., 1987)

increasing temperature between T_C and T_1). Concretely, a thermal hysteresis could exist between T_0 and T_1.

In DKDP crystal, the phase front minimizing the mechanical energy is not a coherent interface as for domain walls: it is an incoherent interface in a plane (001)

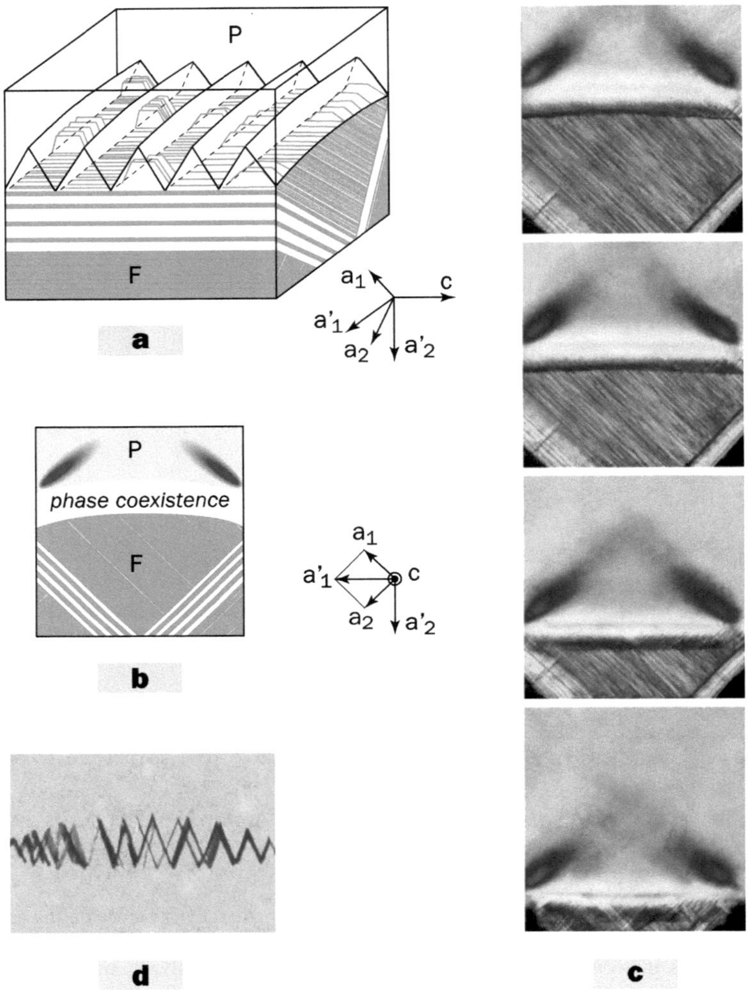

Fig. 5.37 (**a**) DKDP crystal sample with faces in orthorhombic planes. Coexistence takes place during this FP transition with a thermal gradient $\mathbf{G}_i$ directed in the direction a_2'. (**b**, **c**) The photographs in sections (001) with crossed polarizer and analyzer in white light presented in (**c**) highlight, the sets of domains, and the stresses created in the para phase by the intersection between the phase front and the sample edge. The very bright band just above the domains corresponds to the coexistence zone in factory roofs. (**d**) Cross-section photograph (100) of the factory roof phase front. The traces on both faces (100) of the sample are not as well superimposed as when observing with a sample in tetragonal faces. In this more complicated context for interfaces (fronts and domains), the roof ridges are less well aligned following a_1' than they were in Fig. 5.36 following a_2

for a sample homogeneous in temperature without applied conjugate field. If a thermal gradient **G** exists, the front is located between plane (001) and isotherm T_C with an angle β related to the plane (001) depending on the thermal gradient orientation. If **G** is at 90° to (001), the front presents a factory roof shape. Its flat parts have β angle less than 25° for a sample with tetragonal edges and less than 18° for orthorhombic edges. The factory roof edges on the para side are singularity lines that create deformations in the para phase, starting from the intersections between sample edges and the factory roof. Their suppression by decreasing G changes β angle from 25° to 11°.

In this context without applied conjugate field, the domains follow the rules presented in Chap. 4: during a PF cycle, unstable and sparse domain textures, and during a FP cycle, dense and stable textures. But the phase front also conditions the domains which are only dense when they can cross the entire sample along the ferroelectric c-axis. If the ferro part of the coexistence region is limited by two phase fronts, the domain density decreases with the distance between the fronts and even becomes mono-domain if this distance is less than about 200 μm.

The macroscopic dielectric properties are impacted by the coexistence zone, in addition to that of the mono-domain and that of the domains. This contribution grows as the volume of the coexistence zone increases and losses are always the detector of the domain importance.

The results are well explained either by the Khachaturyan inclusion model or by a model of dislocations and disinclinations demonstrating the primary role of mechanical energy in this phase coexistence of DKDP without the action of a conjugate field. Pending questions will be addressed in Chap. 6:

- What happens in the factory roof zone in terms of walls and domains during a PF transition?
- How are the deformations within the coexistence zone in the two phases with factory roof fronts, depending on the distance to the phase front edges?
- What changes does the application of an electric field E that increases the importance of electrical energy make?
- Is it possible to explain with the phase fronts and the domains, the very different results for the dielectric measurements and the hysteresis cycles encountered in the literature especially for the low values of E?
- Are phase diagrams like the one in Fig. 1.22b illuminated by the existence of phase fronts?

Chapter 6
Coexistence of Phases with Conjugate Field

6.1 Phase Coexistence and Competing Energies

Understanding the links between microscopic phenomena and macroscopic experiments and models leads to taking into account often neglected energies. The proposed conclusions cannot be fully applied to all crystals since symmetries and couplings between properties differ. However, the choice of simple and linearly coupled elastic and dielectric properties in our ferro-ferro allows us to imagine what can occur in more complex cases. It is in this logic that the first-order transition of the DKDP was retained.

The application of conjugate fields to a crystal with a first-order transition was illustrated concerning macroscopic properties by Figs. 1.20, 1.21, and 1.22 (the latter, in fact, corresponds to the KDP near the tri-critical regime). Figure 6.1 recalls the achievements for the DKDP of Chaps. 1 and 5 with $E_{//} = 0$ (the notation $E_{//}$ is used in this chapter when an ambiguity could exist between the continuous applied field and the alternating electric field of dielectric measurement): in a crystal of very good thermal homogeneity, coexistence begins at a temperature close to T_C, therefore from point α of the phase diagram in Fig. 1.22b, during a PF transition and a FP transition. Coexistence occurs with a succession of situations (see Fig. 1.21) such as MSM (Metastable, Stable, Metastable), *SΦS* (stable, unstable, stable), etc. Experimentally, phase coexistence does not occupy the entire temperature range from T_1 (at point γ) to T_0 (at point δ). In addition, temperature measurements are often carried out near the sample, the thermal homogeneity of the sample is imperfect, and some thermal and mechanical imperfections may exist. Thus, even if the thermal conditions are of quality, they cannot, during optical observations and dielectric measurements, achieve the performances achieved in neutron diffraction (homogeneity of 10^{-3} K by Zeyen et al.) or in high precision calorimetry of 10^{-4} K

The original version of the chapter has been revised. A correction to this chapter can be found at https://doi.org/10.1007/978-3-031-98099-2_7

J. Bornarel, *Ferroic Crystal Domains, Walls, and Phase Fronts*, Springer Series in Solid-State Sciences 209, https://doi.org/10.1007/978-3-031-98099-2_6

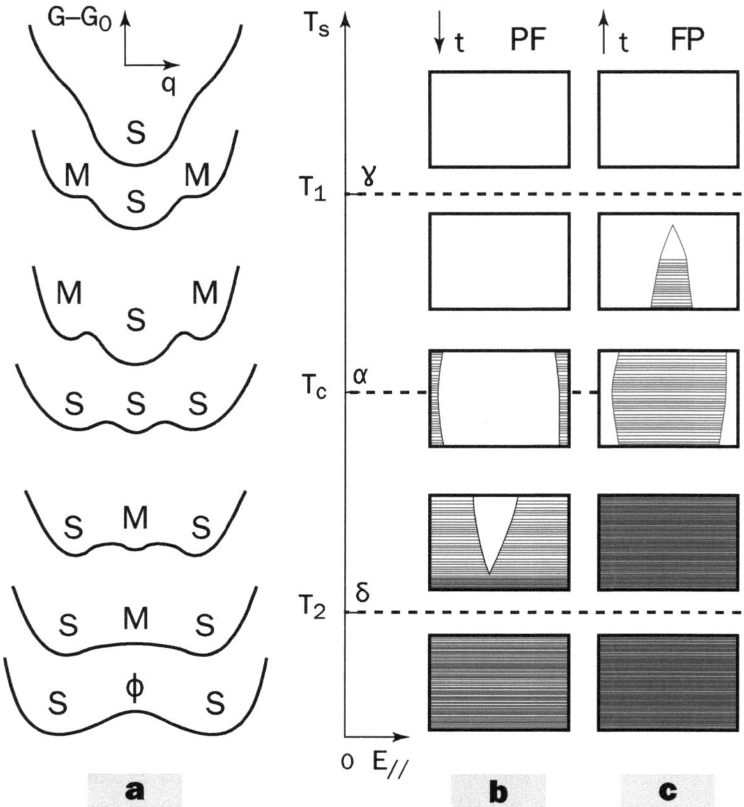

Fig. 6.1 (**a**) In the classical thermodynamic assumption which implicitly assumes an infinite, mono-domain crystal with a homogeneous temperature T_S at all points, the energy density as a function of the order parameter q (in our case polarization or coupled deformation) is illustrated by the graphs conforming to the results of Chap. 1 (Fig. 1.20 and $E = 0$ axis of Fig. 1.22 with the same meanings for S stable, M metastable and ø unstable): above temperature T_1, only the para phase can exist; between T_1 and T_C, three possibilities exist, including the stable one at $P = 0$; three situations are equivalent at T_C; and, below T_C, the two opposite polarizations are more favourable until the disappearance of the one at $P = 0$ below T_0. The temperature scale recalls the names of Fig. 1.22. The experimentally measured temperature T is rarely equal to T_S. (**b** and **c**) The diagrams summarize the situations encountered in terms of phase front and domains in each temperature interval for a sample temperature homogeneity around 10^{-3} K, assuming a phase coexistence that begins at T_C; the arrows for the direction of time t allow to follow the chronology for (**b**): PF and (**c**): FP. The white surfaces correspond to the para phase; the ferro region is represented with striations that provide information on the domain density

by Reese (1967, 1968, 1969a, b). These causes of inaccuracies exist for many published results, with sometimes the difficulty of knowing the temperature measurement process attributed to the crystal. We must remember this possible difference between the temperature T_S of the sample and the measured temperature T.

With the application of a conjugate electric field $E_{//}$, what actually happens inside the γαδεβ region of Fig. 1.22? In this chapter, the same approach is adopted as in Chap. 5: first explore experimental situations of good thermal homogeneity of the crystal (Sect. 6.2), then others with a judiciously oriented constant thermal gradient (Sect. 6.3). Some clues to guess the microscopic phenomena thanks to the macroscopic results are proposed in Sect. 6.4.

Some orders of magnitude of the energies involved can be recalled:

The variation of the chemical energy at the transition compared to the para situation is of the form $S\Delta T$ with ΔT the value of the thermal hysteresis calculated using Landau's theory. We thus find a value between 0.1 and 0.2 K close to some published experimental results (Sidnenko and Gladkii 1973a, b; Chabin and Giletta 1977). But the experimental conditions can be questioned. The choice for S is also delicate since the calculated values diverge: 3.3×10^4 J K^{-1} m^{-3} (Bantle, 1942), 6.8×10^4 J K^{-1} m^{-3} (Strukov et al., 1968), 1.7×10^4 J K^{-1} m^{-3} (Smolenskii et al., 1984). The result in modulus of $S\Delta T$ would thus be between 3.4×10^3 J m^{-3} and 1.36×10^4 J m^{-3}. But the results presented in Chap. 5 demonstrate a measured thermal hysteresis significantly lower than these usually retained theoretical values.

The estimation of the electrostatic energy with an applied field is very different between a mono-domain crystal and a poly-domain situation. In the mono-domain case, the contribution to the free energy of the electrostatic energy is in modulus PE; in the ferro phase, if we consider in first approximation the spontaneous polarization little modified by $E_{//}$ applied, with a value of P_S of 4.2×10^{-2} C/m^2 (Strukov et al., 1972a), we obtain for PE a value of 210 J/m^3 with $E_{//}$ of 50V/cm, of 4.2×10^3 J/m^3 with $E_{//}$ of 1 kV/cm, significantly higher than the coercive field observed in the vicinity of the transition. But, as soon as a domain texture appears, in accordance with relations (2.11) and (2.12), the values of electrostatic energy are significantly lower.

These order-of-magnitude estimates could be continued by considering the contributions listed previously. But to take into account microscopic realities is to use the results of Chap. 5 related to the phase front: without an applied electric field, the elastic energy is decisive. The application of the conjugate electric field could, if sufficient, change this situation. Therefore, the model of an inclusion of phase F or P, extremely thin in the matrix of the other phase (P or F) (Fig. 5.20a), is revisited. The calculation of $B(\beta)$ introduced by relations (5.2) and (5.3) is used in the case of a ferro poly-domain phase with a deformation function of the applied field up to the mono-domain for a very strong field: the average deformation $<u_{12}>$ of the ferro region, compared to that of the para region, will depend on the volume of the domains of each polarity. The numerical values of the u_{ij} and λ_{ijkl} from Figs. 5.18 and 5.19 (Kvitek and Bornarel, 2000) are used again to plot in Fig. 6.2a the curves **a** ($<u_{12}> = 0$) and **b** ($<u_{12}> = +u_{12}$ or $-u_{12}$) with, in both cases, $\lambda_{1212} = 6 \times 10^9$ N/m^2, value of the elastic constant at constant polarization. The result of Fig. 5.18 is confirmed: if the ferro phase is poly-domain with equality of domains + and – (curve **a**), the mechanical energy is minimal for a front perpendicular to c ($\beta = 0$) with a value of B of 1.13×10^3 J/m^3. On the contrary, if the ferro phase is mono-domain (curve **b**) under the action of a strong applied electric field, the energy will be minimal for a front parallel to c ($\beta = 90°$), orientation which minimizes the electrostatic energy, with a value for B of 2.45×10^4 J/m^3. One can question the value of the elastic constant at the transition since at constant electric field, it varies inversely to the

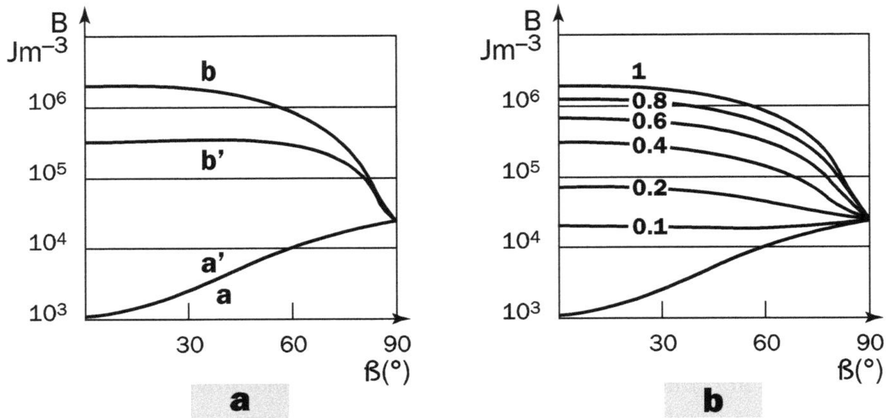

Fig. 6.2 (**a**) Results obtained for DKDP following those presented in Figs. 5.18 and 5.19: the energy density B of an infinite plane inclusion varies as a function of the angle β between its plane and the plane (001), when the ferro region is poly-domain (**a** and **a**′) and when it is mono-domain (**b** and **b**′); the values of λ_{1212} are 6 × 10^9 N m^{-2} for **a** and **b,** and 1 × 10^9 N m^{-2} for **a**′ and **b**′ (which takes into account the piezoelectric coupling and the soft mode at the transition: Fig. 1.27) (Kvitek and Bornarel, 2000). (**b**) The variation of the energy density B of an infinite plane inclusion as a function of the angle β of this plane with the plane (001) depends on the average deformation $<u_{xy}>$ of the ferro region defined by m whose values are indicated on the curves ($<u_{xy}> = m^* u_{xy}$); we can relate m^* to the parameter m introduced in relation (5.1): $m^* = 2m - 1 = (V_+ - V_-)/(V_+ + V_-)$

dielectric constant and drops to 1 × 10^9 N/m^2 (Fig. 1.27). It is with this value of the piezoelectric coupling that the curves **a’** and **b’** are calculated. No change is observed for a ferro poly-domain phase (curve **a’**), but a notable change appears for a mono-domain (curve **b’**), without however questioning the general shape of variation and the minimum at $\beta = 90^\circ$. Figure 6.2b presents the curves B (β) as a function of the parameter m^* varying from 0 to 1 ($m^* = 2$ m^{-1}, m defined in relation (5.1)).

If the other forms of energy were neglected in front of the elastic energy, the phase front would remain perpendicular to c ($B(\beta = 0°)$ being lower than $B(\beta = 90°)$). In the opposite case, the position of the front-most energetically interesting would be parallel to c (in β = 90°). The limit value between the two situations occurs respectively for $<u_{12}>$ = 0.1026 u_{12} without taking into account the piezoelectric coupling at the transition, and for 0.25 u_{12} considering it. Let us recall that these calculations correspond to assumptions of infinite crystal, for an infinitely thin inclusion, without taking into account the energies related to the shape of the front, nor to its singularities.

Thus, the elastic energy in the inclusion model varies with the average deformation of the ferro phase between 1.13 × 10^3 J/m^3 and 2.2 × 10^6 J/m^3 for β = 0°, while it retains the same value 2.45 × 10^4 J/m^3 in all situations for β = 90°. The value of B at β = 0° is therefore very sensitive (over three orders of magnitude) to the average deformation of the ferro phase, which reinforces the interest in studies of microscopic phenomena (phase fronts and domain textures).

6.2 Phase Coexistence with Good Thermal Homogeneity and Applied Conjugate Field

6.2.1 Phase Fronts, Domains, and Dielectric Measurements

For samples of very good thermal homogeneity (thermal gradient G less than a few thousandths of K/mm), the major parameters for phase coexistence are the value of the continuous electric field applied $E_{//}$ and the nature of the transition cycle (PF or FP) (Kvitek and Bornarel, 2000) as illustrated by the results of Figs. 6.3 and 6.4:

If $E_{//}$ is low (for DKDP, <150 V/cm), the domain texture of the ferro phase region can be dense, so the average shear deformation is close to zero; we are in the situation observed without an applied conjugate field, with quasi-plane phase fronts, close to the plane (001) (Fig. 5.3); their inclination is a detector of components of the thermal gradient **G** perpendicular to the *c*-axis. The temperature shifts at the start of coexistence (PF and FP), detected by dielectric measurements (Fig. 6.3b), allow to estimate the thermal hysteresis (about 0.05 K) which integrates the deviations previously highlighted between T_S and T. The loss constant remains the detector of the presence of domains.

For continuous fields applied slightly higher (from 200 to 400 V/cm) (Fig. 6.4a, b), the phenomena are comparable during a PF cycle to those of Fig. 6.3, but completely different during an FP cycle. The front nuclei created in the corners of the sample in a PF cycle are replaced in an FP cycle, from the beginning of coexistence, by a PFPFPF hetero-phase which appears at the hot face of the mono-domain ferro sample as in Fig. 6.4a (bottom) and c. The ferro regions of the coexistence zone are poly-domain with a dense domain texture. Figure 6.5 demonstrates the different nature of the phase front depending on its orientation: parallel to the *c*-axis due to the applied electric field (here in the plane (100)) or close to the plane (001) (orientation which, in the absence of an applied conjugate field, minimizes the mechanical energy due to the importance of the deformation u_{zz}). The dielectric measurements during phase coexistence (Fig. 6.4b) demonstrate the increase in thermal hysteresis with increasing $E_{//}$, contrary to the decrease predicted by the classical Landau-Devonshire theoretical model.

If $E_{//}$ becomes >400 V/cm in the case of the DKDP crystal, the hetero-phase phenomenon disappears, the phase front is a plane (100) imposed by the electrostatic energy, and for even larger values, the transition from para to mono-domain ferro occurs without phase coexistence. Using the phase diagram with isopols of Fig. 1.22, one would follow a path that turns around the point β beyond the critical field E_{cr}.

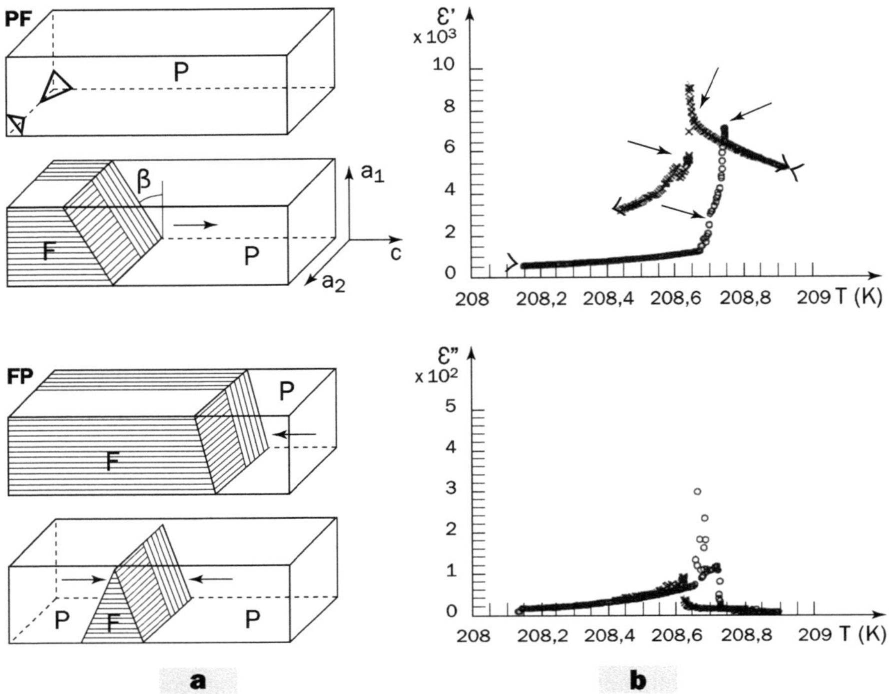

Fig. 6.3 Phase coexistence under very weak conjugate field E. In a DKDP crystal of homogeneous temperature (with a temperature difference between two points less than 5×10^{-3} K) subjected to an electric field applied along the c axis of 150 V cm^{-1}, the PF and FP transitions occur in a way comparable to those presented in Chap. 5, without an external conjugate field applied: the phase fronts created in the corners of the sample on a c face join to form a quasi-plane front close to the plane (001), a front that moves laterally; if there is a small component of thermal gradient perpendicular to the c axis, the front can be slightly inclined, depending on the orientation of this component, following a_1 or a_2 (a_1 in our case). The domain texture created in the ferro phase is usually quite regular, sometimes denser than in the situation without an external field concerning the PF cycle

The schematic representations of (**a**) and the photographs (section a_1 (on the right) and section a_2 (on the left)) during PF transitions in (**c**) and FP in (**d**) illustrate the case of a single front. However, two fronts can be created at the same time on both faces c, and then the process ends with their meeting in a dagger and its withdrawal as illustrated by the example at the bottom of Figure (**a**) and the photographs in Figure (**d**) for the FP transition

The dielectric measurements in Figure (**b**) differ from those at zero conjugate field: the beginning of coexistence in PF and FP occurred at the same temperature (within 2×10^{-3}K), this is no longer the case here. The temperature T is measured on the lower face of the sample. As previously indicated, during coexistence, the temperature T is also a temporal indication as can be shown by changing the rate of temperature change (less than a few 10^{-2} K/h). The arrows indicate on the curves the beginning and end of coexistence; the thermal hysteresis is estimated by the temperature difference between the beginnings of each coexistence (PF and FP); it is close to 0.05 K

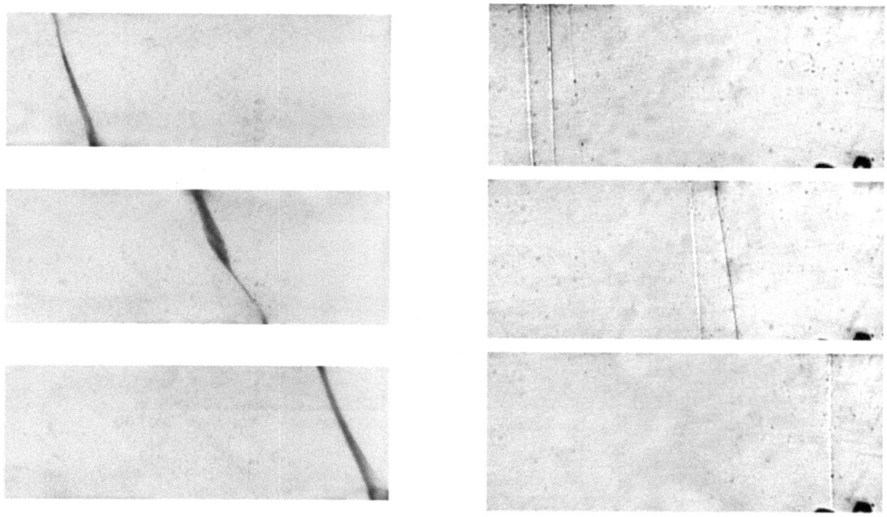

c

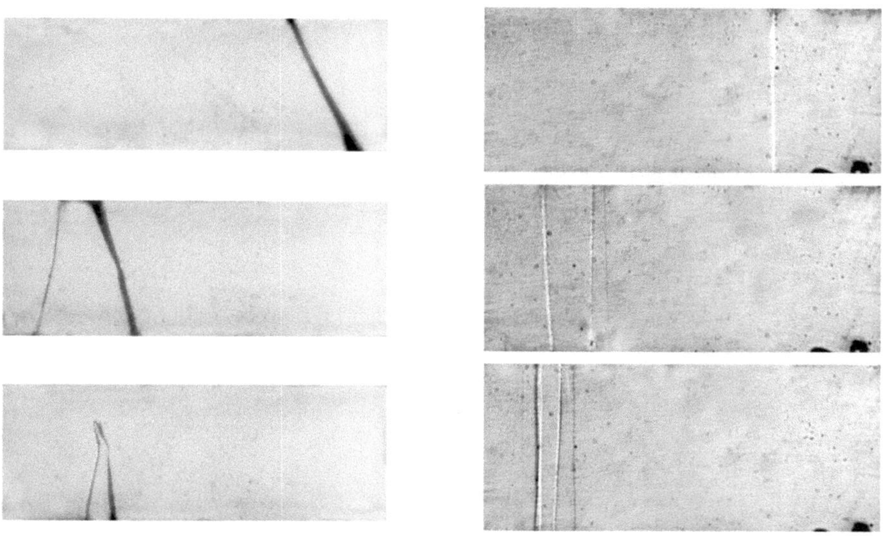

d

Fig. 6.3 (continued)

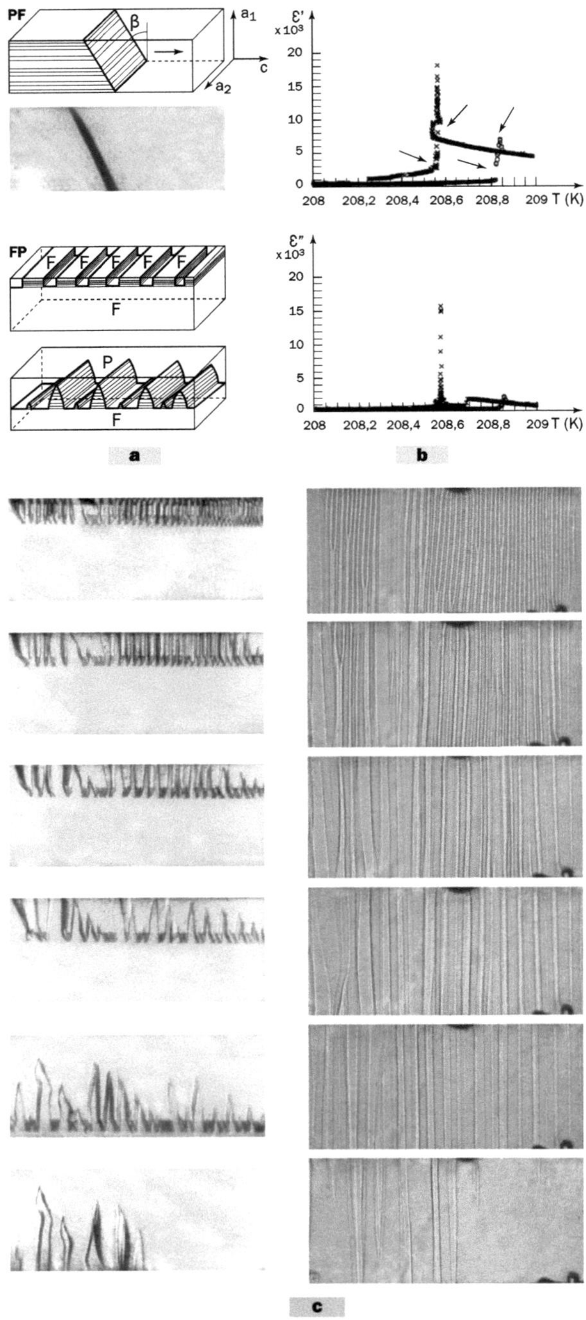
PF
β
a1
c
a2
FP
F
P
ε'
x10³
ε"
x10³
20
15
10
5
0
208
208,2
208,4
208,6
208,8
209
T (K)
a
b
c

Fig. 6.4 Phase coexistence under weak external field (between 150 and 400 V cm^{-1})
·With a stronger electric field than in Fig. 6.3, for the DKDP between 200 V cm^{-1} and 400 V cm^{-1}, the PF transition (Fig. 6.4a top) still occurs with a quasi-planar front almost in the (001) plane unless there are components of G following a_1 or a_2 (the texture in domains remains fine without being very dense). On the other hand, the FP transition is completely different from that with $E = 0$. After remaining at low temperatures for several hours, the sample is practically mono-domain when the FP coexistence begins. From the beginning of the coexistence, in a few seconds, a PFPFPF hetero-phase region appears, bounded by a (100) plane with respect to the ferro phase, with the P and F lamellae of the hetero-phase separated by phase fronts in (001) planes (Fig. 6.4a FP and 6.4c with sections (100) on the right and sections (010) on the left). After this dynamic phenomenon, the coexistence in PFPFPF lamellae can remain stable for several hours, then the hetero-phase lamellae slowly transform into increasingly spaced daggers. Unlike the various situations in Chap. 5, the ferro regions of this coexistence zone are always poly-domains with dense domain textures ($m^* = 0$) as confirmed by the diffraction of a laser beam that passes through the sample along the c axis. The situation at the beginning of coexistence corresponds to well-defined front orientations (internal (001) fronts to the hetero-phase and a (100) boundary between hetero-phase and mono-domain ferro phase ($m^* = 1$ or -1)). This hetero-phase has been imagined in some theoretical publications (Fedosov and Sidorkin, 1979) and observed in fact in synchrotron topographies (Tanner et al., 1983). Then, the shape of the fronts becomes sensitive to the mechanical and thermal imperfections of the sample and the daggers become rare. This ability of the daggers argues for a relatively low energy density per unit of surface of these phase front parts. The dielectric measurements during the phase coexistence (Fig. 6.4b) demonstrate an increase in thermal hysteresis with increasing $E_{//}$, contrary to the decrease predicted by the classical Landau-Devonshire theoretical model. The small sudden step of $\varepsilon'_c(T)$ during the PF cycle observable on Fig. 6.4b proves the very good thermal homogeneity in the sample: the phase front is a heat exchanger and a small variation in its movement is enough to change the local temperature, hence the response in dielectric measurements. This interplay between the moving isotherm and the phase front leads to cascade movements, well studied, for example, in $PbTiO_3$ (Fesenko et al., 1975; Yufatova et al., 1980; Dec, 1988) and in line with the theory of Roitburd (1972). It is also interesting to highlight the fairly close values (in a ratio of 2) of $\varepsilon'_c(T)$ for the two values of $E_{//}$ (between the Fig. 6.3: 150 V cm^{-1} and this Figure at 400 V cm^{-1}), unlike the ratios >5 and sometimes up to 50 between the values of $\varepsilon''_c(T)$ for different values of the applied electric field: the loss constant easily detects the hetero-phase and reflects the characteristics of its dense domain texture within the F lamellae

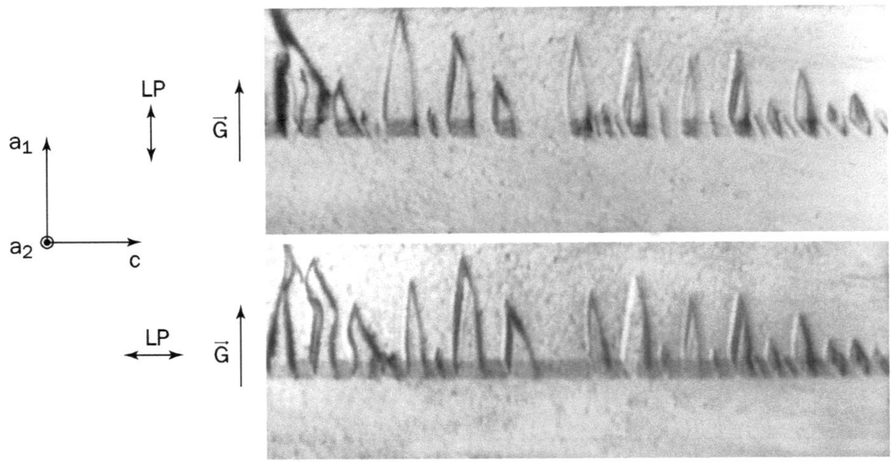

Fig. 6.5 In the context of Fig. 6.4c, the change in polarization of the transmitted light highlights during an FP cycle the different nature of the parts of the phase front, located in a (100) plane and in a (001) plane. Figure 5.23 had illustrated this difference in nature with a model of wedge dislocations

6.2.2 *How to Explain the Shapes of Phase Fronts and the Hetero-phase?*

The results observed with low applied electric fields (Fig. 6.3) are of two types.

First, in the phase coexistence zone, the ferro part, during a PF cycle without an applied electric field, was the site of frequent modifications, instabilities of domains of often different orientations. With a low applied electric field (Fig. 6.3), fairly dense textures are observed, more stable without being comparable to those obtained after several hours at low temperature. The applied electric field seems to bring an anisotropy to the organization of the domains. However, the phase front remains generally close to a plane (001), which shows for these field values, the prominence of the elastic energy over the electrostatic one. This situation is observed during PF cycles (Fig. 6.4) as long as the ferro phase is poly-domain.

The situation changes radically during an FP cycle, after a stay in the ferro phase: the sample is almost mono-domain before the start of phase coexistence (FP cycle of Fig. 6.4a). The electrostatic energy is more important, and the hetero-phase zone (of phase coexistence) can be considered as a special "phase" with its own macro-deformation, as an inclusion in the ferro mono-domain phase in the form of a lamella with an angle $\beta = 90°$ (Kvitek and Bornarel, 2000). As the ferro regions of this hetero-phase are poly-domains, their average shear deformation can be considered as zero. In this model, the macro-deformation values $<u_{kl}>$ of the hetero-phase are calculated in relation to the deformations of the ferro phase in which it appears, introducing the ratio ω between the volume para in the hetero-phase and its total volume. The boundary between these two "phases" (ferro phase and hetero-phase) is a plane parallel to c (minimum obtained for $\beta = 90°$), a situation close to that indicated **b** (or **b'**) in Fig. 6.2a. It is demonstrated (Kvitek and Bornarel, 2000) that the minimum of B for $\beta = 90°$ becomes

$$B^*(90°) = B(90°)\omega^2 \tag{6.1}$$

with at the beginning of the hetero-phase phenomenon, a value of ω of about 1/3 experimentally observed: $B^*(90°)$ is estimated at 2.7×10^3 J/m^3. It is understood that it is easy to pass from the value of 1.13×10^3 J/m^3 of the front perpendicular to c to this situation of a "front" between the phase ferro mono-domain and the hetero-phase parallel to the *c* axis. Arguments support the role of the hetero-phase: the phase front is a heat exchanger optimized by the "finned exchanger" PFPFPF; moreover, the fact of reducing the front surface parallel to c with $B = 2.45 \times 10^4$ J/m^3 by small surfaces with $B^* = 2.7 \times 10^3$ J/m^3 and others with $B = 1.13 \times 10^3$ J/m^{3}can energetically promote this dynamic transition if the geometry is interesting. If the number of lamellae becomes too large and therefore the total surface of the fronts perpendicular to c with $B = 1.13 \times 10^3$ J/m^3 too important, the phenomenon cannot exist.

The model could be improved, for example, by introducing the variation under the effect of the electric field of the deformation in the para parts of the coexistence zone; this does not change the order of magnitude of the results, the value of B^* is only modified by a few per cent.

This model of inclusion of a hetero-phase confirms, as was already the case in situations with $E_{//} = 0$, the major role of elastic energy regarding the shape of the phase front, and what happens at the microscopic level.

6.2.3 *Thermal Hysteresis of the Transition*

The hetero-phase phenomenon also explains the very different results published regarding thermal hysteresis as a function of a continuous field applied. Figure 6.6 summarizes these results. How to explain the growth of thermal hysteresis with the field $E_{//}$ between values close to zero and a few hundred V cm^{-1}, contrary to the predictions of a classic model?

The law of variation of $B(\beta)$ from Fig. 6.1b is again considered: the phase front with the poly-domain ferro phase should be perpendicular to c with $\beta = 0°$ ($B(0) = 1.13 \times 10^3$ J m^3) (Fig. 6.3b); but, if the ferro phase is mono-domain, it cannot maintain this orientation (the value of $B(0)$ would be 2.2×10^6 J/m^3) and the situation with $\beta = 90°$ is energetically more interesting ($B(90) = 2.45 \times 10^4$ J/m^3). The hetero-phase phenomenon facilitates the transition of the front between $\beta = 0°$ and

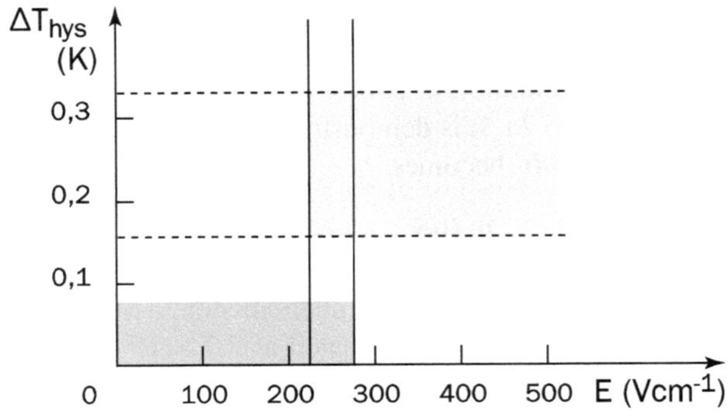

Fig. 6.6 This diagram summarizes the results concerning the thermal hysteresis of the DKDP transition as a function of the applied electric field: two discontinuous horizontal lines between 0.15 and 0.3 K locate the theoretical results using Landau's development relation (1.20) taking into account uncertainties on the parameters used. These results with $\Delta T_{hys} = b^2/3Ac$ are superior to the experimental results. The thin vertical lines delimit the two situations demonstrated in this work:

For applied electric field values less than 250 ± 50 V cm^{-1}, the phase coexistence zone plays a major role between the para and ferro phases with sometimes a complex organization between the shape of the front and domain textures, which optimizes the local and global deformations. The grey rectangle between $E = 0$ and $E = 250$ V cm^{-1} contains high-quality thermal experimental results (measurement uncertainties (10^{-4} to 10^{-3} K)) with ΔT_{hys} values close to zero) (Reese, 1968, 1969a, b, 1973; Zeyen et al., 1976) and the results of some 10^{-2} K arising from interactions at the microscopic level between phase fronts and domains (Kvitek and Bornarel, 2000). The hetero-phase phenomenon explains the variations at low values of the applied field with growth of the thermal hysteresis as a function of E relation (6.3) contrary to the decrease of this hysteresis ΔT^* with E of the classic model (Eberhard and Horn, 1975)

For higher values of the electric field, the phase front is in another situation, confronted with a nearly mono-domain ferro phase region, with the limit of the coexistence zone towards the low temperatures essentially in a plane parallel to c. The grey rectangle beyond 250 V cm^{-1} encompasses direct measurements notably observable in the work with large mono-domain regions in the ferro phase of the sample (Bornarel and Cach, 1999), estimates using Landau's development and macroscopic measurements depending on the experimental conditions (Strukov et al., 1972b; Gladkii and Sidnenko, 1972)

$\beta = 90°$ through an easy thermal exchange between the phases and a phase front with optimized global energy. During this phenomenon, the energy of creation of the hetero-phase ΔF will cause a temperature jump ΔT which can explain the observed hysteresis (Kvitek and Bornarel, 2000):

$$\Delta F = \delta B\omega^2 / 2 + S\Delta T\left(1-\delta\,\omega\right) - \left(1-\delta\right)PE \tag{6.2}$$

with δ proportion in volume of the hetero-phase in the sample and ω proportion of para phase in the hetero-phase. The coefficient 1/2 for $B\omega^2/2$ corresponds to the elastic relaxation (Khachaturyan, 1983). To find the optimal configuration of the hetero-phase, the minimization of ΔF in relation to ω and δ leads to:

$$\omega^* = \left(2PE\right)^{1/2} B^{-1/2} \text{ and } \Delta T^* = \left(2BPE\right)^{1/2} S^{-1} \tag{6.3}$$

If the value of the polarization P is known with good accuracy (4.2×10^{-2} C/m^2) whereas that of S already mentioned in 6.1 varies according to the authors between 1.7×10^4 and 6.8×10^4 J K^{-1} m^{-3}. With the applied external field $E_{//}$ from Fig. 6.4, i.e., 400 V/cm, we obtain for ω^* the value of 0.37 in good agreement with the experimental observation and for ΔT^*, depending on the values of S retained, between 0.13 K and 0.53 K, consistent with the 0.3 K obtained by the dielectric measurements (Fig. 6.4b).

Experimental results obtained during the transitions para-ferro PF and ferro-para FP with good thermal homogeneity of the crystal and with a weak applied electric field, allow to conclude:

- The shapes of the phase fronts, the textures in observed domains, can be explained by a minimization of energy, an important role of the overall mechanical energy, with applied electric fields that change the context of average deformations of the ferro regions at the beginning of phase coexistence (especially during a cycle FP).
- Significant variations in dielectric constant observed at the beginning and end of phase coexistence account for the appearance and disappearance of the phase front. The loss constant is a permanent detector of the domains.
- The thermal hysteresis is much lower than the value calculated using the usual thermodynamic model. Microscopic phenomena allow for easier changes than a macroscopic thermodynamic theory predicted.

6.3 Phase Front and Domains in a Crystal with Thermal Gradient G_e and Applied Conjugate Field $E_{//}$

Section 6.2 showed how the application of a conjugate field of low value could, in a sample of good thermal homogeneity depending on the context, PF or FP, contribute to microscopic phenomena (phase fronts and domains) that facilitate the phase change. What happens if the crystal is subjected simultaneously to a significant thermal gradient and a conjugate field with the values used during the usual macroscopic experiments? Without aiming for exhaustiveness, some answers are provided.

Figures 6.7, 6.8, 6.9 and 6.10 illustrate correlations between the shape of the phase front, domain texture, and dielectric properties, for a sample subjected to a constant $E_{//}$ field (and a constant thermal gradient perpendicular to the ferroelectric axis). The abscissa of the curves is the temperature measured on one face of the sample. This abscissa could be time, during a hysteresis cycle performed at very low frequency, and the results can be recontextualized in this context. From the situations studied, we will retain the following.

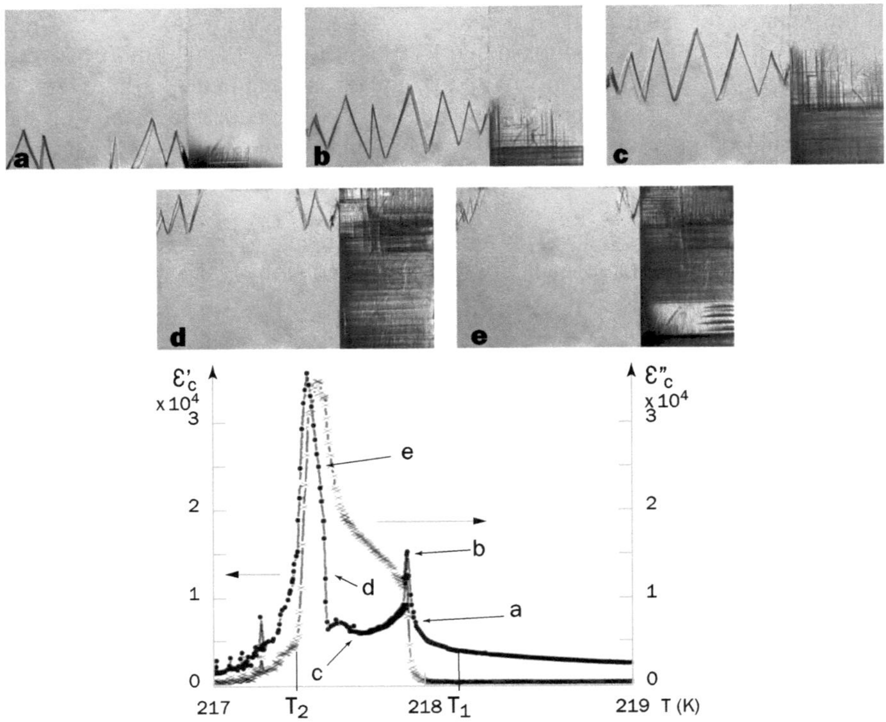

Fig. 6.7 Photographs of phase fronts and domains: those on the left in section a_2 and those on the right in section c during a PF cycle (DKDP sample with $a_1 = 6.66$ mm, $a_2 = 4.41$ mm, $c = 7.11$ mm under an applied electric field $E_{//}$ of 72 V cm^{-1} and a thermal gradient G along the a_1 axis of 6×10^{-2} K mm^{-1}). The situations are identified on the dielectric curves: appearance of the phase front in factory roof shape on the coldest a_1 face of the sample (**a**), coexistence region entirely inside the sample (**b**) and (**c**), beginning of the exit of the coexistence region from the sample on the other a_1 face (**d**), and completed exit (**e**). The temperature interval of coexistence is defined by the temperatures T_1 and T_2 of the glass on which the sample is placed. This PF transition has many points in common with that observed without applied field $E_{//}$ in Fig. 5.11: the ferro regions of the coexistence zone show some unstable domains, and the ferro phase of the sample contains a finely layered domain texture. The ferro part of the coexistence zone is near the mono-domain towards the factory roof edges, close to the para phase. When the coexistence zone exits the sample, small mono-domain regions appear in the ferro region of the sample, due to the action of the electric field

Dielectric constant peaks accompany the entry and exit of the coexistence zone from the sample; the exit peak towards the ferro phase is higher than the entry peak of this PF transition. When the coexistence zone crosses the entire sample, the dielectric constant is sensitive to each incident; the loss constant follows the evolution of the volume occupied by the domains which grows as the photographs following c illustrate, then falls during the withdrawal of the coexistence zone which shows the impact of the unstable domains of the coexistence zone. The maxima of $\varepsilon'_c(T)$ and $\varepsilon''_c(T)$ do not occur at the same measured temperature, therefore not at the same time. When the entire sample is entirely in the ferro phase, rearrangements of domain texture are observed in the few degrees below the transition with abrupt variations of $\varepsilon'_c(T)$ and $\varepsilon''_c(T)$

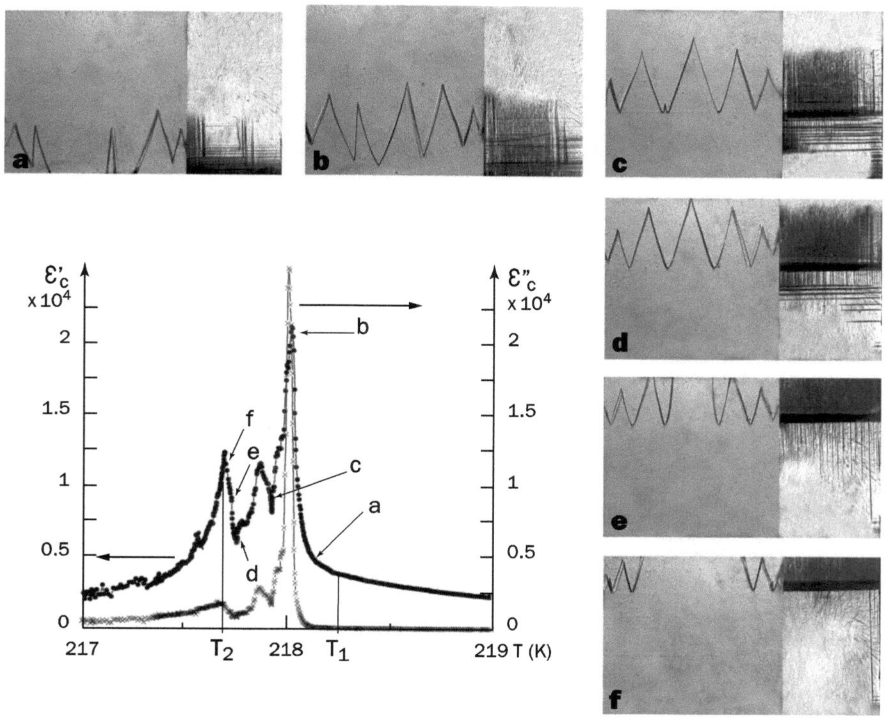

Fig. 6.8 Photographs of phase fronts and domains, from left in section a_2 and from right in section c, characteristic of a PF cycle for a DKDP sample under electric field $E_{//}$ of 150 V cm^{-1} and thermal gradient G following the a_1 axis with a value of 7 × 10^{-2} K mm^{-1}. The situations are identified on the dielectric curves: appearance of the factory roof phase front on one a_1 face of the sample (**a**), coexistence zone entirely within the sample (**b**) and (**c**), start of the coexistence zone exiting from the sample (**d**), (**e**), and exit with a dense domain texture in the ferro regions of the coexistence zone (**f**). The factory roof phase front is still located between two curved envelopes, but the envelope close to the ferro phase approaches a (100) plane as the applied continuous field grows. The ferro phase region of the sample includes only a few domains and becomes mono-domain towards the end of the cycle (**f**). This situation is somewhat the reverse of that of the PF transition of Fig. 6.7 performed with a weaker electric field: a dense domain texture in the ferro phase corresponded to only a few domains in the ferro regions of the coexistence zone, and here, with a higher field and a ferro phase close to the mono-domain, the ferro regions of the coexistence zone show dense domain textures

The variations in dielectric constant and loss constant change compared to those observed in Fig. 6.7: $\varepsilon'_c(T)$ this time present a main maximum at the beginning of the PF cycle rather than at the end. As for $\varepsilon''_c(T)$, a single maximum is observed at the beginning of the cycle, and lower values than in Fig. 6.7 in accordance with the smaller volume of domains throughout the sample and much more stable domains in the ferro daggers of the coexistence zone

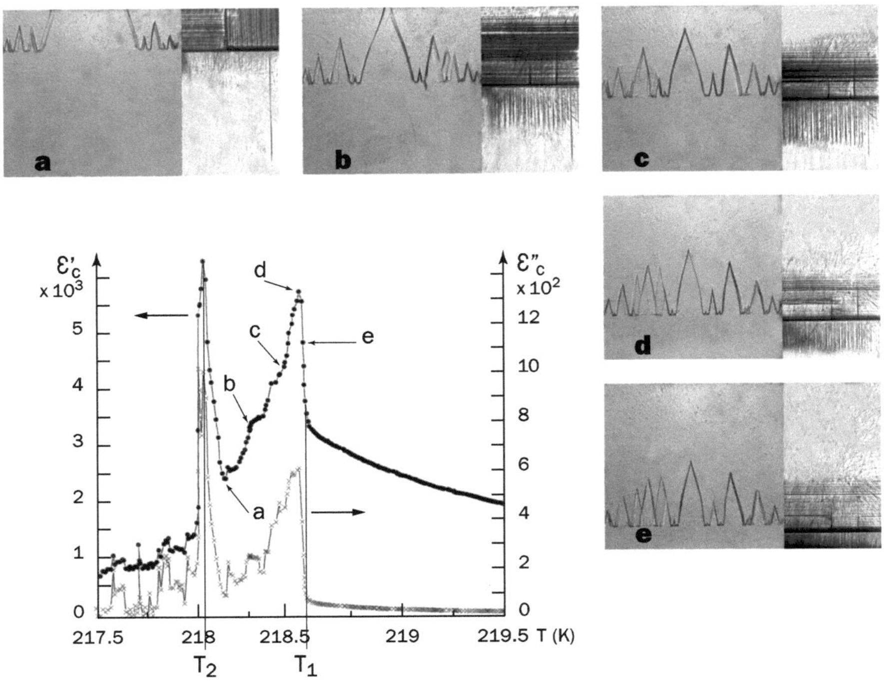

Fig. 6.9 FP cycle of a DKDP sample under a field $E_{//}$ of 72 V cm^{-1} and a thermal gradient G following the axis a_1 with a value of 7×10^{-2} K mm^{-1}: The photographs, those on the left in section a_2 and those on the right in section c, illustrate the characteristic situations identified by arrows on the dielectric curves: appearance of the factory roof phase front on one a_1 face of the sample with relatively dense domain textures within the ferro regions of the coexistence zone (**a**), coexistence region entirely within the sample (**b**), (**c**), start of the coexistence region exiting from the sample (**d**), and exit which occurs with a very lightened domain texture in the ferro daggers of the coexistence zone as in the ferro region close to the coexistence limit (**d**), (**e**). The phase front includes flat parts parallel to c and daggers of various sizes limited by a curved envelope

The situation in terms of domains has completely changed compared to that of the PF cycle at the same value of $E_{//}$: the sample is almost mono-domain. The dielectric constant includes two peaks, at the entrance and exit of the coexistence zone of the sample; the loss constant still illustrates its strong correlation with the density of domains by its rise towards T_1: the domain texture of the ferro region near the front becomes denser. The peaks of dielectric constant and loss constant observed during the entry and exit of the phase coexistence zones in the sample are respectively 10 times and 100 times smaller than they were during the PF cycle with the same applied field value $E_{//}$

6.3.1 Phase Fronts and Domains

- The chemical energy defines the region of the sample in which the coexistence zone can exist. It depends on the temperature distribution outside the sample, the shape of it, and the temperature history. The thermal gradient, for example along an a_1 axis, induces a coexistence zone perpendicular to that of the incoherent interface plane that the phase front of the sample would take without applied

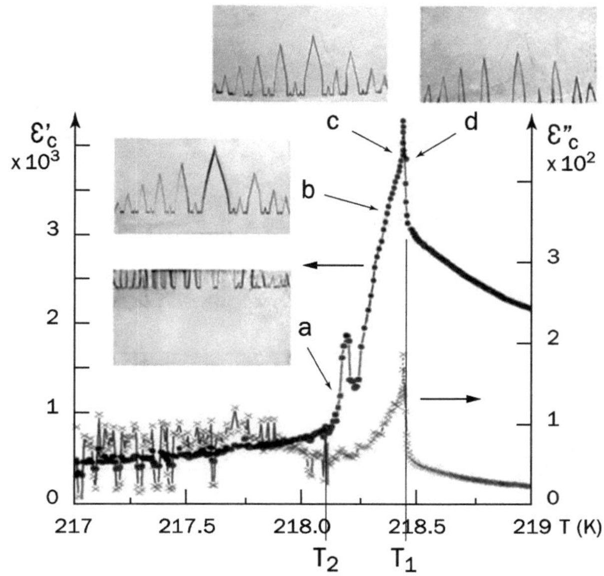

Fig. 6.10 The same sample as that of the three previous figures, remained in ferro phase under applied field, is initially almost mono-domain during this FP cycle (under a field $E_{//}$ of 150 V cm^{-1} and a thermal gradient G following the axis a_1 with a value of 6.5 × 10^{-2} K mm^{-1}). Photographs of the phase front in section a_2 indicate the situations by arrows on the dielectric curves: appearance of the truncated factory roof phase front, thus hetero-phase zone as in Fig. 6.4a (**a**), coexistence region in the middle of the sample (**b**), coexistence zone reaching the edge (**c**), and exit from the coexistence zone of the sample (**d**)

With these values of $E_{//}$ during the FP cycle, the textures observed in the coexistence zone, and in the ferro phase region near the coexistence limit are simplified, sparse. The orientation of the domains in the coexistence zone is often perpendicular to that of the domains that still exist in the ferro phase, and the densities of domain textures in the ferro parts of the coexistence zone and the ferro phase vary inversely, as can be observed in Figs. 6.8, 6.9 and 6.11

The $\varepsilon'_c(T)$ and $\varepsilon''_c(T)$ curves, simpler than those of the previous figures, account for this transition to simplified textures: low value for $\varepsilon''_c(T)$ due to the low density of domains in the sample; detection with $\varepsilon'_c(T)$ in (**a**) of the detachment of the coexistence zone of PFPFPF hetero-phase from the edge a_1 of the sample which becomes para. At the other end of the coexistence in (**d**), peaks of $\varepsilon'_c(T)$ and $\varepsilon''_c(T)$ accompany the departure of the ferro phase; the curves increasingly resemble the theoretical variations of the mono-domain of Fig. 1.16c

conjugate field and without thermal gradient; this coexistence zone, with a factory roof phase front, occupies a volume limited by two curved envelopes, of maximum size in the centre and minimum at the sample edges where the constraints cancel out (Fig. 5.17). The envelope close to the para phase is always more curved than the one close to the ferro phase. This latter approaches a plane parallel to the applied electric field $E_{//}$ as soon as this field exceeds 40 V cm^{-1} during FP transitions and 100 V cm^{-1}during PF cycles.

- The shape of the phase front of this coexistence zone is the result of energy minimization, with obviously, the electric energy favouring a front parallel to c, and the mechanical energy a front perpendicular to this axis as the $B(\beta)$ law of

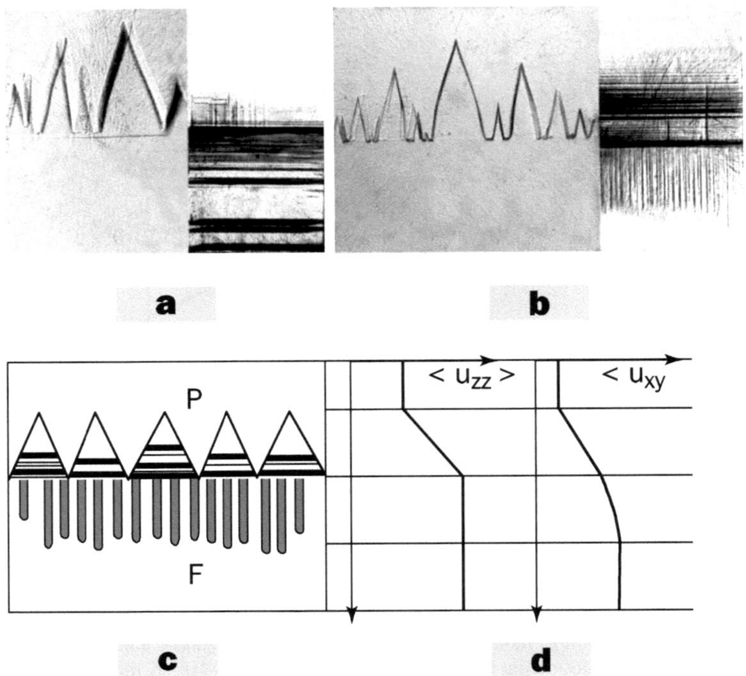

Fig. 6.11 The photographs (section a_2 on the left and section c on the right) correspond to a PF transition (with $E = 175$ V cm^{-1}) for (**a**) and to an FP transition with 72 V cm^{-1} for (**b**) of a sample subjected to a thermal gradient G perpendicular to c which defines a coexistence zone in the shape of a factory roof with its daggers. Regardless of the value of the electric field applied in this context, the ferro part of the coexistence zone is mono-domain near the edges towards P and includes domains in the part close to F. In the ferro phase region near the coexistence zone, domains occupy a volume comparable to that of the coexistence zone between it and the ferro mono-domain region. Very often, the orientation of the domains in this region of the ferro phase is perpendicular to that of the domains in the coexistence zone nearby. (**c**) Schematizes a hypothetical case of domain organization in the ferro coexistence phase region and in the nearby region of the ferro phase. The curves in (**d**) are directly linked to this hypothetical factory roof and propose variations of $<u_{zz}>$ and $<u_{xy}>$ in the sectors delimited by thin horizontal lines. The variations of $<u_{zz}>$ are close to what actually happens, allowing a transition without too much discontinuity, except at the edges of the factory roofs, from P to F. A comparable graph could be drawn for $<u_{xx}>$. The graph concerning $<u_{xy}>$ also indicates that an optimization of local and global deformations exists between the phase front and the region near the ferro phase. As much as the graphs in $<u_{xx}>$ and $<u_{zz}>$ are solid, as much that one in $<u_{xy}>$ varies with the context and the local values of the m parameter introduced in the text. For example, (**a**) includes in phase F a poly-domain zone (very low m), a mono-domain zone ($m^* = 1$ or -1), then again a poly-domain zone. The qualitative representations should not make us forget the orders of magnitude of the contributions to the deformation energy (λ_{1111} $u_{xx}{}^2 = 7.2 \times 10^2$ N m^{-2}, λ_{3333} $u_{zz}{}^2 = 2.3 \times 10^4$ N m^{-2}, λ_{1212} $u_{xy}{}^2 = 3.1 \times 10^5$ N m^{-2}) and those of the chemical and electric energies recalled in Sect. 6.1. More advanced models will need to take into account the role of domain walls with non-zero thicknesses in this transition context as almost para regions, therefore with lower values of $<u_{zz}>$ and $<u_{xx}>$ and values of $<u_{xy}>$ functions of the electric field $E_{//}$

Fig. 6.1b illustrates. The parameter m^* used in this figure, defined by the relation $<u_{12}> = m^* u_{12}$, informs about the average polarization and the average shear deformation of the concerned ferro region: $m^* = 0$ for a fine texture with equal volume for the two polarizations, and $m^* = 1$ or -1 for a mono-domain region.

- For all values of m^*, the average values $<u_{zz}>$ and $<u_{xx}>$ in the ferro regions are the same if we neglect, whatever the sign of u_{12}, the contribution of domain walls, implicitly assumed to be of zero width. The results previously presented validate the impact of domain density, of the average strain-polarization on the β orientation of the front (Fig. 5.5), including within the same "roof" of a factory with changes in slope of the front related to changes in domain density (Fig. 6.9).
- In all situations, when two phase fronts are close, for example at the top of the daggers on the para phase side, the ferro parts of the coexistence zone are mono-domains.
- At the boundaries of the coexistence zone, towards the para region and towards the ferro region, the situations differ. The first is characterized by the singularity lines of the factory roof edges very visible in Fig. 5.21 and in Figs. 5.36 and 5.37 with real layers, black or very bright depending on the orientation of the polarized light. The disappearance of these edges in the faces of the sample (Fig. 5.22) reduces by 30–50% the β orientation angle of the phase front, which allows the calculation of their energy contribution.

The boundary between the phase coexistence zone and the ferro phase zone varies significantly by application of a weak electric field. Part of the phase front is aligned in a plane parallel to c and the other parts depend on the density of the domains in the coexistence zone. The average strain-polarization at the local level allows a regular transition from the coexistence region to the ferro phase region as illustrated in Fig. 6.11 by the variations of $<u_{zz}>$ (and of $<u_{xx}>$) on the one hand, and of $<u_{xy}>$ on the other hand, on a larger scale than the models of Landau's theory predict (Rottman 1988). Although the domain walls in the phase coexistence region are most often perpendicular to those in the ferro phase region do not alter this continuity. This "adaptation" is often encountered in martensitic transformations (Khachaturyan et al., 1991) with their anelastic consequences (Khachin and Solovev, 1975). Figure 6.12 shows a correlation between the orientation of the domains within the coexistence zone and the context of creation of the nuclei of this zone on the a_1 face of the sample. In all situations, the domains, by their orientation and their density, allow to optimize the transition from the para region to the ferro region, taking into account both the electrical aspects and the deformations u_{xy}, u_{xx}, and u_{zz}. The thickness of the adaptation region in ferro phase is approximately comparable to that of the coexistence zone. The result differs depending on the value of $E_{//}$ applied and the type of transition (the E field contributes to simplify the domain texture of the ferro phase up to mono-domain more markedly in an FP cycle than in a PF cycle). The results are obvious with this key to understanding: a coexistence zone has ferro parts without many domains when the ferro phase is in dense domain texture (Fig. 6.7),

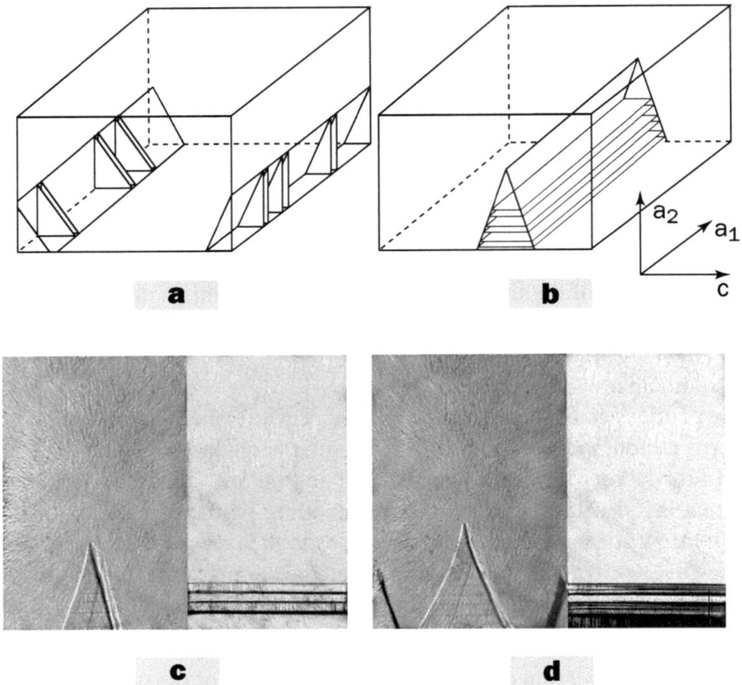

Fig. 6.12 The orientation of the domains in the ferro coexistence zone depends on the context of its creation, therefore on the interaction at the beginning of coexistence, of the phase front with the sample edge at zero constraint. In the case of an FP transition, the effect of the applied field even weak, even without the appearance of notable mono-domain zones in the ferro phase, creates an anisotropy that promotes a front aligned in an a_2 plane with the domain walls almost always, initially, parallel to this plane. In the case of a PF transition, the position of the nuclei of the ferro phase usually favours one orientation of the domain walls rather than the other: nuclei following the sample edges as in (**a**) favour a_1 walls, and nuclei on a face a_2 of the sample favour domain walls which are parallel to it as in (**b**). These orientations can then change during the movement of the front. The photographs (**c**) and (**d**) illustrate the process with, for (**d**), the accumulation of the two situations

ferro parts of dense texture when the ferro phase is almost mono-domain (Fig. 6.8), and finally, a simplified domain texture, both in the ferro parts of the coexistence zone and in the ferro phase region, over comparable thicknesses in intermediate contexts (Fig. 6.9). It is enough to consider the phase coexistence zone as an intermediate region, linking between the para phase and the ferro phase as it is (poly-domain or mono-domain) to predict qualitatively the ferro texture near the coexistence front. In addition to this optimization of global and local energies between front shape and domain textures, the shape of the fronts in an FP context (Figs. 6.9 and 6.10) is a beautiful illustration of the balance local game between front parts with different B values. It is thus possible to use these sets to develop a study of fractals.

6.3.2 Dielectric Results

The macroscopic dielectric responses accumulate the contributions of para phase regions, ferro phase, and coexistence zone with interactions between these regions. The impact of the phase coexistence zone must therefore be evaluated by taking into account these various contributions. The contribution of the coexistence zone to the dielectric constant in Fig. 5.30 for samples without applied field and in various thermal situations illustrated the role of temperature, its distribution at all points of the sample, and its history. The loss constant (or the dissipation factor) remains an optimum detector of domain density.

The application of an electric field E will have several consequences on the dielectric and loss constants: changes in the shape of the phase front and change in the volume of the coexistence zone, domain density in the ferro phase zone and domain density of ferro regions of the coexistence zone. Figure 6.6 allows to globally dissociate two situations, the one with E values sufficient for the temperature variations of the inverse susceptibility and losses to recall those of the theoretical crystal (of infinite dimensions) of Figs. 1.15, 1.20, 1.21, and 1.22, and the one for which the microscopic phenomena (at low E values) of the coexistence zone practically mask the previous ones. The results of Fig. 6.13 correspond to an intermediate situation: the impact of factory roofs and domains of the coexistence zone is evident in the grey areas of Fig. 6.13d, while between these two zones, we recognize the variation of the volumes of zones para and ferro without obvious impact of the phase front specificity.

With conjugate field E values low enough for the interplay between phase front shape and domain textures to decline, the variation of dielectric constant with temperature is far removed from mono-domain theoretical curves. The dielectric constant ε_c' is a detector of instabilities, notably the appearances and disappearances of the phase front in the sample with a value that reflects the context, thus the phase fronts and domain textures in the ferro phase region, and in the coexistence zone. The loss constant is the optimal detector of domains, almost proportional in all situations to their importance (occupied volume and/or wall surface) over two orders of magnitude, as previously illustrated (Fig. 5.7d). The result is obvious when the transitions are crossed slowly enough (in a few hours or tens of hours in the cases presented). This is a well-known result during martensitic transitions quantified by simultaneous acoustic emission and optical observation measurements: domain movements with their dislocations dissipate a very large amount of energy compared to that of a lateral phase front displacement (Bonnot et al., 2008; Vives et al., 1994; Roth et al., 2007). The markers (*a*, *b*, *c*, *d*, etc.) on the curves and the photographs of Figs. 6.7, 6.8, and 6.9 allow in each situation and especially during the peaks of ε_c' and ε_c'' to compare the impact of domain density in the ferro regions of the coexistence zone to that of the domains of the ferro phase region, and also that of texture rearrangements created by the interaction between the coexistence zone and the sample edges…

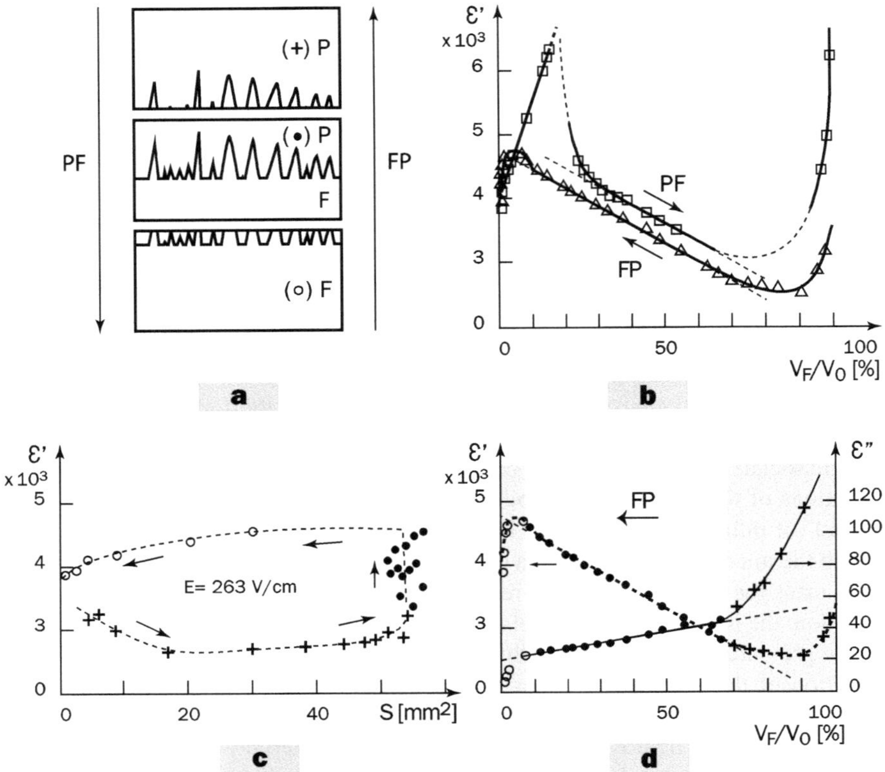

Fig. 6.13 (**a**) In a DKDP sample, during a FP cycle with a thermal gradient **G** perpendicular to c and an applied electric field $E_{//} = 263$ V cm^{-1}, the phase front schematized by its trace in section a_2 appears following the direction of temperature variation in the experiment, either on face a_1 at the bottom the coldest (+) during a PF cycle, or on the face a_1 at the top the hottest during a FP cycle (o). Then, the progress of the front and its disappearance are predictable. The diagram is only indicative because during the FP cycle, the front parallel to c is important as illustrated by Figs. 6.9 and 6.10. (**b**) Variation of the dielectric constant ε'_c as a function of the volumetric proportion of the sample occupied by the ferro phase (V_F/V_0) during PF and FP cycles. The difference in phenomena between these two cycles is clear at the beginning and at the end of the phase coexistence cycles: effect of the entry and exit of the phase front. In both cases, this effect is separable from that of the simple translation of the front in the sample. (**c**) The graph of ε'_c as a function of the total surface of the phase front S uses the marks given in (**a**) to locate the position of the coexistence zone in the sample. The arrows recall that it is the FP cycle of (**b**) whose beginning with the crosses identified in (**a**) starts at $S = 0$. (**d**) The variations of ε'_c and ε''_c as a function of (V_F/V_0) (we recognize the curve of ε'_c already presented in **b**)) and of the total surface of the phase front S (graph **c**)) confirm the direct link between the loss constant and the domain textures, with the dense texture in domains of the hetero-phase at the beginning of the FP cycle (greyed area of right). When the domains are rare under the action of an electric field or a constraint, the variation of the dielectric constant during phase coexistence mainly accounts for the proportion of para phase compared to the ferro phase in the sample and therefore approaches the theoretical "mono-domain" models (white region)

6.4 Macroscopic Results and Microscopic Phenomena

The results published on phase transitions of ferroelectric and ferroelastic crystals are mainly derived from macroscopic measurements with samples of a few mm^3 to a few hundred mm^3. This is the case for polarization and deformation measurements which reflect an average property over all or part of the sample. The same is true of diffraction, scattering measurements: they too reflect an average of integrated responses over the section of a radiation beam, etc. In the case of a first-order transition, even more so than in other situations, the existence of a microscopic phenomenon is systematic and often responsible for properties very different from those predicted by macroscopic models. The ideal is obviously to carry out macroscopic and microscopic studies at the same time, but if this is not possible, how can information about the microscopic phenomena that impact the measurements be deduced from macroscopic measurements?

The most symbolic phenomenon of ferro crystals, the "hysteresis cycle," is performed at frequencies defined by the experimenter. It is thus possible to carry out cycles slow enough for a full declination of microscopic processes, in proximity to the conditions of thermodynamic models. The characteristics of the hysteresis cycle (coercive field, cycle shape) and the microscopic phenomena (phase front, domains) can be correlated during the transition from one phase to another (para-ferro or ferro-para) at fixed applied electric field and variable temperature, or during the transition at variable applied electric field and fixed temperature. Models concerning the similarities and differences between these non-reversible phenomena are published based on "non-equilibrium thermodynamics" with formalisms difficult to relate to the surfaces at the microscopic level that contribute to average energies in volume (phase front and domain walls). The aim here is to enrich reflections on the observed phenomena. The experiments are conducted slowly enough to reduce the effect of characteristic variation speeds of the phenomena, thus without addressing theories related to the kinetics of interfaces.

To do this, Fig. 6.14 reproduces the representation of Fig. 1.22 introduced by Schmidt et al. (1976). Sample diagrams with phase front and domain textures characterize transitions with variation of *T*, thus by vertical displacement in the representation. The assumption of a beautiful thermal homogeneity of the crystal is admitted. Furthermore, phase coexistence is assumed to begin when the $\alpha\beta$ curve is crossed. The arrows indicate the evolution over time, which underlines the limits of the exercise: the history and its major role in a first-order transition do not appear sufficiently with these simplistic diagrams, even when the dynamics are extremely slow.

Figure 6.15 reproduces the previous graph in electric field and temperature, used not by vertical paths by variation of temperature but by horizontal paths with variation of the electric field. This corresponds, for example, to hysteresis cycles performed with a triangular periodic variation of the applied electric field of several hours. The experiments conducted in both contexts (slow variations of *T* or slow variation of *E*) have shown that the comparison between them was qualitatively

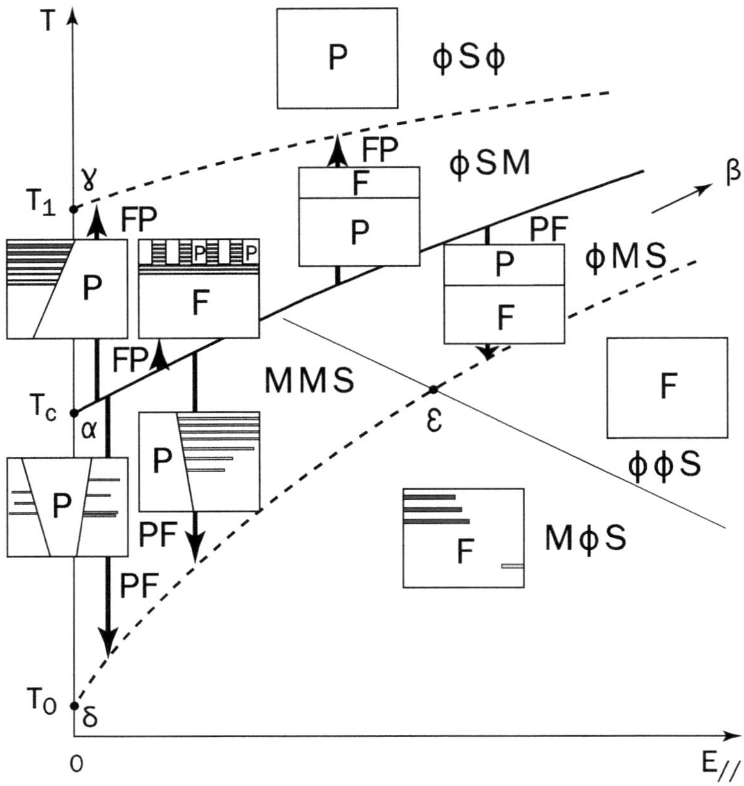

Fig. 6.14 Part of Fig. 1.22b is enlarged. The various regions of this $T(E_{//})$ representation are characterized concerning the function G_2 (P) (see Figs. 1.21a and 1.22b) by stability indications such as *ΦSM*

Synthesis figure concerning PF or FP transitions with constant applied conjugate field $E_{//}$ which would start on the line $T = T_C$. Arrows schematize the variation of temperature T (considered as the homogeneous temperature of the sample T_S). The PF transitions occur from T_C to T_0 and the transitions FP from T_C to T_1; diagrams illustrate what phase fronts and domain textures would be observed in each region in section a_2 of a DKDP sample. For very low values of $E_{//}$, the phase fronts are close to planes (001) while they are clearly in a plane (100) when the value of $E_{//}$ is significant. The domain textures are illustrated. One can observe the hetero-phase PFPFPF created at the beginning of a cycle FP with electric field $E_{//}$ which appears in the region entitled MSM in Fig. 1.22b, for a low value of $E_{//}$ without being very low. It is clear that we can go from P mono-domain to F mono-domain by bypassing point β, which is not addressed here since no interface then exists

acceptable. Three situations are illustrated for temperatures at which phase coexistence could exist. The shapes of hysteresis cycles presented are borrowed from previously published results (Gladkii and Sidnenko, 1972).... If the frequency of the cycle increases too much, even in a system homogeneous in temperature around the sample, the displacements of domains, and especially of the phase front (heat exchanger), disturb the thermal homogeneity and modify the phenomena and the exchange of latent heat. Furthermore, if the frequency is notable in relation to the

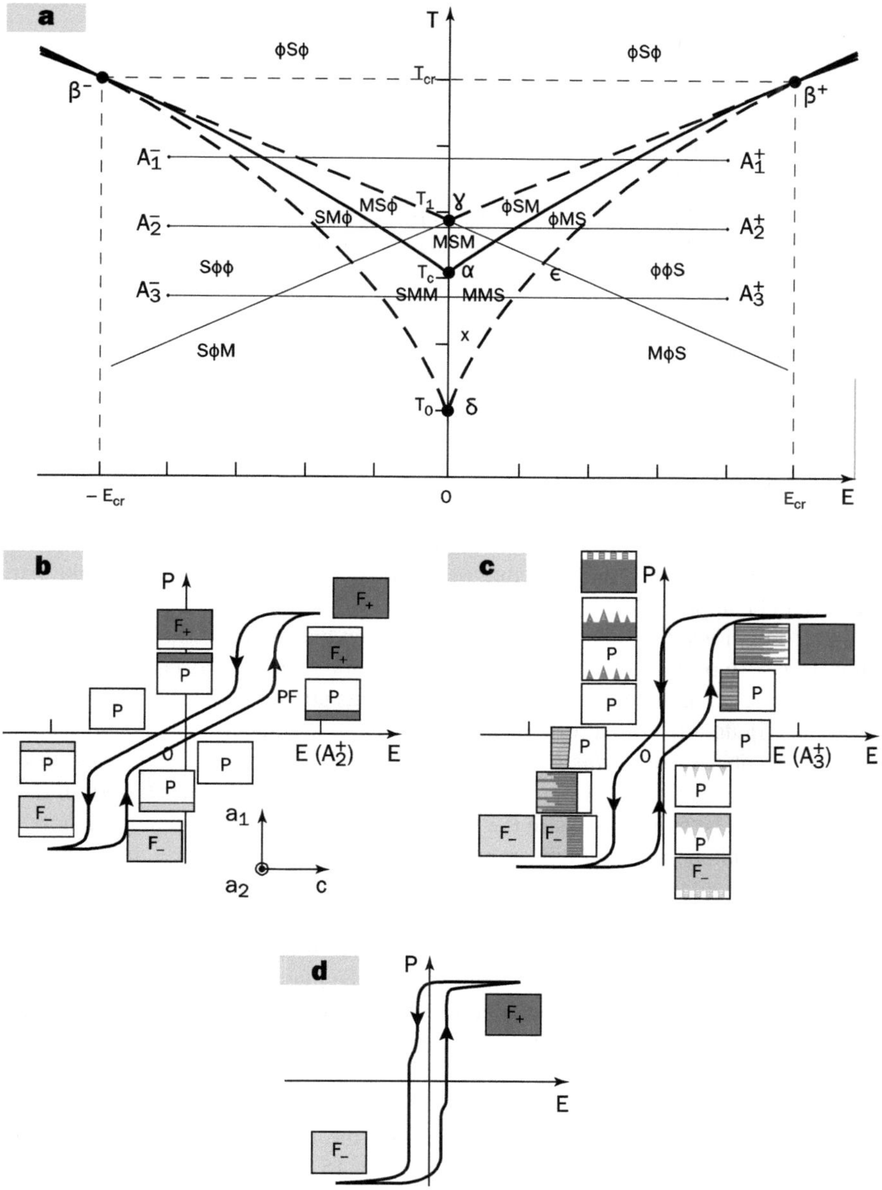

Fig. 6.15 Hysteresis cycles and coexistence zones: correspondence between the shape of very low frequency hysteresis cycles and the possible interfaces in the sample (phase fronts and domains) (**a**) The representation in $T(E_{//})$ of Fig. 6.14 is taken up with $E_{//}$ positive or negative. In the coexistence zones (such $\gamma\delta\beta^+$), the stability criteria (such as ΦSM) defined in Figs. 1.21 and 1.22b) are indicated. By hypothesis, the dynamics to go from one point to another on the plane is very slow. Thus, the movements in both directions between points A_i^+ and A_i^- correspond to hysteresis cycles with $E_{//}$ variable of very low frequency, at the temperature T_i. These cycles are considered at three

(continued)

temperatures of the sample (of very good thermal homogeneity: $T = T_S$). The qualitative shapes represented in (**b–d**) are inspired by the results of our work and those previously published (Gladkii and Sidnenko, 1972). (**b**) When $T_{cr} > T > T_1$, the crystal is in para phase if E is of low value, and in ferro phase when it is more significant. Coexistence occurs for values of E sufficiently large to make the ferro region mono-domain and the phase front parallel to the c axis. On the schematic representations, F^+ and F^- are ferro mono-domain regions of respectively positive and negative polarity, and P a para region of polarity defined by the direction of the applied E field. (**c**) If $T_1 > T > T_C$, phase coexistence occurs differently during PF and FP transitions in terms of phase fronts and domain textures. The effect of electrostatic and mechanical energies on the shape of the front and domains acts as demonstrated in Fig. 6.4. (**d**) If $T_C > T > T_0$, phase coexistence should not be effective. The shape of the hysteresis cycle presented is roughly similar to that of a cycle of the ferro phase, with kinds of stairs. The slope breaks could account for events such as those described in (**c**). It is however prudent to ensure the thermal homogeneity of one's sample before concluding

characteristics of the assembly, the hypothesis on the appearance of phase coexistence when crossing the $\alpha\beta$ line also becomes questionable.

Taking into account these restrictions, the shape of a hysteresis cycle with linear parts, for example, provides information on the speed of the phase front when the front is a single plane moving laterally. Then, the part due to the modification of para and ferro volumes, and that due to the variation in polarizability of the para phase, can be appreciated.

In all situations, the shapes are significant when the slopes of the cycles change. Thus, Fig. 6.16 proposes hysteresis cycles that would qualitatively correspond to the results of Sect. 6.3, therefore with thermal gradients perpendicular to c. It is difficult to go back in detail, from a form of hysteresis cycle to a specific phenomenon concerning the phase front or/and the texture in domains in the ferro region or in the coexistence zone, but the exercise allows to exclude certain models based on risky assumptions. The slope-breaks in the cycles almost always correspond to a notable microscopic phenomenon (appearance or disappearance of phase front, radical change of domain texture in the ferro phase or in the ferro region of the coexistence zone). When a thermal gradient exists, these slope breaks also account for changes in a region about twice the volume of that occupied by the coexistence zone.

In addition to the shape and characteristics of the hysteresis cycles, the dielectric measurements also allow to predict certain microscopic phenomena. Knowing the thermal context is necessary, and the typical experiments illustrated in Fig. 1.15a, b often provide sufficient information concerning potential thermal gradients. The comparison between the inverse susceptibility curves and the theoretical one of a mono-domain crystal then allows to calculate the dielectric contribution of the domains and that of the phase coexistence (Fig. 4.15a, b) as well as the losses. The average values of ε'_{pc} and ε''_{pc} usually confirm the situation of thermal homogeneity or/and the nature of the gradient. Finally, the maxima by their intensity and their situation, at the beginning or at the end of coexistence, in PF or FP cycle, are reliable indicators concerning on the one hand the entry of the phase front and its exit from the sample and on the other hand the density of the domains in the ferro regions of the coexistence zone or in the ferro phase region. A dense texture in the coexistence zone has a higher dielectric impact, especially concerning the losses, than a comparable volume of dense domains in the ferro phase region.

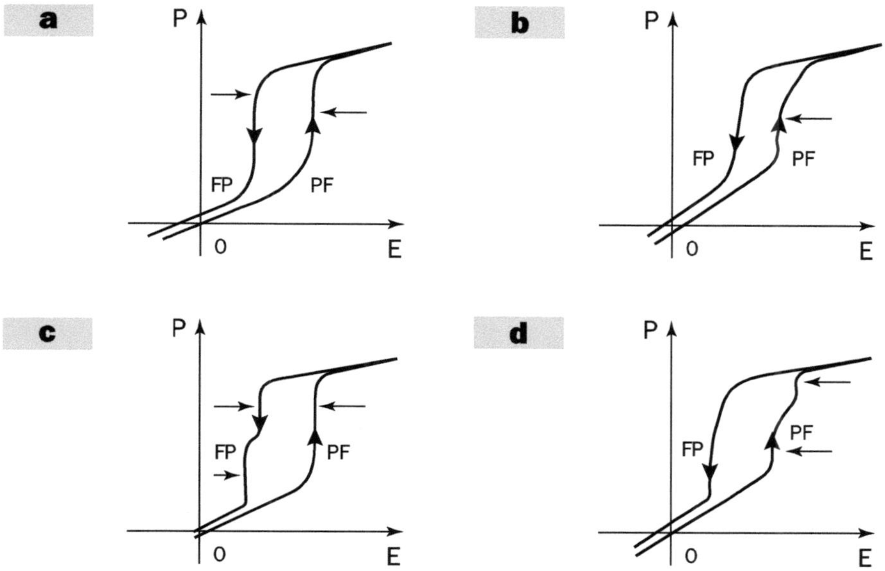

Fig. 6.16 The portions of double hysteresis cycle in the first quadrant (E, P) correspond to a PF transition followed by a FP transition; the frequency of the cycles is $<10^{-4}$ Hz. The four qualitative representations proposed integrate the variation of local polarization under the action of the field in the para regions, and the changes induced by the phase fronts and the domain textures. The arrows indicate situations in which susceptibility is maximum. These qualitative diagrams are inspired by observations and measurements made and notably those presented in Sects. 6.2 and 6.3. Such cycle shapes are not observed when the characteristic times of the assemblies, notably electrical ones, smooth the plots. Some of these cycle shapes can be found during coexistences in martensites (Bonnot et al., 2008). (**a, b**) Very good thermal homogeneity with low values of E in (**a**); in (**b**), the creation of a hetero-phase at the beginning of coexistence of a FP transition is possible. (**c, d**) Thermal gradients exist perpendicular to the polar axis. The fields applied in (**c** and **d**) are the same as those in (**a** and **b**); the creation of phase fronts with complicated shape and their impacts on the domain textures explain the types of stairs observed

This brief report for the busy scientist will obviously not replace the observation of interfaces at the microscopic level carried out simultaneously with macroscopic measurements.

6.5 Games with Phase Front Shapes and Interfaces

It is difficult to conclude this section in which the experimenter manipulates the temperature and the excitation, mechanical stress, or more easily conjugated electric field, without giving a glimpse of the multiple possible phase front shapes when the action of these parameters and the impact of the sample boundaries are considered. One can ensure the role of a break for example with Fig. 6.17 which illustrates that a phase front, such as those created in Figs. 6.4, 6.5, 6.10…, crosses the incident

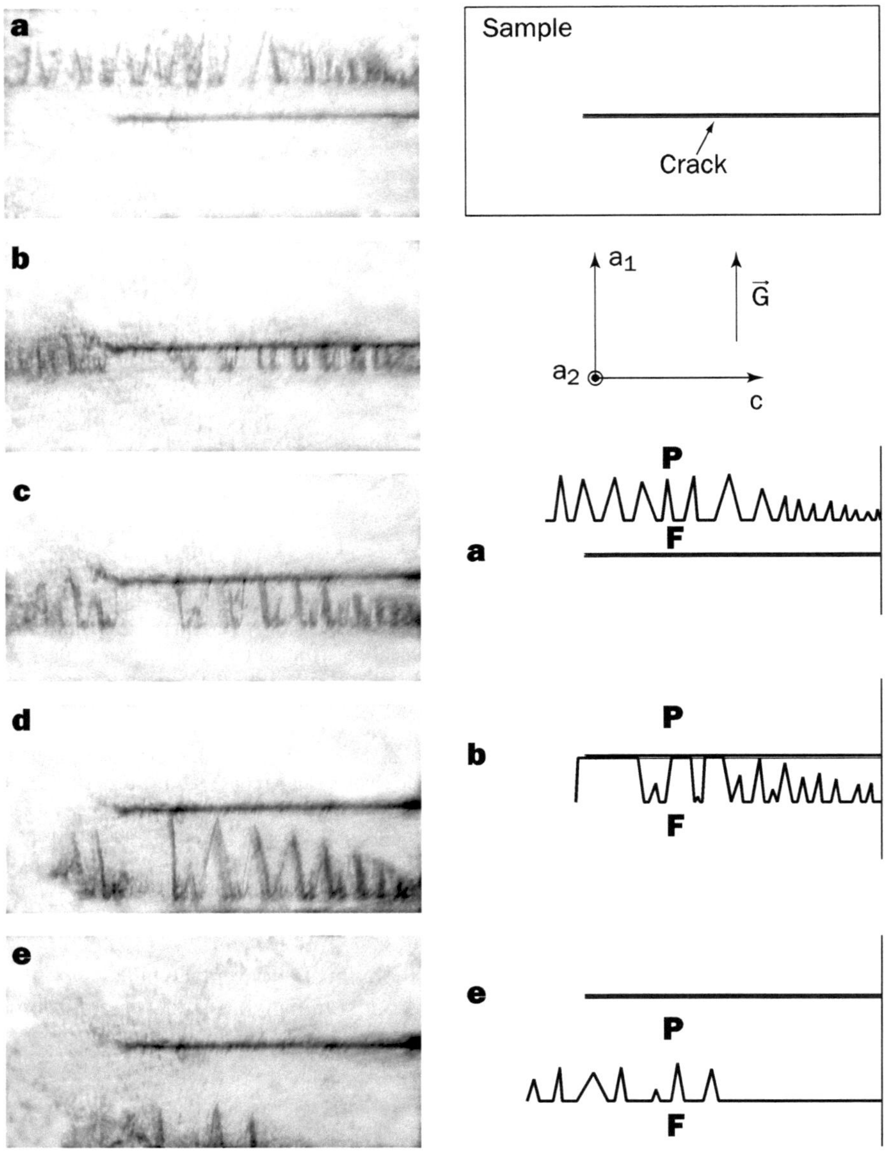

Fig. 6.17 The same context as that of Fig. 6.4c is realized in a sample that includes a clear break inside, located in a tetragonal plane (100): the phase front crosses this plane which releases the constraints and then resumes the same shape

to take back almost the same shape, while a surface scratch can play a much more effective role.

Figure 6.18 gives an overview of the variations in phase front shapes that can be created by following a process alternating small variations in temperature and electric field. The result is easy to present: volumes are created, cylinders of axis a_1 with a rectangular section, of para phase in a ferro matrix with a relative periodicity. Their size and this pseudo-periodicity can be modified on demand.

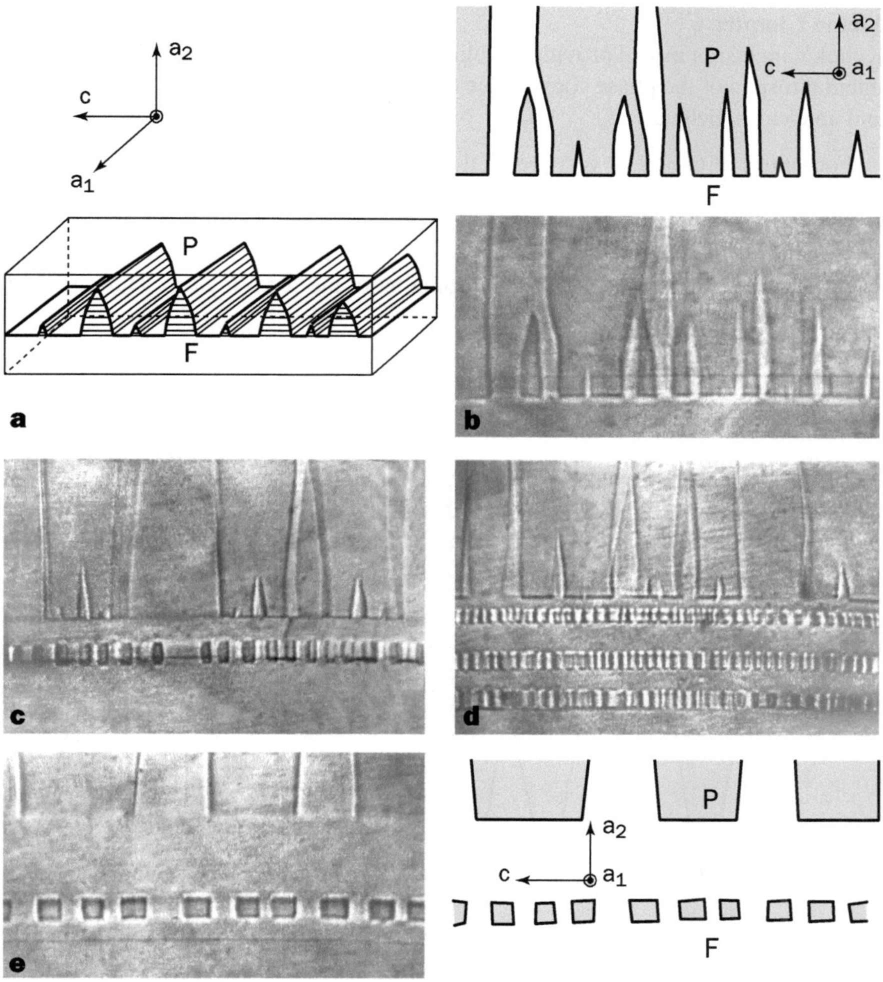

Fig. 6.18 The process of creating a phase front as described in Figure 6.4b is repeated with a DKDP sample on which a small scratch has been drawn on face a_1 (it is very visible in Figure **e**) towards the bottom of the photograph). The alternations of increasing and decreasing temperature and applied electric field allow the creation of cylindrical implants of para phase in the ferro phase matrix of different size and pseudo-periodicity

In the same vein, Fig. 6.19 allows to envisage numerous fractal games, other than the fractal structures of crystal growth (Zolotukhin and Spitsina, 2002), and studies since the relationships between energies per unit surface of phase fronts parallel to c, perpendicular to this axis or whose angles are functions of the density of walls in the ferro daggers. One can delve into the conditions for creating new daggers depending on the sizes of the existing ones. The declination of such effects in the various ferro-ferro compounds with first-order transition opens horizons that may not be just playful…

Memo Chapter 6

Kvitek's inclusion model provides a solid theoretical basis for explaining the experimental results of the phase coexistence of DKDP subjected to a thermal gradient G and an electric field E:

– For values of G null or very weak, the difference between the PF cycles and the FP cycles is significant for weak E fields, of modulus between a few tens of V/cm and 200–300 V/cm concerning the various interfaces: during a PF cycle, the situation is close to those observed with $E = 0$ in Chap. 5 (a plane phase front (001) and a fairly dense texture in domains in the ferro region). During an FP

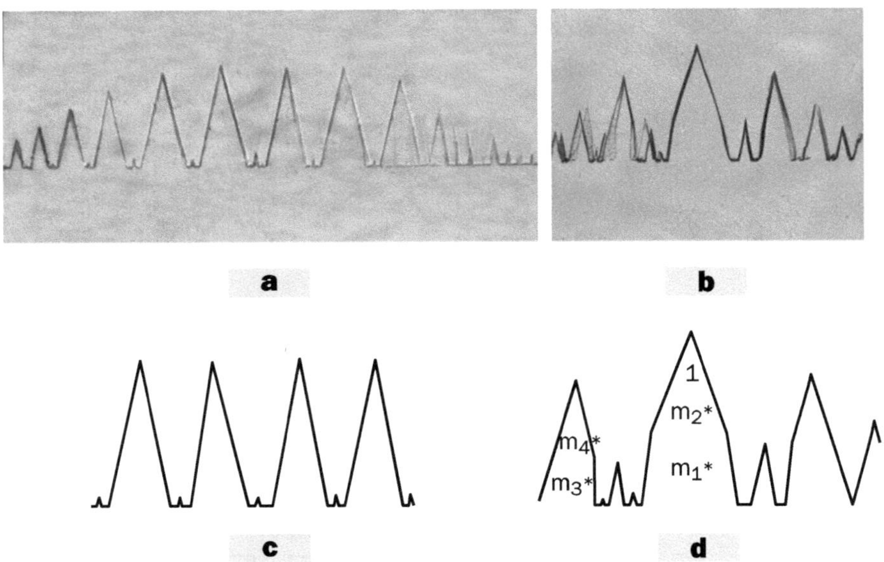

Fig. 6.19 Figure 6.10 defines in which context the phase front takes the form, between two envelopes, of daggers with the creation of fractals, small daggers between the larger ones. The system can be very regular as in (**c**) in a sample of very high thermal homogeneity, in the middle of the sample and with regular domain texture: The daggers have constant angle texture. On the contrary, the angle changes for the same dagger when the domain density changes as in (**d**) ($m^* = 1$ or -1: quasi-mono-domain, $m^* = 0$ dense domain texture and front close to (001)). The angle at the top of the daggers is always smaller for the small daggers of fractals in creation than for the previously created and good-sized daggers

cycle, the ferro phase region is closer to the mono-domain separated from the coexistence zone by a plane (100); but the coexistence zone is a hetero-phase with alternation of para regions and finely layered ferro regions. Its evolution can be achieved with daggers whose angles vary with the texture in domains within them. The notable fact in macroscopic measurements is the thermal hysteresis of the transition that grows with E while classical models predict its decrease, found with values of E beyond the previous fields.

- If G is significant and perpendicular to the ferroelectric axis, the application of an applied field E limits the coexistence zone by a plane (100) towards the ferro region which cuts the "factory roofs" at variable levels depending on the modulus of E. The difference between PF and FP cycles is illustrated by the domain texture in the ferro regions of coexistence: texture close to the mono-domain in PF cycle and dense texture in FP cycle. The volume to consider is about twice the coexistence region, with a region of comparable size in the ferro phase. Thus, all the deformations in this set are harmonized by different densities and orientations of domain walls between ferro region and ferro parts of the coexistence zone (optimization of all deformations notably of u_{xx} and of u_{zz}). The history plays a significant role, for example, for the creation of "factory roof" nuclei on the sample edges or in the middle of a face, which contributes to defining the orientation of the domain walls (between the two possible orientations (100) and (010)).
- Many macroscopic results, often unexplained, are understood as the phenomena at the microscopic level (Dielectric measurements and hysteresis cycles near T_C with double cycle and susceptibility measurements).
- This knowledge could be exploited for fundamental macro-micro-nano studies, but also in applications to create, on demand, structures of inclusions of one phase in the other, and fractal shapes by playing on mechanical and electrostatic energies.

Correction to: Coexistence of Phases with Conjugate Field

Correction to:
Chapter 6 in: J. Bornarel, *Ferroic Crystal Domains, Walls, and Phase Fronts*, Springer Series in Solid-State Sciences 209, https://doi.org/10.1007/978-3-031-98099-2_6

The chapter "Coexistence of Phases with Conjugate Field" was originally published with incorrect Fig. 6.10. The most important curve "ε'_c(T)" has disappeared between T~2~ and T~1~!. The dielectric constant curve versus the temperature for temperatures greater than T~2~ is missing in Fig. 6.10; nevertheless, the arrows that connect points of the curve to the corresponding photographs of the phase front and the domains are present. The error is thus obvious. The correct Fig. 6.10 is provided below:

The updated version of this chapter can be found at https://doi.org/10.1007/978-3-031-98099-2_6

J. Bornarel, *Ferroic Crystal Domains, Walls, and Phase Fronts*, Springer Series in Solid-State Sciences 209, https://doi.org/10.1007/978-3-031-98099-2_7

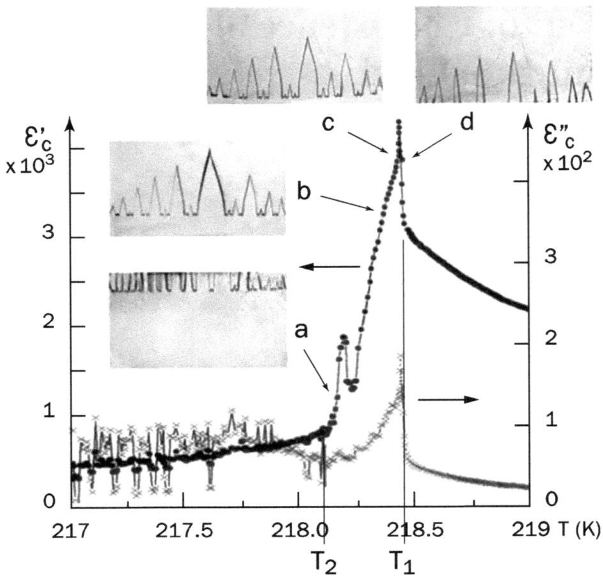

Figure 6.10 has been updated in the chapter.

Annexes

A1 Physical Properties

A1.1 How to Define and Represent Properties of a Homogeneous Crystal?

By convention, the quantities that account for the properties of materials are defined for homogeneous materials, without boundary problems, in a stable thermodynamic situation. This annex recalls these conventions and the proposed references allow to delve deeper into the question. Quantities represented by a number are written in normal straight letters and quantities represented by non-zero rank tensors, including vectors (rank 1), are written in straight letters and bold characters.

The situation of a homogeneous crystal is described by intensive parameters (temperature *T*, electric field **E**, mechanical stress **σ**) and extensive parameters proportional to volume, defined per unit volume (entropy S, polarization **P**, strain **e** or **u**). The simplest physical properties relate the value of an intensive parameter and that of an extensive parameter. These quantities are based on the assumption of a "linear" variation, that is to say, for example, **D** as a function of **E** and not as a function of $\mathbf{E}^2$ or $\mathbf{E}^3$ (Mason, 1950; Burfoot, 1967).

Let's first choose the relations between parameters of the same nature, called conjugates. These relations are called main properties:

Let temperature *T* and entropy *S*, two scalars, hence the expression of the property "calorific capacity" by a single number: the specific heat (calorific capacity per unit mass or per mole). It is written, for a measurement at constant volume or one at constant pressure (and *E* and **σ** constants)

$$C_v = T\left(\delta S / \delta T\right)_v \text{ or } C_p = T\left(\delta S / \delta T\right)_p \quad \text{(A1.1)}$$

J. Bornarel, *Ferroic Crystal Domains, Walls, and Phase Fronts*, Springer Series in Solid-State Sciences 209, https://doi.org/10.1007/978-3-031-98099-2

Then, the conjugate parameters electric field **E** and electric induction (or electric displacement) **D**: if the value of the electric field is changed by d**E**, the induction **D** will be changed by d**D** with the relation:

$$\mathrm{d}\mathbf{D} = \varepsilon\,\mathrm{d}\mathbf{E} \tag{A1.2}$$

ε is the permittivity (or the specific inducting power of the crystal);

We recall the relation:

$$\mathbf{D} = \varepsilon_0\,\mathbf{E} + \mathbf{P} \tag{A1.3}$$

with ε_0 (a number) permittivity of vacuum or specific inductive power of vacuum equal in (SI) to 8.854×10^{-12} F m^{-1} and **P** the polarization (electric moment of the unit volume or polarization charge per unit of surface normal to **P**). Furthermore:

$$\mathrm{d}\mathbf{P} = \chi\,\mathrm{d}\mathbf{E} \tag{A1.4}$$

$\boldsymbol{\chi}$ is the dielectric susceptibility of the material. To obtain dimensionless permittivity and susceptibility, we write: $\boldsymbol{\chi} = \varepsilon_0\,\boldsymbol{\chi}_r$ with $\boldsymbol{\chi}_r$ relative susceptibility of the material, and $\boldsymbol{\varepsilon} = \varepsilon_0\,\boldsymbol{\varepsilon}_r$, $\boldsymbol{\varepsilon}_r$ relative dielectric constant often used in the book.

As **E**, **D**, and **P** are vectors (three components), the susceptibility $\boldsymbol{\chi}$ and the permittivity **ε** are rank two tensors with nine coefficients χ_{ij} and ε_{ij}; *i* and *j* correspond to the three axes of a reference trihedron; the first index (*i*) indicates the row of the tensor and the second (*j*) the column; thus:

$$P_i = \chi_{ij}\,E_j \text{ and } D_i = \varepsilon_{ij}\,E_j \tag{A1.5}$$

Habit leads to abuses of language and writing: the "dielectric constant" (dimensionless) is often the term used for the relative dielectric constant because the physicist knows which tensor coefficient he is measuring. Thus, while $\varepsilon_{ij} = \varepsilon_0\,(\delta_{ij} + \chi_{ij})$, it is the content of the parenthesis that he calls the dielectric constant and often writes ε_c if it corresponds, for example, to a measurement along the main axis *c* (δ_{ij} is the Kröneker symbol).

The conditions for measuring susceptibility (permittivity) must be well defined: is the experiment carried out at constant stress or constant strain? Under isothermal or adiabatic thermal conditions? In the book, unless otherwise indicated, they will most often be obtained at constant stresses and in an isothermal situation. As the dielectric constant is often measured using an alternating electric field, the frequency of this must be low; if it is too high, the situation is adiabatic. Strictly speaking, the following writing should remind it (often forgotten not to weigh down when confusion is excluded):

$$\chi_{ij} = \left(\delta P_i / \delta E_j\right)_{T,\sigma} \tag{A1.6}$$

The dielectric susceptibility tensor will include at most 6 independent non-zero coefficients.

The relationship between the conjugate parameters stresses and strains is more complex since these quantities **σ** and **e** (or **u** or sometimes **ε** not to be confused with the dielectric constant) are expressed by second-order tensors of 9 coefficients. In the case of a linear variation (Hooke's law), the relations between the coefficients e_{ij} of the strain and the coefficients σ_{ij} of the stress are written as:

$$e_{ij} = s_{ijkl}\,\sigma_{kl} \tag{A1.7}$$

the s_{ijkl} are the elastic compliance coefficients of the crystal, and

$$\sigma_{kl} = c_{ijkl}\,e_{kl} \tag{A1.8}$$

the c_{ijkl} are the elastic rigidity constants or rigidities.

These relations introduce tensors **s** and **c** of rank four, with 81 coefficients, but symmetry reasoning allows their number to be limited to 36 and then to 21 independent coefficients. It is demonstrated that each coefficient s_{ijkl} can be expressed in the form of a linear combination of coefficients c_{ijkl}, and the converse is also true.

The choice of the reference axis system is a major factor to obtain the representation at the same time the simplest and the most adapted to the physical problem studied. Tensors that differ according to symmetries can be consulted in specialized works (Nye, 1992; Tagantsev et al., 2010; Malgrange et al., 2014). We will follow the same conventions on mute indices as those chosen by Einstein (when an index appears twice in a monomial term, the summation with respect to this index is automatically implied).

Often used is the "matrix notation" or "Voigt notation," which takes into account symmetric terms with a single digit index notation from 1 to 6 as follows:

$$\text{Index 1 for 11, 2 for 22, 3 for 33} \tag{A1.9}$$

index 4 for 23 and 32, 5 for 13 and 31, 6 for 12 and 21.

We will be able to use this notation in the book (for example, s_{12} for s_{1122}, s_{36} for s_{3312},…). Thus, regarding the elasticity tensor, we cannot encounter more than 21 independent non-zero coefficients. With Voigt notation, we can therefore present results using a two-dimensional matrix, which is pleasant but should not lead to confusion: the matrix tables of deformation and stresses with coefficients at one index, those of elasticity and rigidity, s_{ij} and c_{ij} with coefficients at only two indices are not the elements of a tensor of these ranks and cannot be transformed as such. For example, if we wish to change the axis system, we must return to the tensor notation to do so. The remark obviously applies to other relations, for example to piezoelectric tensors.

Some physical properties, not principal, link parameters described by tensors of different ranks. This is the case of the piezoelectric effect: a stress σ_{kl} applied to the crystal creates a modification of the components of the electric induction or polarization:

$$P_j = d_{jkl}\sigma_{kl} \tag{A1.10}$$

d_{jkl} is one of the coefficients of the piezoelectric module **d** which links the polarization **P** created by the stress **σ**; **d** is a tensor of rank three, with 27 coefficients. Indeed, the previous relation developed for each component of **P** is written as:

$$\begin{aligned} P_1 = {} & d_{111}\sigma_{11} + d_{112}\sigma_{12} + d_{113}\sigma_{13} + d_{121}\sigma_{21} + d_{122}\sigma_{22} \\ & + d_{123}\sigma_{23} + d_{131}\sigma_{31} + d_{132}\sigma_{32} + d_{133}\sigma_{33}, \end{aligned}$$

hence the writing of a rectangular tensor by declining in the same way as P_2 and P_3. The stress tensor being symmetric, $\sigma_{12} = \sigma_{21}$, the same applies to the other non-diagonal terms which correspond to shearings. The tensor representing this property will have at most 18 independent non-zero coefficients.

By thermodynamic reasoning, by deriving a second time in a crossed way the indicated partial derivatives, we demonstrate the numerical equality of the coefficients of the direct piezoelectric effect and of the inverse piezoelectric effect.

This equality is also true between the coefficients of thermal expansion and of the piezo-calorific effect (tensor with at most 6 independent non-zero coefficients) usually called α_i with i varying from 1 to 6 in the matrix notation:

$$\begin{aligned} &e_{ii} \text{ of the tensor} = e_i\left(i = 1{,}2 \text{ or } 3\right) = \alpha_i \Delta T, \\ &e_{ij} \text{ of the tensor} = e_j\left(j = 4{,}5 \text{ or } 6\right) = \alpha_j \Delta T \end{aligned} \tag{A1.11}$$

Similarly, the coefficients of the pyroelectric effect are symmetrically equal to those of the electro-calorific effect (with at most 3 independent non-zero coefficients):

$$D_i = p_i \Delta T \tag{A1.12}$$

These reminders start from the hypothesis of linear relations between excitation and response.

Other coefficients express responses of another order, for example the electrostriction of a coefficient which links the stresses to the products of the components of the polarization:

$$Q_{ijkl} = \delta\left(e_{ij}\right) / \delta\left(P_k P_l\right) \tag{A1.13}$$

The reader can deepen this representation of the crystal physical properties by tensorial relations, and the conventions taken with matrix presentations, by consulting reference works (Mason, 1950; Nye, 1992; Tagantsev et al., 2010; Malgrange et al., 2014).

A1.2 Tensors and Matrices

This book focuses on ferro-ferro crystals of symmetry $\bar{4}2m$ in the para phase and mm^2 in the ferro phase. The term "ferro-ferro" used in the book for convenience summarizes for the reader that with the recognized names, these crystals are ferroelectric, ferroelastic, ferroelastolectric, and ferrobielastic (Aizu, 1972, 1973; Newnham and Cross, 1974a, b; Sirotin and Shakol'skaya, 1979; Amin and Newnham, 1980). In these crystals, the low-temperature ferro symmetry group is a subgroup of that of the para phase. Thus, in the tensors of properties in the reference frame of the para phase, some elements null in the para phase appear as non-null at the transition. The relations between extensive and intensive parameters are written as:

$$\begin{aligned} e_n &= s_{nm}\sigma_m + d_{in}E_i + \alpha_n \Delta T \\ D_i &= d_{im}\sigma_m + \varepsilon_0 \varepsilon_{ij} E_j + p_i \Delta T \\ \Delta S &= \alpha_n \sigma_n + p_j E_j + (C/T)\Delta T \end{aligned} \tag{A1.14}$$

With the Einstein summation convention and the conventions of Nye, those of i and j which change from 1 to 3, of n and m which change from 1 to 6, these properties are summarized by the Voigt matrices in Fig. 1.11 reproduced in Fig. A1.1. The matrices of the properties are detailed: on the left, those which correspond to the symmetry $\bar{4}2m$; in the middle those of the orthorhombic symmetry mm^2 with the same c axis and two axes perpendicular to it, rotated by 45° relative to the tetragonal

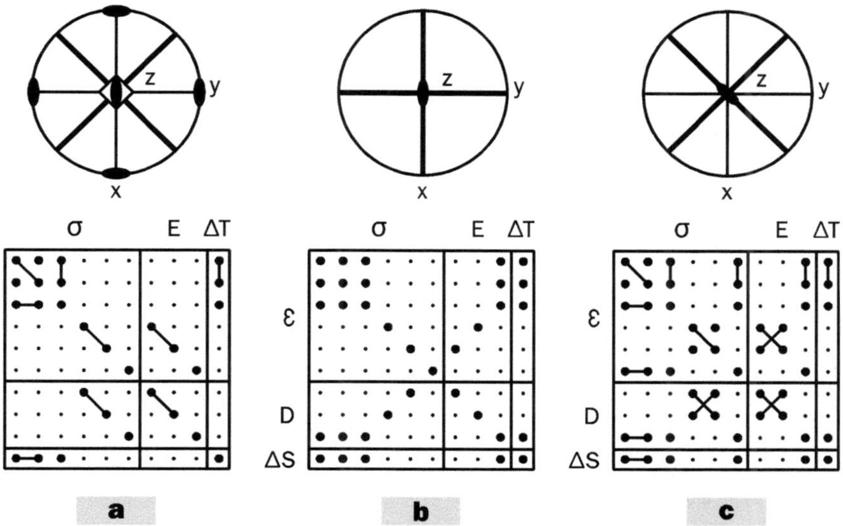

Fig. A1.1 Voigt matrices of properties already provided in Fig. 1.11 for the tetragonal symmetry $\bar{4}$ 2 m (left), the orthorhombic symmetry mm^2 (middle), and considering the symmetry of the ferro orthorhombic phase as a subgroup of the tetragonal symmetry $\bar{4}$ 2 m (right)

axes; on the right, the symmetry of the ferro phase considered as a subgroup of that of the para phase; we still use the coefficients defined in the para phase with different values and some coefficients null in the para phase that become non null. This presentation remains acceptable in many situations because of the limited values of lattice deformations and other properties, especially in the KDP family.

To be complete and avoid any confusion, Fig. A1.2 makes the link in the simple case of the book between the elements of the matrix representation of deformation and what is commonly encountered in the literature, and in our book, which is a "practical shear deformation."

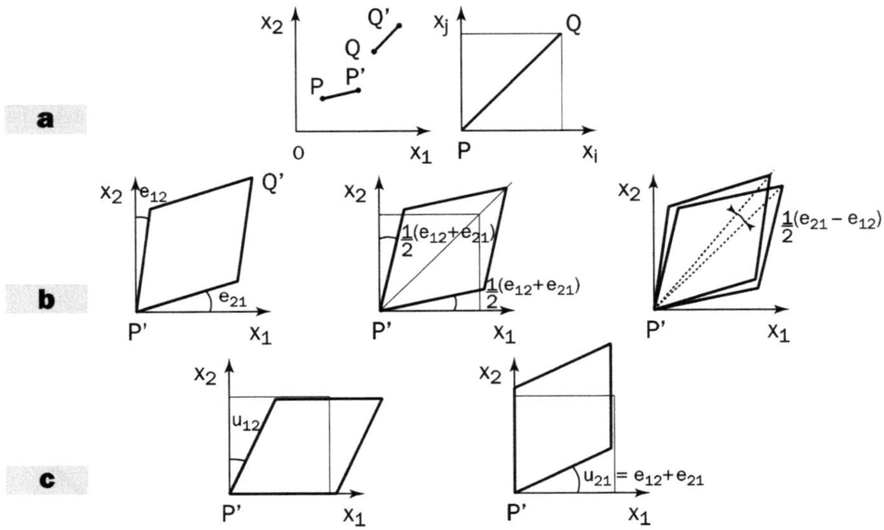

Fig. A1.2 Deformations. These two-dimensional surface deformation diagrams illustrate what also occurs in three dimensions, expressed with the same type of relations: (**a**) Two very close points of the crystal P and Q, defined by the coordinates x_i^P and x_i^Q along the axes ($(\mathrm{OP})_i = x_i^P$), are found after deformation in P' and Q'. We call u_i the displacement in the direction $\mathrm{O}x_i$, for example $(\mathrm{PP})_i = u_i^P$. A particular case is represented on the right, where the points P and Q are located along the diagonal of a square, like the ends of a cubic or tetragonal elementary cell in a section perpendicular to $\mathrm{O}x_3$ in the tetragonal reference frame. (**b**) The previous "cell" deforms as indicated on the left figure. This figure actually corresponds to the sum of a deformation of the square from (**a**) into a diamond as illustrated by the middle figure and a rotation of this figure as illustrated on the right. The values of the angles are indicated in terms of e_{ij}. (**c**) Most of the time, in the context of this book, the domain walls remain in one of the two tetragonal crystallographic planes (perpendicular to $\mathrm{O}x_1$ or to $\mathrm{O}x_2$); we have got used to calling "deformation" or "practical shear deformation" (Nye, 1992; Chap. 6) the values of u_{12} and u_{21}. These are deformations (dimensionless) that cannot be confused with the displacements u_1, u_2 ... which are lengths

$u_{12} = u_{21} = e_{12} + e_{21} = 2e_{12}$

The coefficients e_{ij}, are numerically equal to half of the "practical shear deformations" u_{ij} with i and j different, measured for example in γ diffraction in Annex A3. But if $i = j$, we have $e_{ii} = u_{ii}$

Furthermore, figures in (**b**) remind the experimenter who measures the values of the u_{ij} using diffraction on reticular planes that these can rotate as illustrated in Figs. 4.12 and A3.4.

A1.3 Notations and Symbols, Units and Conversions

Physical quantities are represented by letters which sometimes differ according to scientific communities. Some letters are used for several quantities in different contexts. Below, some recalled relations allow to check the units and conversions between systems. The international unit system (SI) is used, but given the many old publications cited as references, and customs in the various fields of physics, the conversion rules to the units of the so-called Gauss system, "cgs electrostatic(ues)" and "cgs electromagnetic (uem)," are recalled.

Below, for each quantity, are given the letters used in the book with some comments and relations, the unit in SI and its value in units of the Gauss system. Sometimes, other conversions in units of usage are proposed.

Length: *L* or *l*, for the dimensions of the lattice a_1, a_2 and c for the tetragonal symmetry and *a*, *b*, *c* for the orthorhombic symmetry. The metre = 10^2 centimetres(cm) = 10^{10} Angstrom(Å),
Plane angle: Greek letters; the shear deformation u_{12} or u_{xy} is an angle = the radian or seconds, minutes, degrees of arc,
Solid angle : Ω in steradians,
Surface: $S = l^2$. square metre (m^2) = 10^4 square centimetre (cm^2),
Volume: $v = l^3$. cubic metre (m^3) = 10^6 cubic centimetre (cm^3) = 10^3 L = 10^{30} cubic Angstrom (A^{o3}); the volume of a cell in the crystals of the KDP family frequently mentioned in the work is several hundred cubic Angstroms (386.6 for the tetragonal cell of KDP and 767 for its orthorhombic cell towards the transition),
Mass: *m*: The kilogram (kg) = 10^3 grams (g),
Density ρ: kg m^{-3} = 10^{-3} g cm^{-3}.
Time: *t*: second (s), with sometimes as a unit, the period of the phenomenon indicated *T* or τ,
Frequency: $\nu = T^{-1}$ = Hertz (Hz), and the pulsation $\omega = 2\pi\nu$ (Hz),
Speed and acceleration: usually indicated *v* and Υ respectively in m s^{-1} and m s^{-2}.
Temperature: *T* = Kelvin (K) or degree Celsius (°C),
Thermal conductivity: $\lambda = Jm^{-1}\, s^{-1}K^{-1} = 10^5$ erg $cm^{-1}\, s^{-1}\, K^{-1}$,
Pyroelectric coefficient: $p_i = Cm^{-2}\, K^{-1} = 3.10^5$ esu $K^{-1} = 10^5$ emu K^{-1},
Electrocaloric coefficient: $p_i = Cm^{-2}\, K^{-1} = 3.10^5$ esu $K^{-1} = 10^{-5}$ emu K^{-1},
Coefficient of thermal expansion: $\alpha_i = T^{-1} = K^{-1}$,
Piezocaloric coefficient : $\alpha_i = T^{-1} = K^{-1}$,
Specific heat (molar) at constant pressure: $C_p = J\, K\, ^1mol^{-1} = 10^7$ erg $K^{-1}\, mol^{-1}$,
Heat of transition per mole: $\Delta Q_m = Jmol^{-1} = 10^7$ erg mol^{-1},
Transition entropy per mole: $\Delta S_m = JK^{-1}\, mol^{-1} = 10^7$ $ergK^{-1}mol^{-1}$.
Force: **F**. Newton (N)= 10^5 dyn (dynes),
Work, Energy: $W = F \times l$. The joule (J) = 10^7 erg (erg),
Energy per unit area: *W*/*S*. joule per square metre (J m^{-2}) = 10^3 erg cm^{-2},
Energy per unit volume: *W*/*v*. joule per cubic metre (J m^{-3}) = 10 erg cm^{-3},
Power : *p* = W/t = energy per unit of time = watt

Mechanical stress: σ = F/S. newton per square metre (N m^{-2}) = Pascal (Pa) = 10 dynes per square centimetre (dyne cm^{-2}) = 10^{-5} bar (unit of atmosphere),

The definition of the physical properties in Annex A1.1 highlights the difference between "normal" stresses (compression or expansion) σ_{ii} force per unit area parallel to the normal to the considered surface, and shear stresses σ_{ij} applied to a surface that contains directions i and j and therefore cause a rotation around the perpendicular direction k (reference trihedron (**i, j, k**)). In crystals, the stresses are defined by a second-order tensor with 9 coefficients.

Hydrostatic pressure: p = Pa (= N m^{-2}) = 10 dyn cm^{-2} (the pressure of an atmosphere: 1 atm = 101325 Pa = 76 cm Hg)

Deformation. Deformation is a quantity defined in relation to the object considered: the elongation Δl of a length l. Therefore, its definition refers to the lengths, areas, and volumes considered. For example, the elongation per unit length in the Ox_1 direction will be noted e_{11} for this compression or expansion stress (in m/m therefore without dimension); the rotation towards Ox_2 of a linear element parallel to Ox_1, rotation performed around Ox_3 will be noted e_{21} while the rotation towards Ox_1 of a linear element parallel to Ox_2 still around Ox_3 will be noted e_{12}. These deformations are said to be shear. The overall deformation of a crystal volume is defined by a second-order tensor of 9 coefficients.

We will use these usual notations in the book, but often, when there is no ambiguity about the deformation concerned, as in many publications will be used the practical deformation u_{ij} or even simply u (see Fig. A1.2). The doubt between a displacement, often noted u and the deformation does not exist, the displacement is in units of length, the deformation is dimensionless (in the case of shear deformations, they are expressed in units of angle (seconds, minutes, degrees or in radians) (Nye, 1961).

Elastic compliance $s_{ijkl,}$ with the matrix notation s_{ik}: (m^2 N^{-1}) = 10^{-1} cm^2 dyn^{-1},

Stiffness constant ("elastic stiffness constant") c_{ijkl} in matrix notation c_{ik}: N m^{-2} = 10 dyn cm^{-2},

Piezoelectric strain constant (direct effect) d_{ijk} and in matrix notation d_{ik}: CN^{-1} = 3×10^4 esu,

Piezoelectric stress constant (inverse effect) d_{ijk}^* and in matrix notation d_{ik}^*: equal to d_{ik}.

Electric charge: q. The coulomb (C) = 3 $\times10^9$ units ues (exactly 2.998 $\times10^9$)

Electric current I: $I = q/t$. The ampere (A) = 3 $\times10^9$ units ues per second (exactly 2.998 $\times10^9$),

Electric current density: j = Am^{-2} = 3 $\times10^5$ esu = 10^{-5} uem,

Resistivity: ρ = Ωm = $10^{-9}\times 9^{-1}$ esu = 10^{11} uem,

Conductivity: $\sigma = \Omega^{-1}$ m^{-1} = 9 $\times10^9$ esu = 10^{-11} uem,

Electric potential: V. The volt (V) = $(300)^{-1}$ statvolts (exactly $(299.8)^{-1}$),

Electric field E : E = V/l. The volt per metre (V m^{-1}) = 3^{-1} $\times10^{-4}$ statvolts per centimetre (exactly $(2.998)^{-1}\times 10^{-4}$),

Electric dipole moment p: p = le coulomb-metre (Cm)

Dielectric polarization P: $P = ql^{-2}$ defined as the dipole moment per unit volume. The **coulomb** per square metre (Cm^{-2}) = 3×10^5 ues (exactly 2.998×10^5 ues). A "mixed" unit is often used by researchers: the micro-coulomb per square centimetre ($\mu C\ cm^{-2}$),

Dielectric induction ("electric displacement"): $\mathbf{D} = \varepsilon_0\ \mathbf{E} + \mathbf{P}$ knowing that in vacuum $\mathbf{P} = 0$; $\mathbf{D} = \varepsilon_0 \boldsymbol{\varepsilon}_r \mathbf{E}$ with $\boldsymbol{\varepsilon}_r$ dimensionless relative dielectric constant,

Reminder of Maxwell's equations (SI): div $\mathbf{D} = \rho$ free charge density or on conductors and **rot H** $= j + \delta\mathbf{D}/\delta t$ with **j** volumetric current density (Am^{-2}):

The **coulomb** per square metre (Cm^{-2}) = $12\Pi\times 10^5$ ues = $4\Pi\times 10^{-5}$ uem,

Dielectric constant ε = the dimension is that of ε_0, but ordinarily, it is a question of the relative dielectric constant $\boldsymbol{\varepsilon}_r$ without dimension,

Dielectric susceptibility ("electric susceptibility") χ : $Fm^{-1} = 9\times 10^9$ without dimension in ues = 10^{-11} uem,

Relative dielectric susceptibility: χ_r with $\mathbf{D} = \varepsilon_0\ \mathbf{E}\ (1 + \chi_r)$, $\boldsymbol{\varepsilon}_r = 1 + \chi_r$; therefore, χ_r is without dimension,

Electrostrictive constant (for **P**): Q_{ij} in matrix notation = $Q_{ijk,}$ if $i = j$ and $2Q_{ijk}$ if $i \neq j$ = $m^4C^{-2} = 10^{-10}\times 9^{-1}$ ues = 10^{10} uem.

Magnetic induction: $\mathbf{B} = \mu_0(\mathbf{H} + \mathbf{M})$ knowing that in the vacuum $\mathbf{M} = 0$.
Reminder of Maxwell's equations: div $\mathbf{B} = 0$ and **rot E** $= -\delta\mathbf{B}/\delta t$:
The tesla (T) (=Wb m^{-2}) = $10^{-6}\times 3^{-1}$ ues = 10^4 G (gauss)in uem,

Magnetic field: **H**: $A.m^{-1} = 12\Pi\times 10^7$ ues = $4\ \Pi\times 10^{-3}$ oersteds(uem)

Magnetic flux: ϕ = B.S. The weber (Wb) = 10^8 maxwells (or gauss cm^2),

Magnetization: **M**: $Am^{-1} = 3\times 10^7$ esu = 10^{-3} G (uem),

Magnetic polarization J with $\mathbf{B} = \mu_0\mathbf{H} + \mathbf{J}$: The tesla (T) (=Wb m^{-2}) = $10^{-6}\times (12\Pi)^{-1}$ esu = $10^4\times (4\Pi)^{-1}$ G (uem),

Electrical resistance: R. Ohm(Ω) = $9^{-1}\times 10^{-11}$ emu,

A1.4 Fundamental Constants

Avogadro (Number of): N or $N_A = 6.022\times 10^{23}$ molecules per mole of a substance = $mol^{-1} = 6.022\times 10^{26}\ kmol^{-1}$,

Boltzmann (Constant): k_B, derived from the relation of a perfect gas containing N particles with a pressure p and a volume V at a temperature T: $pV = Nk_BT$ with $k_B = 1.3807\times 10^{-23}\ J\ K^{-1} = 1.3807\times 10^{-16}$ erg K^{-1} ues = $8.617\ eVK^{-1}$,

Bohr (radius of): $a_0 = h^2/(4\Pi^2 m_e e^2) = 5.29177\times 10^{-11}$ m = 1 emu;
(magneton of): $\mu_B = eh/8\Pi m_e c = 9.2741\times 10^{-24}\ JT^{-1} = 9.2741\times 10^{-21}\ ergG^{-1} = 5.7884\times 10^{-5}\ eV.T^{-1} = 5.7884\times 1^{-9}\ eV\ G^{-1}$,

Electron (charge of): $-e = -1.60219\times 10^{-19}C = 4.80324\times 10^{-10}$ emu;
(rest mass of) (m_e) = 9.1095×10^{-31} kg = 9.1095×10^{-28} g,

Electron-volt per particle (energy of an electron accelerated by a potential difference of 1 v): the electron volt (eV) = $1.602\times 10^{-19}\ J = 1.602\times 10^{-12}$ erg $\cong$ 2.30 6×10^4 cal $mol^{-1} \cong 2.41796\times 10^{14}$ Hz $\cong 8.0655\times 10^3\ cm^{-1} \cong 1.1604\times 10^4$ K.

Perfect gas (Constant of): $R = 8.314\ J\ K^{-1}\ mol^{-1} = 8.314\times 10^3$ erg $K^{-1}\ mol^{-1}$

Hall (constant of): $R_H = m^3C^{-1} = 10^{-3} \times 3^{-1}$ ues = 10^7 emu.
Kronecker (symbol of): $\delta_{ij} = 1$ if $i = j$ and $= 0$ if $i \neq j$
Light (speed of): $c = 2.997925 \times 10^8$ ms^{-1} = 2.997925×10^{10} cms^{-1}
Magnetic permeability of the vacuum: $\mu_0 = 4\pi \times 10^{-7}$ H m^{-1}; reminder of the relation $\varepsilon_0 \, \mu_0 = c^2$
Permittivity (or dielectric constant) of the vacuum: $\varepsilon_0 = 10^{-9} \times (36\pi)^{-1}$ F m^{-1}
Planck (constant of): h relates in the context of the wave-particle duality, the energy E of the particle with the frequency v of its dual wave: $E = hv$, with $h = 6.6262 \times 10^{-34}$ Js = 6.6262×10^{-27} erg s = 4.1357×10^{15} eVs.
It is often used to express the spin of a particle $\hbar = h/2\pi = 1.05459 \times 10^{-34}$ J.S = 1.05459×10^{-27} erg s = 6.5822×10^{-16} eV s.
Proton (charge of): $e = 1.602 \times 10^{-19}$ C;
(mass of): m_P = 1836,15 mass of the electron = 1.6726×10^{-27} kg = 1.6726×10^{-24} g.
Rydberg (constant of): $R_y = h^2/(8\Pi^2 \, m a_0^2) = 2.18 \times 10^{-18}$ J = 0.5 uem.

A2 Thermodynamic Relations and Descriptions

A2.1 Relations

The purpose of thermodynamic relations is to describe a macroscopic system based on statistical calculations, for example of microscopic results. Let's recall the difference of the so-called main properties between intensive and extensive parameters of the same nature (electrical: **E** and **P**, thermal: T and S, elastic: **σ** and **e**) and the so-called coupled properties (for example between **σ** and **P**).

By principle, the system is assumed to change from a stable state to another close stable state. We do not care about the path, the route taken between the two states but about these states of departure and arrival. Strictly speaking, this concerns reversible and quasi-static processes. Nevertheless, the approach, used with ferro materials, which are hardly reversible, meets with some success. We will not forget that this phenomenological approach should however meet the conditions initially defined to retain all its relevance.

The thermodynamic functions will depend on the parameters (**E**, **σ**, T) and (**P**, **e**, S), but it is demonstrated that it is enough to have three parameters to define the function, one per conjugate pair (**E**, **P**), (T, S), (**σ**, **e**). Depending on the experimental context, we will obviously choose the function and the parameters in the most judicious way. Table A2.1 presents eight useful functions for a "ferro-ferro" with the appropriate parameters. We notice that these parameters are the intensive parameters present by definition in the function and the extensive parameters that are absent from it.

Table A2.1 Thermodynamic functions with their usual names (left column), their definition in terms of intensive and extensive variables (central column), and the three most appropriate independent variables for the function (right column)

Name	Definition	Appropriate variables
Helmholtz function	$A = U - TS$	$T, \mathbf{P}, \mathbf{e}$
Gibbs function	$G = U - TS - \sigma\,\mathbf{e} - \mathbf{EP}$	$T, \mathbf{E}, \sigma$
Gibbs elastic function	$G_1 = U - TS - \sigma\,\mathbf{e}$	$T, \sigma, \mathbf{P}$
Gibbs electric function	$G_2 = U - TS - \mathbf{EP}$	$T, \mathbf{E}, \mathbf{e}$
Energy	U	$S, \mathbf{P}, \mathbf{e}$
Enthalpy	$H = U - \sigma\,\mathbf{e} - \mathbf{EP}$	$\sigma, \mathbf{E}, S$
Elastic enthalpy	$H_1 = U - \sigma\,\mathbf{e}$	$\sigma, \mathbf{P}, S$
Electric enthalpy	$H_2 = U - \mathbf{E}\,\mathbf{P}$	$\mathbf{E}, \mathbf{e}, S$

By respecting the principle of small variations between stable states, we can write the differentials of each of the functions by recalling the first law of thermodynamics:

$$\mathrm{d}U = T\,\mathrm{d}S + \boldsymbol{\sigma}\,\mathrm{d}\mathbf{e} + \mathbf{E}\,\mathrm{d}\mathbf{P} \tag{A2.1}$$

and obtain for example:

$$\mathrm{d}G_2 = \boldsymbol{\sigma}\,\mathrm{d}\mathbf{e} - S\,\mathrm{d}T - \mathbf{P}\,\mathrm{d}\mathbf{E} \tag{A2.2}$$

therefore: $-P = (\delta G_2/\delta \mathbf{E})_{T,\sigma}$; $-S = (\delta G_2/\delta T)_{\mathbf{E},\mathbf{e}}$; $\boldsymbol{\sigma} = (\delta G_2/\delta \mathbf{e})_{T,\mathbf{E}}$.

We obtain 24 equations of this kind between the parameters of the conjugate pairs (with a sign negative when the parameter on the left is extensive), valid for the recalled conditions (T and $\boldsymbol{\sigma}$ constant, etc…). Moreover, Annex A1.1 has defined the essential physical properties (and their conditions of definition), such as for piezoelectricity:

$$\mathbf{d} = (\delta\mathbf{P}/\delta\boldsymbol{\sigma})_{\mathbf{E},T} \text{ with } -\mathbf{P} = (\delta G_2/\delta\mathbf{E})_{T,\sigma}$$

The calculations that follow can be found in various reference works cited; the conditions of the various relations must be compatible (for constant values of…). These calculations allow to establish a number of relations; those that may appear or be underlying in the work are cited below:

$$\mathbf{s}_{\mathrm{E}} - \mathbf{s}_{\mathrm{P}} = (\delta\mathbf{e}/\delta\boldsymbol{\sigma})_{T,\mathbf{E}} - (\delta\mathbf{e}/\delta\boldsymbol{\sigma})_{\mathrm{T},\mathbf{P}} = -(\delta\mathbf{E}/\delta\boldsymbol{\sigma})_{T,\mathbf{P}}.(\delta\mathbf{P}/\delta\boldsymbol{\sigma})_{T,\mathbf{E}} \tag{A2.3}$$

$$\chi_{\sigma} - \chi_{\mathbf{e}} = (\delta\mathbf{P}/\delta\mathbf{e})_{T,\mathbf{E}}.(\delta\mathbf{P}/\delta\boldsymbol{\sigma})_{T,\mathbf{E}} \tag{A2.4}$$

$$\mathbf{1}/\chi_S - 1/\chi_T = q^2 T/C \tag{A2.5}$$

$$\mathbf{1}/\chi_{\mathbf{e}} - 1/\chi_{\sigma} = (\delta\mathbf{E}/\delta\mathbf{e})_{T,\mathbf{P}}.(\delta\mathbf{E}/\delta\boldsymbol{\sigma})_{T,\mathbf{P}} \tag{A2.6}$$

Similarly:

$$\mathbf{d}_T / \mathbf{d}_S = \chi_T / \chi_S \tag{A2.7}$$

$$\mathbf{s}_P / \mathbf{s}_E = \chi_\mathbf{e} / \chi_\sigma \tag{A2.8}$$

A2.2 Different Forms of Developments

The third section of Chap. 1 presents the "thermodynamic and phenomenological approach to the transition" using developments based on even powers of the order parameter in the simplest and most general case. This is quite suitable when the order parameter is simple, for example the polarization in a uniaxial crystal for a second-order transition. Even in this case, the limits for analysing results in which piezoelectric and electrostrictive properties come into play are observed. A good example was proposed by Fousek (1967), and thus, in various contexts, ferroelectric and ferroelastic crystals have been studied (Tagantsev et al., 2010). Regarding the crystals of the KDP family, the authors have retained the incidences between P_3, σ_3, and σ_6 from which a development of the Gibbs energy of the form:

$$\begin{aligned} G = G_0 + \alpha / 2.P_3^2 + \beta / 4.P_3^4 + \Upsilon / 6.P_3^6 \\ -1/2\, s_{33}{}^P \sigma_3{}^2 - 1/2\, s_{66}{}^P \sigma_6{}^2 - b_{36} P_3 \sigma_6 - Q_{33} P_3{}^2 \sigma_3 \end{aligned} \tag{A2.9}$$

from which we deduce relations such as:

$$e_6 = s_{66}{}^P \sigma_6 + b_{36} P_3 \tag{A2.10}$$

$$e_3 = s_{33}{}^P \sigma_3 + Q_{33} P_3^2 \tag{A2.11}$$

which, along with relations from Annex A1, show that KDP is also a ferroelastic with $s_{66}{}^E$ following a Curie law and becoming infinite for $T = T_0$ (or $c_{66}{}^P$ which tends towards zero as demonstrated experimentally in Fig. 1.27):

$$s_{66}{}^E = s_{66}{}^P + b_{36}{}^2 / \alpha_0 \left(T - T_0\right).$$

These reasonings remain in the context of a crystal with the same characteristics at every point ("mono-domain") with either the polarization $+P_3$, or its opposite. The presence of domain walls is not considered. This approach is sometimes used with the order parameter considered as an average value, and it sometimes works in "continuum" regions as discussed in paragraph 4.4. It is prudent to question the relevance of this phenomenological "convenience" and its limits. In the same reference (Tagantsev et al., 2010), a proposal is made for the development of improper uniaxial ferroelectric-ferroelastic crystals such as TMO and GMO in which the order parameter is neither the polarization, nor the deformation. This avoids confusing properties related to symmetry changes, those related to the order parameter, and those specific to domains and their walls, the subject of the book.

A3 Gamma Ray Diffraction and Neutron Diffraction

A3.1 Gamma Ray Diffraction with a Twinned Crystal

Results obtained by gamma diffraction are used in the book; this annex summarizes the characteristics of the setup used (Schneider, 1974a, b; Bastie et al., 1976a, b; Bastie and Bornarel, 1979).

Figure A3.1 illustrates the experimental situation, defines the scanning angle ω and the characteristic reflection angle θ_B. Depending on the position of the γ-ray beam that passes through the crystal, we obtain the scanning curves presented in Fig. A3.2. With a beam divergence greater than the width of the observed peak (for example a few arc minutes) centred on the peak maximum (see around ω_{II} in Fig. A3.1b), it is possible to perform intensity measurements and knowing that of the incident beam, to deduce the integrated reflecting power proportional to the diffracting volume (as shown in Fig. 1.34c) with a precision of 10^{-2}.

A3.2 Differential Gamma Diffractometry

The initial idea of synchronous detection with several counters was first developed in X-rays (Lissalde, 1974); Fig. A3.3 summarizes its transposition to γ diffraction. It allows the measurement of the displacement and shape change of a peak, thus the

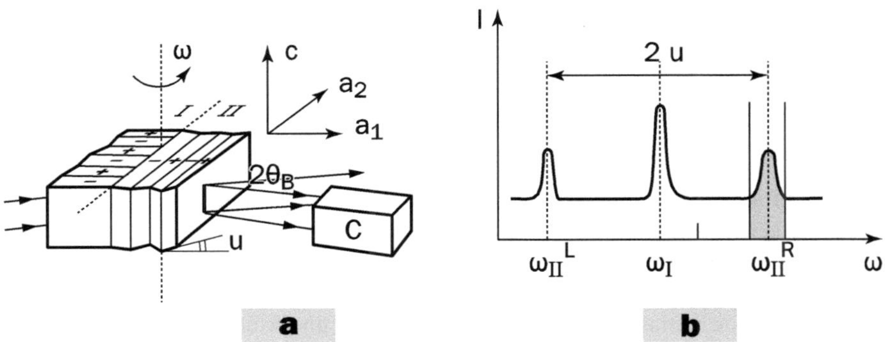

Fig. A3.1 (**a**) The studied crystal (of KDP) is placed in a goniometer which allows a horizontal translation, following a_2 in the diagram and a rotation of 360° of arc around the c axis characterized by the sweep angle ω. Thus, sweep curves can be drawn with γ beams close to the a_1 axis in the diagram, beams whose minimum divergence is 10 arc seconds. (**b**) Such a sweep curve is obtained with steps of 2 arc seconds. The difference between the peaks allows to reach the spontaneous deformation of the twins

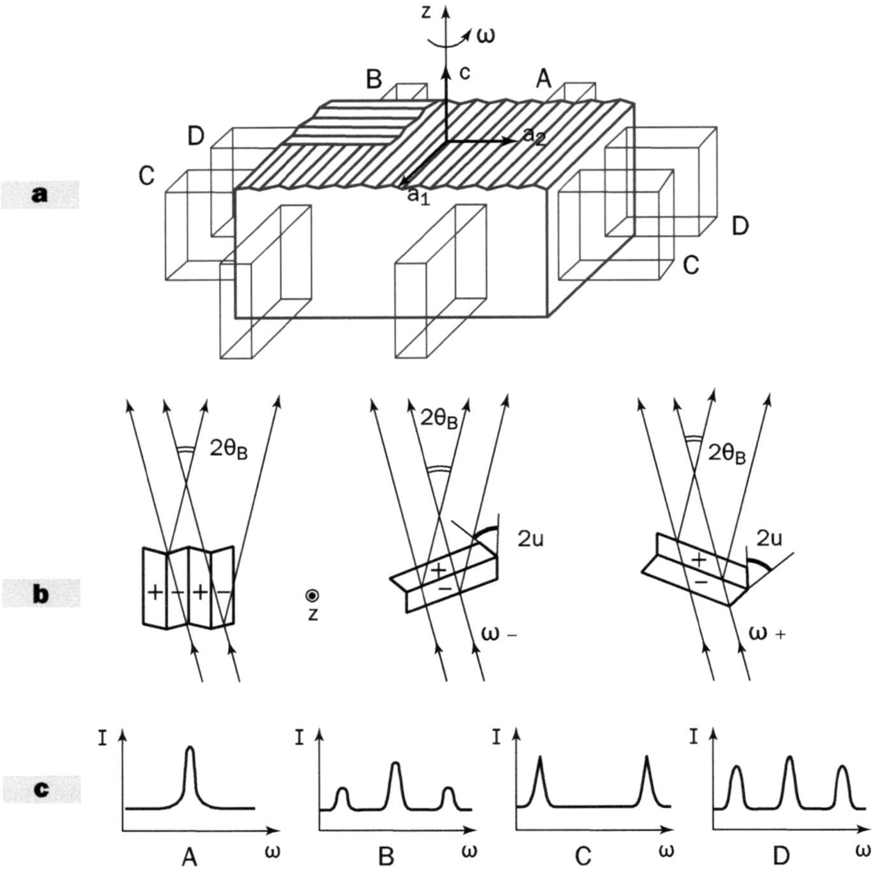

Fig. A3.2 (**a**) The position of the γ-ray beam can be adjusted to cross the twinned sample as indicated by A, B, C, and D, which allows rotation around the *z*-axis parallel to c, to obtain the reflections on the lattice planes illustrated in (**b**) and the scanning curves illustrated in (**c**). The difference in ω obtained between two peaks from opposite deformation domains is a direct measure of twice the practical shear deformation *u*

modification of a domain deformation, under the action of a field of "same sense or opposite sense" polarity. For the experiments presented in Chap. 3, the polarity inversion time was typically 100 μs.

With a large beam divergence, it is also possible to perform measurements of domain volume variations when the conditions of the kinematic regime are met (with a typical precision of 10^{-2} in a few minutes and 10^{-3} in about an hour).

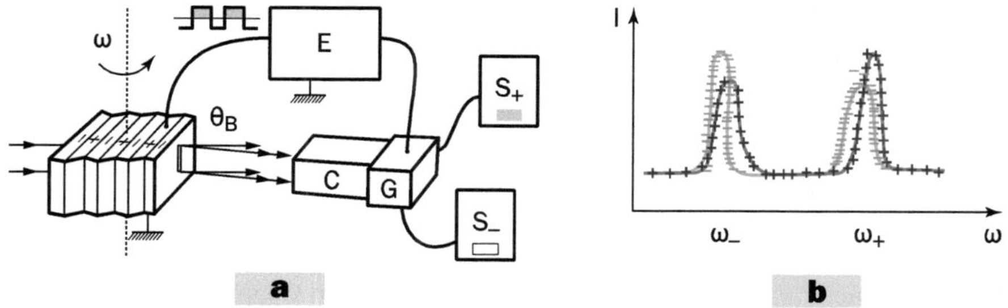

Fig. A3.3 (**a**) The setup presented in Fig. A3.1a is completed by the application of an electric field on the sample equipped with electrodes on the (001) faces. If the polarization is positive, the S_+ counter measures the diffracted intensity and the S_- counter is used for reverse polarization. (**b**) For the same beam, two scanning curves can thus be drawn for the same region of the crystal, each corresponding to a polarization of the applied field. The inverse effects on the + domain peaks and those of the − domains are observed

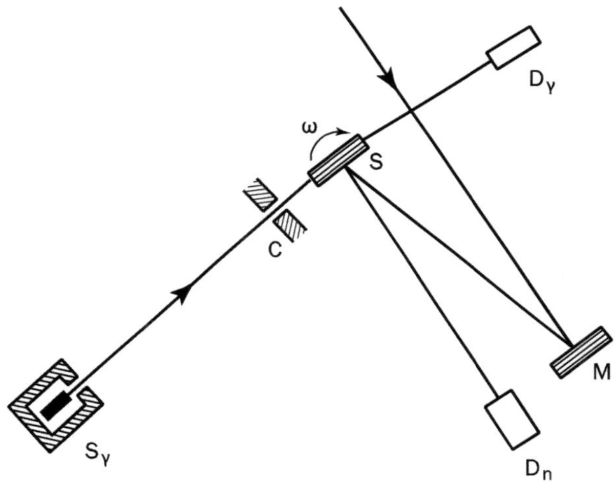

Fig. A3.4 Schematic of the simultaneous γ diffraction experiment setup (source $S\gamma$, detector $D\gamma$, and sample S in the situation of Fig. A3.2 with θ_B of about half a degree of arc) and neutron diffraction (D_n) at two crystals (S and M) with an angle θ_B between 20 and 85° of arc

A3.3 Simultaneous Measurements of Gamma Ray Diffraction and Neutron Diffraction

The γ diffraction presented above can give very precise results, but their interpretation can be skewed if the elementary cell inside the domains deforms at the same time in shear and contraction-dilation. We can know these modifications by accumulating measurements of γ diffraction on the one hand, and X or neutron diffraction on two crystals and large Bragg angles on the other hand. For massive crystals, the experiment with neutron diffraction presented in Fig. A3.4 simultaneously allows the two measurements while minimizing losses by absorption (Freund and

Schneider, 1972; Zeyen, 1975). Thus, variations in lattice parameters are measured with a precision of 10^{-6}. By correctly aligning the beams to obtain information on the same volumes of the sample, we can write:

$$\Delta\omega = \Delta d / d \cdot tg\theta_B + \alpha$$

For γ diffraction, $tg\theta_B$ is approximately equal to 10^{-2} and α is the most important, whereas for neutron diffraction, $tg\theta_B$ of about 10 in value makes the variation of the lattice parameter d predominant. The results proposed in Chaps. 1, 5, and 6 come from such setups (Bastie et al., 1987).

A4 Structural Parameters of the KDP Family Crystals

The results concerning the elementary cells and atomic positions of the compounds in the para and ferro phases are summarized below.

The ferroelectric properties of the KDP crystal, shown in Fig. 1.30, have been known for a long time (West, 1930; Busch, 1938; Busch and Scherrer, 1935; Busch and Ganz, 1942). X-ray diffraction studies have revealed the elementary cell (Fig. 1.7) and the atomic arrangements shown in Fig. 1.31 (De Quervain, 1944; Ubbelohde and Woodward, 1947). The elementary cell contains four XO_4^- (X = P, As, …) tetrahedra linked by hydrogen bonds (with H or D) with, between each tetrahedron along the c axis, alkali ions M^+ (M = K, Rb, Cs, NH_4, …). The tetragonal axes indicated a_1 and a_2 in Fig. 1.31 correspond to the horizontal edges of the parallelepiped in the figure, while the axes a_1' and a_2' are the horizontal traces of the mirror planes m_1 and m_2 in Fig. 1.7. Neutron diffraction (Bacon and Pease, 1953, 1955) provided information about the changes in atomic arrangements between the para and ferro phases (Fig. A4.1): the chemical formula is preserved at the level of each tetrahedron in the ferro phase in a region of homogeneous polarization of the crystal; the position of the hydrogen (or deuterium) ions on their bond and the displacement of the M and X ions in the direction of the c axis are correlated as illustrated in Fig. A4.1d. Numerous studies, notably of neutron diffraction and scattering (Kuhs et al., 1983; Nelmes et al., 1982, 1985), have specified the density of presence of the hydrogens on the bond (or deuteriums in the case of deuterated compounds such as KD_2 PO_4) as a function of temperature, in the para phase and in the ferro phase (Fig. A4.2a–c). The orientations of the segments joining two oxygens and those of the segments between the two positions of the hydrogen on the bond (Kennedy and Nelmes, 1980) have also been determined. Similarly, the link between the variation of this last segment and the rotation of the tetrahedra (Nelmes et al., 1978) has been demonstrated.

In the para phase, the proton (or D) is statistically located in the middle of the hydrogen bond, while in the ferro phase, it is closer to one of the two MO_4^- tetrahedra. If the hydrogens of the two lower bonds (along the c axis) are close to an MO_4^- as in Fig. A4.1d, the dipole moment contributing to the polarization will be along

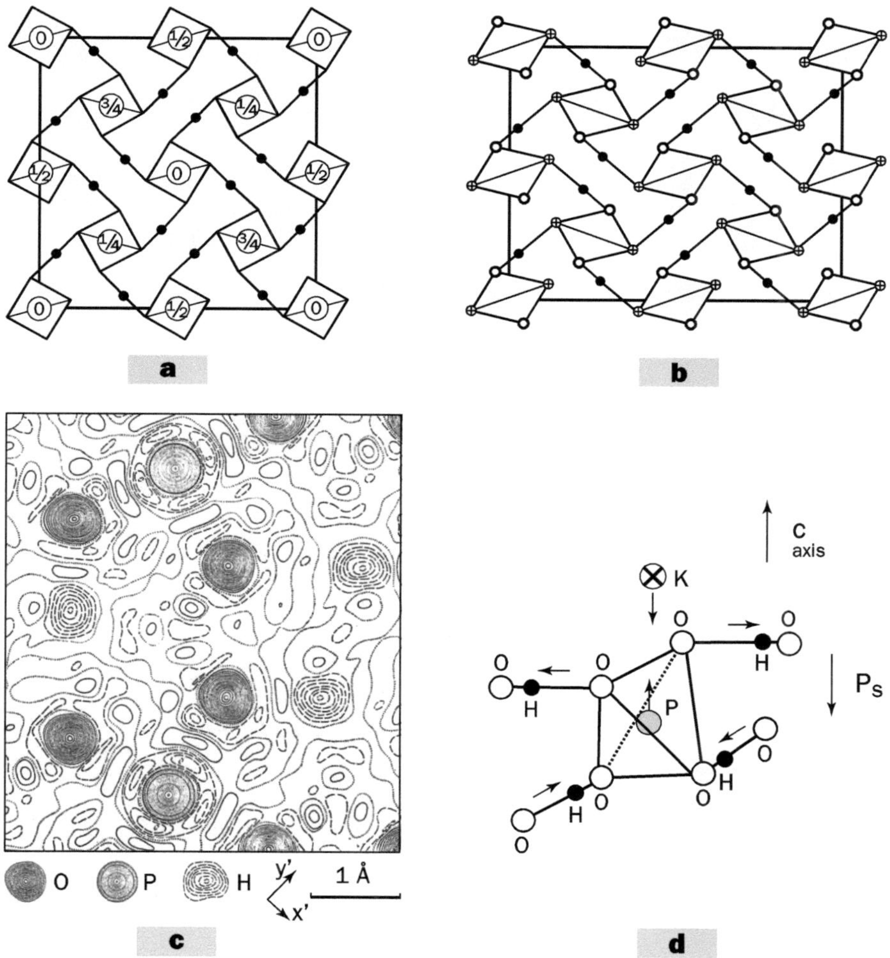

Fig. A4.1 De Quervain (1930) was the first to propose an arrangement between the PO_4^{--} tetrahedra and the hydrogen bonds projected in a plane perpendicular to c with very amplified deformations to make visible the changes from the para (**a**) phase to the ferro (**b**) phase
Neutron diffraction allowed the atoms to be well located for example at 77 K by Bacon and Pease (1955) in (**c**). In fact, the orthorhombic deformation is detectable on this projection: the angle of the figure, which was 90 K at ambient, is 89°33′ for the angle of the figure at the top left; it was thus possible to correlate the positions of H, P, and K: when the spontaneous polarization P_s is oriented in the negative direction of the c axis, the hydrogens of the hydrogen bonds at the bottom of a tetrahedron are close to it as indicated in (**d**).

this *c* axis in the negative direction; if, on the other hand, the hydrogens close to the tetrahedron are those of the highest bonds, the moment will be oriented in the positive direction of *c*. Numerous studies have made it possible to correlate the structural parameters and the transition temperature in all compounds, but we will remain

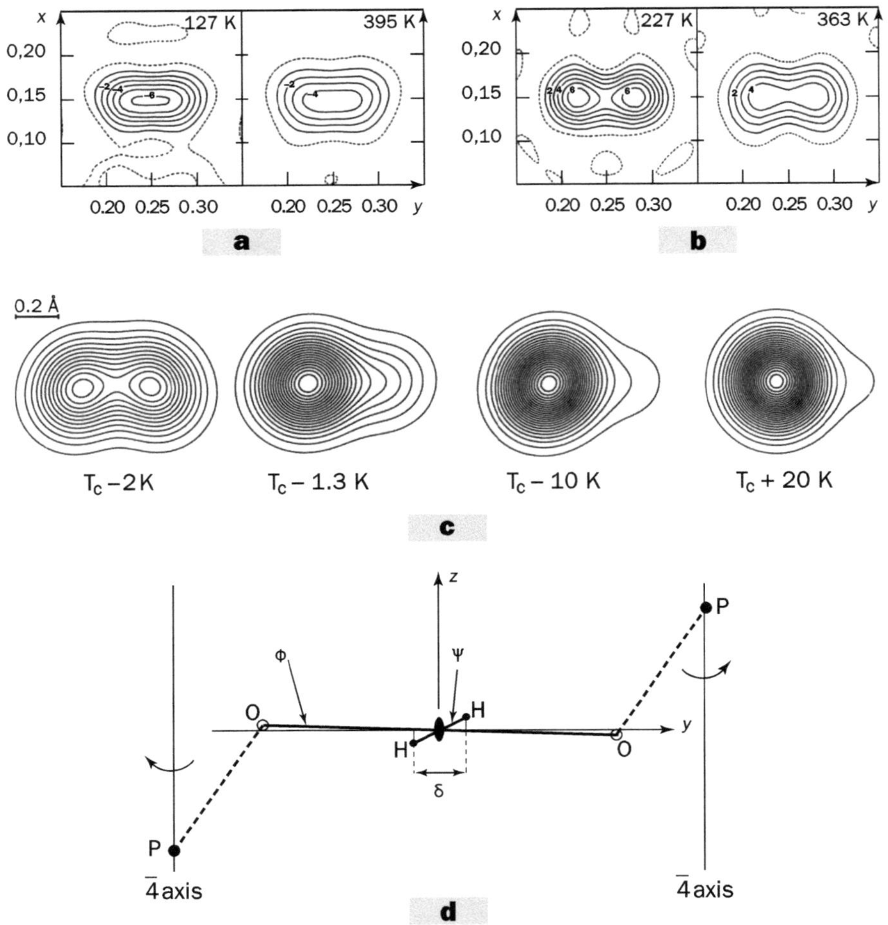

Fig. A4.2 The neutron scattering studies have allowed a better understanding of the positions of the hydrogens (or deuteriums) on the hydrogen bond and its orientation: (**a**) Density of presence of protons in projection in the (001) plane at the indicated temperatures (Nelmes et al., 1982) (KDP). (**b**) Density of presence of deuterons in projection in (001) (Nelmes et al., 1982) (DKDP). (**c**) More detailed results (Nelmes et al., 1985). (**d**) Schematic representation of the O–O and H–H segments in the KDP (Kennedy and Nelmes, 1980) and definition of the distance δ

focused on KH_2PO_4 and KD_2PO_4. The results are less simple than those extrapolated from simple reasoning concerning the dimensions along c and hydrogen bond chains (Albers and Klöpperpieper, 1981). Figure A4.3 proposes a focus on the central tetrahedron of the cell and the hydrogen bonds that connect it to the neighbouring tetrahedra. Figure A4.4a, b defines the following quantities (Ichikawa, 1981; Ichikawa et al., 1987, 2001; Tibbals et al., 1982):

– R distance between the two oxygens of the O–H–O bond, whose orientation is close to that of the tetragonal axes a_i without being along these axes,

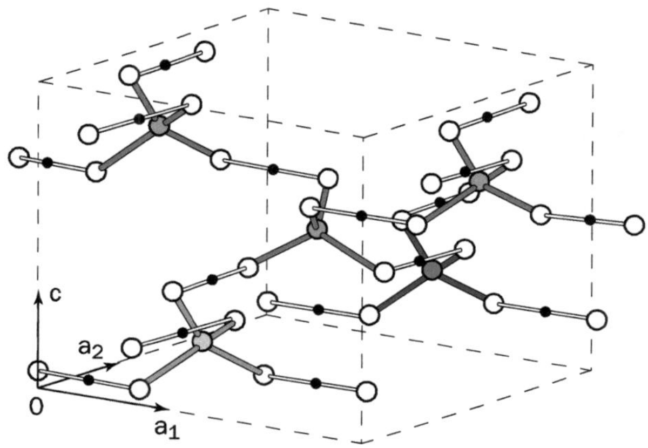

Fig. A4.3 Partial reprise of Fig. 1.31 limited to a tetrahedron with P of coordinates (1/2, 1/2, 1/2) in the reference frame (a_1, a_2, c), and the four tetrahedra directly linked to it by hydrogen bond: the coordinates of their P atoms are (0, 1/2, 3/4), (1, 1/2, 3/4), (1/2, 0, 1/4), and (1/2, 1, 1/4), therefore with ordinates that differ along c by 1/4 of the cell dimension. The four hydrogen bonds are approximately parallel to a_1 and a_2, and the four tetrahedra linked with the central tetrahedron are of the same type, different from this

- α_1 angle between two PO bonds that concern the oxygens of the tetrahedron towards positive c,
- α_2 angle between two PO bonds that concern one oxygen of the tetrahedron towards positive c and the other an oxygen towards negative c,
- s distance between the P and K ions along c,
- In projection in a (001) plane, the angle θ between the segment that joins two neighbouring P and the trace of a PO bond taken as a reference allows to characterize the rotations of the PO_4^{--} tetrahedron around the c axis.

We will add a reference to Fig. A4.2d:

- δ distance on the O–H–O bond between the projections of the two positions that H (or D) can take when it is closer to one oxygen or the other.

X-ray and neutron diffraction experiments have allowed to apprehend the evolution of the previous parameters as a function of temperature and hydrostatic pressure on the one hand, and to demonstrate the link between the transition temperature and these parameters on the other hand. From the extensive bibliography on these subjects (Ichikawa, 1981, Ichikawa et al., 1987, 2001; Tibbals et al., 1982), we will retain the following conclusions:

- As the hydrostatic pressure increases, the distances R and δ decrease while the angle θ increases, and the lattice parameters a and c decrease.
- The transition temperature T_C decreases when the pressure increases, and if δ becomes zero, the transition would no longer be possible. This is observed for a pressure of 17 kbars in the KDP at a transition temperature of 0 K; for the DKDP, calculations indicate that a pressure of 65 kbars would be necessary to achieve this result.

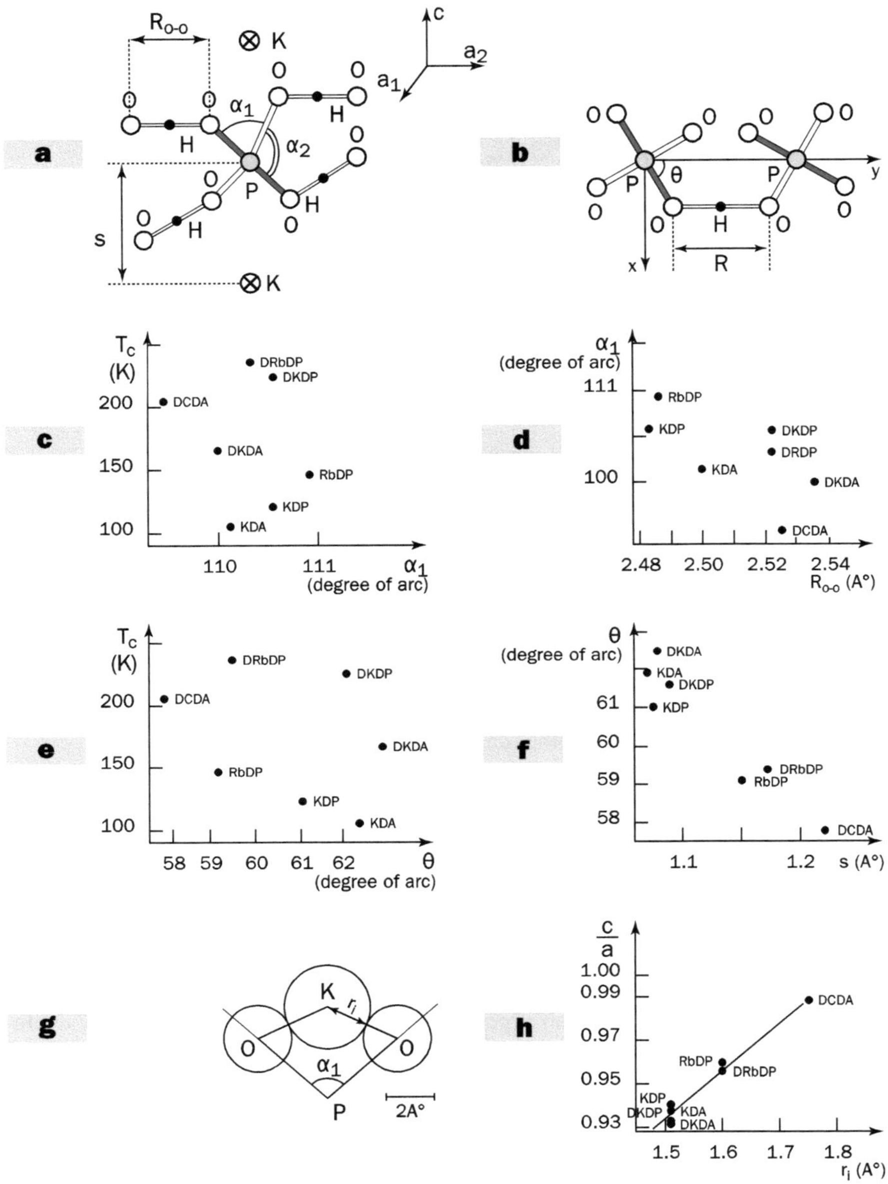

Fig. A4.4 (**a**) The distances and angles used in the text are defined with in (**b**) a projection in the (001) plane. The results for the various compounds are extracted from the publications of Ichikawa et al., (Ichikawa, 1981; Ichikawa et al., 1987, 2001); (**c**) T_C as a function of α_1; (**d**) α_1 as a function of R_{OO}; (**e**) T_C as a function of the angle θ; (**f**) θ as a function of s; (**g** and **h**) define the ionic radius r_i and its correlation with the c/a ratio

– The results of various compounds compiled in Fig. A4.4 c–h show that the distance R is not an independent parameter: the quantities defined previously are linked by the following empirical formulas (Ichikawa et al., 2001):

$$T_C = 2850R + 69.5\alpha_1 - 15.36\theta - 13720,$$
$$R = -\ 2PO(\cos\theta)\sin(\alpha_1/2) + a/2,$$
$$S = c/2 - PO\cos(\alpha_1/2) - r_i,$$

with r_i ionic radius of the K atom. Figure A4.4 h shows that the ratio between the two lattice parameters c and a depends directly on this ionic radius.

The distance δ between the two positions of the hydrogen (or deuterium) on the bond seems an important marker, and for various hydrogen-bonded materials, the transition temperature seems to vary linearly with δ as shown for the KDP: T_C proportional to (δ −0.24 Å) (Nelmes et al., 1991; Stasyuk et al., 1999; Silka and Sharma, 2008).

When the crystal goes from the para phase to the ferro phase, the lattice parameters c and a, like the volume of the elementary cell V, vary in the opposite direction between hydrogenated crystals and their corresponding deuterated (Cook, 1967; Garber and Smolenko, 1969)…:

$\Delta c/c$ of a deuterated crystal > $\Delta c/c$ of a hydrogenated crystal
$\Delta a/a$ of a deuterated crystal < $\Delta a/a$ of a hydrogenated crystal
$\Delta V/V$ of a deuterated crystal < $\Delta V/V$ of a hydrogenated crystal
(Note that these inequalities are reversed when it comes to anti-ferroelectric compounds)

These correlations between the structural parameters and the transition temperature have been decisive in clarifying the question of the tunnel effect and to demonstrate that the transitions of the crystals of the KDP family were of nature, and order-disorder, and displacive.

These solid experimental results are used in the many theoretical publications that propose pseudo-spin models with different approximations (cluster of 4 particles and assumptions on tunnelling and the range of interactions). The simulations and comparisons with the experimental results often depend on the choice of these, and the deviations are usually explained by internal constraints (Bartis, 1971) without simultaneous prospective on interfaces, domains, or/and phase fronts. The richness of these works does not seem to diminish over time.

A5 Transition Models for KDP Family Crystals (without domain and phase front)

The models of this para-ferro transition are numerous. They had stated objectives to answer the following questions:

- is the transition of the "displacive" type or "order-disorder"? In the first case, the transition would occur by small displacements of atoms, ions. In the second case, stable states would be located with equal probability in one or the other of two possible positions above the transition temperature, and preferentially in one rather than the other below this temperature.
- What is the order parameter, the physical quantity that accounts for the transition, the one whose appearance at the transition temperature allows a decrease in energy? The answer seems obvious when only polarization is at stake, for example for non-piezoelectric crystals in their para phase, or in ferro-elastic crystals in which only deformation appears. But, even in such cases, is it by coupling with the true parameter of order that these quantities appear? The answer to the question "proper or improper?" is not so obvious when, as in the chosen materials, deformation, polarization, and other quantities such as birefringence are linearly coupled.
- Given the similarities observed in crystals with hydrogen bonds, what is their role in the transitions of KDP family crystals? How does hydrogen (or deuterium) find itself on the O–H–O bond? Is it statistically in the centre of the bond in the para phase with a centred potential well or equally distributed between two symmetrical wells?
- The study of the transition of KDP and its compounds would justify a complete work; only elements useful for the understanding of the present work are summarized here.

To explain the para-ferro transition of crystals with hydrogen bonding, two approaches were initially used:

- -One was based on molecular field theories (Mason 1946, 1950; Pirenne, 1949, 1955; Blinc, 1960); it accounted for isotopic variations of transition temperatures, without explaining the variations of a crystal's parameters with temperature;
- -One started from an analysis of the situation of protons and the laws of occupation of hydrogen bonds around the tetrahedron PO_4^-. Thus, Slater (1941) suggested a theory of proton order-disorder with local interactions and Pauling's rules (Pauling, 1935) established for ice (only one hydrogen per bond and the conservation of the chemical formula at the level of each tetrahedron), a hypothesis well verified since, including for deuterated compounds (Kind et al., 2002). In this model, the energy of a tetrahedron in the ferro phase with the dipole moment following the c axis is considered equal to zero if the two protons close to the tetrahedron are either on the lower bonds (dipole moment $-\mu_0$ following c) or on the upper bonds (dipole moment $+\mu_0$ following *c*). The energy of a tetrahedron with a dipole perpendicular to the *c* axis (approximately following a_1 or a_2) is taken equal to a value E_0 (configuration called since "Slater defect"). The other positions of protons relative to the tetrahedron were considered of infinite energy by Slater. With these assumptions, Slater proposed a statistical interpretation of the transition. In his model, the arrangements of protons that respect the chemical formula have the same probability in the para phase. The model predicts a first-order transition when $-E_0/kT = \ln(1/2)$ with a spontaneous polarization of

undetermined value, but it does not propose any mechanism to interpret the dynamic properties of the crystal.

Takagi (1948b) assumed, in addition to the previous assumptions, the possible existence of HPO_4^- and $H_3PO_4^+$ ions (called since "Takagi defects") with an energy E_1 positive, and Bjerrum defects (four protons or no protons close to a PO_4^- tetrahedron) of energy E_2 infinite. Takagi defects are supposed to move in the crystal by a simple change of position of hydrogens on their bond. This game of moving protons on hydrogen bond is at the origin of the name "bound charge semiconductors" (Schmidt, 2020). The model allows predicting a second-order transition and proposes a mechanism of response of the crystal to an electric excitation field following a and following c (Takagi, 1948a) and a variation of the polarization as a function of temperature (Ishibashi, 1987).

Slater quickly noted (Slater, 1950) the need for long-distance interactions (assuming dipolar forces on the "local entities"). Ten years later and almost simultaneously, Blinc proposed the tunnelling of protons on hydrogen bonds (Blinc, 1960) and Cochran affirmed the essential role of the network and of all ions (Cochran, 1960; Anderson, 1960). Senko introduced, in addition to interactions between close neighbours, a long-distance interaction between dipole moments (Senko, 1961). Then, for about 20 years, various experiments and models tried to make a choice between the game of protons on their bonds with transfer times of 10^{-12} s (Tokunaga, 1970; Sugimoto and Ikeda, 1994; Sugimoto and Ikeda, 1996), and long or short distance couplings (Blinc and Svetina, 1966; Kobayashi, 1968; Blinc and Zeks, 1982; Tibbals et al., 1982; Tun et al., 1988) (introducing various definitions concerning the local entity, the "quasi-spin" or "pseudo-spin", and the variation of polarizability (non-linear) of this pseudo-spin (Migoni et al., 1976; Blinc and Zeks, 1982; Bilz et al., 1987). Experimental and theoretical arguments have led to the acceptance that it was possible in KDP to have a coexistence of order-disorder type situations (Tokunaga et al., 1985; Schmidt, 1987a, b; Kikuta et al., 1995) and "displacive" type (Bussmann-Holder and Michel, 1998; Dalal et al., 1998; Tokunaga et al., 2000; Zhang et al., 2002). Since these works based on experimental results (technologies and experimental conditions, the models value the games of pseudo-spins, such types of close interactions, the non-linear polarizability of local ions (Bussmann-Holder and Dalal, 2000) and the long-distance interaction with variants on its origin.

With these tools and hypotheses, we can predict both first- and second-order transitions taking into account both local orders and long-distance interactions as proposed by Silsbee et al., (Silsbee et al., 1964a, b) (referred to in the rest of the text as SUS). Retaining the definitions of Fig. 1.32, these authors write for a set of N tetrahedra in a volume V, the polarization P along c, the energy U, and the entropy S in the following form:

$$P = \left(N\mu_0 / V\right) p + \chi_0 E$$

with the first term which takes into account the configurations of the protons ($p = x_2 + x_{-2} + 2(x_1 + x_{-1})$), and the second of field contributions (x_i and E_j defined in Fig. 1.32),

$$U = N(4x_0E_0 + 4(x_1 + x_{-1})E_1 + 2x_4\ E_2 - \mu_1\ Ep - \beta p^2 - \gamma E^2$$

with $\mu_1 = \mu_0 + 2V\chi_0\ \beta/N\mu_0$ in which β depends on the shape of the volume V,

Depending on the assumptions made, we find the results of Slater (with $E_1\ \infty$, $\beta = 0$) and Takagi (with E_1 finite, $\beta = 0$). This framework, with various configurations for the protons, is used every time a statistical approach is adopted, thereby admitting, in fact, in the ferro phase, the existence of a homogeneous medium without taking into account the role of domains and their walls. This approach was supported by conductivity and relaxation time measurements which demonstrated in DKDP that proton, deuteron jumps exist in the para phase from one bond to another only for temperatures much higher than those of para-ferro transitions (Schmidt and Uehling, 1962): between bonds, the lifetime for the deuteron at 70 °C is 15 ms and the activation energy is 0.58 eV, while for values of E_0 of 0.00132 eV and of E_1 of 0.078 eV, the jump times ν^{-1} (with ν jump frequency on the link) between two positions on the same link are about 10^{-11} s.

Other avenues have been explored to explain the differences in properties of KDP compared to DKDP: proton tunnelling (Imry et al., 1965; Blinc and Svetina, 1966; Litov and Garland, 1970), impossible with deuterons; a pseudo-spin model that introduces interaction energies between nearby protons around a tetrahedron (Tokunaga and Matsubara, 1966); the electrostrictive effect with possible long-distance anti-polar interactions (Torstweit, 1979).

The results concerning deuterated crystals are explained quite satisfactorily using a two potential well model with jumps between the two possible positions of the same O–D–O bond concerning the para-ferro transition (Blinc et al., 1977) and the effect of applied external electric fields (Schmidt and Uehling, 1962; Silsbee et al., 1964a, b; Imry et al., 1965; Blinc and Svetina, 1966; Schmidt, 1967, 1987a, b; Bjorkstam, 1967; Strukov and Levanyuk, 1998; Tagantsev et al., 2010).

The situation of KDP remained open for a long time due to the hypothesis of proton tunnelling on hydrogen bonds and doubts about the order of the para-ferro transition. After the experimental demonstration of a first-order transition by calorimetric measurements (Reese and May ,1963) and optical (Vallade 1975) for the KDP and comparison with DKDP (Reese and May, 1967; Reese, 1969a, b), work as a function of hydrostatic pressure on KDP (Schmidt, 1974; Schmidt et al., 1976; Zisman et al., 1980) and on RbDP (Vallade and Dolino, 1982) helped to understand the tri-critical situation of these transitions close to the first order for one (KDP) and the second for the other (RbDP). Numerous studies have also helped to model the anti-ferroelectric properties of compounds from the same family of crystals and the "proton glass" of mixed crystals (Schmidt, 1987a, b).

Measurements as a function of hydrostatic pressure were determining: the Curie temperature decreases as the hydrostatic pressure increases to become zero for a critical value. In KDP, this value is experimentally about 17 kbar, for ADP 33 kbar (Samara, 1971), for RbDP 15.2 kbar (Peercy and Samara, 1973), and a disappearance of the transition at 0 K for DKDP under a pressure of 65 kbar (Samara, 1974). The Curie law remains followed for large temperature intervals with a slight modification of the Curie constant. However, this law is no longer followed at very low temperature before the disappearance of ferroelectricity. The decrease in T_C occurs simultaneously with that of the distance $R_{O\text{-}O}$ between two oxygens of a hydrogen bond. This phenomenon called "quantum paraelectricity" (Muller and Burkard, 1979) also occurs in other ferroelectrics. Attempts at description use the Barrett formula (Barrett, 1952). The models proposed based on this formula use either the hypothesis of tunnelling in the O–H–O bond (Samara, 1979; Tokunaga and Matsubara, 1987) or a model of quantum anharmonic oscillator with coupling by short-distance electron clouds and long-distance electrostatic interactions (Endo et al., 2002a, b). The second hypothesis seems more solid since the phenomenon exists as well for DKDP as for KDP. For these authors, the transition at ambient pressure would be order-disorder and would gradually become displacive at increasing pressure; the coupling between the movement of protons and the torsion of the PO_4^- tetrahedra is proposed.

The transition of KDP was presented as improper type. Following the pseudo-spin model proposed by de Gennes (1963) which describes the order of protons on their hydrogen bond, the transition was attributed to the condensation of a proton mode strongly coupled to the optical vibration mode K-PO4 (Kobayashi, 1968), the whole being itself linearly coupled to the shear acoustic mode by piezoelectric effect, which had been predicted by Cochran on the basis of considerations on network dynamics (Cochran, 1961). As the elastic constant linked to shear in the plane perpendicular to the c axis shows a clear decrease at the transition (Fig. 1.27), it was believed to be a second-order transition for KDP, so close to the tri-critical situation (Brody and Cummins, 1968). This proximity will be retained for certain compounds of the work with tri-critical situations (KDP, RbDP, and mixed crystals with or without deuterium) and on the other hand, the differences in contributions between short-distance interactions (the protons of the other bonds of the "cluster" formed by a tetrahedron and its hydrogen bonds) and those at greater distance. For the future, we will remember the role of Takagi groups in dynamic phenomena, well demonstrated by NMR experiments as well as their creation (in DKDP) in pairs DPO_4–D_3PO_4 (or HPO_4–H_3PO_4 for non-deuterated crystals), followed by a migration of these ions in the crystal (Schmidt and Uehling, 1962; Schmidt 1966; Schmidt, 1967; Schmidt, 1968. It is now accepted that the two aspects, "displacive" (Kweon et al., 2017) and "order-disorder," co-exist in the transition of crystals from the KDP family (Merunka and Rakvin, 2007; Dalal et al., 2007).

For about 20 years, numerous publications have been found in the bibliography, each making certain assumptions about the definitions and properties of "pseudo-spin" and the nature and importance of "long distance" interactions (Abalmassov et al., 2012; Abalmassov 2016; Koval et al., 2010). Similarly, syntheses of this wide variety of studies are proposed (Bussmann-Holder and Dalal, 2007). We suggest to the interested reader to consult those.

For the continuation of our discussion and in the spirit of this book, we will follow the approach of Schmidt (2020) and his statistical approach in a continuous medium ("bound charge semiconductor model"), providing, in the ferro phase, numerical simulations of polarization as a function of temperature. To do this, Slater ion chains are introduced (Schmidt, 1979), terminated at both ends by Takagi ions whose densities vary with temperature (Schmidt, 1988). Figure A5.1 gives an

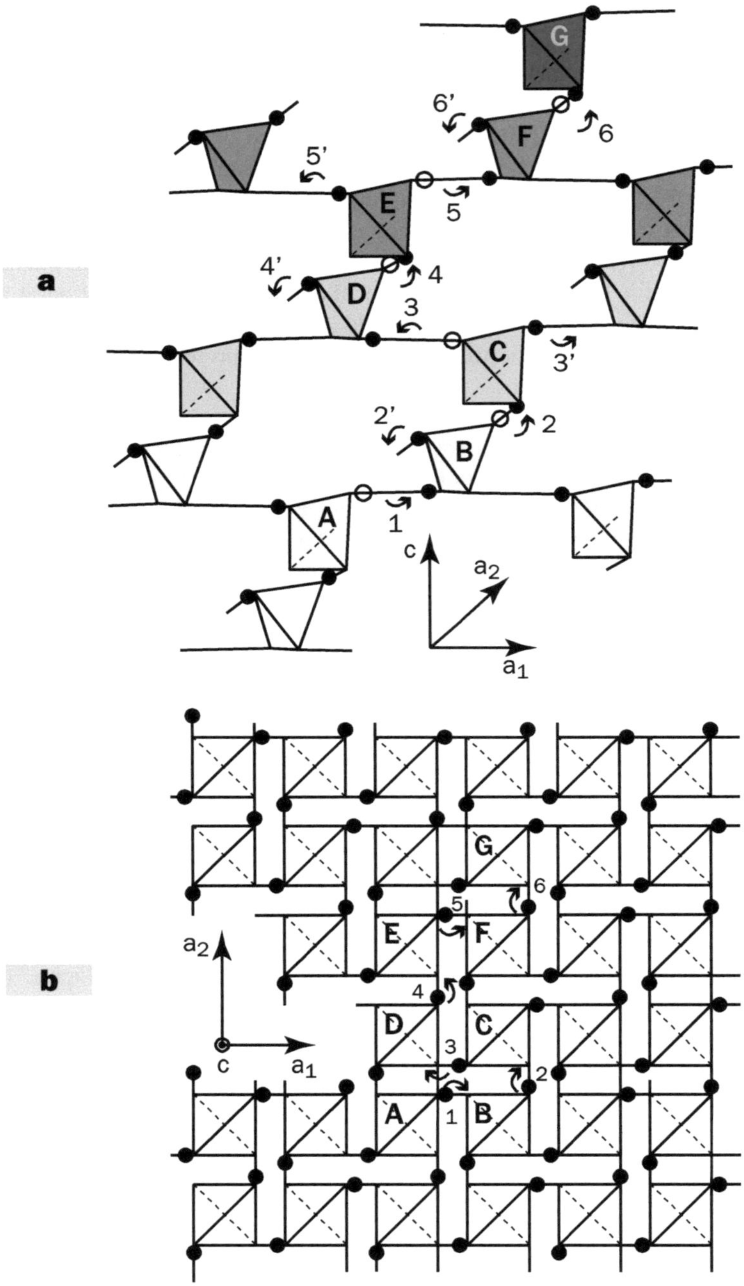

Fig. A5.1 (**a**) Revisiting the 3D perspective representation of West (1930) (Fig. 1.31), the process of creating an ion pair (HPO_4–H_3PO_4), the movement of the $(H_3PO_4)^+$ ion by successive proton movements on their bond with the creation of a series of Slater defects (Schmidt, 2020) are

(continued)

illustrated. The crystal is assumed to be mono-domain, with moments oriented towards positive c; therefore, the protons are close to the tetrahedra on the high bonds. The process begins in A with, in white, the initial position of the proton near the A tetrahedron (in the hypothesis of Schmidt (2020)) and, in black, the position of the proton after jump 1. After this displacement 1, A becomes an ion $(HPO_4)^-$; after the displacement of another proton indicated 2, B becomes a Slater defect and C, an $(H_3PO_4)^+$ ion; after the proton 3 displacement, it is C that has become a Slater defect and D an $(H_3PO_4)^+$ ion…after the displacement 6, the $(H_3PO_4)^+$ ion is in G. The representation highlights the two families of tetrahedra of different orientations in the crystal, the fact that the proton displacements occur alternately along a bond close to the tetragonal axis a_1, then along another close to the tetragonal axis a_2 (and vice versa), and also, that this progression of $(H_3PO_4)^+$ ion takes place towards the positive c axis. The varying shades of grey allow us to see that the progression, in our case, occurs towards the positive a_2 thus towards the bottom of the sheet (if the centre of the axes was placed at the phosphorus of A, we would have for the various phosphorus concerned the coordinates A (0,0,0), C ($a_1/2$, $a_2/2$, $c/2$), E (0, a_2, c), G ($a_1/2$, $3a_2/2$, $3c/2$). In the hypothesis retained with proton movements without return in backwards and thus cancellation of ion pairs, the ions $(H_3PO_4)^+$ progress towards positive c, with numerous possible paths. This is what the arrows 2′, 3′, 4′, 5′, 6′ which could have replaced 1, 2, 3, 4, 5, 6. We take into account the set of possible sequences of Slater defects, functions of the choices of type 2 or 2′, 3 or 3′, etc. With a hypothesis of choice of the same probability (between 2 and 2′…), it is demonstrated that an ion $(H_3PO_4)^+$ created on the surface of a plate perpendicular to c will have a maximum probability of reaching the other surface, in line with c from the starting point. With the same number of movements on hydrogen bonds, it can arrive in the same time at a point defined by a trajectory at 45° from the axis, but with a near-zero probability. We can obviously imagine multiple paths with very long journeys that quickly become unrealistic given the anisotropy of the system, especially in the case where an electric field is applied along c. (**b**) In the simplistic representation used in the book (Figs. 2.10, 2.27, and 2.28 …) in projection in the plane (001), the proton movements of Fig. A5.1a are reproduced. We note that after the displacement of a proton on a bond, it is the displacement of another proton on an approximately perpendicular bond that occurs. The $(H_3PO_4)^+$ ion comes towards us, given the choice of orientation of this mono-domain matrix

example of such chains (Schmidt, 2020). Schmidt calculates in a mono-domain matrix, the rate of creation of Takagi pairs:

$$4\nu N\, e^{2E_1/kT},$$

the proportion of chains with the two Takagi ions separated by a Slater defect:

$$8\, e^{-(2E_1+E_0)/kT},$$

the proportion of chains with many Slater defects between the two Takagi ions:

$$x_0^{\,*} = \left(8\, e^{-(2E_1+E_0)/kT}\right)\left(1-2e^{-E_0/kT}\right)^{-1},$$

and the average length of the chains:

$$L_{\text{chains}} = 4\left(1-e^{-E_0/kT}\right)\left(1-2e^{-E_0/kT}\right)^{-1}$$

which, as in the Slater model becomes infinite at the transition ($-e^{-E_0\, kT} = ½$).

As these chains should become very numerous near the transition, they could intersect. Using a representation like those in Fig. A5.1 or the so-called SUS one, it is observed that at the intersection of two chains in this positive matrix, the intersection tetrahedron has a dipole moment of opposite orientation to that of the matrix. Thus, within a mono-domain of polarization+, there could exist at each intersection of chains a dipole moment of opposite orientation, and there exist Takagi ions with moments of module $\mu_0/2$ and Slater defects without moment in the *c* direction. By calculating the statistical proportion of these situations, Schmidt proposes a variation of the polarization with the temperature which is compared to that of the SUS model (Fig. A5.2, quite satisfactory variation in the few degrees below the transition for DKDP compared to experimental results). When T approaches the transition temperature, the multiplication of intersecting chains obviously leads to the same result as the SUS model: the probability of having dipoles in all directions becomes comparable and we join the para phase. The work carried out on these bases concludes on the instability of the Takagi ions and, on the contrary, on the stability of the Slater ion chains whose density would increase in the vicinity of the transition (Lasave et al., 2005).

We will retain in the continuation of the book the elements concerning the 16 possible arrangements of protons on the 4 hydrogen bonds starting from a tetrahedron with the information on the energies, the associated moments and the proportions of such or such configuration that summarizes Fig. 1.32 (Silsbee et al., 1964a, b). In these types of models, the medium is considered without the presence of domains and walls, which can be imaginable near the transition, but becomes less plausible at lower temperatures. However, one can wonder about the interaction between a domain wall and the Takagi defects that would not be in the cloud constituting the wall, at least in the few degrees below the transition (Fedosov and Sidorkin, 1976a).

A6 Electrostatic Energy of a Periodic Domain Texture

The problem of the equilibrium of a texture in periodic domains was initially addressed for ferromagnetic domains (Landau and Lifshitz, 1935; Kittel, 1949a, b), then adapted to the ferroelectric case (Mitsui and Furuichi, 1953; Kopal et al., 1997; Tagantsev et al., 2010).

The samples are chosen in the form of platelets as in Fig. 2.20a with the smallest thickness along the ferroelectric axis. The perpendicular dimensions are assumed to be very large, in fact infinite, the thickness of the walls assumed negligible compared to the other dimensions, and the polarization in the domains equal to the spontaneous polarization P_S of the homogeneous crystal (at the minima of the curves of Fig. 1.16a).

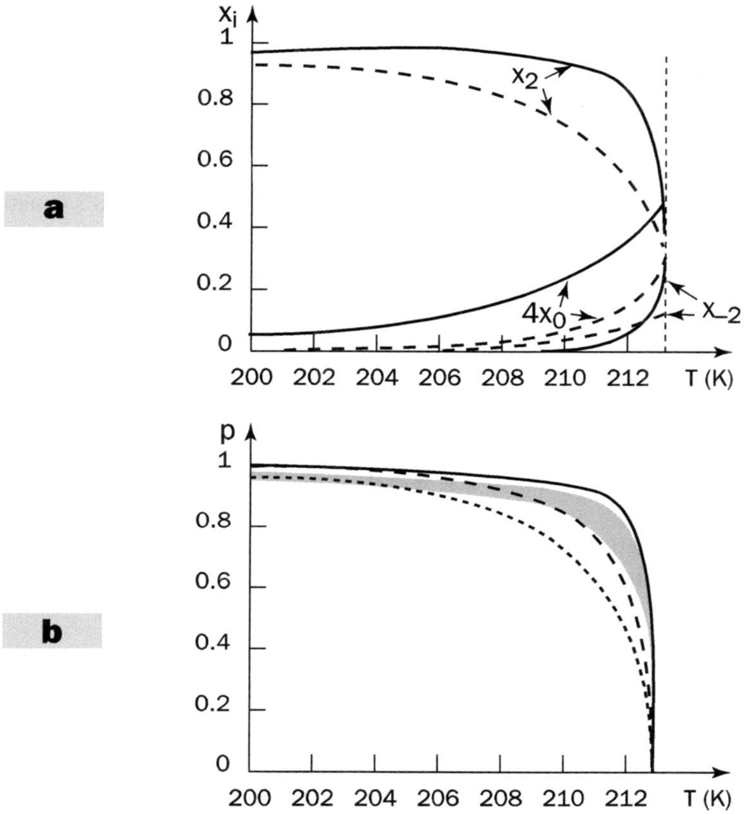

Fig. A5.2 (**a**) The proportions x_i of configurations are represented as a function of temperature in continuous lines for the model with Slater chains and Takagi ions (with E_0/k = 148 K and E_1/k = 900 K) (Schmidt, 2020) and in the SUS model in dashed lines. (**b**) Comparison of polarization measurements with the predictions of continuous line models for the chain model (Schmidt, 2020), discontinuous lines for SUS with E_1/E_0 = 7.0 (1100/154), and dotted lines for E_1/E_0 = 5.9(900/154). The grey area allows the main experimental results to be located (Zwicker and Scherrer, 1944; Samara, 1973; Mayer, 1961; Schmidt and Uehling, 1962)

The effect of the chains is not very clear a few degrees below the transition since the Slater defects do not contribute to P but only a global decrease in other configurations. However, it is very different in the last two degrees and especially the last degree below the transition with the rapid rise of x_{-2} which has a double effect on x_2 and on the sum of the moments.

We see that the "chain" model (Schmidt, 2020) takes into account more precisely the local configurations near the transition even ignoring the long-distance energy parameter β (considered equal to zero) (Senko, 1961). At the transition itself, all models contribute to the disappearance of the polarization.

For the record, regarding DKDP, the values obtained from experiments differ: negligible tunnelling and E_0/k between 92 and 94 K with E_1/k between 900 and 907 K (Reese, 1969a; Reese and May, 1967; Litov and Uehling, 1970).

In this context, a surface charge density σ appears on the faces perpendicular to the ferro axis:

$$\sigma = \mathbf{P_S}\ \mathbf{n}$$

with **n** unit vector normal to the surface ($\boldsymbol{\tau}$ will be the unit tangent vector). The crystal platelet is assumed to be in a vacuum (index v) and Maxwell's equations are written as:

$$div\,\mathbf{D} = \rho; (\mathbf{D_v} - \mathbf{D})\mathbf{n} = \sigma; (\mathbf{E_v} - \mathbf{E})\tau = 0.$$

If the free charges are considered negligible, a depolarizing field is obtained oriented in the opposite direction of the polarization (with ε_c', dielectric constant along the ferro axis):

$$\mathbf{E_d} = -\mathbf{P_S} / \varepsilon_0\ \varepsilon_c'$$

The energy variation due to the depolarizing field in the volume defined by a unit surface of the platelet of thickness t is:

$$U_d = P_S^2 t / 2\varepsilon_0 \varepsilon_c'.$$

The appearance of domains allows to decrease the energy. In the simple case of a single wall orientation, Mitsui and Furuichi chose, by hypothesis, equidistant walls. Kopal and his co-authors, for their part, chose a periodic texture (of periodicity $2w$), but with domain widths that could change according to the direction of polarization ($w_+ - w_- = \Delta w$ with $w_+ + w_- = 2w$). By writing the charge density and the potential in a Fourier series, we obtain for U_d:

$$U_d = \left(P_S^2 t / 2\varepsilon_0 \varepsilon_c'\right).$$
$$[(\Delta w / 2w)^2 + \left(16\varepsilon_c' w / \pi^3 t\right)\Sigma\left(1/n^3\right).\sin^2\left(n\pi\, w_+ / 2w\right)$$
$$\left(1 + \left(\varepsilon_a' \varepsilon_c'\right)^{1/2} \coth\left(n\pi t\, \varepsilon_a'^{1/2} / 2w\, \varepsilon_c'^{1/2}\right)\right]$$

This relationship has also been studied with thin crystals (Kopal et al., 1999). With $\Delta w = 0$ and ($t\ \varepsilon_a'^{1/2}/2w\ \varepsilon_c'^{1/2}$) sufficiently large, we find the formula of Mitsui and Furuichi whose approximations are thus measured:

$$U_d = 0.13 P_S^2\ w / \varepsilon_0 \left(1 + \varepsilon_a'^{1/2} \varepsilon_c'^{1/2}\right)$$

It is observed that the electrostatic energy of a unit surface volume perpendicular to the ferroelectric axis does not depend on the thickness t. This volumetric energy remains concentrated in "layers" of crystal close to the surface, the thickness of which is of the order of the average distance w between the domain walls. The existence or non-existence of surface layers ("surface layers") does not radically change the result (Kopal et al., 1999).

The calculation can also be performed with density planes of stacked charges along the ferroelectric axis, in a vacuum, therefore without anisotropy, without material, by developing the potential and charge densities in a Fourier series, and by using again the Maxwell equations. Taking into account various interactions between the charge planes, the numerical multiplicative coefficients will be modified but, with w very small in front of t, results will always be found in $P_S^2\, w$, therefore without effect of the thickness t of the sample. For example, in the simplest case, with only surface charges, we obtain:

$$U_d = \left(4P_S^2 w / \varepsilon_0 \pi^3\right)\Sigma\left(1/(2n+1)^3\right) = 15.35\times 10^9\ P_S^2\ \ \mathrm{w\,Jm^{-2}}$$

A7 Elastic Energy of a Set of Domain Tips (twins)

A7.1 Context

The first objective of this book is to explain to a fairly wide audience, a number of macroscopic results obtained in ferro-ferro crystals, using the microscopic phenomena observed. We have limited the choice of crystals selected to a simple change of symmetry, and the microscopic models to the objects whose effects are most obvious, essentially dislocations or quasi-dislocations of the wedge type. To analyse in more detail the displacements of nuclei and "knees" in the direction of the ferroelectric axis, the "ridges" and "grooves" of Figs. 2.13, 2.14, and 2.15, the situations between phase fronts and domain textures with applied external electric field, as in Figs. 6.7, 6.8, 6.9, 6.10, and 6.11, would be necessary to call upon screw dislocations, various types of joints, and other microscopic objects.

Thus, the term "disinclination" ("disclination" in English) appears several times in the book to designate the existence of gradients of translation and rotation defects. Such terms have a history that the reader can discover by working on the publications and works of Volterra (1907), Love (1920), Franck (1958), Friedel (1964), Eshelby (1966), Nabarro (1967a, b), De Witt (1970, 1971, 1973a, b), Mura (1972) ... We initially consider a cylinder of material around the line of its axis which we cut between its axis and its edge to then study what happens according to the displacements one with respect to the other of the two cutting surfaces. We find situations of distortions, rotations, and translations. In these exercises, we often consider the environment as infinite, with or without anisotropy, with or without combination between cylinder lines. Given the aforementioned objectives, we do not neglect the existence of these phenomena and objects. Sometimes the objects (screw dislocations, disinclinations...) are mentioned without describing them further and without using them. As in the rest of the book, this annex is limited to dislocations and quasi-dislocations of translation, wedge type (Fig. 2.30) with straight lines, parallel to each other and to the ferroelectric axis. In the description of the cylinder cut between an edge and its axis (dislocation line), the only displacement (Burgers vector) of the cutting surfaces one with respect to the other corresponds to a slip in their plane perpendicular to the axis of the line.

A7.2 Interaction Between Parallel Wedge Dislocations

Figure A7.1 allows us to recall an old result of these straight wedge type dislocations (Cottrell, 1953; Friedel, 1964), very useful: a wedge dislocation of Burgers vector b_1, parallel to the Oz axis (the c axis in our crystals), cuts the plane of our page at the origin of the axes O. Another parallel wedge dislocation also to Oz of Burgers vector b_2 can only move in a glide plane (x, y_0, z) and cuts the page at M. The force that this dislocation undergoes because of the existence of the first, per unit length of line, can be written in polar coordinates (see Fig. A7.1a):

$$F_r = \mu b_1 b_2 / \left(2\pi\left(1-v\right)r\right),$$

$$F_\theta = \mu b_1 \mathrm{b}_2 \sin 2\theta / \left(2\pi\left(1-v\right)r\right)$$

μ shear modulus, v Poisson's ratio; r = OM; $\theta = (\mathbf{Ox}, \mathbf{OM})$. The component of the force that determines the slip at $y = y_0$ is:

$$F_r \cos\theta - F_\theta \sin 2\theta = \cos\theta\left(1-2\sin^2\theta\right)\left(\mu b_1 \mathrm{b}_2 / \left(2\pi\left(1-v\right)r\right)\right.$$

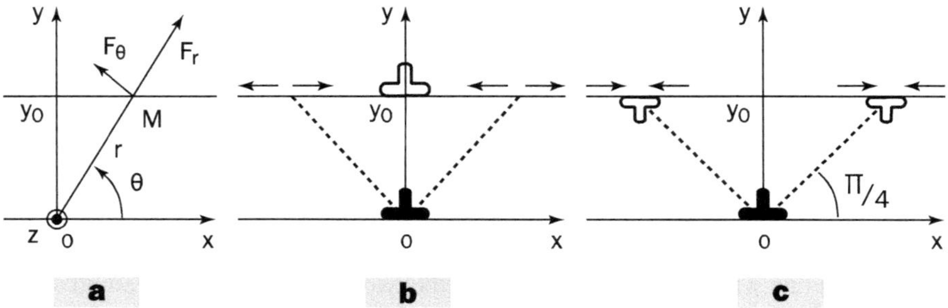

Fig. A7.1 (**a**) A dislocation line of Burgers vector b_1 is along the Oz axis and another, parallel to it of Burgers vector b_2, cuts the page at M and can only slide in the plane (x, y_0, z). (**b**) If $b_1\, b_2 > 0$, the stable situations correspond to $\theta = 0$ (r infinity) and $\theta = \Pi/2$, as illustrated by the small horizontal arrows. (**c**) If $b_1\, b_2 < 0$, the stable situations correspond to $\theta = \Pi/4$ and $\theta = 3\Pi/4$ as illustrated by the small horizontal arrows

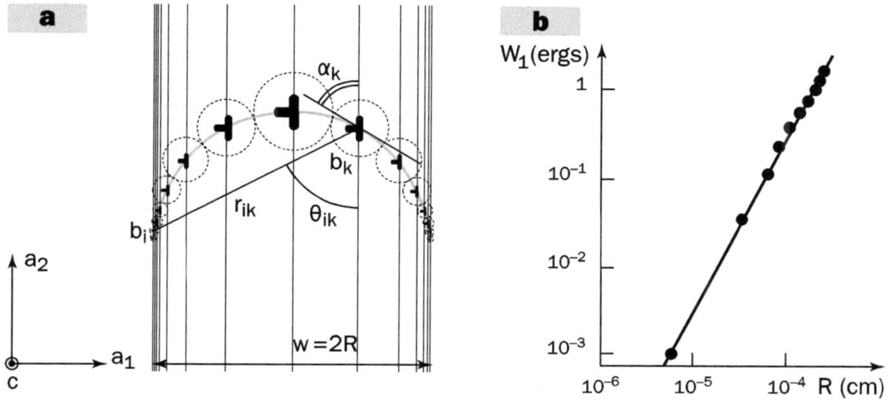

Fig. A7.2 (**a**) Schematic representation of an assembly of quasi-dislocations at a needle domain tip with core radii that delimit dissociable dislocations. (**b**) Variation of the elastic energy W_1 of a semi-circular domain tip as a function of the radius R for a core radius value r_0 of 278 A°. The slope of the line is equal to 1.9

The diagrams of Fig. A7.1b, c illustrate the equilibrium situations that can be observed in the photographic sequences of Figs. 3.27 and 3.31. These relationships intervene in the interaction of a domain tip with the closest ones:

b: when $b_1\ b_2$ is positive, stable equilibriums are observed for $\theta = 0$ and $\theta = \Pi/2$ and unstable equilibriums for $\theta = \Pi/4$ and $\theta = 3\Pi/4$ (see Figs. 3.27 and 3.28).

c: when $b_1\ b_2$ is negative, the situations are reversed compared to the previous ones and the stable equilibrium occurs for $\theta = \Pi/4$ and $\theta = 3\Pi/4$ (see Fig. 3.31).

A7.3 Dislocations Summarizing a Needle Domain Tip

When a domain wall element deviates from one of the allowed wall planes, and if the deformation changes are small, situations can be accounted for using the concept of virtual dislocation density. Many models have thus been developed. The forms of twin ends shown in Fig. 2.29 (Salje and Ishibashi, 1996) are an example.

However, if the incident in the wall is notable and localized, the introduction of a dissociable dislocation and a line of singularity like the one represented in Fig. 2.30a are necessary. In this annex, only the deformation linearly related to the order parameter is taken into account. A tip of needle domains is thus described as an assembly of dissociable dislocations between them, as in Fig. A7.2a. The smallest dislocation will be the elementary quasi-dislocation of Fig. 2.30a corresponding to the smallest possible lateral wall offset: an elementary cell *a* is often assumed, but in the KDP family, one can also imagine *a*/2. The elementary Burgers vector is therefore:

$$b_e = 2a\,tg u_{xy} \tag{A7.1}$$

When the angle α_k of a part of the wall with the permissible wall plane increases (see Figs. 2.30b and A7.2a), a Burgers vector b_k of the singularity can be written

$$b_k = (b_e \ \sin\alpha_k)(2r_0 / a) \tag{A7.2}$$

therefore, with a singularity of maximum importance at the end of the needle domain. The core radius r_0 requires a realistic estimate; this long-studied theme (Gilmann 1969) is a subject in itself, not developed here.

A7.4 Elastic Energy of a Domain Tip

It is assumed, like many authors, that the stress and therefore the energy created by such dissociated dislocations, remain limited by a cut-off radius r_1 of the order of the thickness of the sample following c around each line of the considered dislocation. The approximation of infinite dislocations is also used. The energy W_{ek} per unit length of a dissociated dislocation k can thus be written (Friedel, 1964; Bornarel, 1971):

$$W_{ek} = \left(\mu b_k^2 / 4\pi(1-\nu)\right)\left(\ln(r_1 / r_0) + s\right) \tag{A7.3}$$

with μ shear modulus (equal for KDP to $c_{66}/2$), ν Poisson's ratio equal to (1/2) $(3K-2\mu)/(3K+\mu)$, K inverse of the compression coefficient equal to $c_{33}c_{11}/(2c_{33}+c_{11})$, and s a numerical coefficient which depends on the choice of r_1 and the adhesion energy near the core.

Similarly, one can define W_{ik} energy per unit of length following c of interaction of two dissociable dislocations i and k located at a distance r_{ik} small compared to the thickness of the sample, with θ_{ik} defined in Fig. A7.2a:

$$W_{ik} = \left(\mu b_k b_i / 2\pi(1-\nu)\right).\left(\ln(r_1 / r_{ik}) + 1/2\cos 2\theta_{ik} - 1\right) \tag{A7.4}$$

and then, the energy per unit length following c, of an isolated domain end, assembly of all the dissociable dislocations (Bornarel, 1971):

$$W_1 = A\left(H\Sigma\sin^2\alpha_k + 1/2\Sigma\Sigma\sin\alpha_k\,\sin\alpha_i\left(1/2\cos 2\theta_{ik} + \ln(r_1 / r_{ik}) - 1\right)\right) \tag{A7.5}$$

with $A = \mu\, r_0^2 b_e^2/\pi(1-\nu)a^2$, $H = \ln r_1/(r_0 - 1 + E) > 0$, where E characterizes the core energy and remains close to 1 in KDP (Kamysheva and Zhukov, 1966).

The use of the relation (A7.5) for various forms of ends demonstrates that if the energy values are modified, this is not the case for the dependence on the width of the domain. If we choose for a crystal such as KDP at low temperatures, a form of semi-circular end of radius R ($R = w/2$, w width of the domain), by choosing the energy calculation of the Peierls model (Peierls, 1940) with $E = 1$ (Nabarro, 1967a, b), we obtain the following results:

- For values of w of a few micrometres, a notable variation of r_0 (from 10^{-8} to 10^{-6} m) does not cause a significant variation of W_1.
- For a defined value of r_0, W_1 varies significantly with w (like $w^{1.9}$ in Fig. A7.2b).
- The variation of W_1 as a function of R can be written

$$W_1 = \alpha R - \beta R^2 \ln\left(R / r_1\right) - \gamma R^2 \tag{A7.6}$$

α varies as r_0 Ln R/r_1, β and γ are functions of b_e, μ and ν. Thus, with the values $\mu = 3.5 \times 10^{10}$ dynes cm^{-2}, $\nu = 0.45$ (Barkla and Finlayson 1953) and $b_e = 0.12 \times 10^{8}$ cm (Bornarel, 1971), we find values of W_1, which when brought back to the unit of surface of wall are a hundred times larger than those retained for a permitted wall plane.

A7.5 Interaction Energy of Two Domain Tips

If now, two domain ends are considered, each dissociable dislocation will interact with all the dissociable dislocations of the same tip and with those of the neighbouring tip. The energy of the two tips W_2 can take a simplified form as shown by Fig. 2.31: the action of one of the ends on a dissociable dislocation k of the set of the neighbouring end can be summarized, if r^2/R^2 is sufficiently large, by that of a dislocation of Burgers vector B (said effective),

$$B = \Sigma b_k = 2w\, tgu_{xy} \tag{A7.7}$$

We can thus approximate the elastic energy per unit length following c, W_2 of two domain tips by

$$W_2 = \left(\mu B^2 / 2\pi\left(1-\nu\right)\right)\left(\ln\left(r_1 / d\right) + \%\!\cos 2\theta - 1\right) \tag{A7.8}$$

in which θ is the angle relative to the slip plane of the vector **d** that joins the centres of the two domain tips. For two tips in a situation of minimum energy, on the same perpendicular to the slip plane, we obtain (in cgs) with semi-circular ends, in KDP (at low temperatures)

$$W_2 = 5 \times 10^{6} R^2 \left(2 \ln\left(r_1 / d\right) - 3\right)$$

A7.6 Elastic Energy of a Front with n Needle Domain Tips

The situation represented in Fig. A7.3 corresponds to a periodic distribution of domain walls, with a constant number of dislocations and boundary conditions independent of the number of tips. In addition, the tips are assumed to be of semi-circular shape ($4R = 2w$), arranged in the form of a front, on the same perpendicular to the slip plane. The energy per unit length of this set of n tips ($D = nR$), can be written using the previously cited approximations:

$$W_n = n\left(\alpha R - \beta R^2 \ln\left(R / r_1\right) - \gamma R^2\right) +$$
$$\left(\xi \ln\left(4R / r_1\right) - \delta\right) n\left(n-1\right) R^2 / 2 - \xi R^2 \Sigma\left(n-i\right) \ln\left(i\right);$$

if n is large enough to use Stirling's formula, we write
$\Sigma (n - i) \ln(i) \approx (n^2/2) \ln(n) - 3n^2/4$, hence (Bornarel, 1971):

$$W_n = \psi / n + \eta$$

with $n = D/R$, $\psi = 2 \times 10^6 D^2$ in *cgs*, and $\eta = 3.9 \times 10^6 r_0 \ln(r_1/r_0) - 5 \times 10^6 D \ln(4D/r_1)$.

Thus, we obtain a total elastic energy of the domain ends that decreases as their number increases. This model, general for any assembly of twins, plays in ferro crystals a role comparable to the energy linked to the depolarizing field in electrostatics: the term in ψ/n, or $\psi\, w/2D$ is proportional to the width of the domains. It is used in Chap. 2 to explain the variation in domain width as a function of their length, particularly for needle domains of constant width.

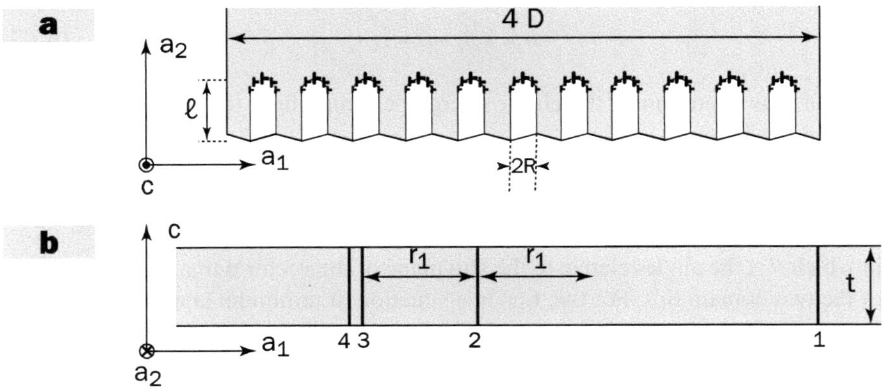

Fig. A7.3 (**a**) Schematic representation in section (001) of a regular domain texture with domain tips in front and semi-circular tip shape. (**b**) The cut-off radius r_1 gives an estimate of the limit distance of interaction between dislocations parallel to c; usually, it is chosen of the order of the sample thickness

The previous calculations, initially proposed for KDP and its tetragonal-orthorhombic context (Bornarel, 1971), were then taken up and adapted to various situations and symmetries, for example in the case of rhombohedral lead phosphate (Torres et al., 1982).

A8 Avalanches and Collective Phenomena

A8.1 Universality of Phenomena

In certain systems, the continuous variation of a parameter (magnetic or electric field, stress, temperature...) leads to that of its major physical properties (magnetization, polarization, deformation...), but we can observe jumps, discontinuities, pulses, which are added or subtracted from the expected continuous evolution, as illustrated in Fig. A8.1. The first observation of this type is due to Barkhausen (1919) during experiments of magnetization reversal of an iron bar. This result was the first attributed to magnetic domains whose existence had been postulated by Weiss (1907). It took about 20 years to have direct measurements of magnetic domain wall

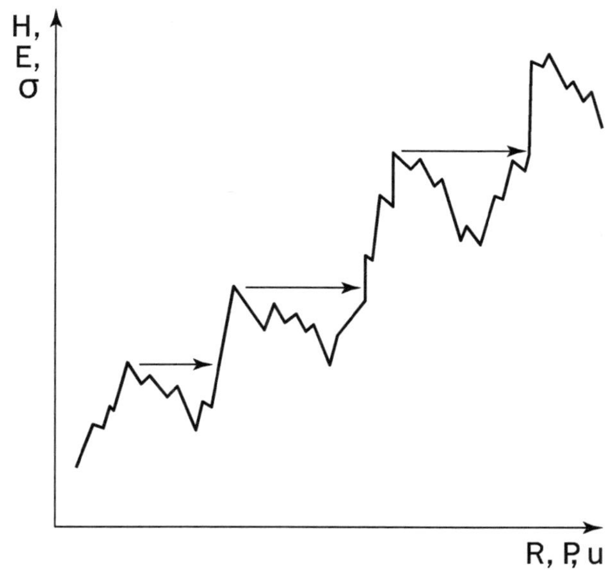

Fig. A8.1 The application of an increasing external excitation (H, E, σ) leads to the growth of magnetization, polarization, and deformation. But, at the level of a wall, the field necessary for its displacement must be sufficient to overcome a local barrier, the result of local forces and longer-distance actions. Each time this happens, the wall moves to the next blockage by a larger barrier. The external excitation must then be increased to trigger a new movement. With each jump, pulses can be observed, for example, Barkhausen pulses displacement on a cobalt crystal (Elmore 1938). Since then, phenomena comparable to "Barkhausen noise" are widely observed (Spasojevic et al., 1996). They can give rise to electrical signals (variations in voltage or current as in Barkhausen noise), brief acoustic signals, brief light signals,... These phenomena can occur during the tracing of hysteresis cycles in ferromagnetism (Bertotti, 1998; Durin and Zapperi, 2006; Durin et al., 2016) in ferroelectricity (Colla et al., 2002), in the responses of granular media (Bretz et al., 1992; Baldassari et al., 2006), in martensitic transitions (Vives et al., 1994), during fractures of materials (Curtin and Scher, 1991), during the creation of dislocations in a plastic deformation (Miguel et al., 2001), in the phenomena of dynamics of superconductors (Ertas and Kardar, 1994) and superfluids (Lilly et al., 1996), but also during earthquakes (Gutenberg and Richter, 1956; Main, 1996; Utsu, 1999), in the fluctuations of financial markets (Bouchaud, 2001), etc.

These so-called avalanche phenomena show a certain universality of properties (Kadanoff et al., 1989); for example, when the sequences of such discontinuous signals are dense, they are of low amplitude, and on the contrary, jumps of great importance are separated by notable intervals. Power law distribution probabilities are also frequently observed for the parameters that characterize the pulses.

After the work on Barkhausen noise in ferromagnetic materials (Colaiori, 2008), studies have multiplied in various sectors of science, and particularly in the last 20 years in ferro-elastic crystals with the realisation of the major role of interfaces on macroscopic properties, and also, due to the growing interest in thin layers in which these phenomena also appear. Various review publications provide an update on these avalanche phenomena in ferroelectric, ferro-elastic, and co-elastic materials (Salje et al., 2019; Nataf and Salje, 2020).

A8.2 How Are Avalanches Described?

The characteristic signals appear in sequences of pulses of varying amplitude, more or less spaced in relation to the continuously varying parameter. One can determine the amplitude A or the surface S of the pulse, its energy E, its duration D or T if the sequence of pulses are plotted as a function of time (see Fig. A8.2). It has also become customary to differentiate the interval between two events, called Φ if they are close, and Ψ if they are very dissociated.

The phenomenon is characterised by calculating the probability P of obtaining a certain value of one of the quantities (A, E, T...) in the sequence of pulses obtained. A satisfactory description of the experimental results is often obtained by representations in the form of a power law, such as

$$P(E) \sim E^{-\varepsilon}, P(A) \sim A^{-\tau'}, P(T) \sim T^{-\alpha}, P(d,\text{waiting time}) \sim d^{-p} \quad \text{(A8.1)}$$

with $p = 1 - \Phi$ for short d and $p = 2 + \psi$ for long d.

A maximum likelihood estimation method allows the merging of as many data series as possible to differentiate phenomena with the best precision and to bring

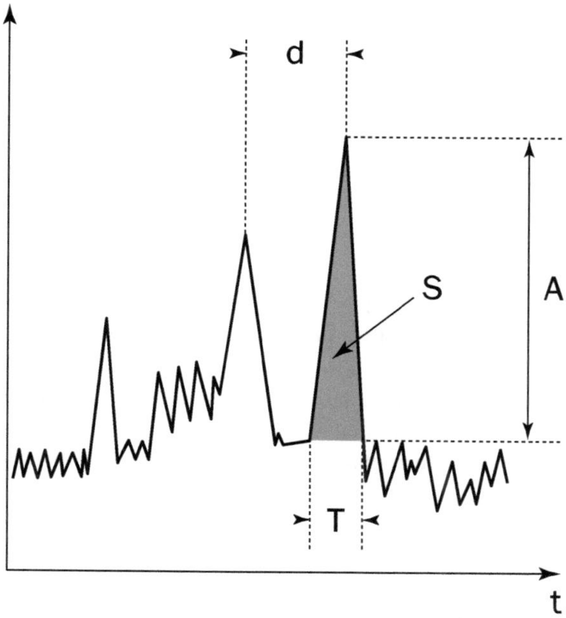

Fig. A8.2 The spectrum of pulses $x(t)$ as a function of time illustrates a general result: the denser the signals, the more their amplitude is limited. The quantities usually used are indicated: for a pulse, its amplitude A, its area S, its duration T. The energy of the pulse often varies as A^2. The distance between two pulses d is here the waiting time

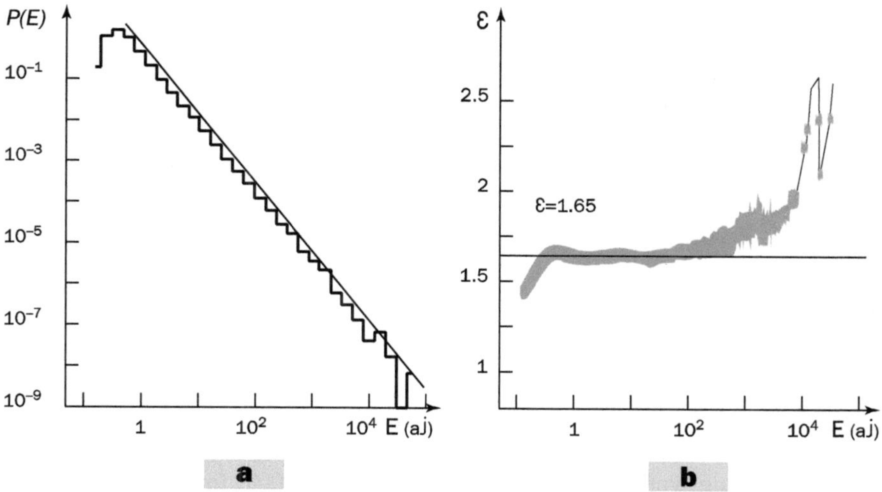

Fig. A8.3 Results of Salje's work on $BaTiO_3$: (**a**) probability $P(E)$ in terms of energy E of the pulses and its power distribution law with an exponent $\varepsilon = 1.65$, and (**b**) graph of maximum probability made with a large number of pulse series, in which the limits of the region with this power law are visible (Salje et al., 2019)

together those of the same universality (Navas-Portella et al., 2019). The results are illustrated in Fig. A8.3, which shows the limits within which the power law works (Salje et al., 2019): for values too small of the quantity studied, it is not possible to extract it due to the low sensitivity of the measurement, and the upper limit usually corresponds to finite size effects.

Experiments carried out on very different systems (limiting ourselves to materials, ferromagnetics, ferroelectrics, ferro-elastics...), and concerning different phenomena themselves (transition with a phase front, wall displacement, crystal growth or fracture...etc.), demonstrate however that it is not simple to link an observed effect of pulses to the phenomenon that is the cause at the nanometric or micrometric level.

The exponents introduced previously vary as illustrated below (Nataf and Salje, 2020):

ε between 1.3 and 2.2: 1.3 for cracks in disordered medium (Salje and Dahmen, 2014), 1.3–1.4 for porous materials subjected to compressions (Soprunyuk et al., 2017), 1.3–1.5 in a phase transition of a ferroelectric with a high value when there is a high level of disorder and various relaxations (Xu et al., 2019), 1.6–1.7 in a global reversal of a ferroelectric (Tan et al., 2019; Salje et al., 2019), 1.8 for a ferroelastic subjected to a uni-axial compression;

τ' between 1.6 and 4: about 1.6 in a martensitic transformation of alloy (Vives et al., 1994), 1.6–1.8 when the domain ends are the cause (Harrison and Salje, 2010; Puchberger et al., 2017), 2.25–2.3 in the case of global reversal of a ferroelectric (Salje et al., 2019; Shur et al., 2002);

α between 2.0 and 4.3: between 2.0 and 2.8 in Stoichiometric Lithium Tantalate (Shur et al., 2002), between 2.7 and 3.0 if we take the globality of a reversal, between 2.8 and 3.2 during martensitic transitions (Vives et al., 1994).

Many of these results do not allow to correlate the situation of the domains and the values of the exponents and only account for phenomena observed macroscopically. However, in some cases, the relative range of interactions in the system can be estimated: the values of τ' and α seem larger when the ranges of interaction are more important (Kuntz and Sethna, 2000).

A8.3 Theoretical Models

The first models proposed aimed to understand the phenomena of reversal of ferromagnetics and to detect defects to improve performance during hysteresis cycles. But the generality of the phenomena led to transpose some analyses to other situations.

An interesting synthesis was proposed by Colaiori (2008) for the movements of magnetic domains. It is based initially on the ABBM model (for Alessandro, Beatrice, Bertotti, and Montorsi) (Alessandro et al., 1990a, b) inspired by the work of Louis Néel (1942, 1943) who was the first to introduce a random energy in a model of hysteresis cycle. The ABBM model assumes that, even in a disordered ferromagnetic, the speed of a "wall" in motion is the solution of a Langevin equation as if its centre of mass were moving in an average field. Thus, it is assumed that the shape of the results does not depend on the "details" at the microscopic level. The results of this theoretical model give for the probabilities P of the various quantities associated with the pulses, relations of the form:

$$P(A) \sim A^{-i} \ F_A \ (A / A_0) \tag{A8.2}$$

with a cut-off function $F_A = \exp(-A/A_0)$ and characteristic values A_0; the exponents i vary according to the rate of change of the excitation but remain well defined under identical experimental conditions.

This model has achieved good results, especially for experimental cases close to the initial assumptions, those of "soft" ferromagnets with flat domain walls that move quite regularly in a context described as "stationary" for the overall set (Bertotti et al., 1981; Kuntz and Sethna, 2000). This is the case, for example, for flat walls, in a situation close to the coercive field as illustrated by Figs. 3.6, 3.7, 3.8, 3.9 and 3.10 of the book.

The reasoning is based on the following: in each basic element (for example a dipole), the probability of change depends on the macroscopic excitation and an energy capable of triggering it; it is considered that a change on one element will facilitate changes on others through interaction and that cascades thus propagate (Salje and Dahmen, 2014). Therefore, the existence of interactions at more or less long distance such as interactions between dipoles, or/and long-distance mechanical interactions that can play the same role, is admitted. Thus, an avalanche field can propagate in the crystal.

The ABBM model, with an analytical treatment expressed by the relation (A8.2), has the advantage of giving meaning to many properties in terms of a stochastic process. Thus, power laws for shapes and durations are derived from return time distributions to the origin. This topic can be further explored by referring to the publication by Colaiori.

However, some characteristics of the pulse distributions were not reproduced by the initial ABBM model. Thus, to improve the correspondence between the results of simulations and experiments concerning the asymmetry of pulse shapes, a negative effective mass was associated with the moving wall. More generally, the asymmetric shape of the pulses would be associated with inertial effects with specific characteristic times. Note that, sometimes, the results can be the sum of several phenomena that overlap. Simulations can demonstrate this by searching for power laws of different exponents.

A8.4 Ferro-elastic, Ferroelectric Crystals, and Avalanches

Avalanches in ferroelectrics and ferro-elastics are numerous during reversals (polarization and/or deformation) and phase transitions. For example, the differences between switchings with "rounded" hysteresis cycles and with "brutal" switching cycles have been studied (Perkovic et al., 1999). All the work confirms the link between the existence of avalanches and that of domains (or phase fronts). However, the correlation between the emission of pulses and the precise phenomenon that causes it at the level of walls and domains is rarely demonstrated experimentally. This explains some contradictions in the interpretation of the results of macroscopic measurements without simultaneous observation of domain textures. To try to clarify this difficulty, correlations between the existence of avalanche signals and the situation of domain texture are proposed in Fig. A8.4 in two simple cases, that of

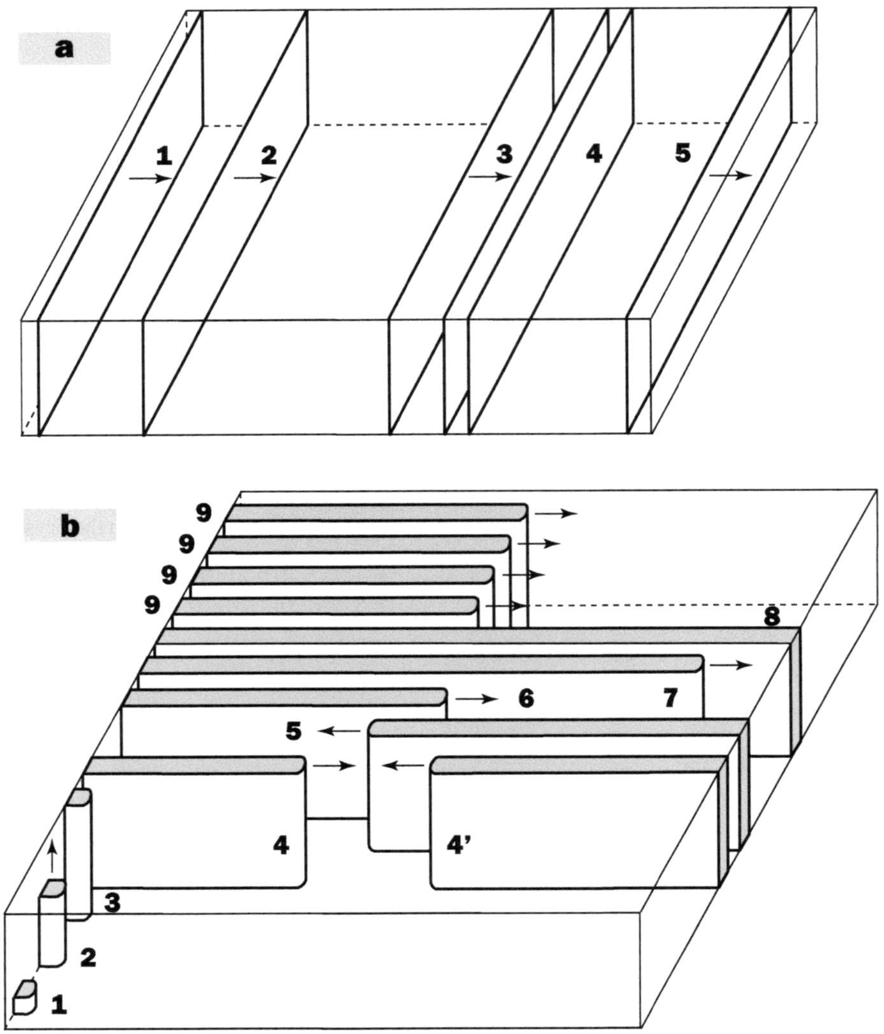

Fig. A8.4 (**a**) Reversal by lateral displacement of plane walls: in this situation of walls with lateral displacements, are represented (the arrows indicate the movement of the concerned wall): (**1**) The creation of the plane wall on one edge (in a plane allowed for our ferro-ferro): Notable pulses are observed in all experiments (Salje et al., 2019) which simulations confirm: atom displacements on surfaces emit energy jumps estimated at a few meV/atom (Salje et al., 2014). (**2**) The lateral displacement of an isolated wall: When the movement is regular and slow, a pulse may not be detected (Viehland and Salje, 2014), or there are pulses j of low amplitude (Shur et al., 2001: GMO) (Udalov et al., 2015: $LiTaO_3$, $LaAlO_3$). Sometimes acoustic emission pulses ac exist preceding electric pulses j (Lupascu et al., 2002: GMO); they become notable at high speeds if the excitations are significant (Shur et al., 2001; Mohamad et al., 1982: Pb_5 Ge_3 O_{11}Zammit-Mangion and Saunders 1984: GMO, TMO; Salje et al., 2017a, b). The estimation of energies during knee movements would give results of a few 10^{-5} eV/atom (Salje et al., 2014). (**3**) A rapidly interacting wall, here with a stationary domain, can also interact with another wall as illustrated in Figs. 3.14 and 3.15.

lateral wall displacements (Fig. A8.4a) and that of dense needle domain textures (Fig. A8.4b). The examples go well beyond ferro-ferro. With the same objective, Fig. A8.5 presents a small monograph of avalanche signals correlated with modes of reversals and shapes of hysteresis cycles. The last example (Fig. A8.5f) underlines the fundamental difference between true reversals and simple modifications of polarization and deformation of a sample that remain of the same sign. This is what often happens during ferroelasticity experiments by varying stresses of the same sign (so as not to lose the mechanical contact between the stress source and the sample). In these situations, the important sources of pulses at the creation and disappearance of domains on the sample edges are almost absent.

When complete switchings are studied, things get complicated and depend a lot on the possible textures. A synthesis of the results of acoustic emissions concerning $BaTiO_3$ has been proposed (Salje et al., 2019), and the displacement of the walls at 90° is, for example, an obvious cause of pulse emission. In this case, it is shown that major shocks are followed by others, and that electrical, acoustic, and optical pulses can be simultaneous. The pulses can result from superpositions of collective phenomena of different origins that need to be differentiated. Thus, with samples of Pb (Zr,Ti)O_3, the superposition of two phenomena was shown with different exponent values: 1.6 and 1.8 (Tan et al., 2019). These superpositions are important if the speed of modification of the excitation is high, during switchings as during rapid passages to phase transitions (Xu et al., 2019). In our examples chosen for their simplicity (only two orientations of walls at 90°), we should be able to dissociate the avalanches due to knees in the walls, to end displacements, to their interactions, to global displacements with rotations and incoherent interfaces, to domain-sample edge interactions...

If a "collapse" of walls is observed, significant pulses j and ac are observed (Shur et al., 2001: GMO; Udalov et al., 2015: $LiTaO_3$; Mohamad et al., 1982: Pb_5 Ge_3 O_{11}; Akhmatkhanov et al., 2020: KTA) with energies of the order of a few 10^{-4} eV/atom (Salje et al., 2014). (**4-5**) The disappearance of nearby domains by creating a needle, and that of a wall on an edge of the sample in strong interaction with the edge, produces effects comparable to the creation **1**. (**b**) Reversal with needle domains: (**1–3**) Creation of a nucleus (**1**) which propagates in the direction of the polarization (**2**) and crosses the entire sample (**3**). Pulses are detected in most experiments during domain creations, whether it is during reversal, or transitions, including a transition from ferroelectric relaxor (Zhang et al., 2017). (**4**) The domain tip has freed itself from the edge and moves freely: ac pulses of low amplitude at regular speed, notable during speed changes (Harrison and Salje, 2010: La AlO_3). (**4–4′**) Opposite domain tips attract each other until they "collapse," emitting notable pulses. These "collapse" phenomena give responses that can be confused with those of texture rearrangements (relative changes in domain sets). (**5, 6**) The domain tips (5) and (6) intersect interacting strongly: the emission of pulses depends on the configuration of the tips between them. (**7**) The domain tip is attracted to the edge of the crystal before disappearing at the edge of the sample; as for the "collapse" of domain tips, very large pulses j are observed (Akhmatkhanov et al., 2020: KTA) and also ac pulses (Lupascu et al., 2003: GMO) with energies of a few meV/atom. (**8**) Stable needle domains emit pulses of very low amplitudes. (**9**) Movement of a set of needle domain tips: a collective moving tip front "softly" emits few pulses *j*, but the shapes of the tps change (Akhmatkhanov et al., 2020: KTA) (Lupascu et al., 2003: GMO)

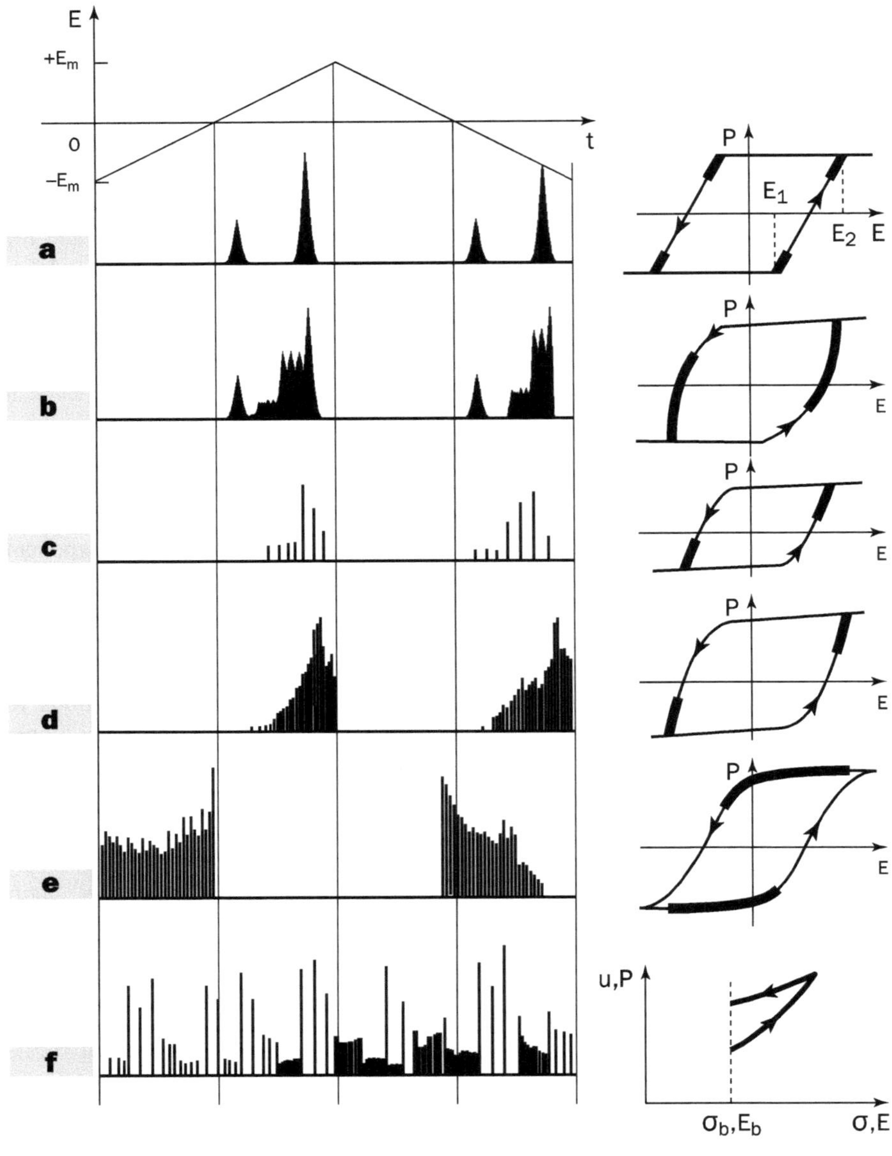

Fig. A8.5 Schematic representation of series of pulses observed in a ferro-ferro crystal during a field cycle with linear variation over time: the cycle starts at $-E_m$, passes through the situation with zero field value, changes sign and grows to $+E_m$ to then decrease to $-E_m$. Typical series of pulses are represented on the left, and on the right, the corresponding hysteresis cycles on which are indicated in bold the parts with observable pulses; depending on the case, the pulses can be electrical, acoustic or optical: (**a**) Complete cycle with lateral displacement of a single wall at almost constant speed. The pulses correspond to the appearance of the wall in E_1 and its disappearance in E_2. (**b**) Similar situation, but the speed of the wall changes with the amplitude of the field. As observed in some studies on GMO, the existence and importance of pulses change with the amplitude of E (Shur et al., 2001, 2002). The slopes and curvatures of the hysteresis cycle change

The experimental emission tests conducted quickly at the DKDP transition of Chaps. 5 and 6 open a field of study that requires an analysis of the sequences of pulses during modifications of domain textures and phase front shapes. Simple beliefs can be quickly shaken: a phase front can appear as a soft interface emitting only very weak dense pulses and domains changing emission types quite drastically in temperature. However, the experimental results of this work demonstrate the opportunity to continue to feed the numerous simulations concerning avalanche laws by lateral displacements of interfaces and by needle domains (Salje et al., 2011).

To conclude this brief illumination on avalanche phenomena, pulse emissions (electrical, acoustic, optical...) constitute an extremely rich field of study, especially if the interfaces in the materials studied are simultaneously observed.

A9 Very Low Temperature Properties of Ferroic Crystals and KDP

The thermal properties (specific heat, conductivity) and their evolutions in temperature and frequency can inform about the phenomena in the materials and allow to analyse their causes. This annex recalls results obtained in ferroelectrics, notably KDP, and the most used models, to lighten the texts concerned such as those on dielectric measurements at LNT in Sect. 4.7.

A9.1 Models Used to Describe the Specific Heat of Solids

A9.1.1 Specific Heat at High Temperature

When the temperature exceeds 300 K, the specific heat at constant volume usually tends towards a constant value. The simplest model considers a monoatomic body with small masses m interacting by springs of stiffness β, forming a one-dimensional chain. The vibrations are assumed to be harmonic, hence equality between

accordingly. (**c**) At a temperature close to the transition, the reversal is easy with a low value of E_m. The set of needle domains changes globally without jerks except during domain species variations (Bornarel and Kvitek, 1993) (Fig. 3.48). (**d**) On the "plateau" of dielectric measurements of a crystal such as KDP, with needle domains, the causes of pulses are numerous as listed in Fig. A8.4; small lateral displacements of the permitted walls must be added (Bornarel and Kvitek, 1993). (**e**) Series of pulses observed during reversals of $BaTiO_3$ (Salje et al., 2019); the particular shape of the series of pulses which reverse over time depending on the direction of polarization remains to be explained. (**f**) Series of almost monotonous pulses during a change in the electric field or the represented constraint, with a larger continuous E_b field or a larger continuous constraint applied. The sign of the crystal's polarization-deformation is not changed; only variations in modulus are produced without real reversal. This is a common situation in ferroelasticity to maintain mechanical contact between the source of stress and the crystal
The above results have been known for a long time; it is at the end of the switching processes that instability is strong, including during dielectric measurements of classic ferroelectric crystals (Fouskova and Janovek 1965; Fouskova 1965)

average kinetic energy per atom and average potential energy per atom. We easily find the classical dispersion law:

$$\omega = \left(4\beta / m\right)^{1/2} \sin\left(ka / 2\right) \tag{A9.1}$$

with the distance a between two adjacent atoms of the chain. Thus, in this one-dimensional material, a wave propagating along this chain would have a speed:

$$v = \left(r / \rho_0\right)^{1/2} \left(\sin ka / 2\right) / \left(ka / 2\right) \tag{A9.2}$$

with r rigidity of the material and ρ_0 its mass density.

Using the Boltzmann theorem, for one degree of freedom, the average kinetic energy of an atom will be $kT/2$, so the total average energy of an atom kT. In three dimensions, the result becomes $3kT$ and, for a gram-atom $3NkT$, or $3RT$; hence, the specific heat at constant volume of a gram-atom:

$$C_v = \mathrm{d}U / \mathrm{d}T = \mathrm{d}\left(3RT\right) / \mathrm{d}T = 3RT = 5.955\,\mathrm{cal\,atom\,g / K} \tag{A9.3}$$

If the material in question contains p different atoms, the specific heat of the gram-molecule will become $3pR$. We find the law of Dulong and Petit of perfect gases which is normal given the similarity of the assumptions made.

A9.1.2 Specific Heat at Medium and Low Temperatures

The experimental results of specific heat of solids show a global decrease in its value for a temperature lower than 250–300 K until it tends towards zero at 0 K. There are dispersions, anomalies depending on the material and its states. Two models are commonly used to describe these properties, due to Einstein and Debye.

The **Einstein model** (1911) is a bias on the quantification: it is assumed that each atom can only vibrate at certain frequencies which depend on the atom and are quantified. In the simple case of a monoatomic material, the energy of an atom will be of the form:

$$E = n\,h\nu = n\hbar\omega \tag{A9.4}$$

With ν the frequency, ω the pulsation, h the Planck constant, n a positive integer quantum number.

To calculate the average energy per atom, we will calculate the ratios of populations between the various levels using the Boltzmann theorem, and we obtain:

$$\bar{E} = \hbar\omega / \left(e^{\hbar\omega/kT} - 1\right) \tag{A9.5}$$

When the temperature is very high, $\bar{E}$ tends towards kT, and we find the law of Dulong and Petit. If, on the contrary, T decreases sufficiently so that the exponential term of the denominator is very large compared to 1 (A9.5) simplifies, and the specific heat at constant volume for a gram-atom tends towards zero according to the relation:

$$C_v = Nk\left(\hbar\omega / kT\right)^2 e^{-\hbar\omega/kT} \tag{A9.6}$$

In the presence of a molecule with p types of atoms, we will take into account of p contributions to C_v. Using the relation (A9.5) for any temperature, it has become customary to introduce a temperature resulting from the calculations carried out with the experimental results, called Einstein temperature θ_E such that:

$$k\theta_E = \hbar\omega_0 = h\nu_0$$

Thus, in the case of a mole of a monoatomic material,

$$C_v = 3R\left(\theta_E / T\right)^2 \left(e^{\theta}{}_{E/T}\right) / \left(e^{\theta}{}_{E/T} - 1\right)^2 = 3RE\left(\theta_{E/T}\right) \tag{A9.7}$$

The function E ($\theta_{E/T}$), called the Einstein function tends towards 1 when T tends towards infinity and towards zero if T tends towards zero.

To better understand the Einstein temperature, let's assume that the experimental adjustments define it at 100 K, the corresponding ν_0 frequency will be about 2.10^{12} Hz, which corresponds, if we consider a sound wave with a propagation speed of 5000 ms^{-1} to a wavelength of 25 Å, a result quite plausible from a physical point of view. However, this model, which works for temperatures above 50 K, becomes increasingly unsuitable as the temperature drops significantly: the values thus calculated are much smaller than those measured experimentally, 10 times too small (approximately) towards 30 K, 10^4 times towards 10 K, and by several tens of orders of magnitude towards 1 K!

A9.1.3 Specific Heat at Very Low Temperatures

Debye (1912) proposed practically at the same time as Einstein a model with very different assumptions; Einstein took a quantum option and Debye chose a classical option:

- The medium is considered as a quasi-continuous set of chains of atoms whose consequence is a spectrum of vibrations in a wide frequency range.
- By assumption, no vibration frequency can exist above a value ν_m whose corresponding wavelength is the interatomic distance.

- -Finally, the distribution of frequencies is the simple result of a classical statistical calculation, with for each interval dν, the number of longitudinal and transverse vibrations

d$n_l = 4\pi V\nu^2$ dν $(1/c_l^2)$ and d$n_t = 4\pi V\nu^2$dν $(1/c_t^2)$ respectively (c_i propagation speed of wave i and V volume); thus, we obtain

$$3N = \int_0^{\nu_m} g(\nu)\mathrm{d}\nu = 4\pi V/3\left(1/c_l^2 + 1/c_t^2\right)\nu^2 \quad \text{with} \quad g(\nu) = 9N/\nu_m^{\ 2}$$

which allows to calculate the energy of a gram atom in the form

$$U = 9R(T/\theta m)^2 \int_0^{\theta m/T} \frac{x^3}{(\exp x - 1)} dx$$

with $x = h\nu/kT$ and $\theta_m = h\nu_m/k$

The specific heat is deduced from this:

$$C_v = 36R(T/\theta_m)^3 \ D(\theta_m/T) \tag{A9.8}$$

The function $D(\theta_m/T)$, called the Debye function, can be simplified if the temperature is sufficiently low, considering that only the lowest frequencies will be excited. We calculate the integral from 0 to infinity with $\theta_m/T < 1/12$ and then find

$$U = \left(3\pi^4/15\right)RT\left(T/\theta_m\right)^3 \text{ hence, } C_v = 12\pi^2 \ R/5\left(T/\theta_m\right)^3 \tag{A9.9}$$

Therefore, with the Debye model, we account for a variation of the specific heat in T^3 at very low temperatures with an initial assumption of a continuous medium.

A9.1.4 Realistic Modelling of the Specific Heat of Crystals

Generally, to account for the specific heat, we note that a crystal with p atoms in the lattice has $3pN$ degrees of freedom; therefore:

- $3N$ degrees for the lattice vibrations in the acoustic spectrum
- $3(p - 1)N$ degrees in the optical spectrum

and, as a result, we usually write for the average specific heat:

$$pC_v = 3R\left(D(\theta_m/T) + \sum_0^{p-1} E(\theta_{Ei}/T)\right) \tag{A9.10}$$

It is by numerical calculation with the experimental results that we deduce the Debye and Einstein temperatures and the coefficients in front of the functions, which offer a number of possibilities for adjustment in the calculations.

There are a number of anomalies, dispersions, which are not easily described by a Debye function added to 2 or 3 Einstein functions. The authors, depending on the case, call upon various imperfections and models to describe them:

- Schottky defects, i.e., the absence of certain positive ions or/and certain negative ions
- Frenkel defects, following the vacancy of an ion which induces an interstitial
- Polarons with electrons that couple to ions and move with high mobility
- Excitons due to an ion whose excitation level is different from that of its counterparts

The great difficulty in this variety of numerical simulations is to succeed, beyond the good simulation, in attributing with a good certainty, a physical phenomenon to the studied anomaly.

A9.2 Specific Heat of Ferroelectrics at Low and Very Low Temperatures

Detailed studies on solids have shown a fairly good relevance of the models recalled above at room temperature and at low temperatures. Figure A9.1 gives a quick insight: the Einstein model accounts quite well for the experimental results for paraelectric and anti-ferroelectric crystals, and the temperature of the maxima of the curves of Fig. A9.1a, b (Lawless, 1976a, b) allows to calculate the temperatures and frequencies of Einstein. For some ferroelectrics, the curves show increases in values at very low temperatures. The authors usually consider the different physical phenomena of the curves with maxima on the one hand and the increases for temperatures lower than T_m, observed minimum temperature, on the other hand. Various plots of the experimental results are chosen to highlight the laws of variation of the specific heat as a function of temperature, as illustrated by the curves of Fig. A9.1.

We obtain confirmation of variation in T^3, therefore as in Debye model, in practically all crystals. At very low temperatures, we observe, in addition to variations in T^3, linear variations in T that are found in glasses and some variations in $T^{3/2}$ which would be the result of surface effects (Lawless, 1981) and the existence of domains (Lawless, 1976a, b). Regarding KDP, the doubt has been lifted for pure crystals of sufficient size demonstrating properties close to those of a Debye model for specific heat (Lawless, 1987). The anomalies observed at very low temperatures, for temperatures below T_m (Fig. A9.1b), are clearly correlated with the existence of domain walls. The excess specific heat is thus attributed to imperfections created by the walls. Depending on the authors and the similarities they detect or demonstrate, the proposed explanations are diverse:

- Existence of spin waves (Lawless, 1976a, b)
- Similarity of the crystal and its walls with a glass therefore with non-localized defects periodically, regularly

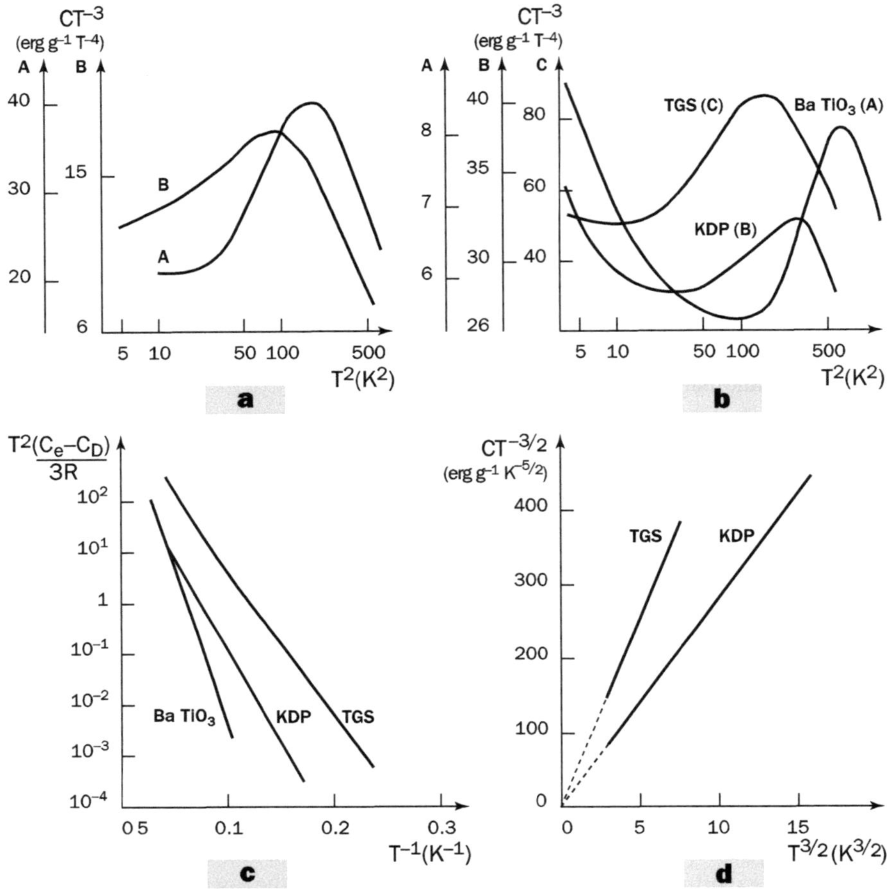

Fig. A9.1 (Lawless, 1976a, b). (**a**) The semi-logarithmic representation highlights the maxima of $C_v T^{-3}$ with precision and allows the calculation of Einstein (optical) modes and Debye (acoustic) modes. The crystals are paraelectric: PbF_2(A) and $KTaO_3$(B). (**b**) In the same representation as in **a**), we observe for three ferroelectric crystals the curve rises correlated with the existence of domains when T decreases to TBT: $BaTiO_3$ (A), KDP (B) and TGS (C). The minima are supposed to represent the acoustic "background" (Burns, 1980). (**c**) Results of the adjustment calculations that validate for the indicated crystals the linear relationship between $\ln(T^2(C_v - C_{Debye}))$ and T^{-1}. (**d**) These linear variations with remarkable adjustments have confirmed the existence of domain walls and their impact on the thermal properties at very low temperatures for KDP and TGS

- Existence of defects, results of tunnelling phenomena between two positions of an atom with the observation for ferroelectrics with hydrogen bonds of more complex acoustic properties than, for example, for perovskites
- Existence of Schottky-type defects

We continue to use the Debye and Einstein models to calculate the specific heat of the "background" to which are added the values of the classical transition model (Chap. 1) and models attributed to suspected defects. This is the case concerning doped KDP to explain the splitting of the peak at the transition or the variations beyond the "background" value of TMO and GMO (Strukov et al., 2008). As the number of adjustment parameters varies, for example in the previously cited publication (between 6 and 8!), one could hope for a better understanding of the phenomena and physical causes, which is often not the case.

A9.3 Conductivity of Solids and Ferro-Ferro Crystals at Low and Very Low Temperatures

The following general relationship summarizes the link between thermal conductivity K_T and specific heat C_v (Kittel, 1949a, b):

$$\mathrm{K}_T = a \cdot C_v \ \ v\Lambda \tag{A9.11}$$

with v speed of phonons and Λ their mean free path (depending on the authors, the arbitrary factor a can be 1/3, 1/4).

In a perfect crystal that only includes harmonic vibrations, the mean free path of phonons is infinite. But with anharmonic interaction terms, Λ takes finite values due to various couplings between lattice vibrations and the distribution of vibration frequencies. Λ will depend on both the importance of the anharmonic terms and the density of phonons.

Some crystals whose transitions are quite diffuse, behave at very low temperatures like glasses with a decrease (in $T^{1.8}$ at very low temperatures (Zeller and Pohl, 1971; Ackermann et al., 1981)): when T decreases, with a marked plateau around 10 K. This is not the case for the selected crystals and notably KDP whose transitions are sharp. At very low temperatures, the previous relationship (A9.11) remains solid with the mean free path constant: it is the variation of C_v that determines that of K_T (De Yoreo et al., 1985).

The measurements of thermal conductivities carried out for a long time for crystals of the KDP family (Suemune, 1967) show a decrease in conductivity above the transition region in which the disorder of protons (or deuterons) on the O–H–O bonds disturbs the network and the PO_4. The suppression of this disorder at the transition changes the context, and under T_C, the conductivity grows very significantly. Even if the forms of temperature variations are the same for crystals such as KDP, DKDP, and KDA, it is noted that the length of the bonds is important: the longer this length is, the less the movement of H (or D) affects the network which explains higher thermal conductivity values for DKDP than for KDP. When the temperature is further reduced, K_T goes through a maximum as in most crystals with the opposite effect versus T of impurities, defects and of dislocations. At very low temperatures, it is the variation of C_v that determines that of K_T (De Yoreo et al., 1985).

The proton (deuteron)-network interactions are major for thermal properties, as they are for the para-ferro transition as illustrated by the correlation between the value of the thermal conductivity K_m at the maximum at very low temperatures and the para-ferro transition temperatures (Suemune, 1967), according to the relationship: $K_m = 2.6 \times 10^{-10}\, T_C^4$ W cm^{-1} K^{-1}.

Conclusion

Simple thermodynamic models allow to justify the possible transitions in crystals assumed to be of infinite dimensions by looking for stable situations, with identical temperature and pressure at all points. We thus define quantities, order parameters corresponding to these stable situations. Knowing the symmetry of the two phases, we can predict the possible elementary cells in the "ferro" phase, and the possible coherent interfaces between the domains thus defined. For example, in this book, the tetragonal planes (100) and (010) constitute the "permissible walls" between the "ferro-ferro" domains of crystals of the same symmetry change tetragonal-orthorhombic at the transition (KDP family, GMO family, Tanane...). Staying within the assumptions of an infinite medium, with gradients of order parameter perpendicular to the permitted walls, we calculate what we are used to calling widths (theoretical) of walls.

These models are valuable but, in real crystals, the physical limits of a sample induce the obligatory existence of walls, domains and the interfaces may not actually be coherent. They are at the basis of certain macroscopic properties of crystals that rely on dynamics; this is obvious for the very symbol of the ferros, the hysteresis cycle, with wall displacements at the coercive field, with a coercive field a hundred times smaller in reality than that predicted by a theoretical model. Interfaces are also major during the coexistence of phases of a first order transition, with the exchange surface of latent heat, the phase front that can interact with the domains.

Moreover, the dynamics of interfaces do not occur in a homogeneous medium, but start, progress, end, at the moment and place energetically most favourable. Hence the importance of sample edges, possible interfaces (phase fronts, walls), and within these interfaces, places that present a favourable imperfection. In the example of the crystals of the book, the systematic importance of quasi-dislocations, notably of wedge type along the ferroelectric axis, is clearly demonstrated at the level of a non-stable permissible wall, a fortiori in any wall outside the planes permitted by symmetry. The theory of dislocations (and disinclinations) explains most of the observed results. The concept can also be successfully used for phase fronts.

J. Bornarel, *Ferroic Crystal Domains, Walls, and Phase Fronts*, Springer Series in Solid-State Sciences 209, https://doi.org/10.1007/978-3-031-98099-2

Depending on the experimental contexts, the properties measured macroscopically in the ferro-ferro phases correspond to phenomena of different natures: global phenomenon sum of continuous variations in a kind of continuum not far from the transition, isolatable domains in strong collective interaction and in metastable equilibrium, walls in short distance interaction, isolated walls, imperfections at the atomic level within the walls. One of the interests of the book is to clearly show the reality of this rather banal statement, and to demonstrate that what essentially matters is the distance at which such an interface will have an effect on another interface or object.

The relative values of the energies involved (chemical, mechanical, electrostatic), here deliberately put in competition, are decisive for the shapes of phase fronts, domains and their textures. But, these values can be misleading. We tend to focus on those directly related to the order parameter, and to somewhat forget such a deformation ten times smaller. For example, the extent of the deformed regions well beyond the coexistence zone of the DKDP, demonstrates the importance of what changes relatively: the deformations u_{xx} and u_{zz} identical in two "opposite" domains but different between the walls and the domains. Their values lower than that of the main deformation u_{xy} at the transition, can nevertheless modify the properties of a set of domains, therefore of the sample. The example of a factory roof phase front with its close domain textures demonstrates how the competition of all energies can have implications throughout the crystal.

The macroscopic properties result from this richness of nested phenomena at various levels, sometimes shielded by a last friction. The contribution of domain walls, that of phase coexistence zones, have been identified within the macroscopic results with losses as a privileged detector of the existence of interfaces and especially of domain walls. The interplay of interfaces explains for example the switching current laws with typical acceleration at the end of the ferro-ferro process.

These dynamic phenomena, like the metastable states created, obviously depend on the crystals and the imposed physical parameters, temperature and pressure, fields conjugated to the order parameter and boundary conditions therefore sample shapes. They also depend on the starting and ending situations, which requires particular attention to time, history.

This book will have achieved its objective if it has convinced the reader of the beauty of physics at all levels: the microscopic hetero-phase of phase coexistence at the beginning of a first order ferro-para transition is an efficient finned heat exchanger. The phenomena of avalanches, those of fractals often modelled theoretically are here concretely illustrated. Let's hope that this link between experiments and models opens horizons for the reader and encourages reflection, including for data processing with large computing powers and many variable parameters thus brought closer to reality…

References

Abalmassov, V.A., Pugachev, A.M., Surovtsev, N.V., Ferroelectrics 440, 113 (2012)
Abalmassov, V.A., Ferroelectrics 501(57)(2016); 538, 1 (2019)
Abe, K., J. Phys. Soc. Jpn. 5(2), 757 (1987)
Ackermann, D.A., Moy, D., Potter, R.C., Anderson, A.C., Lawless, W.N., Phys. Rev. B 23(8), 3886 (1981)
Adams, G., Gibbs, H., J. Chem. Phys. 43, 139 (1965)
Afonikova, N.S., Borovikov, V.V., Shmyt'Ko, I.M., Sov. Phys. Solid State 29(3), 462 (1987)
Aird, A., Salje, E.K.H., J. Phys.: Condens. Matter 10, L337 (1998)
Aird, A., Salje, E.K.H., Eur. Phys. J. B 15, L205 (2000)
Aizu, K.J., J. Phys. Soc. Jpn. 27(2), 387 (1969)
Aizu, K.J., J. Phys. Soc. Jpn. 28, 706 (1970)
Aizu, K.J., J. Phys. Soc. Jpn. 32(5), 1287 (1972)
Aizu, K.J., J. Phys. Soc. Jpn. 34, 121 (1973)
Akhmatkhanov, A.R., Chuvakova, M.A., Dolgushin, N.A., Kolker, D.B., Vedenyapin, V.N., Isaenko, L.I., Shur, V.Ya Ferroelectrics 559, 1 (2020)
Aknazarov, S.K., Shabel'nikov, L.G., Shekhtman, V.S. Sov. Phys. Solid State 17(1), 16 (1975)
Albers, J., Klöpperpieper, A., Ferroelectrics 39, 1071 (1981)
Aleshko-Ozhevkii, O.P., Sov. Phys. Cryst. 27(6), 673 (1982)
Aleshko-Ozhevkii, O.P., Bowen, D.K., Davies, S.T., Ferroelectrics 62, 53 (1985)
Aleshko-Ozhevkii, O.P., Sov. Phys. Solid State 34(3), 499 (1992)
Alessandro, B., Beatrice, C., Bertotti, G., Montorsi, A., J. Appl. Phys. 68, 2901 (1990a)
Alessandro, B., Beatrice, C., Bertotti, G., Montorsi, A., J. Appl. Phys. 68, 2908 (1990b)
Alpay, S.P., Roytburd, A.L., J. Appl. Phys. 83, 4714 (1998)
Amin, A., Newnham, R.E., Phys. Stat. Sol. (a) 61, 215 (1980)
Anderson, P.W. In: Skanavi GI (Ed.), Fizika Dielektrikov. Akad Nauk SSSR Moscow (1960)
Andrews, S.R., Cowley, R.A., J. Phys. C: Solid State Phys. 19, 615 (1986)
Angell, C.A., Non-Cryst. Solids 102, 205 (1988),
Araujo, J.F., Mendes Filho, J., Melo, F.E.A., Letelier, F.V., Chaves, A.S., Phys. Rev. B 57(2), 783(1998)
Arias, I., Catalan, G., Sharma, P., J. Appl. Phys. 131, 020401 (2022)
Authier, A., Petroff, J.F., C. R. Acad. Sci. 258, 4238 (1964)
Avrami, M., J. Chem. Phys. 7, 1103 (1939); 8, 212 (1940); 9, 177 (1941)
Axe, J.D., Dorner, B., Shirane, G., Phys. Rev. Lett. 26, 519 (1971)

J. Bornarel, *Ferroic Crystal Domains, Walls, and Phase Fronts*, Springer Series in Solid-State Sciences 209, https://doi.org/10.1007/978-3-031-98099-2

Bachheimer, J.P., Bastie, P., Bornarel, J., Dolino, G., Vallade, M., Proc. Inst. Conf. on Solid State transformations, Warendale, P.A.: Metallurgical Society of AIME, p. 1533 (1981)
Bacon, G.E., Pease, R.S., Proc. Roy. Soc. 220, 397 (1953)
Bacon, G.E., Pease, R.S., Proc. Roy. Soc. 230, 359 (1955)
Balagurov, A.M., Dutt, I.D., Savenko, B.N., Shuvalov, L.A., I. Akad. Nauk SSSR Ser. Fiz. 43(8), 1740 (1979)
Balagurov, A.M., Damm, I.D., Savenko, B.N., Shuvalov, L.A., Sov. Phys. Solid State 22(9), 1595 (1980)
Balagurov, A.M., Savenko, B.N., Dutt, I.D., Shuvalov, L.A., Ferroelectrics 48, 163 (1983)
Baldassari, A., Dalton, F., Petri, A., Zapperi, S., Pontuale, G., Pietronero, L., Phys. Rev. Lett. 96, 118002 (2006)
Bantle, W., Helv. Phys. Acta 15, 373 (1942)
Barkhausen, H., Phys. Z. 20, 401 (1919)
Barkla, H.M., Finlayson, Phil. Mag. 7(44), 109 (1953)
Baro, J, Vives, E., Phys. Rev. E 85, 6, 066121 (2012)
Barrett, J.H., Phys. Rev. 86, 118 (1952)
Barsch, G.R., Krumhansl, J.A., Phys. Rev. Lett. 53(11), 1060 (1984)
Bartels, M., Hagen, V., Burianek, M., Getzlatt, M., Bismayer, U., Wiesendanger R., J. Phys.: Condens Matter 15, 957 (2003)
Bartis, F.J., Phys. Stat. Sol. (b) 45, K169 (1971)
Bastie, P., Bornarel, J., Lajzerowicz, J., Vallade, M., Phys. Rev. B 12, 5112 (1975)
Bastie, P., Bornarel, J., Lajzerowicz, J., Legrand, J.F., Ferroelectrics 13, 319 (1976a)
Bastie, P., Bornarel, J., Legrand, J.F., Ferroelectrics 13, 455 (1976b)
Bastie, P., Bornarel, J., J. Phys.C:Solid State Phys. 12, 1785 (1979)
Bastie, P.M., Doctoral Thesis, Joseph Fourier University (1980)
Bastie, P., Bornarel, J., Dolino, G., Vallade, M., Ferroelectrics 26, 789 (1980)
Bastie, P., Vallade, M., Vettier, C., Zeyen, C.M.E., Meister, H., J. Physique 42, 445 (1981)
Bastie, P., Bornarel, J., J. Physique 43, 795 (1982)
Bastie, P., Becker, P., J. Phys. C: Solid State Phys. 17, 193 (1984)
Bastie, P., Lajzerowicz, J., Schneider, J.R., J. Phys. C: Solid State Phys. 11, 1203 (1978)
Bastie, P., Lajzerowicz, J., Schneider, J.R., J. Phys. C: Solid State Phys. 11, 1203 (1978a)
Bastie, P., Vallade, M., Vettier, C., Zeyen, C.M.E., Phys. Rev. Lett. 40, 337 (1978b)
Bastie, P., Troussaut, F., Vallade, M., Zeyen, C.M.E., J. Appl. Cryst. 20, 475 (1987)
Baumgartner, H., Helv. Phys. Acta 24, VII, 651 (1950)
Baumgartner, H., Helv. Phys. Acta 24, 326 (1951)
Belouet, C., Dunia, E., Petroff, J.F., J. Crystal Growth 23, 243 (1974)
Benepe, J.W., Reese, W., Phys.Rev. B 3, 3032 (1971)
Ben Ishai, P., de Oliveira, C.E.M., Ryabov, Y., Feldmann, Yu., Agranat, A.J., Phys. Rev. B 70, 132104 (2004)
Bertotti, G., Fiorillo, F., Sassi, M.P., J. Magnetism Mater 23, 136 (1981)
Bertotti, G.H., Hysteresis in Magnetism: For Physicists, Material Scientists and Engineers, Academic Press, San Diego, CA, (1998)
Bhattacharya, S., Nagel, S.R., Fleishman, L., Susman, S., Phys. Rev. Lett. 48, 1267 (1982)
Biedrzycki, K., Ferroelectrics 119, 33 (1991)
Bilz, H., Benedek, G., Bussmann-Holder, A., Phys. Rev. B 35, 4840 (1987)
Binggeli, B., Fatuzzo, E., J. Appl. Phys. 36(1), 1431 (1965)
Binder, A., Knorr, K., Markov, Yu., Phys. Rev. B 61, 190 (2000)
Bjorkstam, J.L., Phys.Rev. 153(3), 599 (1967)
Bjorkstam, J.L., Oettel, R.E., PIMF, Praha V 2, 91 (1966)
Bjorkstam, J.L., Oettel, R.E., Phys. Rev. 159(2), 427 (1967)
Blinc, R., J. Phys. Chem. Solids 13, 204 (1960)
Blinc, R., Svetina, S., Phys. Rev. 147, 423, and 430 (1966)
Blinc, R., Burgar, M., Cizikov, S., Levstik, A., Kadara, P., Shuvalov, L.A., Phys. Stat. Sol. (b) 67, 689 (1975)

Blinc, R., Burgar, M., Rutar, V., Seliger, J., ZupanCic, I., Phys. Rev. Lett. 38(2), 92 (1977)
Blinc, R., Zeks, B., J. Phys.C: Solid State Phys. 15, 4661 (1982)
Blinc, R., Zeks, B., Nelmes, R.J., Phase Transitions 3, 293 (1983)
Bodrov, S.B., Abramovsky, N.A., Paramonov, G.S., Belyaev, S.N., Prokhorov, A.P., Stepanov, A.N., Bakunov, M.I., J. Appl. Phys. 134, 213102 (2023)
Boiko, A.A., Liu Tan Fat, Sov. Phys. Crystallogr. 14(5), 712 (1970)
Bonnot, E., Vives, E., Manosa, L., Planes, A., Phys. Rev. B 78, 094104 (2008)
Borchardt, H.J., Bierstedt, P.H., J. Appl. Phys. 8, 50 (1965)
Borchardt, H.J., Bierstedt, P.H., J. Appl. Phys. 38, 2057 (1967)
Bordeaux, D., Bornarel, J., Capiomont, A., Lajzerowicz-Bonneteau, J., Lajzerowicz, J., Legrand, J.F., Phys. Rev. Lett. 31, 314 (1973)
Bordeaux, D., Capiomont, A., Lajzerowicz-Bonneteau, J., Jouve, M., Thomas, M., Acta Crys. B 30, 2156 (1974)
Born, M., von Karmann, T., Physik. Z. 13, 297 (1912)
Bornarel, J., Fouskova, A., Guyon, P., Lajzerowicz, J. Proceedings of the International Meeting on Ferroelectricity, Praha V 2, 81 (1966)
Bornarel, J. Lajzerowicz, J., J. Appl. Phys. 39, 4339 (1968)
Bornarel, J., Lajzerowicz, J., J. Phys. Soc. Jpn. S2 8, 360 (1970)
Bornarel, J., Thesis Université Joseh Fourier, Grenoble, France (1971)
Bornarel, J., J. Appl. Phys., 43(3), 845 (1972)
Bornarel, J., Lajzerowicz, J., Ferroelectrics 4, 177 (1972)
Bornarel, J., Fousek, J., Glogarova, M., Czech. J. of Phys. B 22, 864 (1972)
Bornarel, J., Lazzerowicz, J., J. Phys. C 2(33), 153 (1972)
Bornarel, J., Lajzerowicz, J., Legrand, J.F., Ferroelectrics 7, 313 (1974)
Bornarel, J., Ferroelectrics 9, 197 (1975)
Bornarel, J., Bastie, P., J. Phys. C: Solid State Phys. 13, 5843 (1980)
Bornarel, J., Kvitek, Z., Ferroelectrics, 141,113 (1993)
Bornarel, J., Legrand, J.F., Ferroelectrics 39, 1127 (1981)
Bornarel, J., Schmidt, V.H., J. Phys. C: Solid State Phys. 14, 2017 (1981)
Bornarel, J., Ferroelectrics 71, 255 (1987)
Bornarel, J., Torche, B., Ferroelectrics 76(1–2), 201 (1987)
Bornarel, J., Ferroelectrics, 106, 21 (1990)
Bornarel, J., Phase Transitions 34, 147 (1991)
Bornarel, J., Torche, B., Ferroelectrics 132, 271 (1992)
Bornarel, J., Ferroelectrics, 172, 53 (1995)
Bornarel, J., Staniorowski, P., Czapla, Z., Phys. Stat. Sol. (b) 203, 59 (1997)
Bornarel, J, Cach, R., Ferroelectrics 124,345 (1991)
Bornarel, J, Cach, R., J. Phys.: Condens. Matter 5, 2977 (1993)
Bornarel, J, Cach, R., J. Phys.: Condens. Matter 6, 1663 (1994)
Bornarel, J, Cach, R., Kvitek, Z., J. Phys.: Condens. Matter 8, 7365 (1996)
Bornarel, J, Cach, R., Phys. Rev. B 60, 3806 (1999)
Bornarel, J, Cach, R, Kvitek, Z., Ferroelectrics 236, 117 (2000)
Bornarel, J, Cach, R., J. Phys.: Condens. Matter 15, 4371 (2003)
Bouchaud, J., Quant. Finance 1, 105 (2001)
Bratkovsky, A.M., Levanyuk, A.P., Phys. Rev. Lett. 85(21), 4614 (2000)
Bretz, M., Cunningham, J.B., Kurczynski, P.L., Nori, F., Phys. Rev. Lett. 69, 2431 (1992)
Brezina, B., Havrankova, M. Mat. Res. Bull. 6 (1971)
Brody, E.M., Cummins, H.Z., Phys. Rev. Lett. 21, 1263 (1968)
Brody, E.M., Cummins, H.Z., Phys. Rev. Lett. 23, 18, 1039 (1969)
Bruce, D.A., J. Phys. C: Solid State Phys. 14, 5195 (1981)
Budin, J.P., Milatos, A., Roufos, Nguyen Due Chinh, Le Roux G., J. Appl. Phys. 46, 2867 (1975)
Bullough, R., Phys. Rev. 110(3), 620 (1958)
Burfoot, J.C., Ferroelectrics, An Introduction to the Physical Principles, Van Nostrand, London (1967)

Burns, G., Solid State Com. 35, 811 (1980)
Bursian, E.V., Vall, A.E., Trunov, N.N., Sov. Phys. Solid State 19(6), 1106 (1977)
Burtsev, E.V., Chervonobrodov, S.P., Ferroelectrics Lett. 44, 293 (1983)
Busch, G., H.P.A., 11, 269 (1938)
Busch, G., Scherrer, P., Naturwiss 23, 737 (1935)
Busch, G., Ganz, E., H.P.A. 15, 501 (1942)
Bussmann-Holder, A., Michel, K., Phys. Rev. Lett. 80, 2173 (1998)
Bussmann-Holder, A., Dalal, N., Ferroelectrics 237, 1 (2000)
Bussmann-Holder, A., Dalal, N., Struct. Bond 124, 1 (2007)
Cach, R., Tomaszewski, P., Bornarel, J., J. Phys. C: Solid State Phys. 18, 915 (1985)
Calleja, M., Dove, M.T., Salje, E.K.H., J. Phys.: Condens. Matter 13, 9445 (2001)
Calleja, M., Dove, M.T., Salje, E.K.H., J. Phys.: Condens. Matter 15, 2301 (2003)
Camara, F., Doukhan, J.C., Salje, E.K.H., Phase Transitions, 71, 227 (2000)
Cao, W. Barsch, G.R., Phys. Rev. B 41, 4334 (1990)
Chabin, M., Giletta, F., Ferroelectrics 15, 149 (1977)
Chaves, A.S., Letelier, F.V., Sampaio, J.F., Gazzzinelli, R., Phys. Rev. B 47, 4880 (1993)
Chervonobrodov, S.P., Roytburd, A.L., Ferroelectrics, 83, 109 (1988)
Chincholkar, V.S., Unruh, H.G., Phys. Stat. Sol. 29, 669 (1968)
Choi, B.K., Lockwood, D.J., Ferroelectrics 106, 15 (1990)
Chynoweth, A.G., Phys. Rev. 110(6), 1316 (1958),
Chynoweth, A.G., Abel, J.L., J. Appl. Phys. 30(7), 1073 (1959)
Cochran, W., Adv. Phys. 9, 387 (1960)
Cochran, W., Adv. Phys. 10,401 (1961)
Colaiori, F., Adv. Phys. 57, 287 (2008)
Colla, E.V., Chao, I.K., Weissman, M.B., Phys. Rev. Lett. 73, 3423 (1994)
Colla, E.V., Chao, I.K., Weissman, M.B., Phys. Rev. Lett. 88(1), 017601 (2002)
Collins, M.A., Blumen, A., Currie, J.F., Ross, J., Phys. Rev. B 19(7), 3630 (1979)
Conti, S., Salje, E.K.H., J. Phys.: Condens. Matter 13, L847 (2001)
Cook, W.R., J. Appl. Phys. 38, 1637 (1967)
Cole, K.S., Cole, R.H.J., Chem. Phys. 9, 341 (1941)
Combs, J.A., Yip, S., Phys. Rev. B 28(12), 6873 (1983)
Cottrell, A.H., Dislocations and Plastic Flow, Oxford University Press, Oxford (1953)
Courtens, E., Phys. Rev. Lett. 52(1), 69 (1984)
Cross, L.E., Fouskova, A., Cummins, S.E., Phys. Rev. Lett. 21, 812 (1968)
Curie, P., J. Phys. Theor. Appli. 3, 393 (1894)
Curtin, W.A., Scher, H., Phys. Rev. Lett. 67, 2457 (1991)
Dabrowska, K., Daszczynska, W., Jaskiewcz, A., Acta Phys. Polonica A 51(4), 539 (1977)
Dalal, N., Klymachyov, A., Bussmann-Holder, A., Phys. Rev. Lett. 81, 5924 (1998)
Dalal, N.S., Gunaydin, O., Bussmann-Holder, A., Struct. Bond 124, 23 (2007)
Damjanovic, D., Demartin, M., J. Phys. D: Appl. Phys. 29, 2057 (1996)
Damjanovic, D., Phys. Rev. B 55(2), 649 (1997)
Daniel, V.V., Dielectric relaxation, Academic Press, London and New York (1967)
Daraktchjev, M., Salje E.K.H., Lee, W.T., Redfern, S.A.T., Phys. Rev. B 75, 134102 (2007)
Darinskii, B.M., Sidorkin, A.S., Ferroelectrics 71, 269 (1987)
Darinskii, B.M., Sidorkin, A.S., Lazarev, A.P., Ferroelectrics 98, 245 (1989)
Darinskii, B.M., Sidorkin, A.S., Ferroelectrics 497, 107 (2016)
Debye, P., Ann. Physik 39,789 (1912)
Dec, J., J. Phys. C: Solid State Phys. 21, 1257 (1988)
Dec, J., Phase Transitions 45, 35 (1993)
De Gennes, P.G., Solid State Comm. 1, 132 (1963)
Deguchi, K., Nakamura, E., Kohno, S., J. Phys. Soc. Japan, 48(6), 2007 (1980)
Delgado-Sanchez, J.M., Martin-Olalla, J.M., Gallardo, M.C., Ramos, S., Koralewski, M., del Cerro, J., Phys. Rev. (2005)
Deluca, M., Taniguchi, H., J. Phys. Soc. Jpn. 91, 091001 (2022)

De Quervain, Helv. Phys. Acta 17, 509 (1944)
Devonshire, A.F., Phil. Mag. Suppl. 3(10), 85 (1954)
De Witt, R., Phys. Stat. Sol. 20, 567 and 575 (1967)
De Witt, R., Fundamentals Aspects of Dislocations Theory, U.S. Gov. Print. Office, Washington (1970)
De Witt, R., J. Appl. Phys. 42, 3304 (1971)
De Witt, R., Theory of Disclinations 3, 359 (1973a)
De Witt, R., J. Res. National Bureau of Standards 77A, 1 (1973b)
De Yoreo, J.J., Pohl, R.O., Burns, G., Phys.Rev. B 32(9), 5780 (1985)
De Yoreo, J.J., Burnham, A.K., Whitman, P.K., Int. Mater. Rev. 47(3), 113 (2002)
Diamant, H., Drenck, K., Pepinsky, R., Rev. Scient. Inst. 28(1), 30 (1951)
Dimmler, K., Parris, M., Butler, D., Eaton, S., Pouligny, B., Scott, J.F., Ishibashi, Y., J. Appl. Phys. 61, 5467 (1987)
Domenichini, P., Pasquini, G., Capellita, M.G., Phys. Rev. B 108, 224434 (2023)
Dorner, B., Axe, J.D., Shirane, G., Phys. Rev. B 6, 1950 (1972)
Drougard, M.E., Landauer, R., J. Appl. Phys. 30(11), 1663 (1959)
Duiker, H.M., Beale, P.D., Phys. Rev. B 41(1), 490 (1990)
Durand, E., Electrostatics, Mason et Cie, Paris (1964)
Durin, G, Zapperi, S., The Science of Hysteresis. In: Bertotti, G. and Mayergoyz, I.D. (eds.) vol. 2, Elsevier, Amsterdam, p. 181 (2006)
Durin, G, Bohn, F., Corrêa, M.A., Sommer, R.L., Le Doussal, P., Wiese, K.J., Phys. Rev. Lett. 117, 087201 (2016)
Dvorak, V., Phys. Stat. Sol. (b) 46, 147 (1971a)
Dvorak, V., Phys. Stat. Sol. (b) 46, 763 (1971b)
Dvorak, V., Czec. J. Phys, 28(9), 989 (1978)
Eberhard, J.W., Horn, P.M., Solid State Comm. 16, 1343 (1975)
Einstein, A., Ann. D. Physik 22, 180 (1907); 34, 170 (1911)
Ehrenfest, P., Suppl. 75b, Proc. Roy. Acad. Amsterdam 36, 153 (1933)
Elmore, W.C., Phys. Rev. 53, 757 (1938)
Endo, S., Sawada, T., Tsukawake, T., Kobayashi, Y., Ishizuka, M., Deguchi, K., Tokunaga, M., Solid State Commun. 112, 655 (1999)
Endo, S., Deguchi, K., Tokunaga, M., Phys. Rev. Lett. 88(3), 035503 (2002a);
Endo, S., Deguchi, K., Tokunaga, M., J. Phys.: Condens. Matter 14, 11275 (2002b)
Erhart, J., Phase Transitions 77, 12, 989 (2004)
Erhart, J., Franclikova, M., Rusin, L., Ferroelectrics 376, 99 (2008)
Ertas, D., Kardar, M., Phys. Rev. Lett. 73, 1703 (1994)
Eshelby, J.D., Brit. J., Appl. Phys. 17, 1136 (1966)
Falk, F., Z. Phys. B 51, 177 (1983)
Fally, M., Kubinec, P., Fuith, A., Warhanec, H., Filipic, C., J. Phys.C: Condens. Matter 7, 2195 (1995)
Fally, M., Kubinec, P., Fuith, A., Warhanec, H., Ferroelectrics 190, 43 (1997)
Fally, M., Fuith, A., Schranz, W., Müller, V., Phys. Rev. B 64, 026101 (2001)
Fang, Z., Wang, Yu.I., Tang, Y.L., Zhu, Y.L., Ma, X.L., J. Appl. Phys. 135, 044104 (2024)
Fatuzzo, E., Helv. Phys. Acta 33, 21 (1960a)
Fatuzzo, E., Proc. Phys. Soc. (London) 76, 797 (1960b)
Fatuzzo, E. Merz, W.J., Phys. Rev. 116, 61 (1959)
Fatuzzo, E. Merz, W.J., Phys. Rev. 127(6), 1999 (1962)
Fatuzzo, E. Merz, W.J., Ferroelectricity, North-Holland, Amsterdam (1966)
Fedosov, V.N., Sidorkin, A.S., Sov. Phys. Solid State 18(6), 964 (1976)
Fedosov, V.N., Sidorkin, A.S., Sov. Phys. Crystallogr. 21(6), 644 (1976a)
Fedosov, V.N., Sidorkin, A.S., Sov. Phys. Solid State 18(6), 964 (1976b)
Fedosov, V.N., Sidorkin, A.S., Sov. Phys. Solid State 19(6), 952 (1977a)
Fedosov, V.N., Sidorkin, A.S., Sov. Phys. Solid State 19(8), 1359 (1977b)
Fedosov, V.N., Sidorkin, A.S., Sov. Phys. Solid State 21(7), 1211 (1979)

Fedosov, V.N., Sov. Phys. Solid State 22(11), 1913 (1980)
Fesenko, E.G., Martynenko, M.A., Gavrilyachenko, V.G., Semenchev, A.F., Izv. Akad. Nauk Ser. Fiz. 39, 762 (1975)
Fetisov, Y.K., Srinivasan, G., J. Appl. Phys. 135, 024102 (2024)
Fisher, M.E., Szpilka, A.M., Phys. Rev. B 36(1), 644 (1987)
Flippen, R.B., Haas, C.W., Solid State Commun. 13, 1207 (1973)
Fomitchev, H.H., Soviet Acad. Nauk 29(6), 962 (1965)
Fousek, J., Safrankova, M., Jpn. J. Appl. Phys. 4(6), 403 (1965)
Fousek, J., Jpn. J. Appl. Phys. 6, 950 (1967)
Fousek, J., Janovec, V., J. Appl. Phys. 40, 135 (1969)
Fousek, J., Proc. of the European Meeting on Ferroelectricity, Saarbrücken 1969, 147 (1970)
Fousek, J., Koväk, C., Phys. Stat. Sol. B 52, K13 (1972)
Fousek, J., Glogarova, M., Kürsten, H.D., Ferroelectrics 11, 469 (1976)
Fouskova, A., Janovek, V., J. Phys. Soc. Jpn 20(9), 1619 (1965)
Fouskova, A., J. Phys. Soc. Jpn 20(9), 1625 (1965)
Franck, F.C., Disc.Faraday Soc. 25, 19 (1958)
Freund, A., Schneider, J.R., J. Cryst. Growth 13/14, 247 (1972)
Friedel, J., Dislocations, International Series of Monographs on Solid State Physics 3, Pergamon Press (1964)
Fukami, T., Akahoshi, S., Hukada, K., Jpn.J. Appl. Phys. 24(Suppl. 2), 929 (1985)
Frondel, C., Am. Mineral. 30, 447 (1945)
Fulcher, Gordon S., J. Am. Ceramic Soc. 8(6), 339 (1925)
Gallardo, M.C., Manchado, J., Romero, F.J., J. del Cero, Salje, E.K.H., Planes, A., Vives, E., Phys. Rev. B 81, 174102 (2010)
Gan, Y.C., Cao, W.Q., Ferroelectrics Letters 47(4), 96 (2020)
Garber, S.R., Smolenko, L.A., Sov. Phys. JETP 28(6), 1072 (1969)
Garcia-Colin, L.S., del Castillo, L.F., Goldstein, P., Phys. Rev. B 40(10), 7040 (1989)
Gavrilova, I.V., Kuznjecov, Crys. Grown Ed. Sov. Acad. Sci. 4, 85 (1964)
Gilchrist, J., J. Phys. Chem. Solids 50(8), 857 (1989)
Gilchrist, J., Solid State Comm. 108, 12, 939 (1998)
Gilchrist, J., Private Communication (1999)
Gilletta, F., Phys. Stat. Sol. (a) 11, 721 (1972)
Gilmann, J.J., J. Appl. Phys., 36, 3195 (1965)
Gilmann, J.J., Comments on Solid State Phys. 11(2), 37 (1969)
Ginzburg, V.L., Sov. Phys. Solid State 2, 1123 (1960)
Gladkii, V.V., Sidnenko, E.V., Sov. Phys. Solid State 13(10), 2592 (1972)
Glazer, A.M., Groves, P., Smith, D.T., J. Phys. E Sci. Instrum. 17, 95 (1984)
Goian, V., Borodavka, F., Repcek, D., Savinov, M., Misek, M., Kastil, J., Skoromets, V., Ondrejkovic, P., Petzelt, J., Hlinka, J., Kuzel, P., Kamba, S., Phys. Rev. B 108, 224101 (2023)
Goncalves-Ferreira, L., Redfern, S.A.T., Artacho, E., Salje, E.K.H., Phys. Rev. Lett. 101, 097602 (2008)
Gonzalo, J.A., Phys. Rev. 144, 662 (1966)
Gordon, A., Physica 122B, 321 (1983)
Gordon, A., Physica 138B, 239 (1986)
Gridnev, S.A., Kamisheva, L.N., Romanova, N.F., Sov. Phys. Crys. 18(6), 773 (1974)
Gridnev, S.A., Kamisheva, L.N., Postnikov, V.S., Izv. Akad. Nauk SSSR ser. Fiz. V 39(4), 827 (1975)
Gridnev, S.A., Darinskii, B.M., Phys. Stat. Sol. (a) 47, 379 (1978)
Gridnev, S.A., Darinskii, B.M., Kudryash, V.I., Prasolov, B.N., Shuvalov, L.A., Fiz. Tverd. Tela 24(1), 217 (1982)
Gridnev, S.A., Kravchenko, S.A., Phys. Stat. Sol. (a) 181, R4 (2000)
Gridnev, S.A., Ferroelectrics 266, 171 (2002)
Gridnev, S.A., Kosilov, A.T., J. Phys. D: Appl. Phys. 35, 1538 (2002)

Gridnev, S.A., Ferroelectrics 360, 1 (2007)
Gridnev, S.A., Trukhachev, S.E., Ferroelectrics 443, 63 (2013)
Griffits, R.B., Phys.Rev. Lett. 24, 715 (1970)
Gröger, R., Lookman, T., Saxena, A., Phys. Rev. B 78, 184101 (2008)
Gureev, M.Y., Tagantsev, A.K., Setter, N., Phys. Rev. B 83, 23 (2011)
Gutenberg, B., Richter, C.F., Ann. Geophys. 9, 1 (1956)
Guyon, P., Lajzerowicz, J., Phys. Stat. Solid. 16, 525 (1966)
Ha, D.H., Kim, J.J., Jpn. J. of Appl. Phys. 24(S24), 556 (1985)
Haas, C.W., Jaep, W.F., Phys. Lett. 49A(1), 77 (1974)
Hadni, A., Thomas, R., Ungar, S., Gerbaux, X., Ferroelectrics 47, 201 (1983)
Hall, D.A., Ben-Omran, M.M., J. Phys.: Condens. Matter 10, 9129 (1998).
Hammermesh, M., Group Theory, Adison Wesley Publ. Co., U.S.A. (1964)
Hamzaoui, S., Labrune, M., Puchalska, I.B., Appl. Phys. Lett. 45(11), 1246 (1984)
Harrison, R.J., Redfern, S.A.T., Phys. Earth Planet. Inter. 134, 253 (2002)
Harrison, R.J., Redfern, S.A.T., Buckley, A., Salje, E.K.H., J. Appl. Phys. 95(4), 1706 (2004)
Harrison, R.J., Salje, E.K.H., Appl. Phys. Lett. 97(2), 021907 (2010)
Harrison, R.J., Salje, E.K.H., Appl. Phys. Lett. 99, 151915 (2011)
Hatano, J., Suda, F., Futana, H., J. Phys. Soc. Japan 43(6), 1933 (1977)
Hatano, J., Watanabe, T., Le Bihan, R., Ferroelectrics 126, 311 (1992)
Hayashi, M., J. Phys. Soc. Jpn. 33(3), 616 (1972)
Hayashi, K., Deguchi, K., Nakamura, E., J. Phys. Soc. Jpn. 57(10), 3594 (1988)
Heaney, P.J., Veblen, D.R., Am. Mineral. 76, 1459 (1991)
Hecht, R., Cieszymski, S.F., Colla, E.V., Weissman, M.B., Phys. Rev. Mater. 1, 044403 (2017)
Highland, M.J., Fister, T.T., Richard, M.I., Fong, D.D., Fuoss, P.H., Thompson, C., Eastman, J.A., Streiffer, S.K., Stephenson, G.B., Phys. Rev. Lett. 105, 167601-1 (2010)
Hill, R.M., Ichiki, S.K., Phys. Rev. 130(1), 150 (1963a)
Hill, R.M., Ichiki, S.K., Phys. Rev. 132(4), 1603 (1963b)
Hill, R.M., Ichiki, S.K., Phys. Rev. 135(6), 1640 (1964)
Hill, R.M., Herrmann, G.F., Ichiki, S.K., J. Appl. Phys. 36(11), 3672 (1965)
Höchli, U.T., J. Phys.C: Solid State Phys. 8, 3896 (1975)
Holste, J.C., Lawless W.N., Samara, G.A., Ferroelectrics 11, 337 (1976)
Houchmandzadeh, B., Lajzerowicz, J., Salje, E.K.H., J. Phys.: Conden. Matter 3, 5163 (1991)
Houchmandzadeh, B., Lajzerowicz, J., Salje, E.K.H., J. Phys.: Conden. Matter 4, 9779 (1992)
Hu, Z.W., Jiang, S.S., Huang, X.R., Feng, D., Wang, J.Y., Li, L.X., Appl. Phys. Lett. 64, 55 (1994)
Huang, Y.N., Wang, Y.N., Shen, H.M., Phys. Rev. B 46(6), 3290 (1992)
Huang, X.R., Jiang, S.S., Hu, X.B., Xu, X.Y., Zeng, W., Feng, D., Phys. Rev. B 52(14), 9932 (1995)
Huang, Y.N., Li, X., Ding, Y., Shen, H.M., Zhang, Z.F., Wang, Y.N., Fang, C.S., Zhuo, S.H., J. de Phys. IV, C8, 815 (1996)
Huang, Y.N., Li, X., Ding, Y., Wang, Y.N., Shen, H.M., Zhang, Z.F., Fang, C.S., Zhuo, S.H., Fung, P.C.W., Phys. Rev. B 55(24), 16159 (1997)
Huang, X., Tao, J., Tang, Z., Liu, L., Zhang, F., Chen, W., Zheng, Y., J. Appl. Phys. 134, 014101 (2023)
Ichikawa, M., Motika, K., Yamada, N., Phys. Rev. B 36(1), 874 (1987)
Ichikawa, M., Chem. Phys. Lett. 79(3), 583 (1981)
Ichikawa, M., Amasaki, D., Gustafson, T., Olovson, I., Phys. Rev. B 64, 100101 (2001)
Ikeda, S., Physica B 226, 207 (1996)
Imry, Y., Pelah, I., Wiener, E., J. Chem. Phys. 43(7), 2332 (1965)
Indenbom, V.L., Chamrov, V.A., Sov. Phys. Crys. 25(2), 123 (1980)
Ioffe, L.B., Vinokur, V.M., J. Phys. C: Solid State Phys. 20, 6149 (1987)
Ishibashi, Y., Takagi, Y., J. Phys. Soc. Jpn. 31, 506 (1971)
Ishibashi, Y., Ohya, S., Takagi, Y., J. Phys. Soc. Jpn. 33(6), 1545 (1972)
Ishibashi, Y., Suzuki, I., J. Phys. Soc. Jpn. 53(3), 1093 (1984)
Ishibashi, Y., J. Phys. Soc. Jpn. 56(6), 2089 (1987)

Ishibashi, Y., Iwata, M., Ferroelectrics, 354, 8 (2007)
Ishibashi, Y., Iwata, M., Ferroelectrics, 433, 41 (2012)
Jaccard, C., Känzig, K., Peter, M., Helv. Phys. Acta 26, 521 (1953)
Jacobs, A.E., Phys. Rev. B 31, 5984 (1985)
Jaskiewicz, A., Daszczynska, W., Dabrowska, K., Bull. Acad. Pol. 25(9), 929 (1977)
Jaskiewicz, A., Przeslawski, J. Phys. Stat. Sol. (a) 56, 365 (1979)
Jona, F., Pepinsky, R. Phys. Rev. 105, 861 (1957)
Jung, D.J., Dawber, M., Scott, J.F., Sinnamon, L.J., Gregg, J.M., Integ. Ferroelec. 48, 59 (2002)
Kadanoff, L.P., Nagel, S.R., Wu, L., Zhou, Su-min, Phys. Rev A 39(12), 6524(1989)
Kamysheva, L.N., Zhukov, O.Z., PIMF Prague 2, 20 (1966)
Kamysheva, L.N., Drozdhin, S.N., Ferroelectrics 71, 281 (1987)
Känzig, W., Sommerhalder, R. Helv. Phys. Acta 26, 603 (1953)
Kennedy, N.S.J., Nelmes, R.J., Thornley, F.R., Rouse, K., Ferroelectrics 14, 591 (1976)
Kennedy, N.S.J., Nelmes, R.J., J. Phys. C: Solid State Phys. 13, 4841 (1980)
Keve, E.T., Abrahams, S.C., Nassau, K., Glass, A.M., Solid State Comm. 8, 1517 (1970)
Khachaturyan, A.G., Sov.Phys. Solid State 8, 2163 (1967)
Khachaturyan, A.G., Phys. Stat. Sol. (b) 60, 9 (1973)
Khachaturyan, A.G., Shapiro, S.M., Semenovkaya, S., Phys. Rev. B 43(13), 10832 (1991)
Khachaturyan, A.G., Theory of Structural transformations in Solids, New York, Wiley (1983) (1993)
Khachin, V.N., Solovev, L.A., Phys. Stat. Sol. (a) 30, 671 (1975)
Kikuta, T., Onodera, A., Takama, T., Fujita, M., Yamashita, H., Ferroelectrics 170, 17 (1995)
Kim, B.G., Kim, J.J., Phys. Rev. B 56(10), 6343 (1997)
Kim, B.G., Kim, J.J., Phys. Rev. B 59(21), 13509 (1999)
Kim, S.H., Oh, B.H., Han, J.H., Lee, CH.E., Choi, J.Y., Hahn, K.D., Jang, C.Y., Youn, B.H., Kim, J., Curr. Appl. Phys. 9, 1307 (2009)
Kind, R., Cereghetti, P.M., Jeitziner, C.A., Zalar, B., Dolinsek, J., Blinc, R., Phys. Rev. Lett. 88(19), 196501 (2002)
Kipenko, I.A., Akhmatkhanov, A.R., Chunakova, M.A., Shur, V. Ya, Ferroelectrics 604, 40 (2023)
Kirikov, V.A., Gladkii, V.V., Magataev, V.K., Sov. Phys. Crys. 21(6), 701 (1976)
Kitaeva, G.Kh., Kulik, S.P., Penin, A.N., Belinsky, A.V., Phys. Rev. B 51(6), 3362 (1995)
Kittel, C., Phys. Rev. 75(6), 972 (1949a)
Kittel, C., Red. Mod. Phys. 21, 541 (1949b)
Kittel, C., Solid State Commun. 10, 119 (1972)
Kityk, A.V., Sopronyuk, V.P., Schranz, W., Fuith, A., Warhanek, H., Phys. Rev. B 53, 6337 (1996)
Kityk, A.V., Schranz, W., Sondergeld, P., Havlik, D., Sopronyuk, V.P., Warhanek, H., Phys. Rev. B 61(2), 946 (2000)
Klapper, H., Fishman, Yu. M., Lutsau, V.G., Phys. Stat. Sol. (a) 21, 115 (1974)
Kleman, M., Schlenker, M., J. Appl. Phys. 43(7), 3184 (1972)
Kobayashi, K.K., J. Phys. Soc. Japan 24(3), 497 (1968)
Kobayashi, J., Uesu, Y., Mitsutani, I., Enomoto, Y., Phys. Stat. Solidi (a) 3, 63 (1970)
Kobayashi, J., Uesu, Y., Enomoto, Y., Phys. Stat. Solidi B 45, 293 (1971)
Kolmogorov, A.N., Tzv. Acad. Nauk USSSR, Ser.Math. 3, 355 (1937)
Konak, C., Chapelle, J., Matras, J., Phys. Stat. Sol. (b) 67, K47 (1975)
Kopal, A., Bahnik, T. Fousek, J., Ferroelectrics 202, 267 (1997)
Kopal, A., Mokry, P., Fousek, J., Bahnik, T. Ferroelectrics 223, 127 (1999)
Kopal, A., Mokry, P., Fousek, J., Bahnik, T., Ferroelectrics, 238, 203 (2000)
Kosevich, A.M., Boiko, V.S., Sov. Phys. Uspekhi 14, 286 (1971)
Kosevich, A.M., Boiko, V.S., Mater. Sci. Eng. A 164, 111 (1993)
Koval, S., Lasave, J., Kohanoff, J., Migoni, R., Ferroelectrics 401, 103 (2010)
Krainyuk, G.G., Otko, A.I., Ferroelectrics 111, 189 (1990)
Krupa, F., J. Phys. C3, 27,154 (1966)
Kubinec, P., Fally, M., Fuith, A., Kabelka, H., Filipic, C., J. Phys.: Condens. Matter 7, 2205 (1995)
Kuhs, W.F., Nelmes, R.J., Tibbals, J.E., J. Phys. C: Solid State Phys. 16, L1029 (1983)

Kumada, A., Phys. Lett. 30A, 186 (1969)
Kumada, A., Yumoto, H., Ashida, S., J. Phys. Soc. Jpn, S 28, 351 (1970)
Kumada, A., Ferroelectrics 3,115 (1972)
Kumar, J., Awasthi, A.M., Appl. Phys. Lett. 103, 132903 (2013)
Kuntz, M.C., Sethna, J.P., Phys. Rev. B 62, 11699 (2000)
Kuramoto, K., Motegi, H., Nakamura, E., J. Phys. Soc. Japan 53(12), 4413 (1984)
Kuramoto, K., J. Phys. Soc. Japan 56(5), 1859 (1987)
Kvitek, Z., Bornarel, J., Ferroelectrics 172, 271 (1995)
Kvitek, Z., J. Phys.: Condens. Matter 9, 127 (1997)
Kvitek, Z., Bornarel, J., J. Phys.: Condens. Matter 12, 7819 (2000)
Kvitek, Z., Bornarel, J., Eur.Phys. J. B 30, 153 (2002)
Kvitek, Z., Thesis University Joseph Fourier, Grenoble (1995)
Kweon, J.J., Fu, R., Choi, E.S., Dalal, N.S., J. Phys. : Condens. Matter 29, 16LT01 (2017)
Lajzerowicz, J., Legrand, J.F., Phys. Rev. B 17(3), 1438 (1978)
Lajzerowicz, J., Ferroelectrics 24, 179 (1980)
Lajzerowicz, J., Ferroelectrics 35, 219 (1981)
Lajzerowicz, J., Legrand, J.F., J. Phys. Paris, 31, 1375 (1980)
Landau, L.D., Lifshitz, E., Phys.Z. Sov. U. 8, 653 (1935)
Landau, L.D., On the theory of Phase Transitions, Pergamon, Oxford (1965)
Landau, L.D., Lifshitz, E., Statistical Physics, Edition Mir, Moscow, Chap. 14 (1967)
Landauer, R. J. Appl. Phys. 28(2), 227 (1957)
Landolt-Börnstein New Series Ferroelectrics lll, Springer Berlin (1981)
Lasave, J., Koval, S., Dalal, N.S., Migoni, R., Phys. Rev. B 72, 104104 (2005)
Lawless, W.N., Phys. Rev. Lett. 36(9), 478 (1976a)
Lawless, W.N., Phys. Rev. B 14(1), 134 (1976b)
Lawless, W.N., Phys. Rev. B 23(5), 2421 (1981)
Lawless, W.N., Ferroelectrics 71, 149 (1987)
Lee, W.T., Salje, E.K.H., J. Phys. : Condens. Matter 11, 7385 (1999)
Lee, W.T., Salje, E.K.H., Bismayer, U.J., J. Phys.: Condens. Matter 14, 7901 (2002)
Lee, W.T., Salje, E.K.H., Bismayer, U., Phase Transitions 76(1), 81 (2003a)
Lee, W.T., Salje, E.K.H., Bismayer, U., J. Appl. Phys. 93(12), 9890 (2003b)
Lee, W.T., Salje, E.K.H., Bismayer, U.J., J. Phys.: Condens. Matter 15, 1353 (2003c)
Lee, W.T., Salje, E.K.H., Bismayer, U., Ferroelectrics 303, 3 (2004)
Lee, W.T., Salje, E.K.H., Bismayer, U., Phys. Rev. B 72, 104116 (2005)
Legrand, J.F., Phys. Rev. Lett., 31, 314 (1973)
Legrand, J.F., Thesis, Joseph Fourier University (1981)
Levanyuk, A.P., Misirlioglu, I.B., Okatan, M.B., Ferroelectrics 569, 310 (2020)
Li, J.C.M., Surface Science 31, 12 (1972)
Li, S., Bismayer, U., Ding, X., Salje, E.K.H., Appl. Phys. Lett. 108, 022901 (2016)
Lifchitz, I.M., Zn. Eksp. Teor. Fiz. 18, 1134 (1948)
Lilly, M.P., Wootters, A.H., Hallock, R.B., Phys. Rev. Lett. 77, 4222 (1996)
Lines, M.E., Glass, A.M., Principles and Applications of Ferroelectrics and Related Materials (Oxford University Press, Oxford 1977)
Lissalde, F. Thesis AO9435, Grenoble (1974)
Litov, E., Uehling, E.A., Phys. Rev. Lett. 21, 809 (1968)
Litov, E., Uehling, E.A., Phys. Rev.B 1, 9, 3713 (1970)
Litov, E., Garland, C.W., Phys. Rev. B 2, 11, 4597 (1970)
Locherer, K.R., Chrosch, J., Slje, E.K.H., Phase Transitions 67, 51(1998)
Loh, K.K., Ong, L.H., Chew, K.H., Ferroelectrics, 445, 136 (2013)
Love, A.E.H., Treatise on the Mathematical Theory of Elasticity, Cambridge U.P. (1920)
Lu G., Li, S., Ding, X., Sun, J., Salje, E.K.H., Phys. Rev. Mater. 3, 114405 (2019)
Lundgreen, L., Svedlindh, P., Beckmann, O., Mag. Magn. Matter 31, 1349 (1983)
Lupascu, D.C., Shur, Y.S., Shur, A.G., Appl. Phys. Lett. 80, 2359 (2002)

Lupascu, D.C., Utschig, T., Shur, Y.S., Shur, A.G., Ferroelectrics 290, 207 (2003)
McCrum, N.G., Anelastic and Dielectric Effects in Polymer Solids, Wiley NY (1967)
Main, I., Rev. Geophys. 34, 433 (1996)
Malgrange, C., Ricolleau, C., Schlenker, M., Symmetry and Physical Properties of Crystals, Grenoble Sciences, Springer (2014)
Manaka, H., Uetsubara, K., Korogi, S., Miura, Y., J. Phys. Soc. Jpn. 91, 084704 (2022)
Mansingh, A., Prasard, S.R.J., J. Phys. D: Appl. Phys. 9, 1379 (1976)
Mason, W.P., Phys. Rev. 68(5–6), 173 (1946)
Mason, W.P., Piezo-electric Crystals and their Applications to Ultrasonics, Van Nostrand, Princeton (1950)
Matsuda, T., Abe, R., J. Phys. Soc. Jpn. 36(3), 765 (1974)
Mayer, R.I., M.S.Thesis, University of Washington, Seattle, W.A. (1961)
Meeks, S.W., Auld, B.A., Maccagno, P., Miller, A., Ferroelectrics, 245 (1983)
Meeks, S.W., Auld, B.A., Appl. Phys. Lett. 47(2), 102 (1985)
Meeks, S.W., Auld, B.A., Adv. Electron. Electr. Phys. 71, 251 (1988)
Megaw, H.D., Proc. Phys. Soc. London 58, 133 (1946)
Menchon, R., Colizzi, G., Johnston, C., Torresi, F., Lasave, J., Koval, S. Kohanoff, J., Migoni, R., Phys. Rev. B 98, 104108 (2018)
Merunka, D., Rakvin, B. Struct. Bond 124, 149 (2007)
Merz, W.J., Phys. Rev. 95, 690 (1954)
Merz, W.J., J. Appl. Phys. 27(8), 938 (1956)
Metrat, G., Ferroelectrics, 26, 801 (1980)
Migoni, R., Bilz, H., Bäuerle, D., Phys. Rev. Lett. 37, 1155 (1976)
Miguel, M.C., Vespignani, A., Zapperi, S., Weiss, J., Grasso, J., Nature 410, 667 (2001)
Miller, R.C., Phys. Rev. 111(3), 736 (1958)
Miller, R.C., Savage, A., Phys. Rev. 112(3), 755 (1958)
Miller, R.C., Savage, A., J. Appl. Phys., 31(4), 662 (1960)
Miller, R.C., Savage, A., J. Appl. Phys. 32(4), 714 (1961)
Miller, R.C., Weinreich, G., Phys. Rev. 117(6), 1460 (1960)
Miltat, Kleman M., Phil. Mag. 28, 1015 (1973)
Mitsui, T., Furuichi, J., Phys. Rev. 90, 193 (1953)
Mitsui, T., Furuichi, J., Phys. Rev. 95, 558 (1954)
Mohamad, I.J., Mangion, L.Z., Lambson, E.F., Saunders, G.A., J. Phys.Chem Solids 43(8), 749 (1982)
Mokry, P., Kopal, A., Fousek, J., Ferroelectrics 257, 211 (2001)
Mokry, P., Ferroelectrics 375, 40 (2008)
Molotskii, M., Phil. Mag. Lett. 83, 12, 763 (2003)
Moravets, F., Konstantinova, V.P., Sov. Phys. Crystallogr. 13, 221 (1968)
Motegi, H., Kuramoto, K., Nakamura, E., Jap. J of Appl. Phys. 19(7), L433 (1983a)
Motegi, H., Kuramoto, K., Nakamura J. Phys. Soc. Jpn. 52(4), 1131 (1983b)
Motegi, H., Kuramoto, K., Nakamura, E., Hayashi, K., J. Phys. Soc. Jpn. 54(7), 2735 (1985)
Mueller, V., Shchur, Y., Beige, H., Ferroelectrics 269, 201 (2002a)
Mueller, V., Shchur, Y., Beige, H., Mattauch, S., Glinnemann, J., Heger, G., Phys. Rev. B 65, 134102 (2002b)
Mueller, V., Beige, H., Shchur, Y., Europhys. Lett. 57, 107 (2002c)
Mueller, V., Beige, H., Shchur, Y., Ferroelectrics 290, 151 (2003)
Mueller, V., Beige, H., Shchur, Y., Ferroelectrics 303, 75 (2004)
Muller, K.A., Burkard, H., Phys. Rev. B 19, 3593 (1979)
Mura, T., Advances in Materials Science Vol. 3, Wiley, Interscience Publishers (1968)
Mura, T., Arch.Mech. Stos. 24, 449 (1972)
Nabarro, F.R.N. Theory of crystal dislocations, Clarendon Press, Oxford (1967a)
Nabarro, F.R.N., Dislocations and plastics flows in Crystals, the International series of monographs on physics, Clarendon Press, Oxford (1967b)

Nakamura, T., J. Phys. Soc. Japan 9, 425 (1954)
Nakamura, E., Ihara, H., Phys. Lett. 30A(1), 75 (1969)
Nakamura, E., Ihara, H., J. Phys. Soc. Jpn. 28, 204 (1970)
Nakamura, E., Tsukamoto, M., Itoh, K., Phys. Lett. 48A(6), 439 (1974)
Nakamura, E., Ushio, S., Abe, K., J. Phys. Soc. Jpn. 53(1), 403 (1984)
Nakamura, E., Kuramoto, K. J. Phys. Soc. Japan 57, 2182 (1988)
Nakamura, E., Kuramoto, K., Deguchi, K., Hayashi, K., Ferroelectrics 98, 51 (1989)
Nakatani, N. Jpn. J. Appl. Phys. 24(7), L528 (1985)
Nassau, K., Levistein, H.J., Loicano, G.M., J. Phys. Chem. Solids 26, 1805 (1965)
Nataf, G.F., Salje, E.K.H., Ferroelectrics, 569, 82 (2020)
Nataf, G.F., Guennou, M., J. Phys.: Condens. Matter 32, 183001 (2020)
Nattermann, T., Shapir, Y., Vilfan, I., Phys. Rev. B 42, 13, 8577 (1990)
Navas-Portella, V., Gonzales, A., Serra, I., Vives, E., Corral, A., Phys. Rev. E, 100, 062106 (2019)
Navirian, H., Enquist, H., Nüske, R., Jurgilaitis, A., Korff Schmissing, C., Sondhauss, P., Larsson, J., Phys. Rev. B 81, 024113 (2010)
Nazario, I., Gonzalo, J.A., Solid State Commun. 7, 1305 (1969)
Néel, L. Cahiers Phys. 12, 1 (1942)
Néel, L. Cahiers Phys. 13, 18 (1943)
Néel, L., Ann. Geophys. 5, 99 (1949)
Nelmes, R.J., Kennddy, N.S.J., Baharie, E., Ferroelectrics, 21, 439 (1978)
Nelmes, R.J., Meyer, G.H.M., Tibballs, J.E., J. Phys. C: Solid State Phys. 15, 59 (1982)
Nelmes, R.J., Kuhs, W.F., Howard, C.J., Tibballs, J.E., Ryan, T.W. J. Phys. C: Solid State Phys. 18, L711 (1985), 18, L1023 (1985)
Nelmes, R.J., Mc Mahon, M.I., Piltz, R.O., Wright, N.G., Ferroelectrics 124, 355 (1991)
Nettleton, R.E., J. Appl. Phys. 38(7), 2775 (1967)
Newnham, R.E., Cross, L.E., Mater. Res. Bull. 9, 927 (1974a)
Newnham, R.E., Cross, L.E., Mater. Res. Bull. 9, 1021 (1974b)
Nogami, G., Imasaki, M., J. Phys. Soc. Jpn. 35(6), 1694 (1973)
Novak, J., Salje, E.K.H., Eur. Phys. Journal B, 4, 279 (1998)
Novak, J., Salje, E.K.H., Eur. Phys. Journal B, 4, 279 (1998a)
Novak, J., Salje, E.K.H., J. Phys.: Condens. Matter 10, L359 (1998b)
Novak, J., Bismayer, U., Salje, E.K.H., J. Phys.: Condens. Matter 14, 657 (2002)
Novik, O.D., Malyshkina, I.A., Gavrilova, N.D., Ferroelectrics 606, 39 (2023)
Nye, J.F. Physical properties of crystals, O.U.P., London (1960) and (1992), Physical properties of crystals, Dunod (1961)
Oettel, R.E., Thesis, University of Washington, Seattle, Washington USA (1965)
Okada, K., Sugié, H. Phys. Lett. 37A(4), 337 (1971)
Okada, K., Sugié, H., Ferroelectrics, 17, 325 (1977)
O'Keeffe, M., Perrino, C.T., J. Phys. Chem. Solids 28, 211 (1967)
Osborn, J.A., Phys. Rev. 67, 11 and 12, 351 (1945)
Ozaki, T., Nakamura, E., J. Phys. Soc. Jpn. 51(1), 213 (1982)
Ozaki, T., Senju, T., Nakamura, E., J. Phys. Soc. Jpn. 62(9), 3027 (1993)
Ozaki, T., Ohgami, J., J. Phys.: Condens. Matter 7, 1711 (1995)
Ozaki, T., Fujii, K., Ohgami, J., J. Phys. Soc. Jpn. 64(7), 2282 (1995)
Ozaki, T., Miyoshi, N., Adyagi, S. Ferroelectrics 269, 183 (2002)
Pakulski, G., Breczewski, T., Mroz, B., Krajewski, T., Ferroelectrics 74, 375 (1987)
Paul, B., Albers, J., Müser, H.E., Ferroelectrics 14, 707 (1976)
Pauling, L., J. Am. Chem. Soc. 57, 2680 (1935)
Peercy, P.S., Samara, G.A., Phys. Rev. B 8(5), 2033 (1973)
Peierls, R.E., Proc. Phys. Soc. London 52, 34 (1940)
Perkovic, O., Dahmen, K.A., Sethna, J. P., Phys. Rev. B 59(9), 6106 (1999)
Pertsev, N.A., Novak, J., Salje, E.K.H., Phil. Mag. A 80(9), 2201 (2000)
Petzelt, J., Dvorak, V., Phys. Stat. Sol. (b) 46, 413 (1971)

Pierre, Dufour, Remoissenet, Solid State Comm. 9, 1493 (1971)
Pirc, R., Blinc, R., Phys. Rev. B 76, 020101 (2007)
Pirc, R., Blinc, R., Ferroelectrics 379(1), 30 (2009)
Pirenne, J., Physica XV, 11–12, 1019 (1949)
Poprawski, R., Kolarz, A., J. Phys. Chem. Solids 50(7), 693 (1989)
Pöttker, H., Salje, E.K.H., J. Phys.: Condensed Matter 26, 342201 (2014)
Préjean, J.J., Souletie, J.J., J. Physique 41, 1335 (1980)
Puchberger, S., Soprunyuk, V., Schranz, W., Tröster, A., Roleder, K., Majchrowski, A., Carpenter, M.A., Salje, E.K.H., APL Materials 5, 046102 (2017)
Puchberger, S., Soprunyuk, V., Schranz, W., Phys. Rev. Mater. 2, 013603 (2018)
Pulveri, C.F., Kuebler, W., J. Appl. Phys. 29(12), 1742 (1958)
Pytte, E., Solid State Comm. 8, 2101 (1970)
Raj, S., Perumal, R.N., Ferroelectrics 598, 61 (2022)
Rashkovich, L.N., KDP Family of Crystals, New York, Adam-Hilger (1991)
Redfern, S.A.T., Sarbaeva, A., GeoCHIM. Cosmochim. Acta 66, 1627 (2002)
Reese, W., May, L.F., Phys. Rev. 161(2), 504 (1963)
Reese, W., May, L.F., Phys. Rev. 162(2), 510 (1967)
Reese, W., May, L.F., Phys. Rev. 167(2), 504 (1968)
Reese, W., Phys. Rev. 181, 905 (1969)
Reese, W., Phys. Rev. 181, 905 (1969a)
Reese, W., Phys.Rev. 182(2), 646 (1969b)
Reese, W., Fritz, I.J., Cummins, H.Z., Phys.Rev.B 7(9), 4165 (1973)
Remoissenet, M., Desvignes, M., Marecek, V., Kris. und Teck. 5, 535 (1970)
Ren, X., Xu, D., Xue, D., J. Crystl Growth 310, 2005 (2008)
Ricinschi, D., Okuyama, M., Ferroelectrics, 349, 111 (2007)
Robels, U., Arlt, G., J. Appl. Phys. 73, 3454 (1993)
Roetschi, H., J. Sci. Instrum. 39, 152 (1962)
Rosova, A., Boulesteix, C., Vavra, I., Physica C 214, 247 (1993)
Roth, M., Mojaev, E., Dul'kin, E., Gemeiner, P., Dkhil, B., Phys. Rev. Lett. 98, 264701 (2007)
Rottman, C., Phys. Rev. B 38, 16, 12031 (1988)
Roucau, C., Tanaka, M., Torres, J., Ayroles, R., J. Microsc. Spectrosc. Electron. 4, 603 (1979)
Roytburd, A.L., Sov. Phys. Usp. 17, 3, 326 (1974)
Roytburd, A.L., Kurdjumov, G.V., Material Sci. and Engineering 39, 141 (1979)
Roytburd, A.L., Izv. Akad. Nauk SSSR Ser. Fiz. 47, 435 (1983)
Roytburd, A.L., Phase Transitions 45, 1 (1993)
Roytburd, A.L., J. Appl. Phys. 83, 228 (1998a)
Roytburd, A.L., J. Appl. Phys. 83, 239 (1998b)
Rumyantsev, E.L., Ferroelectrics 560, 70 (2020)
Saint-Paul, M., Gilchrist, J., J. Phys. C 19, 2091 (1986)
Salje, E.K.H., Hoppman, G., Mat. Res. Bull. 11, 1545 (1976)
Salje, E.K.H., Phys. Chem. Min. 15, 336 (1988)
Salje, E.K.H., Phase Transitions in Ferroelastic and co-elastic Crystals, Cambridge U.P. (1990)
Salje, E.K.H., Neues Jahr. Mineral. Abh. 163, 43 (1991)
Salje, E.H.K., Wruck, B., Thomas, H.Z., Z. Phys. B Condensed Matter 82, 399 (1991)
Salje, E.H.K., Physics Reports (Review Section of Physics Letters) 215(2), 49 (1992)
Salje, E.H.K., Ishibashi, Y., J. Phys. : Condens. Matter 8, 44, 8477 (1996)
Salje, E.K.H., Aird, A., Locherer, K., Hayward, S., Novak, J., Chrosch, J., Ferroelectrics, 223, 1 (1999)
Salje, E.H.K., Ferroelectrics, 267, 113 (2002)
Salje, E.K.H., Hayward, S.T., Lee, W.T., Acta Cryst. A 61, 3 (2005)
Salje, E.H.K., Zhang, H., Phase Transition 82, 6, 452 (2009)
Salje, E.H.K., Ding, X., Zhao, Z., Lookman, T., Saxena, A., Phys. Rev. B 83, 104109 (2011)
Salje, E.H.K., Ding, X., Zhao, Z., Lookman, T., Appl. Phys. Lett. 100, 222905 (2012)

Salje, E.H.K., Phase Transitions 86, 1 (2013)
Salje, E.K.H., Dahmen, K.A., Annu. Rev. Condens. Matter Phys. 5(1), 233 (2014)
Salje, E.K.H., Wang, X., Ding, X., Sun, J., Phys. Rev. B 90, 064103 (2014)
Salje, E.K.H., Wang, X., Ding, X., Scott, J.F., Adv. Funct. Mater. 27, 1700367 (2017a)
Salje, E.K.H., Planes, A., Vives, E., Phys. Rev. E 96, 4, 042122 (2017b)
Salje, E.K.H., Xue, D., Ding, X., Dahmen, K.A., Scott, J.F., Phys. Rev. Mat. 3, 014415 (2019)
Salvino, D.J., Rogge, S., Tigner, B., Osheroff, D.D., Phys. Rev. Lett. 73(2), 268 (1994)
Samara, G.A., Phys. Rev. Lett. 27(2), 103 (1971)
Samara, G.A., Ferroelectrics 5(1), 25 (1973)
Samara, G.A., Ferroelectrics 7, 221 (1974)
Samara, G.A., Ferroelectrics 22, 925 (1979)
Samara, G.A., J. Phys.: Condens. Matter 15, R367(2001)
Sapriel, J., Phys. Rev. B 12, 5128 (1975)
Savage, A., Miller, R.C., J. Appl. Phys. 31(9), 1546 (1960)
Sawyer, C.B., Tower, C.H., Phys. Rev. 35, 269 (1930)
Schmidt, V.H., J. Sci. Instrum. 42, 889 (1965)
Schmidt, V.H., J. PIMF, V 2, 97 Praha (1966)
Schmidt, V.H., Phys. Rev. 164, 2, 749 (1967)
Schmidt, V.H., Phys. Rev. 173, 630 (1968)
Schmidt, V.H., Bull. Am. Phys. Soc. 19, 649 (1974)
Schmidt, V.H., Phys. Rev. 30(5), 2795 (1984)
Schmidt, V.H., Ferroelectrics, 72, 157 (1987a)
Schmidt, V.H., Uehling, E.A., Phys. Rev. 126(2), 447(1962)
Schmidt, V.H., Western, A.B., Baker, A.G., Phys. Rev. Lett. 37(13), 839 (1976)
Schmidt, V.H., Phys. Rev. B 20(11), 4397 (1979)
Schmidt, V.H., Ferroelectrics 72, 157 (1987b)
Schmidt, V.H., Ferroelectrics 78, 207 (1988)
Schmidt, V.H., Bohannan, G., Arbogast, D., Tuthill, G., J. Phys.Chem. Sol. 61, 283 (2000)
Schmidt, V.H., Ferroelectrics 569, 70 (2020)
Schneider, J.R., J. Appl. Cryst. 7, 541 (1974a)
Schneider, J.R., J. Appl. Cryst. 7, 547 (1974b)
Schranz, W., Havlik, D., Phys. Rev. Lett. 73(19), 2575 (1994)
Schranz, W., Phys. Rev. B 83, 094120 (2011)
Schranz, W., Kabelka, H., Sarras, A., Burock, M., Appl. Phys. Lett. 101, 141913 (2012)
Schubling, N.W., Nolta, J.P., Dolk, R.A., Rev. Mod. Inst. 35(11), 1517 (1964)
Scott, J.F., Ferroelectrics, 503, 117 (2016)
Sedlakova, L., Petzilka, V., Mikula, P., Chalupa, B., Michalec, R., Cech, J., Phys. Stat. Sol. (a) 27, 309 (1975)
Sekido, T., Mitsui, T., J. Phys. Chem. Solids 28, 967 (1967)
Senko, M.E., Phys. Rev. 121(6), 1599 (1961)
Shandarov, S.M., Savchenkov, E.N., Borodin, M.V., Mandel, A.E., Akhmatkhanov, A.R., Shur, V. Ya. Ferroelectrics 542, 58 (2019)
Shapovalov, K., Yudin, P.V., Tagantsev, A.K., Eliseev, E.A., Morozovska, A.N., Setter, N., Phys. Rev. Lett. 113, 207601 (2014)
Shekhman, V. Sh., Shabel'nikov, L.G., Shmyt'ko, I.M., Aknazarov, S. Kh., Sov. Phys. Solid State 14(11), 2690 (1973)
Sheperd, I.W., Barkley, J.R., Solid State Comm. 10, 123 (1972)
Shiskina, E.V., Chivakova, M.A., Yuzhalov, V.V., Akhnatkhanov, A.R., Pelegova, E.V., Nebogatikov, M.S., Ushakov, A.D., Linker, E.A., Ivleva, L.I., Shur, V.YaJ. Appl. Phys. 132, 184101 (2022)
Shur, V., Rumyantsev, E.L., Makarov, S., J. Appl. Phys. 84(1), 445 (1998)
Shur, V.Y., Kozhevnikov, V.L., Pelegov, D.V., Nikolaeva, E.V., Shishkin, E.I., Phys. Solid State 43(6), 1128 (2001)

Shur, B.Y., Rumyantsev, E.L. Pelegov, D.V., Kozhevnikov, V.L., Nikolaeva, E.V., Shiskin, E.I., Chermykh, A.P., Ivanov, R.K., Ferroelectrics 267, 347 (2002)
Shur, V.Y., Nikolaeva, E.V., Shishkin, E.I., Baturin, I.S., Shur, A.G., Utschig, T., Schleger, T., Lupascu, D.C., J. Appl. Phys. 98, 074106 (2005)
Shur, V., Rumyantsev, E.L., Shiskina, E., Ferroelectrics 443, 105 (2013)
Shuvalov, L.A., Gridnev, S.A., Kudryash, V.I., Prasolov, B.N., Phys Stat. Sol. (a) 83, 131 (1984)
Sidnenko, E.V., Gladkii, V.V., Sov. Phys. Cryst. 18, 83 (1973)
Sidnenko, E.V., Gladkii, V.V., Sov. Phys. Cryst. 17(5), 861 (1973a)
Sidnenko, E.V., Gladkii, V.V., Sov. Phys. Cryst. 18, 83 (1973b)
Sidorkin, A.S., Fedosov, V.N., Sov. Phys. Solid State 23(9), 1667 (1981)
Sidorkin, A.S., Burdanina, N.A., Lazarev, A.P., Sov. Phys. Solid State 26(5), 861 (1984)
Sidorkin, A.S., J. Appl. Phys. 83(3), 3762 (1998)
Silka, S.H., Sharma, S.M., Phase transitions 81(10), 907 (2008)
Sikora, R., Parlinski, K., Phase Transitions 70, 41 (1999)
Silsbee, H.B., Uehling, E.A., Schmidt, V.H., Phys. Rev. Lett. 37, 839 (1964a)
Silsbee, H.B., Uehling, E.A., Schmidt, V.H., Phys. Rev. 133(1A), 165 (1964b)
Singh, S., Draegert, D.A., Geusic, J.E., Phys. Rev. 2709 (1970)
Sinha, J.K., J. Sci. Instrum. 42, 696 (1965)
Sinitski, A., Schmidt, V.H., Phys. Rev. B 54, 842 (1996)
Sirotin, Y.I., Shakol'skaya, M.P., Foundations of Crystal Physics, Nauka, Moscow (1979)
Siun Wenyan, Shen Huimin, Wang Yening, Lu Baosheng, J. de Phys. C 10, 46(Suppl. 12), 609 (1985)
Slater, J.C., J. Chem. Phys. 9, 16 (1941)
Slater, J.C., Phys. Rev. 78, 748 (1950)
Slonczewski, J.C., Thomas, H., Phys. Rev. 21, 3599 (1970)
Sluka, T., Bednyakov, P., Yudin, P., Crassous, A., Tagantsev, A., Topological Structures in Ferroic Materials, Springer Series in Material Science 228, 103 (2016)
Smolenskii, G.A., Bokov, V.A., Isupov, V.A., Krainik, N.N., Pasynkov, R.E., Sokolov, A.I., Ferroelectrics and Related Materials, Gordon and Breach, New York (1984)
Solanski, M.B., Patel, P., Shinde, S., Parekh, B.B., Joshi, M., Ferroelectrics 571, 1 (2021a)
Soprunyuk, V et al J. Phys. Condens. Matter 29, 22, 224002 (2017)
Spasojevic, D.F., Bukvic, S., Milosevic, S., Stanley, H.E., Phys. Rev. E 54(3), 2531 (1996)
Stadler, H.L., J. Appl. Phys. 37(5), 1947 (1966)
Stadler, H.L., Zachmanidis, P.J., J. Appl. Phys. 34(11), 3255 (1963)
Stadler, H.L., Zachmanidis, P.J., J. Appl. Phys. 35(9), 2625 (1964a)
Stadler, H.L., Zachmanidis, P.J., J. Appl. Phys. 35(10), 2895 (1964b)
Staines, M.P., Talloon, I.L., Robinson, W.H., Flower, W.E., Appl. Phys. Lett. 53, 161560 (1988)
Staniorowski, P., Bornarel, J., J. Phys.: Condens. Matter 12, 669 (2000)
Stankovska, J., Acta Phys. Polon. A 43, 603 (1973)
Solanski, M.B., Patel, P., Shinde, S., Parekh, B.B., Joshi, M. Ferroelectrics 571, 1 (2021b)
Stasyuk, I.V., Levitskii, R.R., Moina, A.P., Phys. Rev. B 59(13), 8530 (1999)
Strukov, B.A., Amin, M., Kopchik, V.A., Phys. Status Solidi 27, 741 (1968)
Strukov, B.A., Korzhuev, M.A., Baddur, A., Koptsik, V.A., Soviet Phys. Solid State 13, 1569 (1972)
Strukov, B.A., Korzhuev, M.A., Baddur, A., Koptsik, V.A., Soviet Phys. Solid State 13, 1569 (1972a)
Strukov, B.A., Baddur, A., Velichko, I.A., Soviet Phys. Solid State 13,2085 (1972a)
Strukov, B.A., Baddur, A., Velichko, I.A., Soviet Phys. Solid State 13,2085 (1972b)
Strukov, B.A., Baddur, A., Koptsik, V.A., Velichko, I.A., Soviet Phys. Solid State 14, 885 (1972b)
Strukov, B.A., Baddur, A., Zinenko, V.N., Mishchenko, A.V., Kpotsik, V.A., Soviet Phys. Solid State 15, 939 (1973)
Strukov, B.A., Levanyuk, A.P., Ferroelectric Phenomena in Crystals: Physical Foundations, Springer, Heidelberg (1998)
Strukov, B., Shnaidshtein, I., Onodora, A., Ferroelectrics 363, 27 (2008)
Styrkowiec, R., Phys. Stat. Sol. (a) 87, K135 (1985)

Suemune, Y., J. Phys. Soc. Jpn 22(3), 735 (1967)
Sugié, H. Okada, K., Kan'no, K.I., J. Phys. Soc. Jpn. 33, 1727 (1972)
Sugimoto, H., Ikeda, S., J. Phys.: Condens. Matter 6, 5561 (1994)
Sugimoto, H., Ikeda, S., J. Phys.: Condens. Matter 8, 603 (1996)
Sujar, B., Biedrzycki, K., Jpn. J. Appl. Phys. 24, 81 (1985)
Suzuki, K., Solid State Comm. 11, 937 (1972)
TACM van der Put, Phase transitions 84, 11, 999 (2011)
Tagantsev, A.K., Fousek, J., Ferroelectrics 221, 193 (1999)
Tagantsev, A.K., Ferroelectrics 375, 19 (2008)
Tagantsev, A.K., Cross, L.E., Fousek, J., Domains in Ferroic crystals and Thin Films, Springer (2010)
Tagantsev, A.K., Shapolvalov, K., Yudin, P.Y., Ferroelectrics 503, 163 (2016)
Takagi, Y., J. Phys. Soc. Jpn. 3, 271 (1948a)
Takagi, Y., J. Phys. Soc. Jpn 3, 273 (1948b)
Takagi, Y., Shigenari, T., J. Optical Soc. America 63(8), 995 (1973)
Tan, M., Flannigan, C., Gardner, J., Morrison, F.D., Salje, E.K.H., Scott, J.F., Phys. Rev. Mater 3, 034402(2019)
Tanner, B.K., Clark, G.F., Goddard, P.A., Boxen, D.K., Davie's, S.T., Aleshko-Ozhevsky, O.P., Nuclear Instruments and Methods 208, 713 (1983)
Taylor, G.W., IEEE Trans. Electron. Comput. 14, 881 (1965)
Taylor, G.W., J. Appl. Phys. 37, 593 (1966)
Tibbals, J.E., Nelmes, R.J., McIntyre, G.J., J. Phys. C: Solid State Phys. 15, 37 (1982)
Tikhomirov, O.A., J. Appl. Phys. 80(4), 2358 (1996)
Tikhomirov, O.A., Ferroelectrics 190, 37 (1997)
Tokunaga, M., Matsubara, T., Prog. Theor. Phys. 35, 581 (1966)
Tokunaga, M., Ferroelectrics 1, 195 (1970)
Tokunaga, M., Tominaga, Y., Tatsuzaki, I., Ferroelevtrics 63, (1985)
Tokunaga, M., Matsubara, T., Ferroelectrics 72, 175 (1987)
Tokunaga, M., Endo, S., Sawada, T., Tsukawake, T., Deguchi, K., Ferrolectrics 237, 169 (2000)
Tomashevski, P.E., Phase Transitions 38,127 (1992a)
Tomashevski, P.E., Phase Transitions 38, 221(1992b)
Toledano, J.C., Annales des Télécommunications 29, 249 (1974)
Toledano, J.C., Toledano, P., The Landau theory of phase Transition. World Scientific, Singapore (1988)
Toledano, J.C., Toledano, P., Phys. Rev. B 21, 3, 1139 (1980)
Tonkov, E.Y., High Pressure Phase Transformations, Gordon and Breach, Philadelphia (1992)
Torche, B., State Thesis, Joseph Fourier University (1986)
Torres, J., Roucau, C., Ayrolles, R., Phys. Stat. Sol. (a) 70, 193 (1982)
Torstweit, S., Phys. Rev. B 20, 4431 (1979)
Toshev, S.D., Soviet. Crystallogr. 8, 541 (1963)
Troussaut, F., Vallade, M., J. Phys. C: Solid State Phys. 13, 4649 (1980)
Troussaut, F., Bastie, P., Bornarel, J., Vallade, M., Ferroelectrics 39, 1053 (1981)
Troussaut, F., Thesis, Joseph Fourier University, Grenoble, France (1987)
Tsatskis, I., Salje, E.K.H., Heine, V., J. Phys.: Condens. Matter 6, 11027 (1994)
Tsukamoto, T., Futama, H., Phase transitions 45, 59 (1993)
Tsunekawa, S., Ishibashi, Y., Takagi, Y., J. Phys. Soc. Jpn. 27, 919 (1969)
Tufatova, S.M., Sindeyev, Yu G., Gavrilyachenko, V.G., Fesenko, E.G., Ferroelectrics 26, 809 (1980)
Tun, Z., Nelmes, R.J., Kuhs, W.F., Stansfield, R.D.F., J. Phys. C 21, 245 (1988)
Ubbelohde, A.R., Woodward, I., Proc. Roy. Soc. A 188, 358 (1947)
Udalov, A.R., Korzhenevskii, A.L., Shur, V. Ya., Ferroelectrics 476, 17 (2015)
Utsu, T., Pure Appl. Geophys. 155, 509 (1999)
Vaks, V.G., Sov. Phys. JETP 27,486 (1968)
Valla, Ph., D.E.S., University of Grenoble, France (1967)

Vallade, M., Phys. Rev. B 12(9), 3755 (1975)
Vallade, M., Dolino, G., J. de Phys. C4, S12, 43, 401 (1982)
Van Aert, S., Turner, S., Deville, R., Schryvers, D., Van Tendeloo, G., Ding, X., Salje, E.K.H., Phase Transitions 86, 1052 (2013)
Vasilevskaya, A.S., Kaznacheev, A.V., Slepkov, I.A., Sonin, A.S., Sov. Phys. Solid State 29(11), 1881 (1987)
Velichkina, T.S., Golubeva, O.N., Shustin, D.A., Yakovlev, I.A., JETP LetT. 9(5), 153 (1969)
Verma, S., Phase Transitions 83(9), 714 (2010)
Viehland, D.D., Salje, E.K.H., Adv. In Phys. 63(4), 267 (2014)
Vives, E., Ortin, J., Manosa, L., Rafols, I., Perez-Magrane, R., Planes, A., Phys. Rev. Lett. 72, 1694 (1994)
Vogel, H., Physikalische Zeitschrift 22, 645 (1921)
Volterra, V., AnScient. Ec.norm. sup. Paris 3, 24,400 (1907)
Von Arx, A., Bantle, W., Helv. Phys. Acta 16, 211 (1943)
Von Arx, A., Bantle, W., Helv. Phys. Acta 17, 298 (1944)
Von Hippel, A., Breckenridge, R.G., Chesley, F.G., Tisza, L., Ind. Engn. Chem. 38, 1097 (1946)
Voronkova, V.I., Wolf, T. Physica C 218, 175 (1993)
Wadhavan, V.K., Phase Transitions, 3, 3 (1982)
Wang, Y.N., Huang, H.N., Shen, H.M., Zhang, Z.F., J. Physique IV C 8(6), 505 (1996)
Wang, Y.N., Huang, H.N., Shen, H.M., Ferroelectrics 197, 75 (1997)
Wang, C., Fang, Q.F., Shi, Y., Zhu, Z.G., Mater. Res. Bull. 36, 2657 (2001)
Wang Y.L., Tagantsev, A.K., Damjanovic, D., Setter, N., Appl. Phys. Lett. 91, 62905 (2007)
Weilert, M.A., Msall, M.E., Anderson, A.C., Wolfe, J.P., Phys. Rev. Lett. 71(5), 735 (1993)
Weiss, P. J. Physique 6, 661 (1907)
West, J., Z. Krist. 74, 306 (1930)
Western, A.B., Schmidt, V.H., Solid State Comm. 19, 885 (1976)
Western, A.B., Baker, A.G., Bacon, C.R., Schmidt, V.H., Phys. Rev. B 17(11), 4461 (1978)
Wieder, H.H., J. Appl. Phys. 28(3), 367 (1957)
Wieder, H.H., Proceed. IRE, 1094 (1958a)
Wieder, H.H., Phys. Rev. 110, 1, 29 (1958b)
Wieder, H.H., J. Appl. Phys 31, 180 (1960)
Wieder, H.H., Collins, D.A. Phys. Rev. 120(3), 725 (1960)
William, R., Cook, J.R., J.Appl. Phys. 38 (1967)
Xu, H.W., Heaney, P.J., Am. Mineral. 82, 99 (1997)
Xu, Y., Xue, D., Zhou, Y., Su, T., Ding, X., Sun, X., Sun, J., Salje, E.K.H., Appl. Phys. Lett. 115, 022901 (2019)
Yamamoto, N., Yagi, K., Honjo, G., Phys. Stat. Sol. (a) 42, 257 (1977a)
Yamamoto, N., Yagi, K., Honjo, G., Phys. Stat. Sol. (a) 41, 523 (1977b)
You, J.G., Kim, J.J., Ferroelectrics 190, 19 (1997)
Yudin, P., Gureev, M.Y., Sluka, T., Tagantsev, A.K., Setter, N., Phys. Rev. B 91, 060102 (2015)
Zammit-Mangion, L.J., Saunders, G.A., J. Phys. C: Solid State Phys. 17, 2825 (1984)
Zeller, R.C., Pohl, R.O., Phys. Rev. 4, 6, 2029 (1971)
Zeyen, C.M.E., Diplomarbeit, Munich, Germany (1975)
Zeyen, C.M.E., Meister, H., Proc. of Conf. Neutron Scattering 1, 445, Gatlingburg Tennessee, USA (1976)
Zeyen, C.M.E., Meister, H., Proc. of Conf. Neutron Scattering 1, 445, Gatlingburg Tennessee, USA (1976a)
Zeyen, C.M.E., Meister, H., Kley, W., Solid State Comm. 18, 621 (1976)
Zeyen, C.M.E., Meister, H., Ferroelectrics, 14, 731 (1976)
Zeyen, C.M.E., Meister, H., Ferroelectrics, 14, 731 (1976b)
Zhang, Q., Kioussis, N., Demos, S.G., Radousky, H.B., J. Phys. : Condens. Matter 14, L89 (2002)
Zhang, X., Mellinger, C., Colla, E.V., Weissman, M.B., Viehland, D.D., Phys. Rev. B 95, 144203 (2017)

Zhang, Z., Wen, Z., Li, T., Wang, Z., Liu, Z., Liao, X., Ke, S., Shu, L., J. Appl. Phys. 133, 054102 (2023)
Zhirnov, V.A., Zh. Eksp. Teor. Fiz. 35, 1175 (1958)
Zhirnov, V.A., Sov. Phys. JETP 35,5, 822 (1959)
Zhong, W.L., Wang, Y.G., Zhang, P.L., Qu, B.D., Phys. Rev. B 50, 2, 698 (1994)
Zimmer, H.E., Engert, J., Hegenbarth, E., Ferrolectrics Lett: 6, 165 (1987),
Zimmermann, J., Schranz, W., J. Phys.: Condens. Matter 8, 7085 (1996)
Zimmermann, J., Schranz, W., Ferroelectrics 190, 99 (1997)
Zisman, A.N., Kachinskii, V.N., Stishov, S.M., Zh. Eksp. Teor. Fiz. 31, 172 (1980)
Zolotukhin, I.V., Spitsina, S.V., Ferroelctrics 265, 139 (2002)
Zwicker, B., Scherrer, P., Helv. Phys. Acta 17, 346 (1944)
Zwicker, B., H. P. A. 20, VIII, 523 (1946)